3-Membered Heterocycle Synthesis

3-Membered Heterocycle Synthesis

Navjeet Kaur
Department of Chemistry and Division of Research and Development,
Lovely Professional University, Phagwara, Punjab, India

ELSEVIER

Elsevier
Radarweg 29, PO Box 211, 1000 AE Amsterdam, Netherlands
50 Hampshire Street, 5th Floor, Cambridge, MA 02139, United States

Notices
Knowledge and best practice in this field are constantly changing. As new research and experience broaden our understanding, changes in research methods, professional practices, or medical treatment may become necessary.

Practitioners and researchers must always rely on their own experience and knowledge in evaluating and using any information, methods, compounds, or experiments described herein. In using such information or methods they should be mindful of their own safety and the safety of others, including parties for whom they have a professional responsibility.

To the fullest extent of the law, neither the Publisher nor the authors, contributors, or editors, assume any liability for any injury and/or damage to persons or property as a matter of products liability, negligence or otherwise, or from any use or operation of any methods, products, instructions, or ideas contained in the material herein.

ISBN: 978-0-443-13583-5

For information on all Elsevier publications visit our website at https://www.elsevier.com/books-and-journals

Publisher: Candice Janco
Acquisitions Editor: Gabriela D. Capille
Editorial Project Manager: Lena Sparks
Production Project Manager: Kumar Anbazhagan
Cover Designer: Miles Hitchen

Typeset by TNQ Technologies

Contents

Author biography

Dr. Navjeet Kaur was born in Punjab, India. She received her BSc from Panjab University, Chandigarh (Punjab, India) in 2008. In 2010, she completed her MSc in Chemistry from Banasthali Vidyapith. She was awarded with PhD in 2014 by the same university under the supervision of **Prof. D. Kishore**. Being a meritorious student, she won a merit scholarship for her master's and doctorate degrees. She worked as an Assistant Professor in the Department of Chemistry, Banasthali Vidyapith, and has started a specialized research career focusing on the synthesis of 1,4-benzodiazepine-based heterocyclic compounds (Organic Synthetic and Medicinal Chemistry). At present, she is working as an Associate Professor in the Department of Chemistry and Division of Research and Development at Lovely Professional University, Punjab. **Dr. Kaur** has guided four PhD students and many MSc dissertation students. Currently, she is supervising one research scholar as their PhD supervisor. With 12 years of teaching experience, she has published more than 185 scientific research papers, review articles, books, book chapters, and monographs in the field of organic synthesis in national and international journals. She has published thirteen authored books: *Palladium Assisted Synthesis of Heterocycles* and *Metals and Non-metals: Five-Membered N-Heterocycle Synthesis* with CRC Press, Taylor & Francis group; *Metal- and Nonmetal-Assisted Synthesis of Six-Membered Heterocycles*, *Raney Nickel-Assisted Synthesis of Heterocycles*, *Synthesis of Azetidines from Imines by Cycloaddition Reactions*, *8-Membered Heterocycle Synthesis*, *Iodine-Assisted Synthesis of Six- and Seven-Membered Heterocycles*, *Five-Membered Heterocycle Synthesis Using Iodine*, "Oxetane Synthesis", "4-Membered Heterocycle Synthesis", "Synthesis of 5-Membered Heterocycles – Role of DMF as Solvent, Reagent, Catalyst, and Stabilizer", and "Synthesis of 6- and 7-Membered Heterocycles: the Role of DMF" with Elsevier; and *Lawesson's Reagent in Heterocycle Synthesis* with Springer Nature. She has featured consecutively for six years in the World Ranking of Top 2% Scientists in subject-wise analyses conducted by a team of scientists at Stanford University, United States (2020–2023: Full career and Single year ranking; 2019: Full career ranking; 2018: Single year ranking). Numerous newspapers have reported on her achievement of top ranks in such a short span (5–6 years) of her career. She was presented the Prof. G. L. Telesara Award in 2011 by the Indian Council of Chemists (Agra, Uttar Pradesh) at Osmania University (Hyderabad), and the Best Paper Presentation Award in National Conference on "Emerging Trends in Chemical and Pharmaceutical Sciences" (Banasthali Vidyapith, Rajasthan). In 2020, she has been cited as the most productive and distinguished author of Banasthali Vidyapith in "Scientometrics Profile of the Banasthali Vidyapith: A Deemed University of Rajasthan, India," *Library Philosophy and Practice* (e-journal). In 2021, Dr. Kaur was selected as a candidate for membership by the Royal Society of Research, London, England, United Kingdom, with full waiver of application fees. She has been certified for her contribution to Science and Technology on Women's Day by DST, Govt. of Rajasthan. She is currently serving as an Editor-in-Chief for *Advanced Chemicobiology Research* journal and Editorial Board Member for *Universal Journal of Green Chemistry* journal. She has attended about 50 conferences, workshops, seminars, and refresher courses. Using these platforms, she has disseminated her research work through oral and poster presentations. She has delivered many invited lectures and radio talks. She acted as a reviewer for many papers in different journals. She has served as a Program Co-Chair and member of organizing committee of conferences conducted by her department and acted as a member of national advisory committee in an international conference organized by Akal University, Punjab, in collaboration with the Indian Chemical Society. In addition, she has worked as a National Service Scheme (NSS) Program Officer from 2016 to 2022 and a member of Unnat Bharat Abhiyan (UBA) from 2018 to 2022, involving in the basic daily needs, health care, education, environment sustainability, and agriculture activities in rural areas. **Dr. Navjeet** finds interest in Sikh literature and has completed a 2-year Sikh missionary course from Sikh Missionary College (Ludhiana, Punjab).

Preface

The chemistry of small ring compounds has played an important role in the development of modern organic chemistry. Leading among these reactive molecules are the small ring heterocyclic compounds. Three-membered heterocyclic compounds are invested with a special allure that is derived from their apparent simplicity and spartan architecture. Yet these systems are multifaceted, and the literature continues to afford a proof of their diversity both in terms of synthetic pathways and consequent conversions. These smallest heterocyclic compounds also show a synthetically very beneficial balance between reactivity and stability. Therefore, they are frequently used as selective and versatile intermediates. Three-membered heterocyclic rings offer an unusual combination of synthetic flexibility, reactivity, and atom economy. Uncountable classic and modern procedures exist for their synthesis from simple starting compounds.

The 2*H*-azirines have attracted a great deal of attention from experimental and theoretical chemists. The chemistry of highly reactive 2*H*-azirines has been discovered widely for several synthetic objectives. The research on the chemistry of azirines is emerging greatly and is healthy. The 2*H*-azirines have been widely studied for their presence in the natural products and high synthetic potential in the formation of functionalized amino derivatives and *N*-heterocyclic compounds. These heterocyclic compounds undergo reactions in which they can act either as an electrophile or as a nucleophile.

The diaziridines have also been examined as possible high-energy compounds and show psychotropic activity, especially antidepressant nature and monoamine oxidase inhibition. Many reviews have been published on the formation and chemical properties of diaziridines. The diaziridines are prone to ring-enlargement reactions with electrophilic reagents, which led to the development of novel easy approaches for the formation of both known and earlier unknown heterocyclic systems.

In organic synthesis, the epoxides serve as useful intermediates. They are important functional groups of natural and medicinal products, polymers, and supramolecular compounds. An epoxidation of terminal alkenes to epoxides is especially interesting because epoxides are appreciable intermediates for the formation of fine chemicals like paints, epoxy resins, surfactants, vital intermediates in organic synthesis, agrochemicals, and pharmaceuticals. For the formation of epoxides, an epoxidation of alkenes has been an important key pathway and a number of researchers are paying attention to it.

The thiiranes are beneficial and important building blocks for the formation of medicines, herbicides, pesticides, and polymers. Due to their industrial and synthetic importance, a number of approaches have been developed for the formation of thiiranes. The general process for the formation of thiiranes is the transformation of epoxides by O—S exchange reaction. The sulfur transferring agents like thiourea, inorganic thiocyanates, silica-supported potassium thiocyanate or thiourea, polymer-supported thiocyanates, and Dowex-50WX8-supported $(NH_2)_2CS$ have been utilized for this purpose.

Because of the importance of decreasing environmental pollution in chemical reactions, green procedures have attracted great attention in both industrial and laboratory synthesis. Discovery of innovative approaches to decrease waste and hazard has been the main motive of various organic chemists. In this book, I have given special attention on synthetic methods available for the synthesis of three-membered heterocycles.

Abbreviations

acac acetylacetonate
ADAs adenosine deaminases
ADPM aza-di-π-methane
Aoe (2*S*,9*S*)-2-amino-8-oxo-9,10-epoxydecanoic acid
ARFs ADP ribosylation factors
bmim 1-butyl-3-methylimidazolium
Boc *tert*-butoxycarbonyl
COD cyclooctadiene
COFGTs comprehensive organic functional group transformations
CSA camphor sulfonic acid
DABCO 1,4-diazabicyclo[2.2.2]octane
DADS 4,4′-diaminostilbene-2,2′-disulfonic acid
DBN 1,5-diazabicyclo[4.3.0]non-5-ene
DBU 1,8-diazabicyclo[5.4.0]undec-7-ene
DCA dicyanoanthracene
DCC *N*,*N*′-dicyclohexylcarbodiimide
DCE dichloroethane
DCM dichloromethane
2D COSY-LR two-dimensional correlation spectroscopy low resolution
DEA diethanolamine
DEPC diethyl pyrocarbonate
DFT density functional theory
DHP dihydropyran
DIAD diisopropyl azodicarboxylate
DIBAL-H diisobutylaluminum hydride
DIC diisopropylcarbodiimide
DIMSOY dimethylsulfoxonium methylide
DIMSY dimethylsulfonium methylide
DMAP dimethylaminopyridine
DMD dimethyldioxirane
DMDO dimethyldioxirane
DMF dimethylformamide
DMI 1,3-dimethylimidazolidin-2-one
DMP Dess–Martin periodinane/dimethoxypyridine/2,9-dimethyl-1,10-phenanthroline
DMSO dimethylsulfoxide
DMSY dimethylsulfonium methylide
DMTF dimethylthioformamide
DMTrCl 4,4′-dimethoxytriphenylmethyl chloride
DPPA diphenylphosphorazidate
DPPCl diphenylphosphinyl chloride
DSC disuccinimide carbonate
EDC 1-ethyl-3-(3-dimethylaminopropyl)carbodiimide
EDTA ethylenediaminetetraacetic acid
EJB epoxy jatropha *n*-butyl esters
EWG electron-withdrawing group
FAM 6-carboxyfluorescein

FC flash chromatography/fractional crystallization
Fmoc fluorenylmethoxycarbonyl
FVP flash vacuum pyrolysis
GC-MS gas chromatography-mass spectrometry
GMA glycidyl methacrylate
HASA hydroxylamine-*O*-sulfonic acid
HMDS hexamethyldisilazane
HOBt 1-hydroxy-1*H*-benzotriazole
HPLC high-performance liquid chromatography
IL ionic liquid
IR infrared
ISC intersystem crossing
LDA lithium diisopropylamine
LFP laser flash photolysis
LiHMDS lithium hexamethyldisilamide
LR Lawesson reagent
LTMP lithium tetramethylpiperidide
***m*-CPBA** *meta*-chloroperoxybenzoic acid
MCT 4-methyl-5-carboethoxy-1,2,3-thiadiazole
MTs methylene thiiranes
MW microwave
MWI microwave irradiation
NBS *N*-bromosuccinimide
NCS *N*-chlorosuccinimide
NIS *N*-iodosuccinimide
NMO *N*-methylmorpholine-*N*-oxide
NMP *N*-methylpyrrolidone
NMR nuclear magnetic resonance
NOE nuclear Overhauser effect
OPG osteoprotegerin
OTBDPS *tert*-butyldiphenylsilyl
OTf trifluoromethanesulfonate
PAA poly(allylamine)
PDC pyridinium dichromate
PEG polyethylene glycol
PIDA phenyliodine diacetate
PMB *para*-methoxybenzyl
PPTS pyridinium *para*-tolylsulfonate
PTC phase-transfer catalyst/catalysis/conditions
PTSA *para*-toluenesulfonic acid
SET single-electron transfer
S_N2 bimolecular nucleophilic substitution
TBAF tetrabutylammonium fluoride
TBDMS *tert*-butyldimethylsilyl
TBDPS *tert*-butyldiphenylsilyl
TBS *tert*-butyldimethylsilyl
TEA triethylamine
TEG triethylene glycol

TES triethylsilyl
TFA trifluoroacetic acid
TFAA trifluoroacetic anhydride
TFDO/TMDO (trifluoromethyl)methyldioxirane
TFMAD 3-(trifluoromethyl)-3-phenyl-diazirine
TFMPA 3-(trifluoromethyl)-3-phenyl-azirine
THF tetrahydrofuran
THP tetrahydropyran
TIS triisopropylsilyl
TLC thin layer chromatography
TmdPhe 3-(trifluoromethyl)-3-phenylalanine-diazirine
TMEDA tetramethylethylenediamine
TMS trimethylsilyl/tetramethylsilane
TOSCl *para*-toluenesulfonyl chloride
TPD 3-(trifluoromethyl)-3-phenyl-diazirine
TPP triphenylphosphine/tetraphenylporphyrin
TPS triphenylsilyl
TRIR time-resolved mid-infrared
TTP tetra-*para*-tolylporphyrin
UV ultraviolet
UVLED ultraviolet light-emitting diode

CHAPTER

Synthesis of azirines

1

Chapter outline

1.1 Introduction

The antibiotics such as cephalosporins and penicillins, alkaloids such as morphine, vinblastine, reserpine, and ellipticine, and cardiac glycosides such as the class of digitalis are heterocyclic natural products and are very important for the health of humans and animals. The pharmacological scientists have continuously designed and synthesized better medicines for a better living. In the similar light, insecticides, pesticides, weed killers followed natural models and rodenticides, and a significant portion of such biologically active compounds are heterocycles [1–4].

Since the mid-1960s, the structure, synthetic chemistry, and biological uses of these heterocyclic compounds are widely studied. Various reviews have been published on azirines during this period [5]. Various natural products contain azirine ring. The first isolated azirine bearing natural product is azirinomycin [6,7]. The azirinomycin, isolated from *Streptomyces aureus*, and its methyl ester are found to show broad-spectrum antibiotic activity in vitro against both Gram-negative and Gram-positive bacteria [8,9].

The research on the chemistry of azirines is emerging greatly and is healthy although the level of activity does not match that of the aziridines. The researchers are mainly directed to two excellent reviews by Palacios et al. [10,11]. The 2*H*-azirine is an unsaturated cousin of aziridine ring, also known as azacyclopropene (and confoundingly as 1-azirine), which might well be explained metaphorically as a cyclic imine. It is a weaker base than aziridine, which is analogous to the trend in open-chain imines and amines [12–14].

A class of highly strained and reactive molecules, 2*H*-azirines, with a C=N bond incorporated into a three-membered ring, has been widely studied for their presence in the natural products [15,16] and high synthetic potential in the formation of functionalized amino derivatives and *N*-heterocyclic compounds. Particularly, the 2*H*-azirines having an aryl group are impressive intermediates for the

3-Membered Heterocycle Synthesis. https://doi.org/10.1016/B978-0-443-13583-5.00008-6

formation of a number of substituted indoles; this unique method began from an alkyl-substituted arene and joins the nitrogen functionality on the side chain to the benzene ring through an aryl carbon–hydrogen amination at the final synthetic stage [17–22].

1.2 Synthesis of azirines

The enamine substrates are prepared through the condensation by known processes [23,24]. An intramolecular azirination of a number of substituted enamine compounds [25] was performed under the optimized conditions to study the generality and scope of the procedure. Various enamino nitriles with diverse substitution patterns could be utilized for the synthesis of 2-aryl-2*H*-azirine-2-carbonitriles in 65%–83% yield. The substrate was synthesized by exchanging the position of Ph and Me group, and it also provided 3-aryl-2*H*-azirine-2-carbonitrile in 85% yield (Scheme 1). This consequence showed that the aryl group in ketonitriles was not essential for the azirination [26].

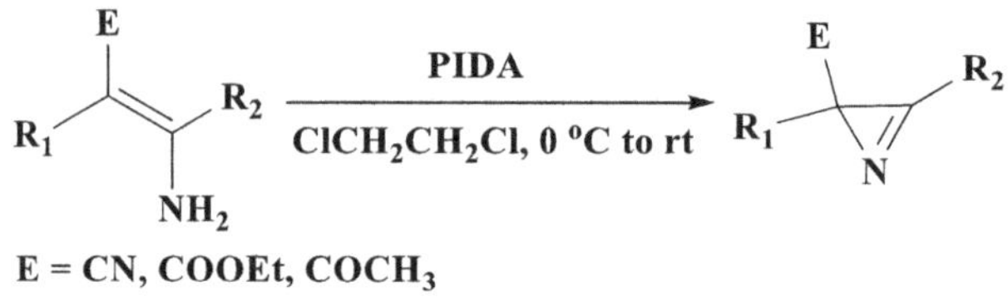

SCHEME 1

The 1*H*-azirines were formed along with isomeric 2*H*-azirines only in minor yield (15%) by an addition of nitrenes to alkynes. This was an outcome of a 1,3-sigmatropic shift from an unstable intermediate 1*H*-azirine (Scheme 2) [27–29]. An oxidation of hydrazine derivatives with lead(IV) acetate provided nitrenes. An intramolecular version of this reaction is also known to provide the 2*H*-azirine derivative in an improved yield (78%) [11,30].

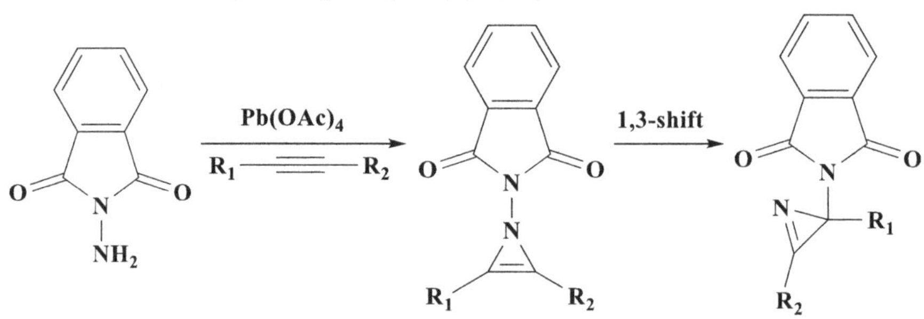

SCHEME 2

The azirine synthesis includes the reaction of nitriles and carbenes of which only a small number of successful syntheses are known. The photochemically synthesized 1-naphthylcarbene was reacted with nitriles to synthesize the products via trapping of intermediate nitrile ylides instead of the desired 2*H*-azirine [31]. Better results have been reported by a [1 + 2]-cycloaddition reaction of phosphinocarbene and benzonitrile to synthesize the 2*H*-azirine in good yield (Scheme 3) [32,33]. The reaction was favored by heteroatom-substituted carbene [11].

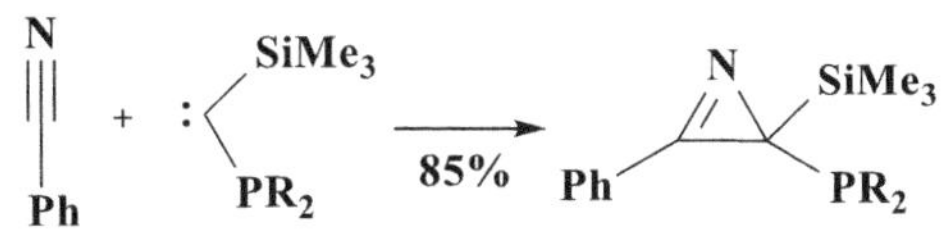

SCHEME 3

Bartoli et al. [34] synthesized amino alcohol with >99% enantiomeric excess and in quantitative yield (depending on *t*-butyl carbamate) by reacting racemic propylene oxide with *t*-butyl carbamate (0.45 eq.) under the aminolytic kinetic resolution conditions [35,36]. The volatile (*R*)-aziridine was obtained in a single flask conversion by alcohol tosylation and after that potassium hydroxide-assisted aziridine ring closure [37]. The aziridinyl ester as a 1:19 *cis*/*trans* mixture, formed by reacting LTMP under the optimized migration conditions, was purified to provide the *trans*-aziridinyl ester in 70% yield. Because of the reduced steric demand of aziridine methyl substituent, the *cis*-diastereomer was formed in traces. Finally, the *t*-butyl ester of natural (*S*)-azirinomycin was prepared by a Swern oxidation (Scheme 4) as a single regioisomer from racemic propylene oxide in only four steps [38].

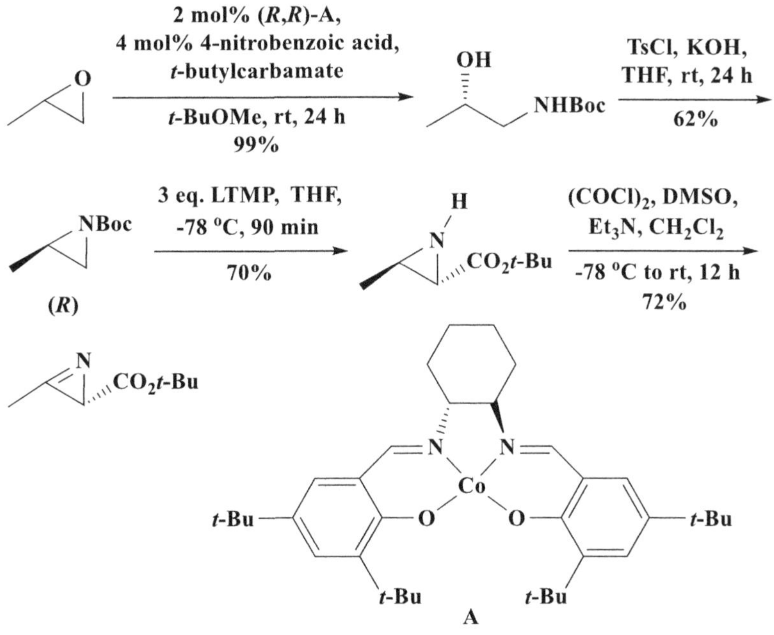

SCHEME 4

The ring contraction of four- and five-membered heterocyclic compounds to three-membered heterocyclic compounds is an alternative pathway to 2*H*-azirines. The 2*H*-azirines could be prepared from azete derivative by an oxidation with DMSO or an intermediate bicyclic adduct, synthesized from 1-diazo-1-phenylethane, rearranged photochemically or thermally to synthesize the 2*H*-azirinyl azine (Scheme 5) [11,39].

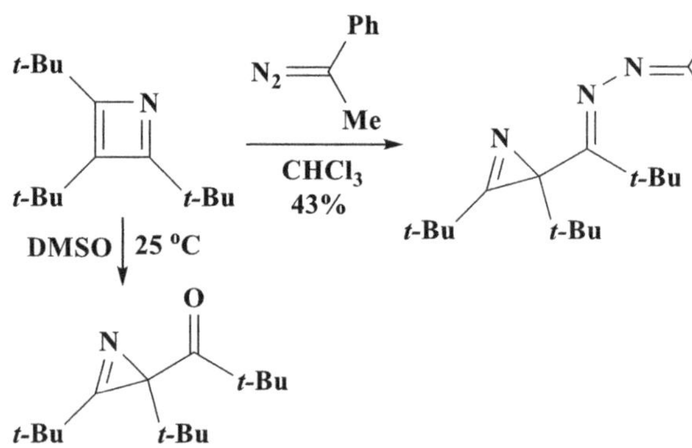

SCHEME 5

1.3 Synthesis of azirines from imines

This method was tolerant to other leaving groups on the imino nitrogen. One commonly met group of compounds in this regard included quaternary hydrazonium functionality [40]. For instance, the hydrazonium salt prepared from propiophenone was transformed to azirine (Scheme 6) in good yield on reaction with NaH in dimethylsulfoxide [41], and an excellent yield of azirine was obtained from the similar unsaturated system (Scheme 7) under the identical conditions [42]. In these cases, the regiochemistry was unambiguous, because there was only one set of α-protons. However, the procedure could be selective even when two regioisomers were possible. Thus, the naphthylacetone derivative afforded azirine (Scheme 8) via hydrazone synthesis, exhaustive methylation, and deprotonation, where the product was provided via benzylic anion [43]. The reasonably good yield of azirine was obtained by a subtle reaction sequence of pregnenolone hydrazonium salt (Scheme 9) corresponding to the ring closure through more substituted azaenolate species. The striking example of azirine stability was provided by later case because the azirine could be stored in C_2H_5OH for 2 weeks at rt without decomposition [44,45].

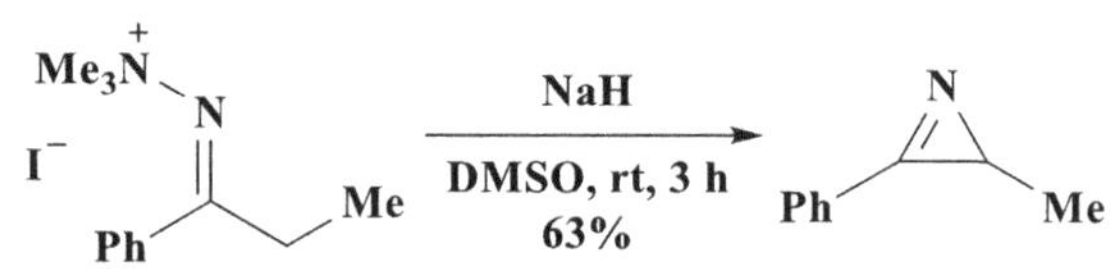

SCHEME 6

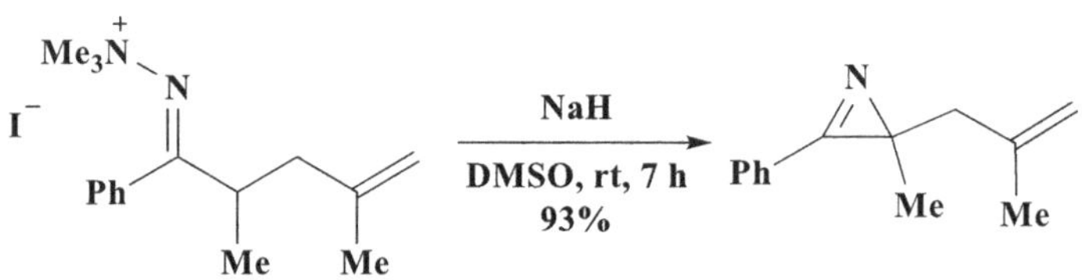

SCHEME 7

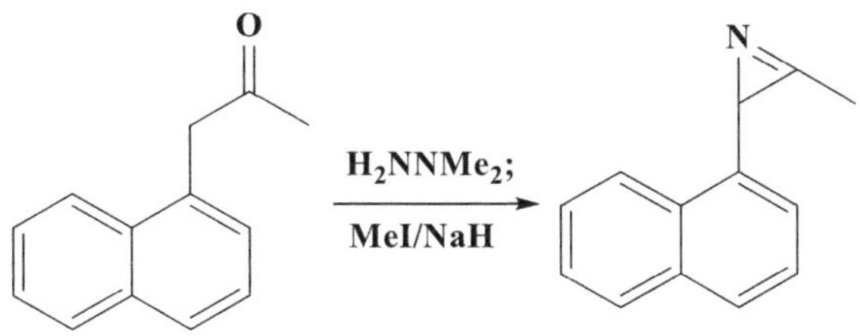

SCHEME 8

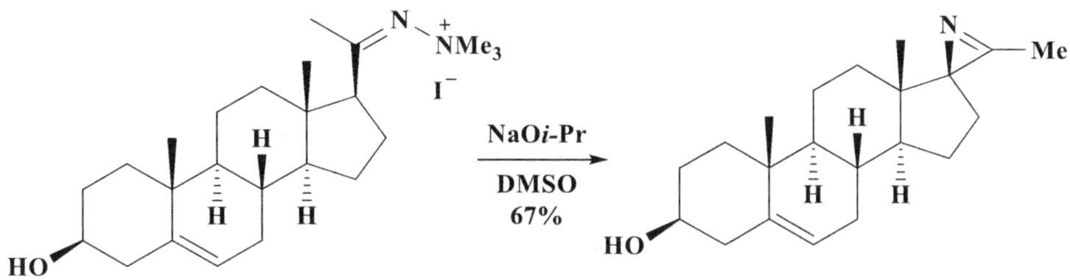

SCHEME 9

The ketone trimethylhydrazonium halides were utilized as an alternative of oxime sulfonate esters in a modified Neber rearrangement, thus allowing the formation of various substituted 2*H*-simple azirines in good yields, as given in Scheme 10 for allyl azirines [11,40,44].

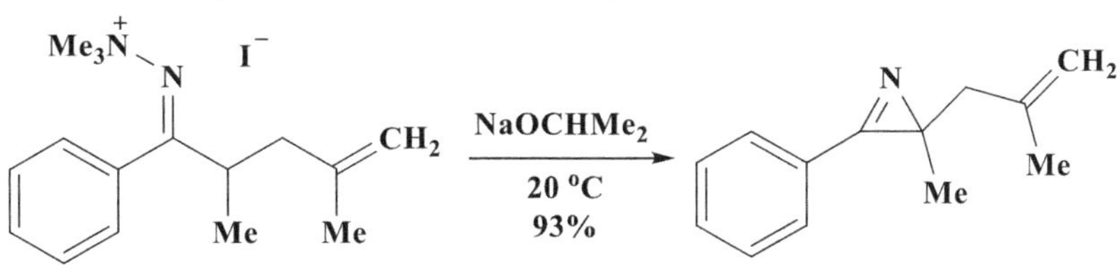

SCHEME 10

The 2*H*-azirines were formed when methyl iodides of dimethyl hydrazones were reacted with bases such as C_3H_7ONa (Scheme 11) [46–49].

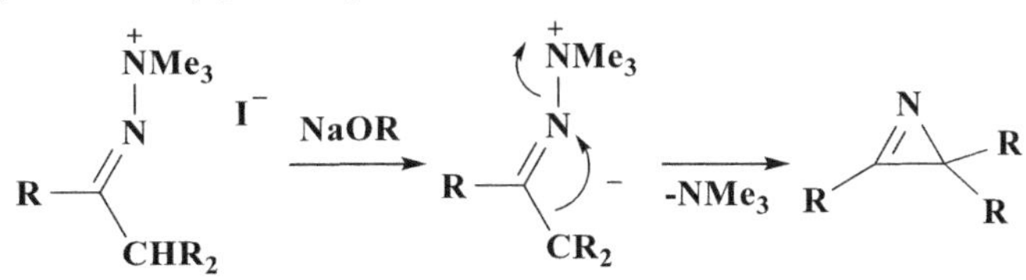

SCHEME 11

The 2,3-trisubstituted 2*H*-azirines were synthesized by reacting methiodides of ketone dimethylhydrazones with alcoholates (Scheme 12) [11,50].

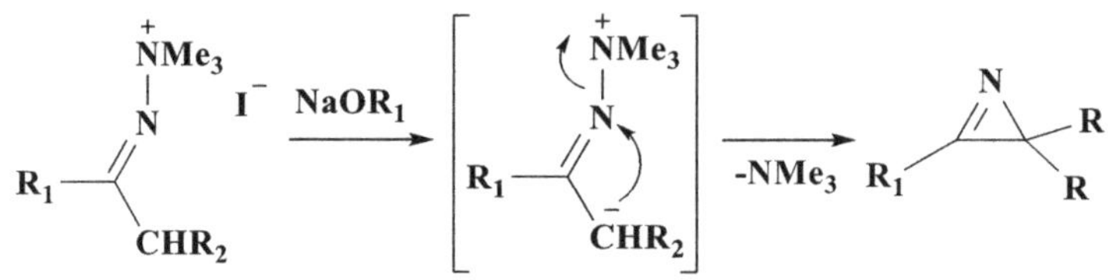

SCHEME 12

The optically active 2*H*-azirine was synthesized by the Neber reaction of *O*-mesyl amidoxime derivative bearing a chiral phenylglycine (Phg) ester as a chiral auxiliary. The reaction of *O*-mesyl amidoxime derivative bearing a chiral phenylglycine (Phg) ester with base afforded 3-amino-2*H*-azirine in 96:4 stereoselectivity and good yield (Scheme 13) [10,51].

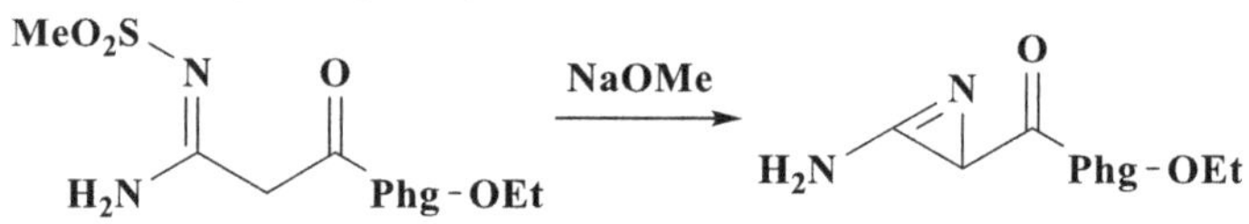

SCHEME 13

It was not until 1993 that Piskunova and coworkers [51] synthesized the synthetic optically active 2*H*-azirine by the reaction of *N*-acyl phenylglycine derivatives with base synthesizing azirines with high diastereomeric ratio (96:4) (Schemes 14 and 15).

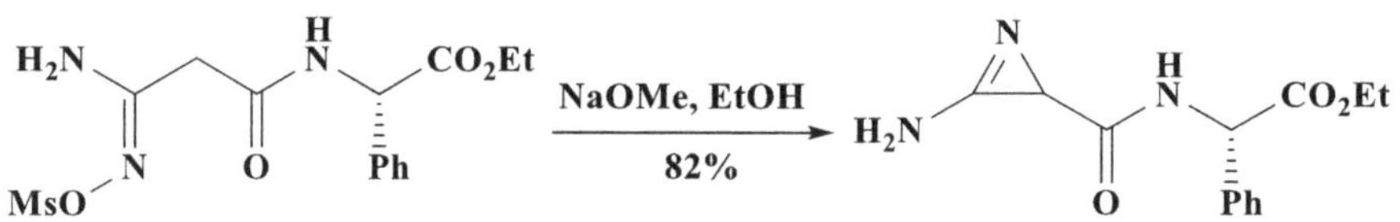

SCHEME 14

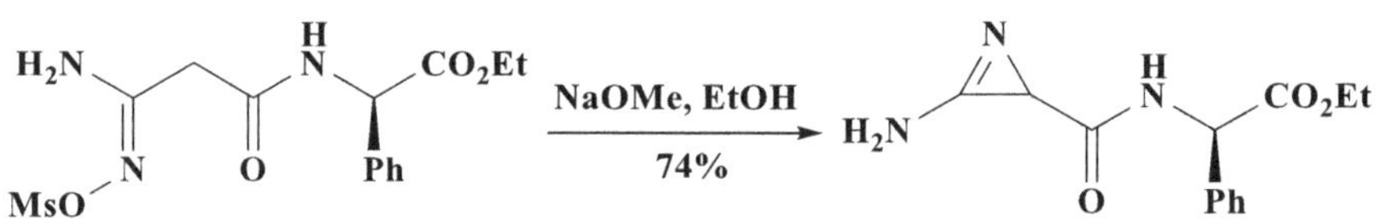

SCHEME 15

Various alternative approaches have been reported for the formation of chiral azirines [52], and this has subsequently resulted in a renewed interest in the utilization of these intermediates for aziridine formation. Davis and coworkers [53] synthesized (−)-dysidazirine, which is a natural product antibiotic, from an unsaturated chiral *N*-sulfinyl aziridine by an asymmetric aza-Darzens procedure (vide infra) revolving around the removal of methyl tolylsulfinite as a key azirine forming reaction (Scheme 16). The regiochemistry of elimination was intriguing, proceeding by the removal of less acidic proton (apparently).

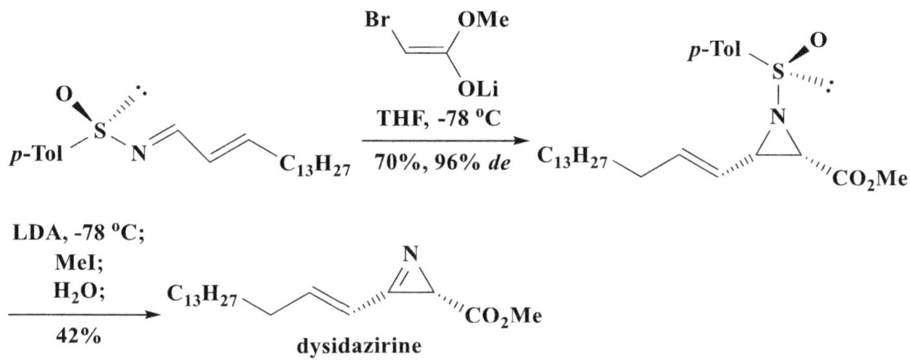

SCHEME 16

Davis et al. [53,54] reported a pathway to highly enantiopure 2*H*-azirines that included the removal of sulfenic acid from chiral, enantiopure *N*-sulfinylaziridine-2-carboxylate esters. The reaction of Andersen's reagent ((1*R*,2*S*,5*R*)-(−)-menthyl (*S*)-*p*-toluenesulfinate) with lithium bis(trimethylsilyl) amide followed by C_6H_5CHO provided *N*-*p*-toluenesulfinimine (Scheme 17). The *N*-*p*-toluenesulfinylaziridine was synthesized by the reaction of *N*-*p*-toluenesulfinimine with (*E*)-enolate of methyl bromoacetate. The removal of sulfinic acid with lithium diisopropylamide provided 2*H*-azirine-2-carboxylate ester in modest yield (42%) and ≥95% enantiomeric excess. An addition of methyl iodide in an elimination step improved the yield of 2*H*-azirine-2-carboxylate ester to some extent and allowed an isolation of *p*-TolylS(O)Me in high yield, presumably obtained by the trapping of removed sulfinic acid by methyl iodide. The isomeric 2*H*-azirine-3-carboxylate ester was not obtained even in traces.

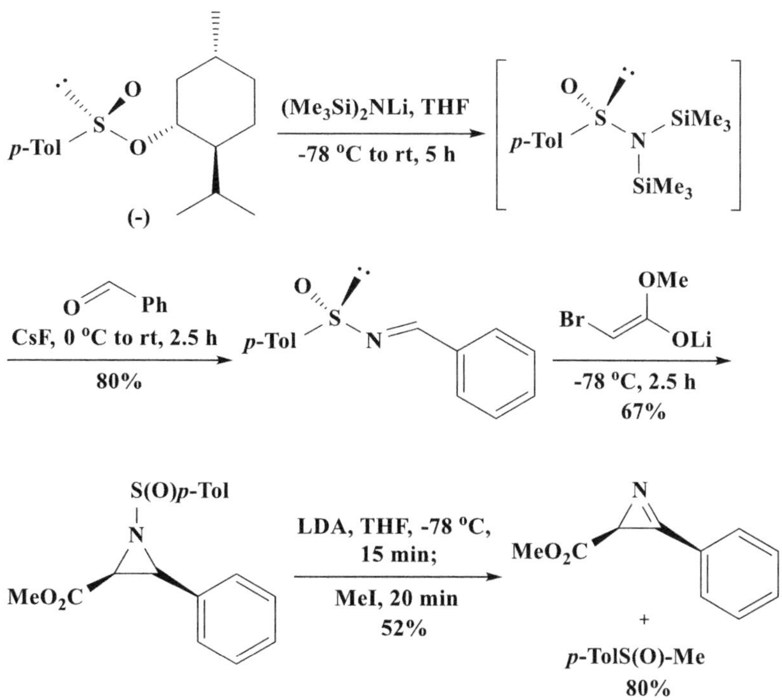

SCHEME 17

This improved process was employed to the first total formation of (−)-(2*R*)-(*E*)-dysidazirine. The *N*-*p*-toluenesulfinimine was prepared in good yield from (+)-(1*R*,2*S*,5*R*)-(−)-menthyl (*S*)-*p*-toluenesulfinate (Scheme 18). The aziridine synthesis proceeded readily to provide the aziridine, which was reacted with trimethylsilyl chloride and lithium diisopropylamide to synthesize the (−)-(2*R*)-(*E*)-dysidazirine. Nuclear magnetic resonance analysis with chiral shift reagent Eu(HFC)$_3$ indicated that the (−)-(2*R*)-(*E*)-dysidazirine was obtained with ≥95% enantiomeric excess [3].

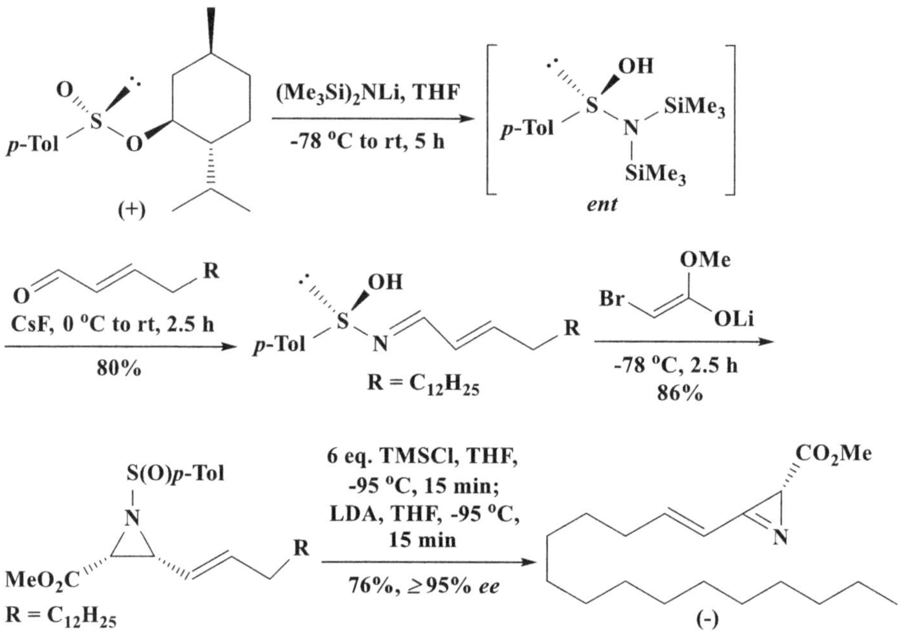

SCHEME 18

1.4 Synthesis of azirines from isoxazoles and oxazoles

The aryl- or alkyl-substituted isoxazoles have comparatively high thermal stability. The isoxazoles were stable for 10 d on heating at 280°C [55]. However, the isoxazoles containing heteroatom (oxygen, sulfur, and nitrogen) substituent at C-5 underwent a ring cleavage at lower temperatures. The 2*H*-azirine was formed on heating 5-alkoxy-3-arylisoxazole at 200°C (Scheme 19) [56]. The 3-amino-2*H*-azirines were formed in high yields through both photolysis and thermolysis of 3,5-bis(dimethylamino)isoxazoles (Scheme 20) [57].

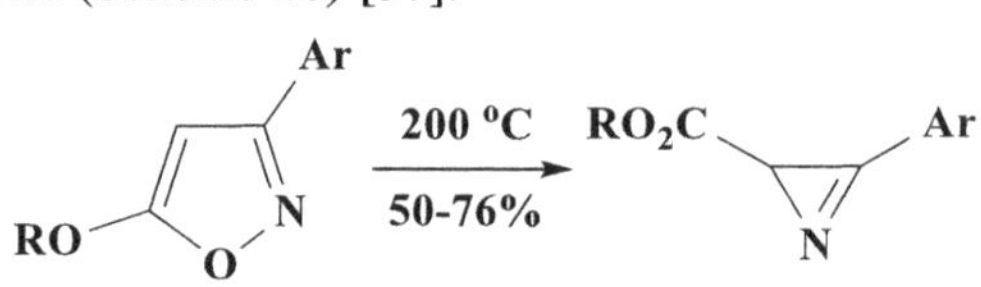

SCHEME 19

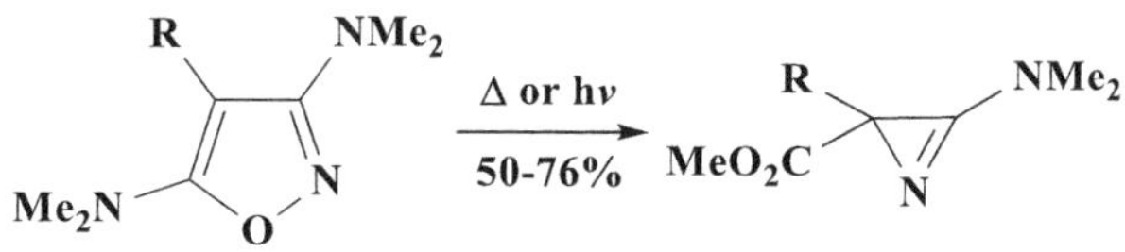

SCHEME 20

The acyl 2*H*-azirines were synthesized by the photochemical or thermal ring contraction of isoxazoles. They were rearranged to synthesize other heterocyclic compounds such as oxazoles [58–60]. These conversions were reversible by changing an irradiation wavelength or at high temperature. The thermolysis of isoxazoles afforded azirines in synthetically beneficial yields as shown by the thermal rearrangement of aminoisoxazole to azirinyl carboxamide under the Ar at high temperature (Scheme 21). The neat thermolysis was the preferred approach as lower yields were obtained in the presence of solvent [61]. The 5-alkoxy substituent was crucial as the 5-aryl or 5-alkyl substituents afforded enaminoketones under the same conditions [62]. This method was of limited synthetic value because of the need of high temperatures although the thermal rearrangement of isoxazoles provided various azirines in good yields. The theoretical and experimental studies have proposed that the mechanism of both the photochemically and thermally induced ring contraction of isoxazoles included a diradical intermediate generated by nitrogen–oxygen bond cleavage [63]. The 2*H*-azirine, once formed, opened to synthesize the oxazoles or nitriles, depending on the ring substitution pattern. The ring contraction of isoxazoles to azirines could also be promoted by base treatment or Fe(II) catalyst at rt. The 5-amino- and 5-alkoxyisoxazoles isomerized to 2*H*-azirine-2-carboxylic carboxamides and esters in approximately quantitative yield, respectively, by treatment with equimolecular or catalytic amounts of iron(II) chloride (Scheme 22). A regiochemically different approach of thermal rearrangement was shown by benzisoxazole derivatives. Instead of bonding to the internal α-carbon, the nitrogen synthesized azirine ring with the adjacent carbon of the substituent so that the aromaticity was preserved [64]. The functionalized benzisoxazoles containing electron-releasing substituents at the α-position underwent a ring contraction in the presence of strong bases such as sodium hydride, potassium *t*-butoxide, or sodium methoxide to synthesize the 2*H*-azirine-2-carboxylates in good yields (Scheme 23). Similar to isoxazoles, both pyrolysis and photolysis of 1,2,3-triazoles also delivered 2*H*-azirines, although in low yields because of the formation of other heterocyclic compounds [11,65].

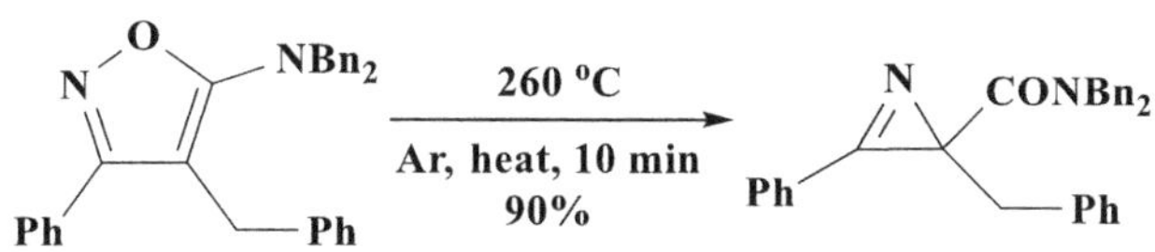

SCHEME 21

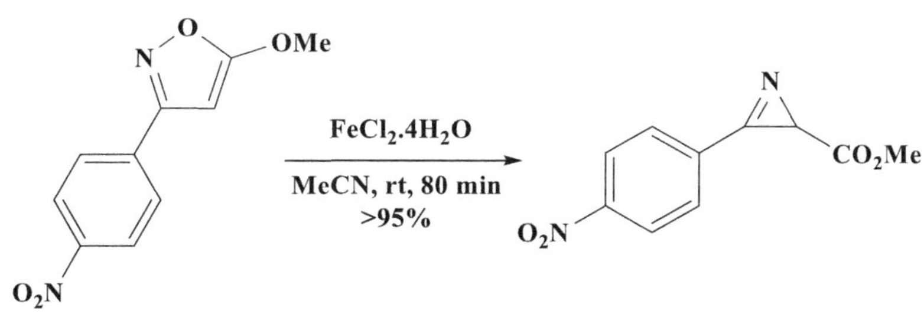

SCHEME 22

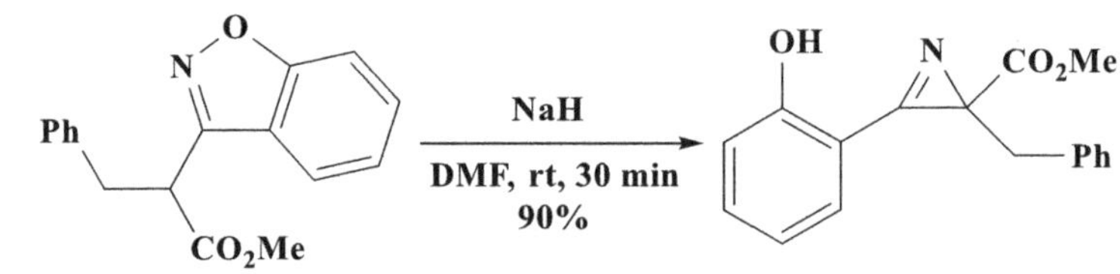

SCHEME 23

The isoxazole was heated to synthesize the acylazirine in 67% yield. The thermally initiated reactions are not terminated at the stage of acylazirine (Scheme 24) [66].

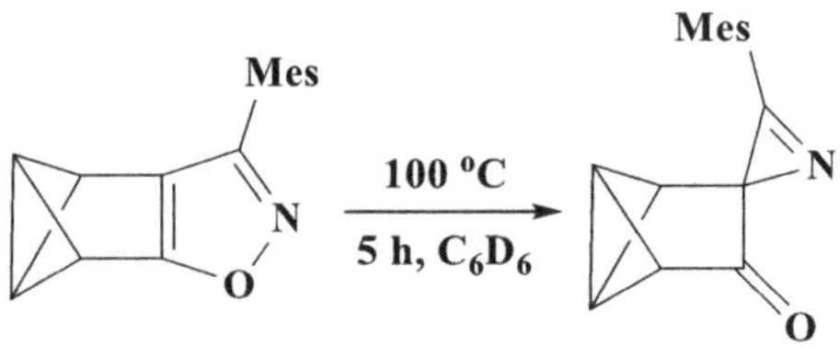

SCHEME 24

Another type of ring contraction affording 2*H*-azirines was thermally induced extrusion of phosphine oxide from 1,2,5- and 1,3,5-oxazaphosphole heterocyclic compounds [67–69]. The electron-withdrawing substituents at C-4 in 1,2,5-oxazaphospholes favored the formation of ketene-imine instead of 2*H*-arizine. Both products were the result of initial cleavage of phosphorus–carbon bond and successive removal of PPh_3O (Scheme 25). The 1,2,5-oxazaphospholes were prepared by the 1,3-dipolar cycloaddition of nitrile oxides to alkylidenephosphoranes [70] or by the reaction of α-bromoketoximes with PPh_3 and after that base-induced cyclization of an intermediate phosphonium salt [71]. The latter technique afforded 2*H*-azirines not easily accessible through a vinyl azide pathway [11].

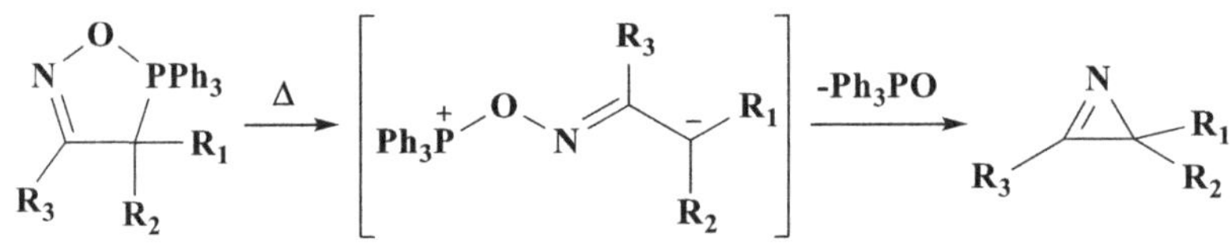

SCHEME 25

The azirine was formed when oxazaphosphole was subjected to pyrolysis under high vacuum at 400°C (Scheme 26), which was trapped at low temperature and characterized by infrared spectroscopy. The nitrile ylides were synthesized by the pyrolysis at higher temperatures (700°C) [67]. A rather more synthetically favorable process was reported for oxazaphospholine (available in four steps from α-bromoketoxime), which decomposed at atmospheric pressure and 120°C to synthesize the methyl azirine as a distillate in good yield (Scheme 27) [71].

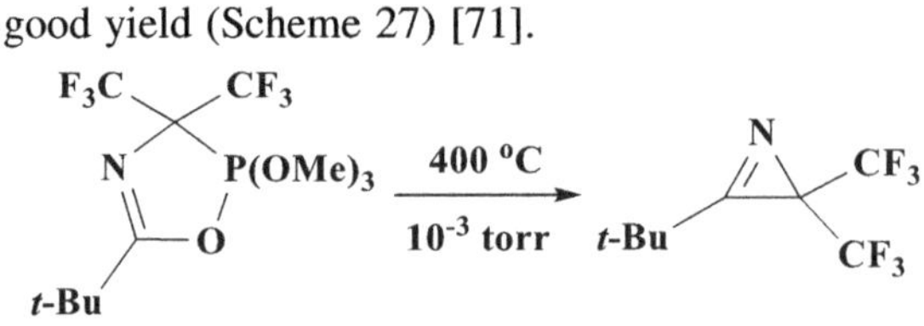

SCHEME 26

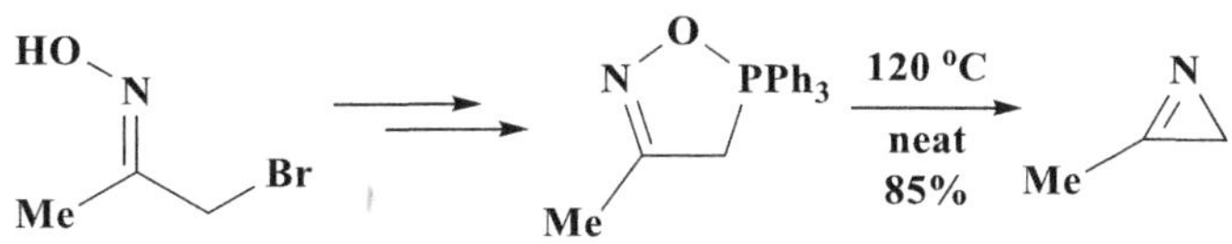

SCHEME 27

1.5 Synthesis of azirines from oximes

The first formation of enantiomerically enriched 2-phosphinyl-2*H*-azirines (R_1 = OEt, phenyl) was based on Neber rearrangement. The alkyl- and aryl-substituted azirines were obtained in excellent enantiomeric excess and chemical yields up to 82% (Scheme 28) [72]. The tosyloximes were obtained by the tosylation of oximophosphine oxides, easily prepared by an addition of hydroxylamine to allenes [73] in the case of alkyl-substituted oximes, and by the condensation reaction of bicarbonyl phosphine oxides with hydroxylamines in the case of aryl-substituted oximes. The aryl and alkyl azirines containing a phosphonate group in the 2-position were synthesized from tosyloxime employing an analogous approach [74,75]. This process could not be used for the formation of unsubstituted azirines (R_2=H) due to the inaccessibility of tosyloximes [48]. The 2*H*-azirine-2-carboxylic and the isosteric phosphoric esters prepared by this technique are of specific interest because they constitute a pathway for the formation of unnatural amino acids [59] and their isosteric phosphorus analogs through a reduction to aziridine and after that ring opening [11,76].

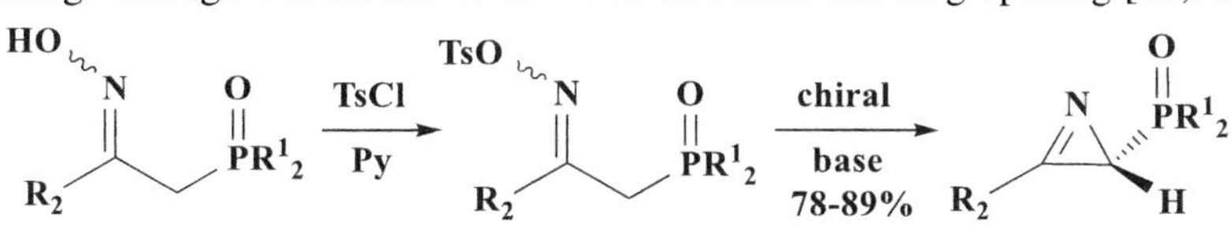

SCHEME 28

The acidity of protons in the α-position was increased by the presence of strong electron-withdrawing groups in the α-position to the C—N double bond, thus favoring cycloelimination reaction under the milder conditions. The Neber reaction occurred via either an internal concerted nucleophilic displacement (route A) [77] or through an electrocyclization of vinyl nitrene (route B) (Scheme 29) [11,78].

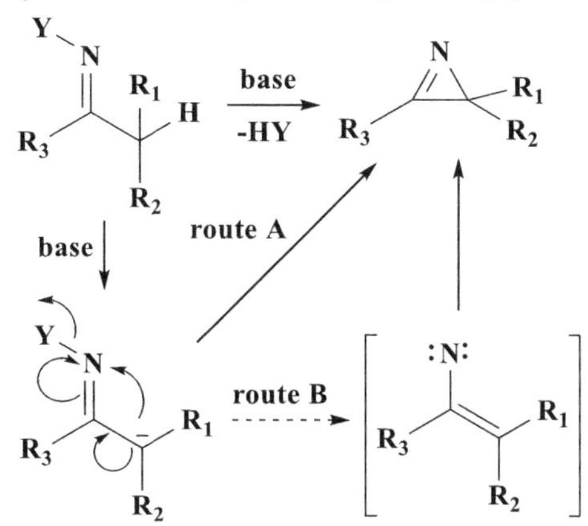

SCHEME 29

A stoichiometric amount of quinine or dihydroquinidine as chiral tertiary base was utilized for an asymmetric synthesis of azirine-2-carboxylates with 44%−82% *ee*. The catalytic amount (10 mol%) of quinidine afforded good results. The OH group of base was proved to be fundamental for a good stereoselectivity. Indeed, other chiral tertiary bases without such OH group, brucine, strychnine, and sparteine do not give any optically active azirine [79]. Later, this approach has also been used for the formation of enantiomerically enriched 2-phosphinyl-2*H*-azirines (Scheme 30) [72].

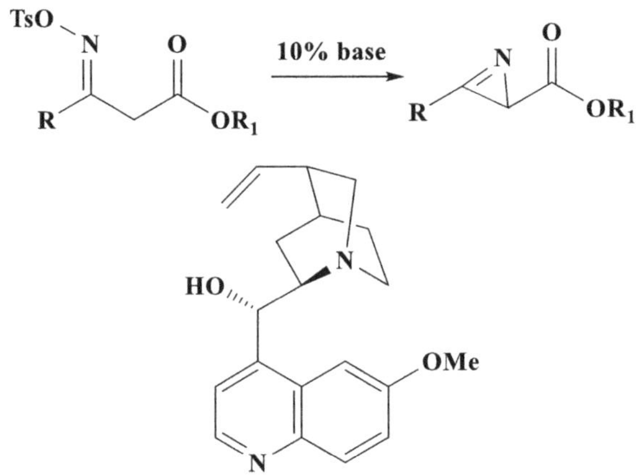

base

SCHEME 30

Neber and coworkers [80,81] synthesized 2*H*-azirines. The Neber reaction probably proceeded either via an internal concerted nucleophilic displacement (route A) or through an electrocyclization of vinyl nitrene (route B), a reactive species synthesized by a base-assisted loss of leaving group on the nitrogen atom of oxime sulfonates and hydrazonium halides (Scheme 31) [82].

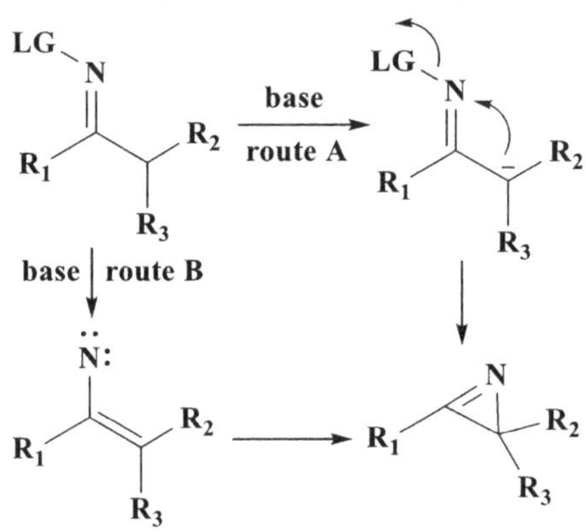

SCHEME 31

The optically active 2*H*-azirines substituted in 2-position with phosphonate or in 3-position with phosphine oxide group were synthesized with modest enantiomeric excess by the Neber reaction starting from easily accessible oximes utilizing chiral polymer-bound amines (Scheme 32) [83].

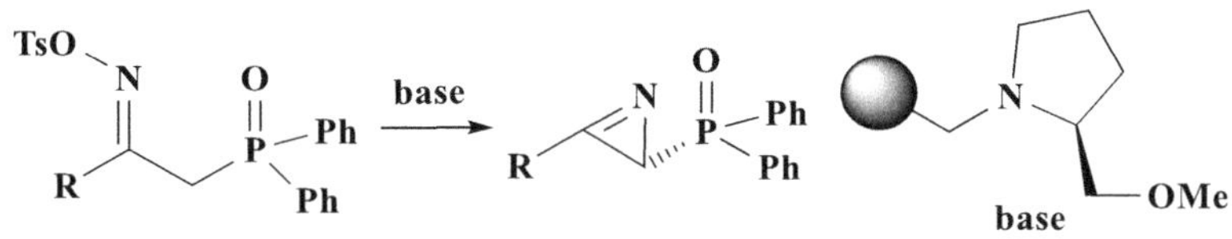

SCHEME 32

The 2*H*-azirines are thermally active and therefore react further, decreasing the yields and causing difficulty in their isolation. To prevent these difficulties, the Neber reaction [81] pathway has also been widely used and could be promoted by an introduction of electron-withdrawing groups at the α-position of *N*-functionalized imine (Scheme 33). An asymmetric synthesis of 2*H*-azirines having a carboxylic ester group [79] or a phosphorous substituent [84] by this method has been reported.

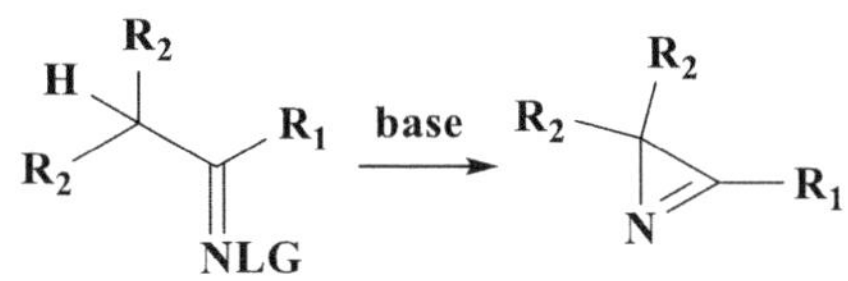

LG = leaving group (i.e. OTs, $\overset{+}{N}(Me)_3$ etc.)

SCHEME 33

The alkaloid-assisted Neber reactions of β-keto tosyloximes resulted in an asymmetric synthesis of 2*H*-azirines having phosphinyl [72] and phosphonyl [75] substituents. Likewise, the utilization of chiral polymer-supported bases [85] provided 2*H*-azirines in high yields and regioselectively (Scheme 34) [86].

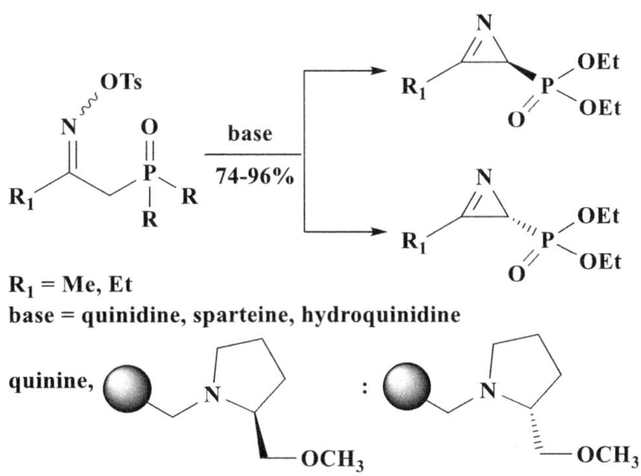

SCHEME 34

The reaction of cyanoacetanilide with NH_2OH using NaOMe as a basic catalyst afforded oxime, which was reacted with *p*-toluene sulfonyl chloride to provide the imine. The 3-amino-*N*-phenyl-1*H*-azirine-2-carboxamide was synthesized by an intramolecular nucleophilic substitution of imine in NaOMe (Scheme 35) [87,88].

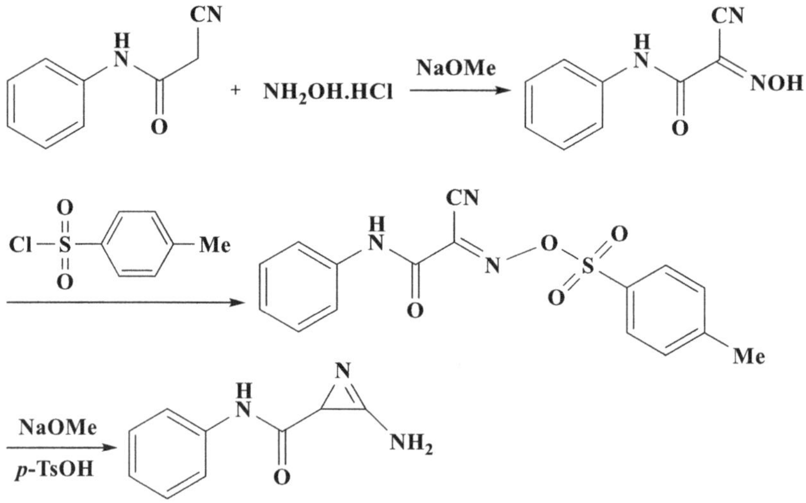

SCHEME 35

The optically active azirines were synthesized by the Neber rearrangement of *o*-methanesulfonyl derivative of amidoxime in which a chiral phenylglycine had been incorporated as a chiral auxiliary. The reaction of this derivative with base provided 3-amine-2*H*-azirine containing two stereocenters, one endocyclic (C-2 of azirine) and one exocyclic, in good stereoselectivity (96:4) and yield (Scheme 36) [11,89].

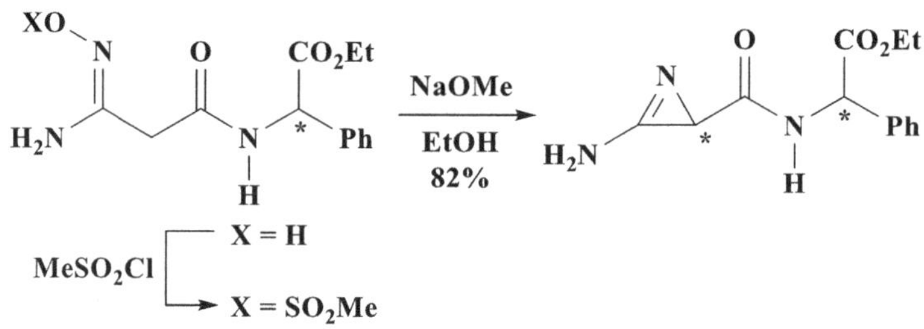

SCHEME 36

There seems to be no generally appropriate method to chiral nonracemic azirines; however, strides are being synthesized. In one instance, the chiral amido oxime derivative underwent transformation into azirine with exclusive diastereoselectivity (Scheme 37) [51]. It was also interesting to report here the ability to isolate the azirine from the type of protic media generally met in the Neber rearrangement, which seemed promising for adapting these types of procedures to azirine formation. The utilization of chiral bases has encountered with marginal achievement. In one situation, moderate enantioselectivities (up to 70% enantiomeric excess) were attained utilizing a chiral phase-transfer agent, which was supposed to give a tight ion pair with hydroxide and thus impose an asymmetric atmosphere around the proton-transfer transition state [90]. The chiral base could also be utilized for good results as shown by the formation of azirine using quinidine (Scheme 38). Best results were observed utilizing stoichiometric amounts of organic base. However, a 20 mol% loading could be utilized using K_2CO_3 with only a small decrease in selectivity [79]. This approach has also been used to synthesize the azirines derived from phosphine oxides (Scheme 39) [72].

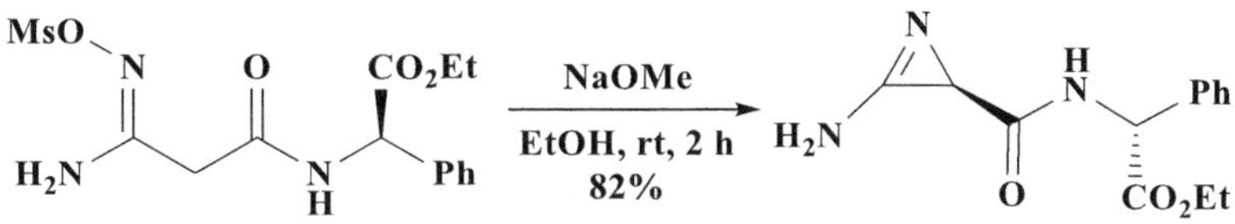

SCHEME 37

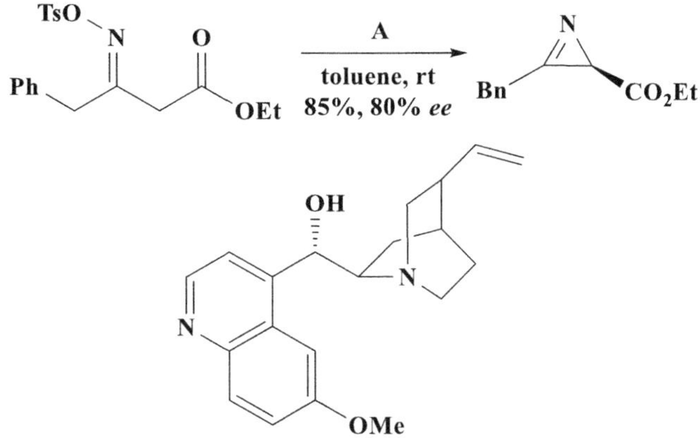

A

SCHEME 38

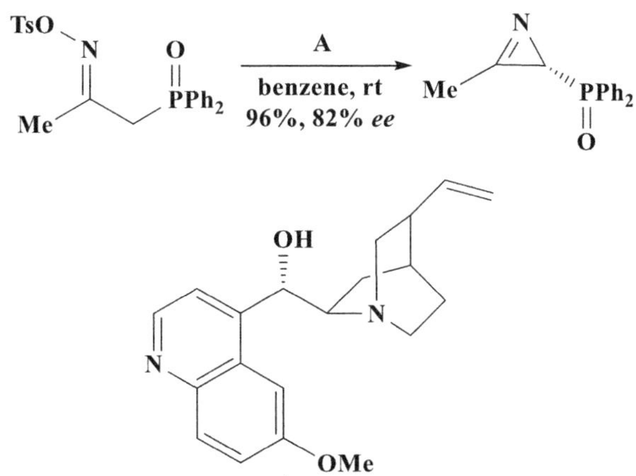

A

SCHEME 39

Palacios and coworkers [72,75] extended this method to the synthesis of azirines substituted at C-2 with low % enantiomeric excess with either phosphine oxides or phosphonates (Schemes 40 and 41).

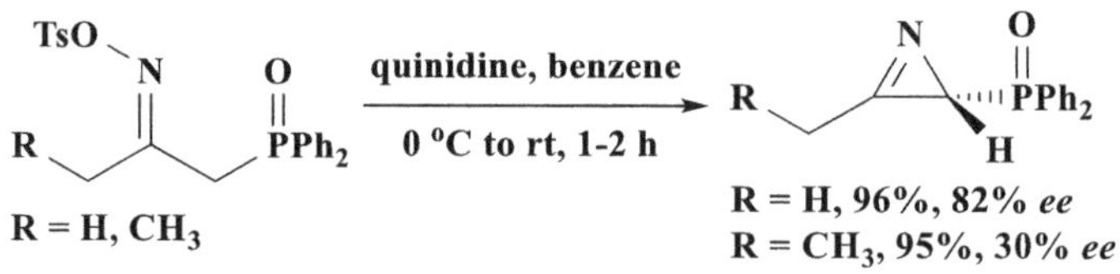

SCHEME 40

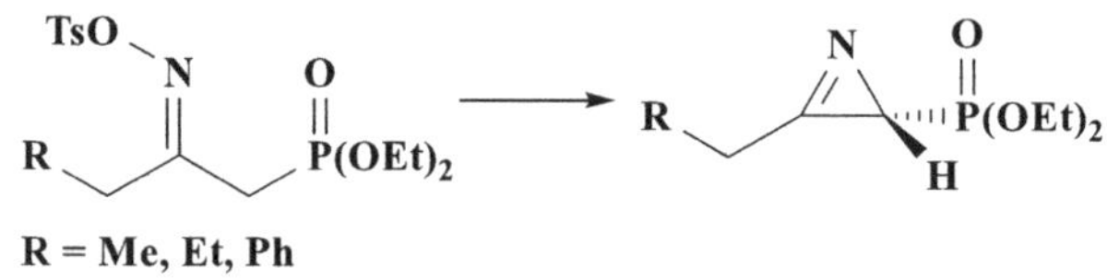

base	yield (%)	*ee* (%)
a: quinidine	90	20 (*S*)
(-)-sparteine	87	2 (*S*)
b: quinidine	95	24 (*S*)
hydroquinidine	92	22 (*S*)
quinine	94	8 (*R*)
c: quinidine	85	52 (*S*)

SCHEME 41

The azirines (three-membered cyclic imines) are related to aziridines by a single redox step, and these reagents could serve as precursors for the synthesis of aziridines by addition reactions. An addition of carbon nucleophiles has been known for some time [91] and attracting the interest of various research groups. The Neber reaction [92], involving the cyclization of 2-(*O*-tosyl)oximino carbonyl compounds, is the oldest known azirine formation, and asymmetric variants have been reported. For instance, Verstappen and coworkers [79] synthesized nonracemic chiral azirines from oximes of 3-ketoesters in the presence of *Cinchona* alkaloids as catalysts (Scheme 42) [90].

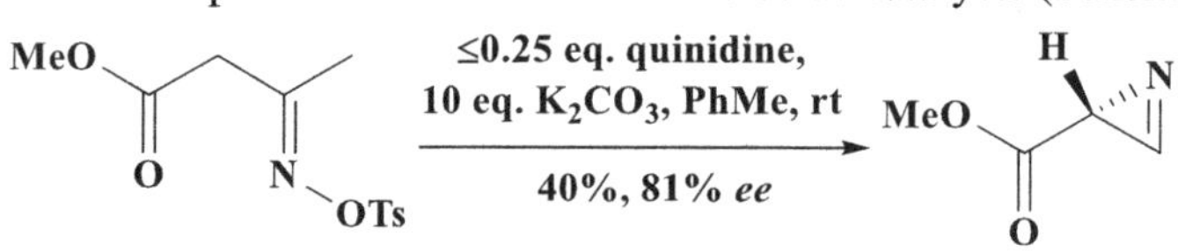

SCHEME 42

The azirines were synthesized, enantioselectively if needed utilizing a natural alkaloid as base, from *O*-tosyl derivatives of oximes of 1,3-keto esters. The carbon was nucleophilic center, and nitrogen was attacked with the departure of tosylate (Scheme 43) [79].

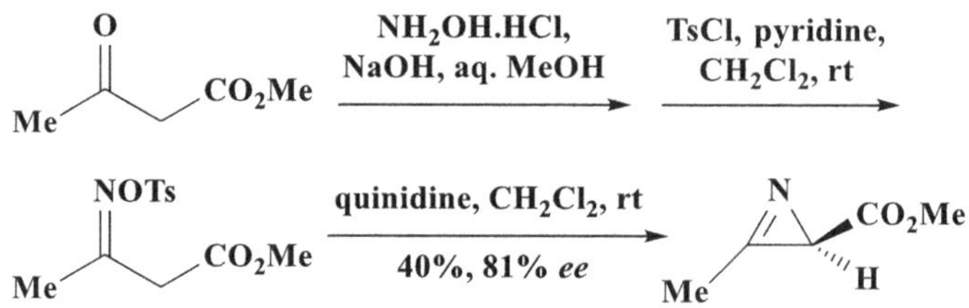

SCHEME 43

The azirines have attracted attention as they represent the smallest nitrogen-containing unsaturated heterocycles [11,93–96]. Most interest stems from the effect of ring strain of the azirine on the potential for elaboration to other heterocycles. As with the formation of pyrazolo[1,5-*a*]pyridines, azirine synthesis can be categorized into inter- and intramolecular classes. The intramolecular variants (reaction of *N*-functionalized imine) are generally named as Neber reaction based on the first synthesis of an azirine in a cycloelimination reaction under the basic conditions [81]. Other intramolecular syntheses of azirines start with vinyl azides, isoxazoles [61], and oxazaphospholes [97]. The less common intermolecular method for the formation of azirines generally depends on a cycloaddition between nitriles and carbenes [27]. Useful yields in this method usually need stabilized phosphanyl carbenes, which afforded phosphorous-substituted azirines (Scheme 44) [32,98].

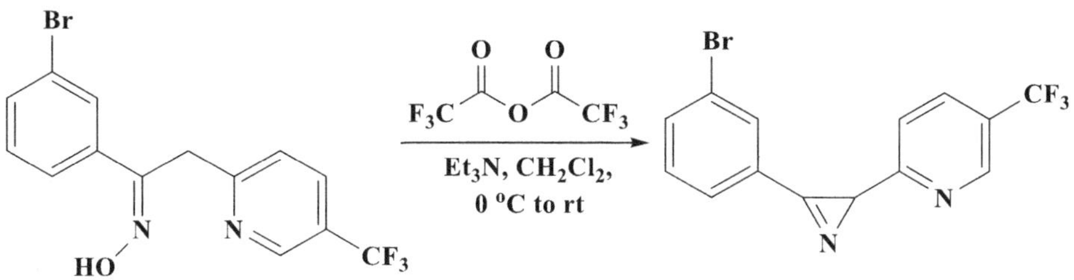

SCHEME 44

A modified Neber reaction of oxime tosylates is an attractive method for azirine esters synthesis (Scheme 45) [99]. An asymmetric synthesis of azirine esters was performed with chiral base quinidine. The nature of solvent is essential in this asymmetric type of reaction. The enantiomeric excess of azirine esters was 80% in toluene, while enantiomeric excess in CH_3CN, ether, or CH_2Cl_2 was 8%–47%. Also, the catalytic version of this asymmetric Neber reaction was reported utilizing solid K_2CO_3 as a regenerating auxiliary base [100].

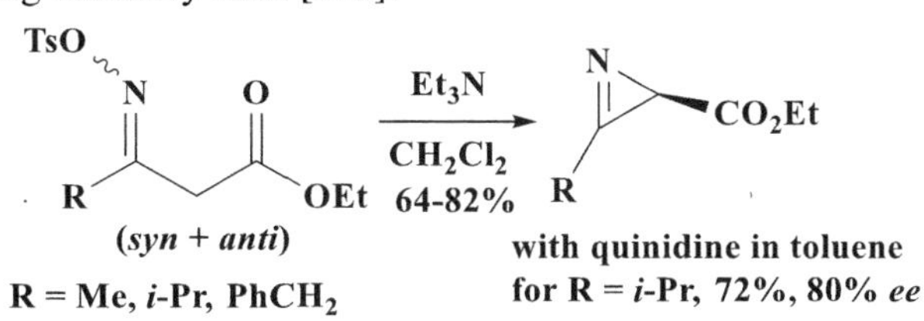

SCHEME 45

Generally, there are two regiochemical possibilities for an amination, in turn, via two regioisomeric azirines. House and Berkowitz [78] reported that the regiochemistry of Neber rearrangement was driven not by oxime geometry (as with the Beckmann rearrangement) but rather by the relative acidities of the α-protons. In other words, an initial tosylate displacement was influenced by more readily formed carbanion. The active methylene almost always ends up as the saturated carbon of the azirine in Neber precursors with electron-withdrawing groups. For instance, the reaction of phosphonatoketoxime with TEA at rt afforded azirine as an exclusive regioisomer (Scheme 46) [75]. Equally satisfactory results could be observed utilizing a solid-supported TEA (Scheme 47) [85]. An unusual set of oxidative conditions has been reported for the thioacetal oxime carbamate, which underwent an oxidation of only one sulfide linkage with $KMnO_4$ and further cyclized to azirine, presumably under the effect of hydroxide liberated during an oxidation (Scheme 48) [77].

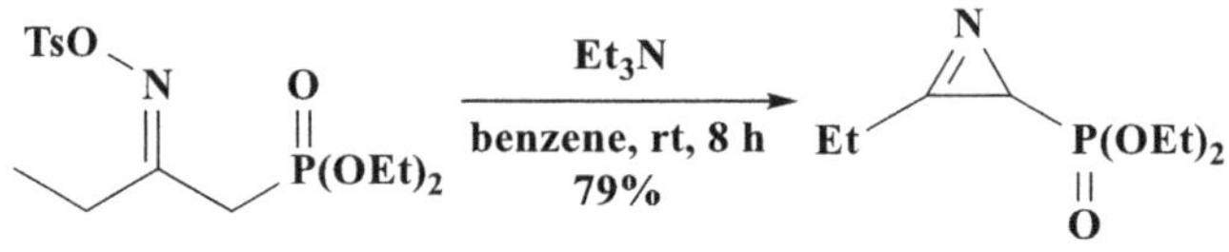

SCHEME 46

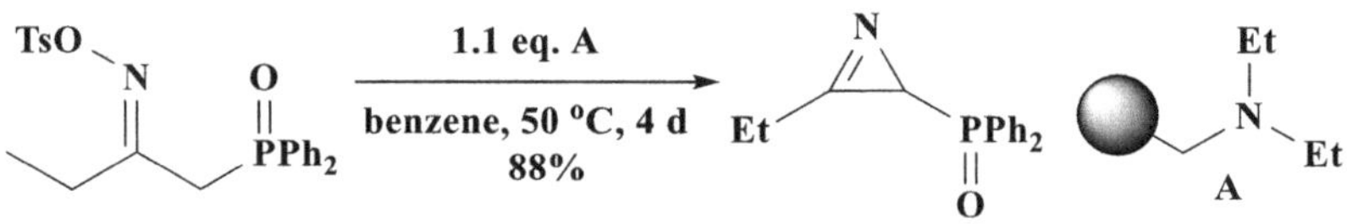

SCHEME 47

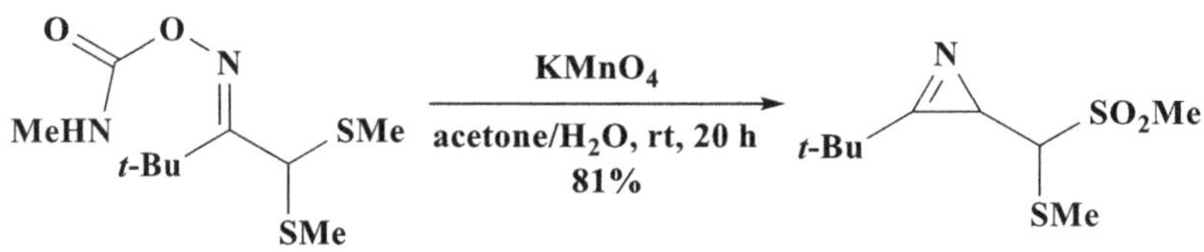

SCHEME 48

The oximes were obtained in 91%–93% yield from aryl benzotriazol-1-ylmethyl ketones, $NH_2OH.HCl$, and sodium hydroxide in refluxing C_2H_5OH/H_2O. They could be transformed to their tosylates, but a large excess of potassium hydroxide transformed them directly into 2*H*-azirines (58%–66% yield). The benzotriazolyl functionality in azirines was easily substituted by nucleophiles (sodium thiophenoxide, potassium phthalimide, or organomagnesium reagents) to synthesize the disubstituted azirines in 50%–79% yield (Scheme 49) [101,102].

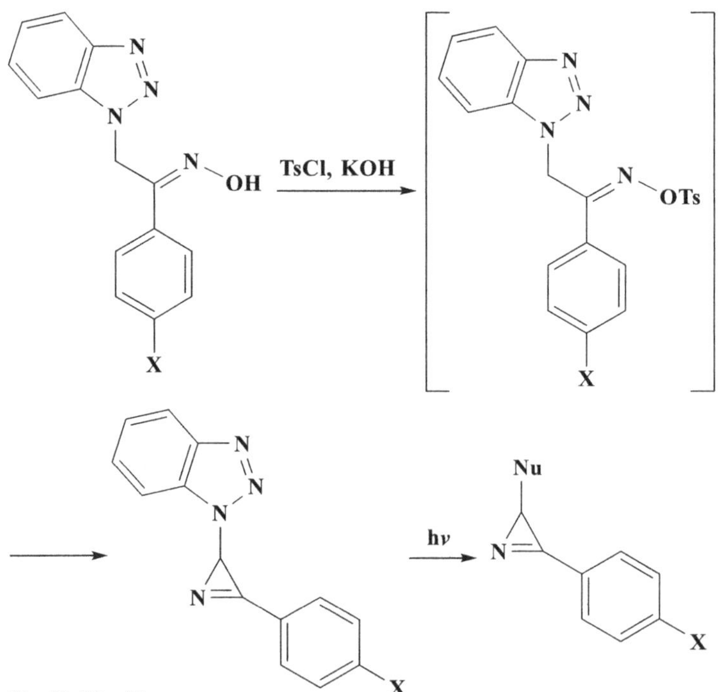

X = H, Me, Cl

Nu = $PhCH_2MgBr$, $4\text{-}MeC_6H_4CH_2MgBr$, PhSNa, K-phthalimide

SCHEME 49

The reaction of oximes with TsCl and aqueous potassium hydroxide resulted in the formation of 2-(benzotriazol-1-yl)-2*H*-azirines, which were further reacted with 4-methylbenzylmagnesium bromide or benzylmagnesium bromide using $ZnCl_2$ to provide the 2-benzyl-2*H*-azirines. The 2-(benzotriazol-1-yl)-2*H*-azirines were transformed into 2*H*-azirines in the presence of sodium salt of benzenethiol and potassium phthalimide (Scheme 50) [101]. The *N*-acylmethylbenzotriazoles were prepared either by the reaction of acid chlorides with 1-(trimethylsilylmethyl)benzotriazole [103] or by the reaction of 2-chloroacetophenone and benzotriazole in toluene under the reflux [104]. The reaction of *N*-acylmethylbenzotriazoles with $NH_2OH.HCl$ afforded oximes (90%) as single isomers, which were reacted with TsCl and aqueous potassium hydroxide in a mixture of $CHCl_3$ and Et_2O at 0°C using tetra-*n*-butyl ammonium hydrogen sulfate as a PTC to synthesize the 2-(benzotriazol-1-yl)-2*H*-azirines. The amount and concentration of aqueous potassium hydroxide and the reaction time influenced the yields considerably. The optimal conditions were found to be 10 eq. 10% aqueous potassium hydroxide and 5–6 h at 0°C; this had prevented any isolation of oxime tosylates. Most recognized reactions of 2*H*-azirines with nucleophiles include nucleophilic addition to the imine bond [9,105]. No earlier nucleophilic replacement of 2-substituent of 2*H*-azirines with carbon nucleophile has been reported. The reported successful halide displacements of 2-halo-2*H*-azirines with nucleophiles have all involved an additional EWG, for example, carboxylate at the 2-position [106]. The reaction of 2-(benzotriazol-1-yl)-2*H*-azirines and 4-methylbenzylmagnesium bromide or benzylmagnesium chloride using $ZnCl_2$ provided modest to good yields of desired products [107]. Unfortunately, the reactions of 2-(benzotriazol-1-yl)-2*H*-azirines with other Grignard reagents, both alkyl (e.g., pentyl and phenylethyl) and aryl (e.g., *p*-chlorophenyl, *p*-tolyl, and phenyl), in the presence of $ZnCl_2$ failed. In some cases, the reactions not proceeded but allowed the recovery of starting 2-(benzotriazol-1-yl)-2*H*-azirines, while in other cases, the disappearance of starting 2-(benzotriazol-1-yl)-2*H*-azirines provided a complicated mixture. Following a process earlier utilized for the substitution of 2-halo-2*H*-azirines [108], the reaction of 2-(benzotriazol-1-yl)-2*H*-azirines with potassium phthalimide at rt in dimethylformamide provided 2*H*-azirine derivatives in good yield. The reaction with sodium salt of benzenethiol was less successful. The azirines provided an oil that consisted largely of phenyl 2*H*-azirines, a compound that decomposed at 20°C and could not be purified completely.

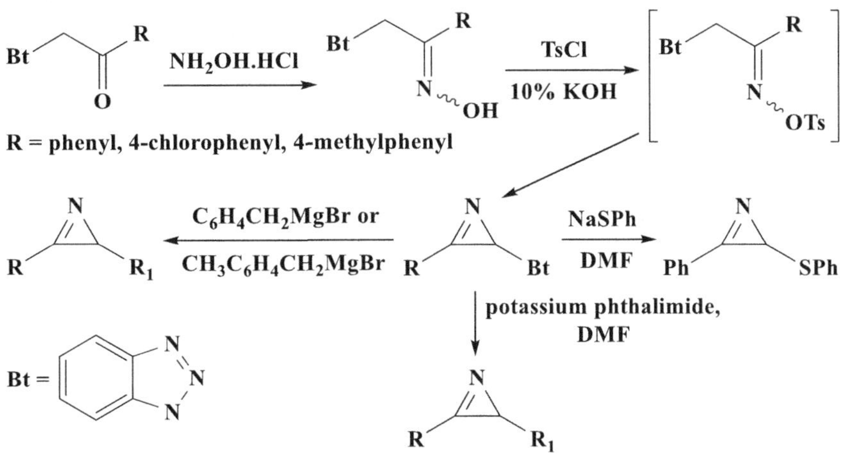

SCHEME 50

The reaction of α-benzotriazolyl oximes was beneficial as a benzotriazolyl unit in the product could be displaced by nucleophiles (Scheme 51) [101].

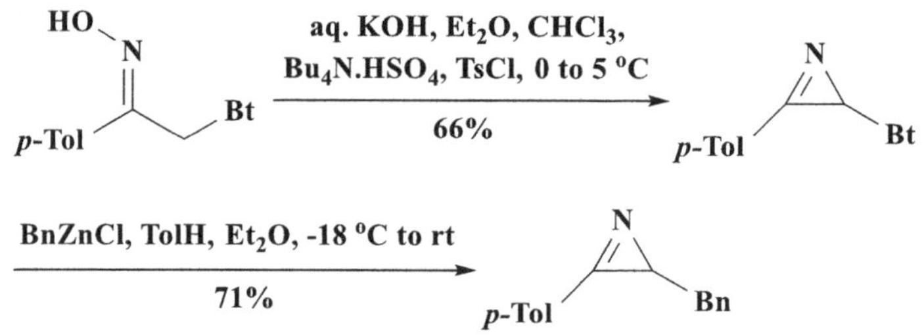

SCHEME 51

The 2*H*-azirine (81%) was prepared stereospecifically by a modified Neber reaction of (*E*)-oxime carbamate with potassium permanganate or *m*-CPBA (Scheme 52) [77]. An initial oxidation of sulfur increased the acidity of methine proton, resulting in the cyclization through an intermediate.

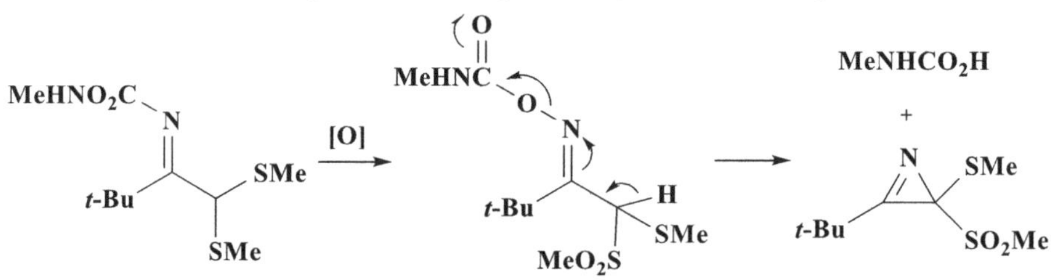

SCHEME 52

The Neber reaction used a chiral tertiary base such as quinine or dihydroquinidine not covalently bonded to the substrate (Scheme 53) [79]. No optically active heterocyclic compounds were formed using other chiral tertiary bases lacking the OH group such as brucine, strychnine, and sparteine or utilizing hydroxylic solvents such as C_2H_5OH. Therefore, the stereoselectivity obtained utilizing quinidine or dihydroquinidine could be explained by the formation of a hydrogen bond between the OH group of the base and one of the S=O functionalities of the ketoxime tosylate during an abstraction of a methylene proton. The 44%–82% enantiomeric excess was obtained when an equimolecular amount of base was used, but excellent results were also found when 10 mol% quinidine was utilized and 10–20 eq. K_2CO_3 was added to regenerate the alkaloid base in situ. Finally, it was remarkable that the opposite antipodes of product were obtained from pseudo-enantiomers of alkaloid bases [11].

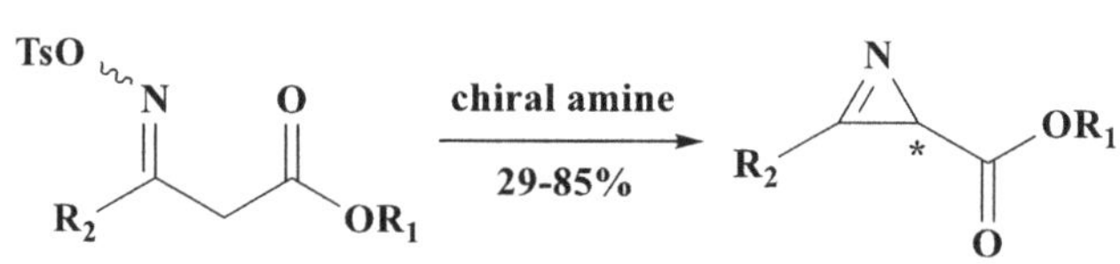

SCHEME 53

Neber and coworkers [80,81] reported the first 2*H*-azirine synthesis as intermediates in the formation of aminoketones by reacting oxime *p*-toluenesulfonates with base (Scheme 54). Since then, various azirines have been synthesized by this method including highly functionalized derivatives such as sulfonyl azirines [109]. This method can also be extended to other substituted oxime derivatives [110] such as carbamate oximes or *o*-benzoyl oximes [77]. An introduction of double bond into a strained three-membered ring was thought to preclude the existence of azirine, and this assignment was encountered with skepticism at the time. Hatch and Cram [111] conducted an extensive study and confirmed the Neber's original assignment [11].

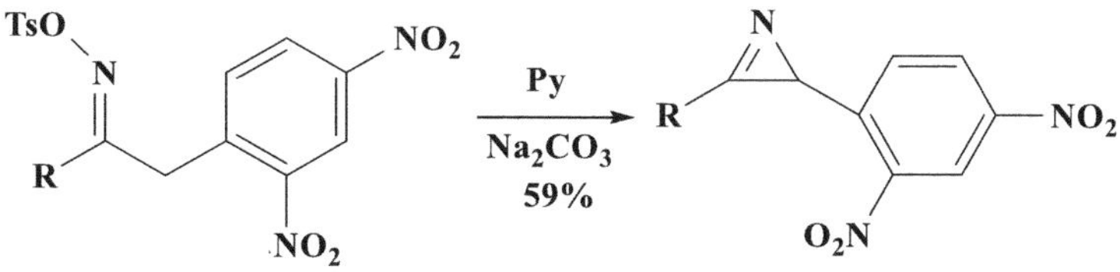

SCHEME 54

References

1. (a) N. Kaur. 2018. Mercury-catalyzed synthesis of heterocycles. Synth. Commun. 48: 2715-2749. (b) N. Kaur, N.K. Jangid and V. Rawat. 2018. Synthesis of heterocycles through platinum-catalyzed reactions. Curr. Catal. 7: 3-25. (c) N. Kaur, Y. Verma, P. Grewal, P. Bhardwaj and M. Devi. 2019. Application of titanium catalysts for the syntheses of heterocycles. Synth. Commun. 49: 1847-1894. (d) N. Kaur and D. Kishore. 2014. Synthesis of 2-(oxadiazolo, pyrimido, imidazolo, and benzimidazolo) substituted analogues of 1,4-benzodiazepin-5-carboxamides linked through a phenoxyl bridge. J. Chem. Sci. 126: 1861-1867. (e) N. Kaur. 2015. Six-membered *N*-heterocycles: microwave-assisted synthesis. Synth. Commun. 45: 1-34. (f) N. Kaur. 2019. Synthesis of five-membered heterocycles containing nitrogen heteroatom under ultrasonic irradiation. Mini Rev. Org. Chem. 16: 481-503. (g) N. Kaur. 2013. Solid-phase synthetic approach to the synthesis of azepine heterocycles of medicinal interest. Int. J. Pharm. Biol. Sci. 4: 357-372.
2. (a) N. Kaur. 2015. Polycyclic six-membered *N*-heterocycles: microwave-assisted synthesis. Synth. Commun. 45: 35-69. (b) N. Kaur. 2015. Application of microwave-assisted synthesis in the synthesis of fused six-membered heterocycles with *N*-heteroatom. Synth. Commun. 45: 173-201. (c) N. Kaur, R. Tyagi, M. Srivastava and D. Kishore. 2014. Application of dimethylaminomethylene ketone in heterocycles synthesis: synthesis of 2-(isoxazolo, pyrazolo and pyrimido) substituted analogs of 1,4-benzodiazepin-5-carboxamides linked through an oxyphenyl bridge. J. Heterocycl. Chem. 51: E50-E54. (d) N. Kaur. 2015. Microwave-assisted synthesis of fused polycyclic six-membered *N*-heterocycles. Synth. Commun. 45: 273-299. (e) N. Kaur. 2015. Review on the synthesis of six-membered *N,N*-heterocycles by microwave irradiation. Synth. Commun. 45: 1145-1182. (f) N. Kaur. 2018. Ultrasound-assisted green synthesis of five-membered *O*- and *S*-heterocycles. Synth. Commun. 48: 1715-1738. (g) N. Kaur. 2018. Solid-phase synthesis of sulfur-containing heterocycles. J. Sulfur Chem. 39: 544-577.
3. (a) N. Kaur. 2015. Applications of microwaves in the synthesis of polycyclic six-membered *N,N*-heterocycles. Synth. Commun. 45: 1599-1631. (b) N. Kaur. 2015. Greener and expeditious synthesis of fused six-membered *N,N*-heterocycles using microwave irradiation. Synth. Commun. 45: 1493-1519. (c) N. Kaur, P. Grewal, P. Bhardwaj, M. Devi, N. Ahlawat and Y. Verma. 2020. Synthesis of five-membered *N*-heterocycles using silver metal. Synth. Commun. 49: 3058-3100. (d) N. Kaur, N. Ahlawat, Y. Verma, P. Grewal and P. Bhardwaj. 2019. A review of ruthenium catalyzed C-N bond formation reactions for the synthesis of

five-membered *N*-heterocycles. Curr. Org. Chem. 23: 1901-1944. (e) N. Kaur, N. Ahlawat, P. Bhardwaj, Y. Verma, P. Grewal and N.K. Jangid. 2020. Synthesis of five-membered *N*-heterocycles using Rh based metal catalysts. Synth. Commun. 50: 137-160. (f) N. Kaur. 2018. Ultrasound assisted synthesis of six-membered *N*-heterocycles. Mini Rev. Org. Chem. 15: 520-536. (g) N. Kaur. 2017. Ionic liquids: promising but challenging solvents for the synthesis of *N*-heterocycles. Mini Rev. Org. Chem. 14: 3-23.

4. (a) N. Kaur. 2018. Copper catalysts in the synthesis of five-membered *N*-polyheterocycles. Curr. Org. Synth. 15: 940-971. (b) N. Kaur, Y. Verma, P. Grewal, N. Ahlawat, P. Bhardwaj and N.K. Jangid. 2020. Palladium acetate assisted synthesis of five-membered *N*-polyheterocycles. Synth. Commun. 50: 1567-1621. (c) N. Kaur. 2015. Palladium catalysts: synthesis of five-membered *N*-heterocycles fused with other heterocycles. Catal. Rev. 57: 1-78. (d) N. Kaur. 2015. Review of microwave-assisted synthesis of benzo-fused six-membered *N,N*-heterocycles. Synth. Commun. 45: 300-330. (e) N. Kaur and D. Kishore. 2014. Nitrogen-containing six-membered heterocycles: solid-phase synthesis. Synth. Commun. 44: 1173-1211. (f) N. Kaur and D. Kishore. 2014. Solid-phase synthetic approach toward the synthesis of oxygen-containing heterocycles. Synth. Commun. 44: 1019-1042. (g) N. Kaur, P. Bhardwaj, M. Devi, Y. Verma, N. Ahlawat and P. Grewal. 2019. Ionic liquids in the synthesis of five-membered *N,N*-, *N,N,N*- and *N,N,N,N*-heterocycles. Curr. Org. Chem. 23: 1214-1238.
5. (a) T.F. Molinski and C.M. Ireland. 1988. Dysidazirine, a cytotoxic azacyclopropene from the marine sponge *Dysidea fragilis*. J. Org. Chem. 53: 2103-2105. (b) C.E. Salomon, D.H. Williams and D.J. Faulkner. 1995. New azacyclopropene derivatives from *Dysidea fragilis* collected in Pohnpei. J. Nat. Prod. 58: 1463-1466. (c) C.K. Skepper and T.F. Molinski. 2008. Long-chain 2*H*-azirines with heterogeneous terminal halogenation from the marine sponge *Dysidea fragilis*. J. Org. Chem. 73: 2592-2597. (d) K. Isomura, K. Uto and H. Taniguchi. 1977. Palladium(II)-catalysed formation of indoles from 2,2-diphenyl-2*H*-azirines. J. Chem. Soc. Chem. Commun. 19: 664-665. (e) H. Alper and J.E. Prickett. 1976. Reaction of 2-aryl-azirines with rhodium(I) complexes. J. Chem. Soc. Chem. Commun. 12: 483-483.
6. S. Chiba, G. Hattoti and K. Narasaka. 2007. Rh(II)-Catalyzed isomerization of 2-aryl-2*H*-azirines to 2,3-disubstituted indoles. Chem. Lett. 36: 52-53.
7. G.A. Russell, C.-F. Yao, H.I. Tashtoush, J.E. Russell and D.F. Dedolph. 1991. Addition, substitution and deoxygenation reactions of alpha-phenyl-beta-nitrostyrenes with the anions of thiols and diethyl phosphite: formation of indoles by reaction with ethyl phosphites. J. Org. Chem. 56: 663-669.
8. H. Inui and S. Murata. 2005. Photochemistry of 2-(1-naphthyl)-2*H*-azirines in matrixes and in solutions: wavelength-dependent C-C and C-N bond cleavage of the azirine ring. J. Am. Chem. Soc. 127: 2628-2636.
9. M.J. Alves, T.L. Gilchrist and J.H. Sousa. 1999. Reactions of methyl 2-aryl-2*H*-azirine-3-carboxylates with nucleophiles. J. Chem. Soc. Perkin Trans. 1 10: 1305-1310.
10. F. Palacios, A.M.O. de Retana, E.M. de Marigorta and J.M. de los Santos. 2001. ChemInform Abstract: 2*H*-Azirines as synthetic tools in organic chemistry. Eur. J. Org. Chem. 13: 2401-2414.
11. F. Palacios, M.O. de Retana, M.M. de Marigorta and J.M. de los Santos. 2002. Preparation, properties and synthetic applications of 2*H*-azirines: a review. Org. Prep. Proced. Int. 34: 219-269.
12. M. Alcami, O. Mo and M. Yaniez. 1993. Stabilization of nitrogen-containing three-membered rings by proton and lithium ion association in the gas phase. J. Am. Chem. Soc. 115: 11074-11083.
13. P.M. Mayer, M.S. Taylor, M.W. Wong and L. Radom. 1998. Thermochemistry of CH_3CN, CH_3NC, and their cyclic isomers and related radicals, cations, and anions: some curious discrepancies between theory and experiment. J. Phys. Chem. A 102: 7074-7080.
14. O. Mo, J.L.G. de Paz and M. Yanez. 1987. Protonation of three-membered ring heterocycles: an ab initio molecular orbital study. J. Phys. Chem. 91: 6484-6490.
15. T.W. Miller, E.W. Tristram and F.J. Wolf. 1971. Azirinomycin. II. J. Antibiot. 24: 48-50.
16. E.O. Stapley, D. Hendlin, M. Jackson, A.K. Miller, S. Hernandez and J.M. Mata. 1971. Azirinomycin. I. J. Antibiot. 24: 42-47.
17. D.F. Taber and W. Tian. 2006. The Neber route to substituted indoles. J. Am. Chem. Soc. 126: 1058-1059.

18. D. Knittel. 1985. Verbesserte synthese von α-azidozimtsäure-estern und 2*H*-azirinen. Synthesis 2: 186-187.
19. L. Henn, D.M.B. Hickey, C.J. Moody and C.W. Rees. 1984. Formation of indoles, isoquinolines, and other fused pyridines from azidoacrylates. J. Chem. Soc. Perkin Trans. 1 0: 2189-2196.
20. L.A. Wendling and R.G. Bergman. 1976. Thermal decomposition of 2*H*-azirines. Formation of products resulting from carbon-carbon bond cleavage. J. Org. Chem. 41: 831-836.
21. T.L. Gilchrist, G.E. Gymer and C.W. Rees. 1971. Mechanism of the pyrolysis of 1,2,3-triazoles. 1*H*-Azirines as intermediates. J. Chem. Soc. Chem. Commun. 23: 1519-1520.
22. K. Isomura, S. Kobayashi and H. Taniguchi. 1968. Indole formation by pyrolysis of β-styrylazides. Tetrahedron Lett. 9: 3499-3502.
23. Y. Ji, W.C. Trenkle and J.V. Vowles. 2006. A high-yielding preparation of β-ketonitriles. Org. Lett. 8: 1161-1163.
24. S.R. Jagtap, M.J. Bhanushali, N.S. Nandurkar and B.M. Bhanage. 2007. Ultrasound-assisted synthesis of β-enaminonitriles in the presence of base. Synth. Commun. 37: 2253-2258.
25. S. Treppendahl, P. Jakobsen and C. Berg. 1983. The structure of the stereoisomeric 3-amino-2,4-diphenyl-2-butenenitriles. Acta Chem. Scand. B 37: 645-646.
26. X. Li, Y. Du, Z. Liang, X. Li, Y. Pan and K. Zhao. 2009. Simple conversion of enamines to 2*H*-azirines and their rearrangements under thermal conditions. Org. Lett. 11: 2643-2646.
27. D.J. Anderson, T.L. Gilchrist, G.E. Gymer and C.W. Rees. 1973. Reactive intermediates. Part XXII. Formation of 2*H*-azirines by oxidation of *N*-aminophthalimide in the presence of alkynes. J. Chem. Soc. Perkin Trans. 1 0: 550-555.
28. J. Meinwald and D.H. Ane. 1966. The photochemical addition of methyl azidoformate to 2-butyne. J. Am. Chem. Soc. 88: 2849-2850.
29. R. Huisgen and H. Blaschke. 1965. 1.3-Dipolare cycloadditionen, XX: 2-Äthoxy-oxazole aus äthoxycarbonyl-azen und alkinen. Chem. Ber. 98: 2985-2997.
30. R.S. Atkinson and M.J. Grinshire. 1986. Intramolecular reactions of *N*-nitrenes with alkynes: conformational anchoring in spiro-fused 2*H*-azirines. J. Chem. Soc. Perkin Trans. 1 0: 1215-1224.
31. R.L. Barcus, L.M. Hadel, L.J. Johnston, M.S. Platz, T.G. Savino and J.C. Scaiano. 1986. 1-Naphthylcarbene: spectroscopy, kinetics, and mechanisms. J. Am. Chem. Soc. 108: 3928-3937.
32. G. Alcaraz, U. Wecker, A. Baceiredo, F. Dahan and G. Bertrand. 1995. Synthesis of a 2*H*-azirine by [1+2] cycloaddition of a phosphinocarbene with a nitrile and its ring-expansion to a 1,2λ^5-azaphosphete. Angew. Chem. Int. Ed. 34: 1246-1248.
33. V. Piquet, A. Baceiredo, H. Gomitzka, F. Dahan and G. Bertrand. 1997. The stable (phosphino)(silyl)carbene as a useful building block: synthesis and reactivity of 2-phosphorus-substituted 2*H*-azirines. Chem. Eur. J. 3: 1757-1764.
34. G. Bartoli, M. Bosco, A. Carlone, M. Locatelli, P. Melchiorre and L. Sambri. 2004. Asymmetric catalytic synthesis of enantiopure *N*-protected 1,2-amino alcohols. Org. Lett. 6: 3973-3975.
35. J.U. Jeong, B. Tao, I. Sagasser, H. Henniges and K.B. Sharpless. 1998. Bromine-catalyzed aziridination of olefins. A rare example of atom-transfer redox catalysis by a main group element. J. Am. Chem. Soc. 120: 6844-6845.
36. S.K. Kim and E.N. Jacobsen. 2004. General catalytic synthesis of highly enantiomerically enriched terminal aziridines from racemic epoxides. Angew. Chem. Int. Ed. 43: 3952-3954.
37. P. Wessig and J. Schwarz. 1977. A convenient one-pot conversion of *N*-Boc-β-aminoalcohols into *N*-Boc-aziridines. Synlett 8: 893-894.
38. D.M. Hodgson, P.G. Humphreys, S.M. Miles, C.A.J. Brierley and J.G. Ward. 2007. Dimerization and isomerization reactions of α-lithiated terminal aziridines. J. Org. Chem. 72: 10009-10021.
39. U.J. Vogelbacher, M. Ledermann, T. Schach, G. Michels, U. Hess and M. Regitz. 1988. Reaktivität von tri-*tert*-butylazet. Angew. Chem. 100: 304-306.

40. P.A.S. Smith and E.E. Most. 1957. Quaternary hydrazones and their rearrangement. J. Org. Chem. 22: 358-362.
41. V. Nair. 1968. The reaction of azirines with diazomethane to produce allylic azides. J. Org. Chem. 33: 2121-2123.
42. A. Padwa and P.H.J. Carlsen. 1977. Photochemical transformations of small ring heterocycles. 81. Carbenic reactions of nitrile ylides. An example of a stepwise 1,3-dipolar cycloaddition. J. Am. Chem. Soc. 99: 1514-1523.
43. R.L. Barcus, B.B. Wright, M.S. Platz and J.C. Scaiano. 1983. Chemical, kinetic, and spectroscopic evidence for the reaction of 1-naphthylcarbene with acetonitrile to form a nitrile ylide. Tetrahedron Lett. 24: 3955-3958.
44. D.F. Morrow, M.E. Butler and E.C.Y. Huang. 1965. The synthesis of 17β-amino-17-isoprogesterone. J. Org. Chem. 30: 579-587.
45. D.F. Morrow and M.E. Butler. 1964. Stereoselectivity in the Neber rearrangement: synthesis of a steroidal spiroazirine. J. Heterocycl. Chem. 1: 53-54.
46. P. Wipf and H. Heimgartner. 1988. Anwendung der azirin/oxazolon-methode in der peptid-chemie: synthese von modell-tripeptiden. Helv. Chim. Acta 71: 140-154.
47. H. Heimgartner. 1991. 3-Amino-2*H*-azirines. Synthons for α,α-disubstituted α-amino acids in heterocycle and peptide synthesis. Angew. Chem. Int. Ed. 30: 238-264.
48. C.B. Bucher and H. Heimgartner. 1996. Optisch aktive 3-amino2*H*-azirine als bausteine für enantiomer-enreine α,α-disubstituierte α-aminosäuren: synthese von isovalin-synthonen und einbau in ein trichotoxin-A-50-segment. Helv. Chim. Acta 79: 1903-1915.
49. C. Alonso, J.M. de los Santos, J. Vicario and F. Palacios. 2011. Phosphorus substituted hydroxylamine and hydroxamic acid derivatives: synthesis and reactivity. ARKIVOC (iii): 221-253.
50. S. Brahma and J.K. Ray. 2008. Synthesis of azirines containing aldehyde functionality and their utilization as synthetic tools for five membered oxazoles and isoxazoles. J. Heterocycl. Chem. 45: 311-317.
51. I.P. Piskunova, A.V. Eremeev, A.F. Mishnev and I.A. Vosekalnaa. 1993. Synthesis and structure of optically active 3-amino-2*H*-azirines. Tetrahedron 49: 4671-4676.
52. M. Alajarín, R.-Á. Orenes, Á. Vidal and A. Pastor. 2003. A convenient and expedient synthesis of 3-aryl-2*H*-azirine-2-carboxaldehydes. Synthesis 1: 49-52.
53. F.A. Davis, G.V. Reddy and H. Liu. 1995. Asymmetric synthesis of 2*H*-azirines: first enantioselective synthesis of the cytotoxic antibiotic (*R*)-(-)-dysidazirine. J. Am. Chem. Soc. 117: 3651-3652.
54. F.A. Davis, H. Liu, C.-H. Liang, G.V. Reddy, Y. Zhang, T. Fang and D.D. Titus. 1999. Asymmetric synthesis of 2*H*-azirine 2-carboxylate esters. J. Org. Chem. 64: 8929-8935.
55. A. Padwa, E. Chen and A. Ku. 1975. Photochemical transformations of small ring heterocyclic compounds. 70. Thermal and photochemical valence isomerizations of 4-carbonyl-substituted isoxazoles. J. Am. Chem. Soc. 97: 6484-6491.
56. G. Szeimies, K. Mannhardt and W. Mickler. 1977. Synthese und thermisches verhalten einiger 3-aryl-3-azidoaziridin-2-carbonsäure-ethylester. Chem. Ber. 110: 2922-2938.
57. G.J. Voghel, T.L. Eggerichs, B. Clamot and H.G. Viehe. 1976. Facile synthesis of 2-dimethylamino-delta-1-azirine derivatives by thermal or photo-isomerization of isoxazoles. Chimia 30: 191-192.
58. D.A. Wunderlin, G.E. Davico and J.D. Perez. 1992. Evidences on the intermediacy of furoxan in the flash vacuum thermolysis of some 4-nitroisoxazoles. Inf. J. Chem. Kinef. 24: 31-40.
59. R.R. Sauers, L.M. Hadel, A.A. Scimone and T.A. Stevenson. 1990. Photochemistry of 4-acylisoxazoles. J. Org. Chem. 55: 4011-4019.
60. G. Himbert, H. Kuhn and M. Barz. 1990. Cycloadditionen, 18. 5-Aminoisoxazole durch cycloaddition von nitriloxiden an inamine. Liebigs Ann. Chem. 4: 403-407.

61. B.H. Lipshutz and D.C. Reuter. 1988. Cyclopeptide alkaloid model studies. A two-step conversion of 5-aminoisoxazoles to amino acid bis-amides. Tetrahedron Lett. 29: 6067-6070.
62. S. Auricchio, A. Biri, E. Pastormerlo and A.M. Truscello. 1997. Iron dichloride induced isomerization or reductive cleavage of isoxazoles: a facile synthesis of 2-carboxy-azirines. Tetrahedron 53: 10911-10920.
63. G.E. Davico and J.D. Perez. 1990. Theoretical study of the gas-phase thermal isomerization of isoxazoles. Part I. Isoxazole. J. Phys. Org. Chem. 3: 611-619.
64. S. Ueda, S. Naruto, T. Yoshida, T. Sawayama and H. Uno. 1988. Novel base-induced reactions of substituted (1,2-benzisoxazol-3-yl)acetates. J. Chem. Soc. Perkin Trans. 1 5: 1013-1021.
65. C.P. Hadjiantoniou-Maroulis, A. Charalambopoulos and A.J. Maroulis. 1998. Pyrolysis and photolysis of 1-aroylamino-4,5-diaryl-1,2,3-triazoles: generation and thermal transformations of 4,5-diaryl-1,2,3-triazolyl radicals. J. Heterocycl. Chem. 35: 891-894.
66. M. Christl, S. Krimm and A. Kraft. 1990. Some valenes of benzannelated five-membered heteroarenes - synthesis and NMR spectra. Angew. Chem. Int. Ed. 29: 675-677.
67. A. Ruhmann and C. Wentrup. 1994. Synthesis of a photoactivatable 9-*Z*-oleic acid for protein kinase C labeling. Tetrahedron 50: 3785-3796.
68. R. Huisgen and J. Wulff. 1967. Cycloadditionen von nitriloxiden an methylenphosphorane. Tetrahedron Lett. 8: 917-920.
69. H.J. Bestmann and R. Kunstmann. 1966. Reactions of benzonitrile oxide with alkylidenephosphoranes. Angew. Chem. Int. Ed. 5: 1039-1040.
70. H.J. Bestmann and R. Kunstmann. 1969. Reaktionen mit phosphinalkylenen, XXIII. Reaktionen von phosphinalkylenen mit nitriloxiden. Eine neue synthesemöglichkeit für azirine, ketenimine und α.β-ungesättigte oxime. Chem. Ber. 102: 1816-1832.
71. A. Hassner and V. Alexanian. 1979. Synthesis of some 1-azirines from alpha-bromo ketoximes via oxazaphospholines. J. Org. Chem. 44: 3861-3864.
72. F. Palacios, A.M. Ochoa de Retana, J.I. Gil and J.M. Ezpeleta. 2000. Simple asymmetric synthesis of 2*H*-azirines derived from phosphine oxides. J. Org. Chem. 65: 3213-3217.
73. F. Palacios, D. Aparicio, J.M. de los Santos and E. Rodriguez. 1996. An improved and general method for the synthesis of α,β-unsaturated oximes from phosphine oxide allenes. Tetrahedron Lett. 37: 1289-1292.
74. F. Palacios, D. Aparicio, J.M. de los Santos and E. Rodriguez. 1998. An improved and effective method for the preparation of α,β-unsaturated oximes and isoxazole derivatives. Tetrahedron 54: 599-614.
75. F. Palacios, A.M. Ochoa de Retana and J.I. Gil. 2000. Easy and efficient synthesis of enantiomerically enriched 2*H*-azirines derived from phosphonates. Tetrahedron Lett. 41: 5363-5366.
76. M. Haddach, R. Pastor and J.G. Riess. 1993. Synthese et reactivite de nouveaux synthons perfluoroalkyles. Comportement atypique des azirines et aziridines F-alkylees. Tetrahedron 49: 4627-4638.
77. H.G. Corkins, L. Storace and E. Osgood. 1980. Stereospecific 2*H*-azirine formation from the modified Neber reaction of oxime carbamates. J. Org. Chem. 45: 3156-3159.
78. H.O. House and W.F. Berkowitz. 1963. The stereochemistry of the Neber rearrangement. J. Org. Chem. 28: 2271-2276.
79. M.M.H. Verstappen, G.J.A. Ariaans and B. Zwanenburg. 1996. ChemInform Abstract: Asymmetric synthesis of 2*H*-azirine carboxylic esters by an alkaloid-mediated Neber reaction. J. Am. Chem. Soc. 118: 8491-8492.
80. P.W. Neber and A. Burgard. 1932. Über den reaktionsverlauf einer neuen art von umlagerung bei ketoximen. III. Justus Liebigs Ann. Chem. 493: 281-294.
81. P.W. Neber and G. Huh. 1935. Eine neue, allgemeine methode zur gewinnung von α-aminoketonen. I. Justus Liebigs Ann. Chem. 515: 283-296.
82. T.M.V.D. Pinho e Melo and A.M.A.R. Gonsalves. 2004. Exploiting 2-halo-2*H*-azirine chemistry. Curr. Org. Synth. 1: 275-292.

83. F. Palacios, D. Aparicio, A.M.O. de Retana, J.M. de los Santos, J.I. Gil and R.L. de Munain. 2003. Asymmetric synthesis of 2*H*-aziridine phosphonates, and α- or β-aminophosphonates from enantiomerically enriched 2*H*-azirines. Tetrahedron: Asymmetry 14: 689-700.
84. A. Padwa and T. Stengel. 2005. Grubb's and Wilkinson catalyzed reactions of 2-phenyl-3-vinyl substituted 2*H*-azirines. ARKIVOC (v): 21-32.
85. F. Palacio, D. Apricio, A.M. Ochoa de Retana, J.M. de los Santos, J.I. Gil and J.M. Alonso. 2002. Asymmetric synthesis of 2*H*-azirines derived from phosphine oxides using solid-supported amines. Ring opening of azirines with carboxylic acids. J. Org. Chem. 67: 7283-7288.
86. A. Lemos. 2009. Addition and cycloaddition reactions of phosphinyl- and phosphonyl-2*H*-azirines, nitrosoalkenes and azoalkenes. Molecules 14: 4098-4119.
87. G.V. Shustov, O.N. Krutius, V.N. Voznesensky, I.I. Chervin, A.V. Eremeev, R.G. Kostyanovskii and F.D. Polyak. 1990. Asymmetric nitrogen - LXXI. Asymmetric synthesis and lactonization of 1-β-hydroxyalkylaziridine-2-carboxylic esters into 4-oxa-1-azabicyclo[4.1.0]heptan-5-ones. Tetrahedron 46: 6741-6752.
88. A.A. Fadda, S. Bondock, R. Rabie and H.A. Etman. 2008. Cyanoacetamide derivatives as synthons in heterocyclic synthesis. Turk. J. Chem. 32: 259-286.
89. A.V. Eremeev and I.P. Piskunova. 1990. Synthesis and properties of substituted 3-amino-2*H*-azirines (review). Chem. Heterocycl. Compd. 26: 719-738.
90. T. Ooi, M. Takahashi, K. Doda and K. Maruoka. 2002. Asymmetric induction in the Neber rearrangement of simple ketoxime sulfonates under phase-transfer conditions: experimental evidence for the participation of an anionic pathway. J. Am. Chem. Soc. 124: 7640-7641.
91. R.M. Carlson and S.Y. Lee. 1969. Olefin synthesis the stereospecific deamination of aziridines. Tetrahedron Lett. 10: 4001-4004.
92. P.W. Neber and A. Friedolsheim. 1926. Über eine neue art der umlagerung von oximen. Justus Liebigs Ann. Chem. 449: 109-134.
93. F. Palacios, A.M.O. de Retana, E.M. de Marigorta and J.M. de Los Santos. 2003. Targets Heterocycl. Systems 7: 206-245.
94. A. Padwa and S. Murphree. 2003. Chapter 4.1 Three-membered ring systems. Progress in Heterocycl. Chem. 15: 75-99.
95. B. Zwanenburg and P. ten Holte. 2001. The synthetic potential of three-membered ring aza-heterocycles. Top. Curr. Chem. 216: 93-124.
96. T.L. Gilchrist. 2001. Activated 2*H*-azirines as dienophiles and electrophiles. Aldrichimica Acta 34: 51-55.
97. C. Wentrup, S. Fischer, H. Berstermann, M. Kuzaj, H. Luerssen and K. Burger. 1986. Nitrile ylides and azirines: gas-phase generation from 2,3-dihydro-1,4-λ^5-oxazaphospholes and matrix isolation. Angew. Chem. Int. Ed. 25: 85-86.
98. S. Greszler and K.L. Stevens. 2009. Synthesis of pyrazolo[1,5-*a*]pyridines via azirines: preparation of 2-(3-bromophenyl)-6-(trifluoromethyl)pyrazolo[1,5-*a*]pyridine. Org. Synth. 86: 18-27.
99. (a) L. Gentilucci, Y. Grijzen, L. Thijs and B. Zwanenburg. 1995. Convenient synthesis of optically active 2*H*-azirine-2-carboxylic esters by Swern oxidation of aziridine-2-carboxylic esters. Tetrahedron Lett. 36: 4665-4668. (b) J. Legters, L. Thijs and B. Zwanenburg. 1992. Conversion of aziridine-2-carboxylic esters into 2*H*-azirine-2-carboxylic esters. Recueil Des Travaux Chimiques Des Pays-Bas 111: 75-78.
100. B. Zwanenburg and L. Thijs. 1996. Aziridine and azirine carboxylic esters. Pure Appl. Chem. 68: 735-738.
101. A.R. Katritzky, M. Wang, C.R. Wilkerson and H. Yang. 2003. A novel approach to substituted 2*H*-azirines. J. Org. Chem. 68: 9105-9108.
102. A.R. Katritzky and S. Rachwal. 2010. Synthesis of heterocycles mediated by benzotriazole. 1. Monocyclic systems. Chem. Rev. 110: 1564-1610.

103. A.R. Katritzky, S. Zhang and Y. Fang. 2000. $BtCH_2TMS$-Assisted homologation of carboxylic acids: a safe alternative to the Arndt-Eistert reaction. Org. Lett. 2: 3789-3791.
104. A.R. Katritzky and J. Wu. 1994. A simple, versatile synthetic route to *N*-1-aryl-, -heteroaryl-, -acylmethyl-, -carboxymethyl- and -alkyl-benzotriazoles via regiospecific or highly regioselective substitutions of benzotriazole. Synthesis 6: 597-600.
105. T.L. Gilchrist and R. Mendonca. 2000. Addition of pyrimidine and purine bases to benzyl 2*H*-azirine-3-carboxylate. Synlett 12: 1843-1845.
106. T.M.V.D. Pinho e Melo, C.S.J. Lopes and A.M.A.R. Gonsalves. 2000. Synthesis and reactivity of 2-halo-2*H*-azirines towards nucleophiles. Tetrahedron Lett. 41: 7217-7220.
107. K. Banert, M. Hagedorn, C. Liedtke, A. Melzer and C. Schoffler. 2000. Azido-1,2,3-triphenylpropenes of varying stabilities: a corrigendum of structure assignment. Eur. J. Org. Chem. 2: 257-267.
108. T.M.V.D. Pinho e Melo, C.S.J. Lopes, A.M.A.R. Gonsalves, A.M. Beja, J.A. Paixao, M.R. Silva and L.A. DaVeiga. 2002. Reactivity of 2-halo-2*H*-azirines. 1. Reactions with nucleophiles. J. Org. Chem. 67: 66-71.
109. B. Zwanenburg, H.J.F. Philipse, M.M.H. Verstappen and R.G. Gieling. 1997. Sulfur substituted small-ring heterocycles. Phosphorus, Sulfir, Silicon Relat. Elem. 120-121: 453-454.
110. W.B. Renfrow, J.F. Witte, R.A. Wolf and W.R. Bohl. 1968. Reactions of *O*-benzoyl oximes with sodium hydride. Substituted isoxazoles and the Neber rearrangement. J. Org. Chem. 33: 150-153.
111. M.J. Hatch and D.J. Cram. 1953. The problem of the unsaturated three-membered ring containing nitrogen. J. Am. Chem. Soc. 75: 33-38.

CHAPTER

Synthesis of azirines from azides

2

Chapter outline

2.1 Introduction

The *N*-containing heterocyclic compounds are present in many chemotherapeutic agents and natural products. Their attachment to sugar molecules as either nucleoside or thioglycoside analogs plays an essential role in various biological processes and in the synthetic organic chemistry [1a–o].

The 2*H*-azirines are the smallest of unsaturated *N*-containing heterocyclic compounds. The azirine is the term utilized to define the smallest unsaturated *N*-heterocyclic system with two carbon atoms and one double bond in a three-membered ring. The enormous chemistry of these compounds is because of the unique array of activities that renders possible to break each of the three bonds of ring by controlling experimental conditions [1p].

The 2*H*-azirines are important for their versatile biological and chemical behavior [2]. The 2*H*-azirines are natural antibiotics [3]. The functionalized aziridines were prepared using substituted and functionalized azirines as versatile precursors [4,5]. The azirines with phosphorus substituents control many important biological functions [6].

Since first reported by Neber and coworkers [7,8], the 2*H*-azirines have attracted a great deal of attention from experimental and theoretical chemists [9–11]. The highly strained imine bond confined in a three-membered ring confers unique properties to these heterocyclic compounds [12,13]. They can serve as electrophiles, nucleophiles, and also as dipolarophiles and dienophiles in cycloaddition procedures. As a result of this reactivity, they act as precursors for a number of heterocycles. The stabilized 2*H*-azirine containing systems have also been isolated from natural sources [14]. The asymmetric syntheses of two of them, azirinomycin and dysidazirine, have been reported [15–18]. On the other hand, even more reactive 2-halo-2*H*-azirines have been much less studied [19].

Some azirines are synthesized because of their biological interest [20,21], and others are synthesized as intermediates in synthesis to synthesize the amino acid derivatives [22] and heterocyclic compounds [23a–c].

3-Membered Heterocycle Synthesis. https://doi.org/10.1016/B978-0-443-13583-5.00015-3

Because of their ability to react both with retention and opening of three-membered ring, the 2*H*-azirines are unique strained compounds, which have several uses in organic synthesis. Even though each of the three bonds in the azirine ring can be cleaved under certain conditions, the majority of synthetic uses of these compounds imply cleavage of either the N–C3 or N–C2 bond. In these syntheses, the 2*H*-azirines act as C–C–N three-atomic building blocks for the formation of both acyclic compounds such as nonproteinogenic amino acids and peptides [24] and several four- to seven-membered heterocyclic compounds [25a–e]. Many general reviews have been published on azirines in the past [26].

2.2 Synthesis of azirines from azides

The 2-formyl-3-phenyl-2*H*-azirine was prepared by an addition of iodine azide to dimethyl acetal of cinnamaldehyde followed by dehydrohalogenation, thermolysis, and aqueous hydrolysis. Further reaction of aldehyde with aniline and Wittig reagents provided substituted 2*H*-azirines in good yield (Scheme 1) [27–29].

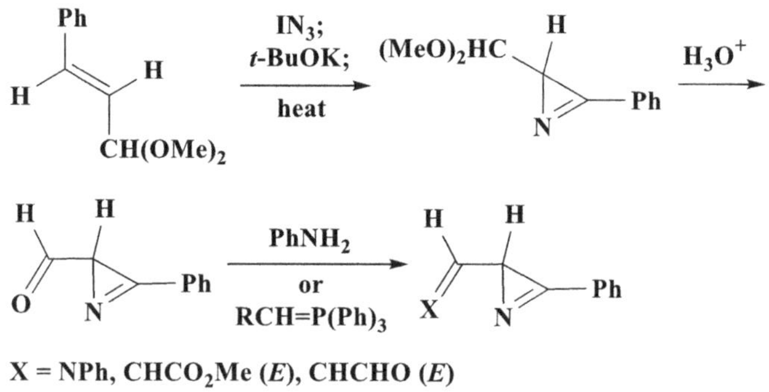

SCHEME 1

An efficient formation of 3-amino-2*H*-azirines occurred by reacting α-mono- or disubstituted amides and an azide source via a nonisolable vinyl azide intermediate. The amide was reacted with phosgene/TEA to synthesize the α-chloroenamine, which was further reacted with sodium azide to afford the 3-amino-2*H*-azirines through the azidoenamine [30,31]. The utilization of highly toxic $COCl_2$ could be avoided by an alternative process involving the reaction of amide enolate with diphenylphosphorochloridate (DPPCl) followed by reaction with sodium azide [32,33]. This technique also has the benefit that it is not essential to isolate the sensitive α-chloroenamine intermediate and it has been utilized to synthesize the heterospirocyclic 3-amino-2*H*-azirines, synthons for heterocyclic amino acids [34]. The DPPA (diphenylphosphorazidate) was utilized as an azide source; thus, 3-amino-2*H*-azirines could be formed in one pot in very good yields (Scheme 2) [35]. This reaction was the technique of choice for the formation of 2,2-disubstituted 3-amino-2*H*-azirines [36].

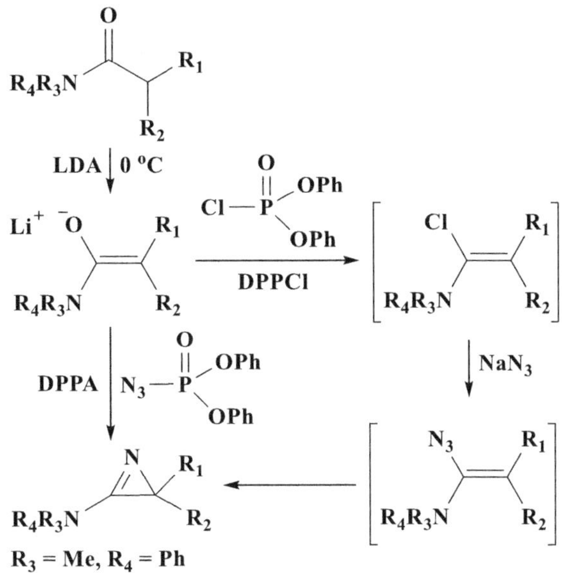

SCHEME 2

A 3:2:1 mixture of methyl azidocinnamate, 3-carbomethoxy-2-phenyl-1-azirine, and methyl α-iodocinnamate was obtained when iodo azide was reacted with DABCO (1,4-diazabicyclo[2.2.2] octane) in CH_3COCH_3 at rt (Scheme 3). The azirine synthesis was entirely suppressed if an elimination was performed at 5°C, and in addition to some iodo azide, only vinyl azide and methyl α-iodocinnamate were formed [37]. Since the reaction was very slow at 5°C, an elimination for preparative purposes was performed at rt. The transformation of residual vinyl azide to azirine occurred by the pyrolysis of reaction mixture in refluxing skellysolve B. No isoxazole was obtained when benzalacetophenone adduct was reacted with 1,4-diazabicyclo[2.2.2]octane. This was consistent with the reduced basicity of carbonyl oxygen in vinyl azide in comparison with the keto vinyl azide from benzalacetophenone. The removal of HN_3 forming α-iodocinnamate was possibly promoted in this case due to the greater acidity of proton α to the carbonyl function. Both HI and HN_3 were removed when the elimination was performed with KOH in CH_3OH, and only phenylpropiolic acid was isolated [38].

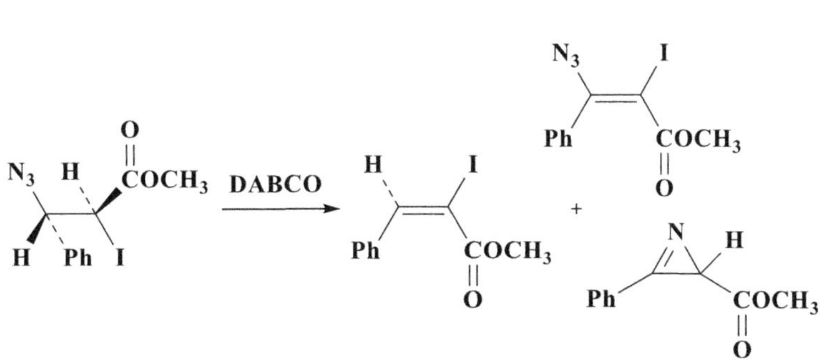

SCHEME 3

The vinyl azide was used to synthesize the heterospirocyclic 3-amino-2*H*-azirines, which are beneficial synthons for heterocyclic amino acids [34]. The 3-amino-2*H*-azirines were obtained in "one pot" and high yield when diphenylphosphorazidate was utilized as azide source and by the replacement of oxygen atom of amide enolates [35]. The modification of this method with a chiral substituent at the amino group of thioamide was utilized for the formation of optically active 3-amino-2*H*-azirines (Scheme 4) [39].

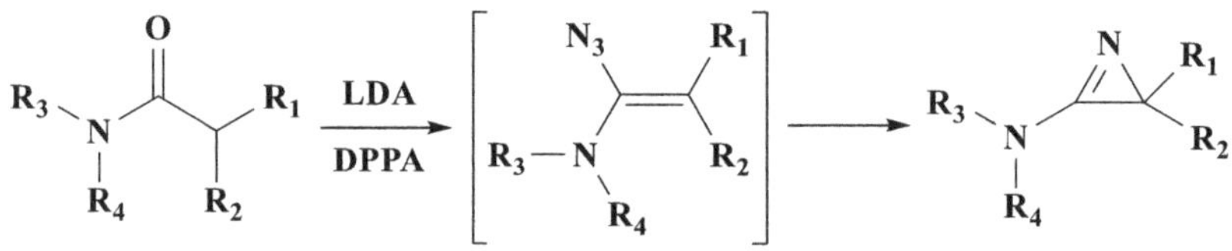

SCHEME 4

The halo 2*H*-azirine system was synthesized starting from oxophosphorus ylides. These ylides were reacted with bromine, chlorine, and electrophilic halogen donor reagents using nucleophiles to provide the substituted alkenes by the removal of PPh_3O. The haloazidoalkenes were formed in good yield (Y=N_3) when oxophosphorus ylides were reacted with *N*-halosuccinimide using azidotrimethylsilane. These vinyl azides were transformed to 2-halo-2*H*-azirines on heating in heptane for 2–3 h (Scheme 5) [40–42].

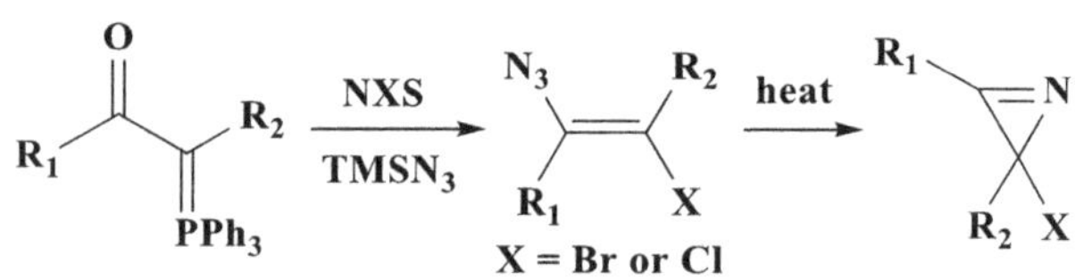

SCHEME 5

The diphenylphosphinyl 2*H*-azirine was isolated in good yield by the thermolysis of vinyl azide (Scheme 6) [43], but this approach was not appropriate for the formation of enantiopure azirines [44].

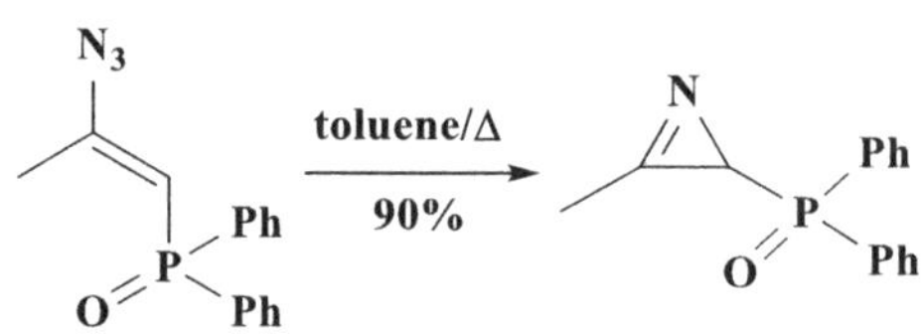

SCHEME 6

The halogenated alkenyl azides were prepared from α-oxophosphonium ylides upon reacting with $TMSN_3$ and *N*-halosuccinimides. These haloalkenyl azides can be converted into equivalent 2-halo-2*H*-azirines by heating to 98°C (Scheme 7) [42,45].

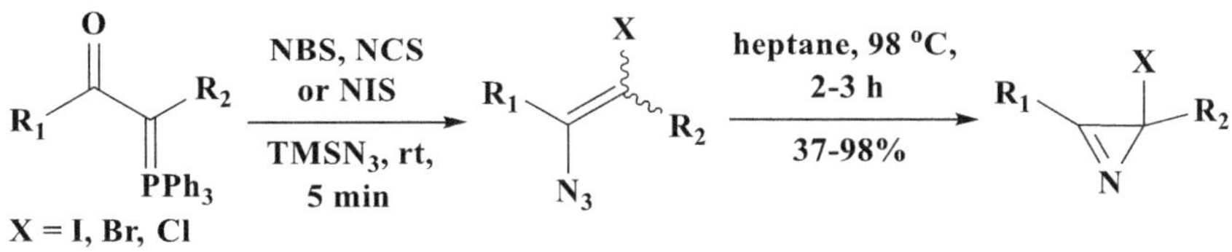

SCHEME 7

The 1,4-diazido-1,3-dienes acted as precursors for biazirinyls. Only conrotatory opening of *trans*-diazidocyclobutanes afforded biazirinyls in satisfactory yields (Scheme 8). The *trans*-diazidocyclobutanes were obtained by the reaction of *trans*-dihalocyclobutanes with tributylhexadecylphosphonium azide [45,46].

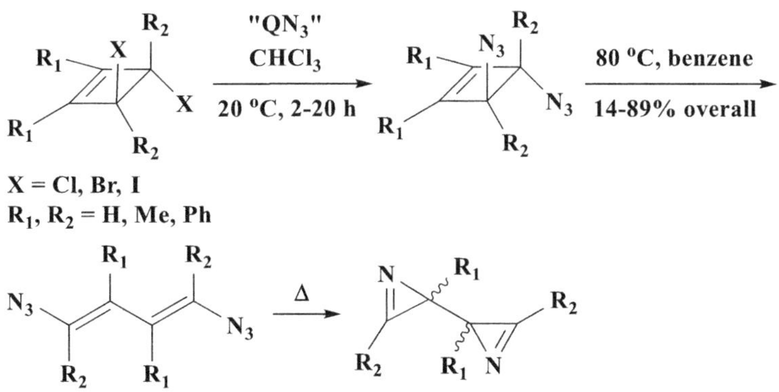

SCHEME 8

The triazole is synthesized by an intramolecular azide cycloaddition to an alkyne. The triazoles are much more stable than triazolines and are most often directly isolable. The azide cyclized to triazole in 89% yield (Scheme 9) [47]. An intramolecular cycloaddition of azido enyne afforded triazole where the carbon–carbon triple bond acted as dipolarophile due to geometric constraints [48]. The alkynyl β-lactam was cyclized to synthesize the azirine in 12% yield and triazolocephem in 49% yield [49]. The triazolocephem was derived from (*Z*)-isomer, whereas the azirine was derived from (*E*)-isomer through the nitrene. The triazolopenam was prepared in an analogous manner.

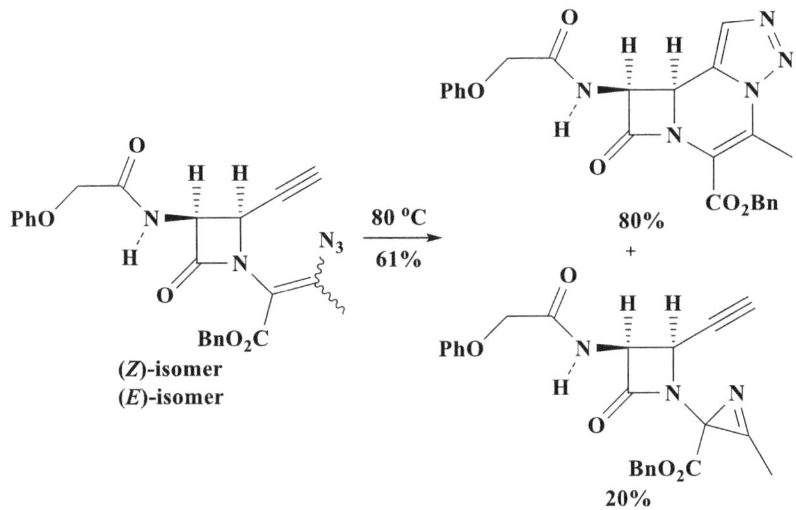

SCHEME 9

Smolinsky [50,51] reported that the vapor-phase thermolysis of α-aziridinylstyrene afforded substantial amounts of phenyl azirine (Scheme 10). Then, the procedure was adapted to be more amenable to the preparative scale as demonstrated by the formation of trifluoromethylphenylazirine, which was performed on a 10-g scale [52]. The vinyl azide in this case was prepared from alkene by sequential bromination, azide displacement, and dehydrobromination, a procedure applied by Gilchrist et al. [53,54] to access the azirinyl esters. Timen et al. [55] synthesized azirine in much improved yield (>95%) by thermolysis at 150°C in dichloromethane for 20 min in a sealed tube (Scheme 11). The procedure could be unusually tolerant to other moiety and quite high yielding even at atmospheric pressure. Thus, the hexopyranose-derived vinyl azide was heated in toluene for 20 h for the smooth preparation of azirine (Scheme 12) [56]. Possibly, the most suitable route was the neat thermolysis of aryl vinyl azides in an open container under the MWI that occurred in very good to excellent yields within few minutes (Scheme 13) [57].

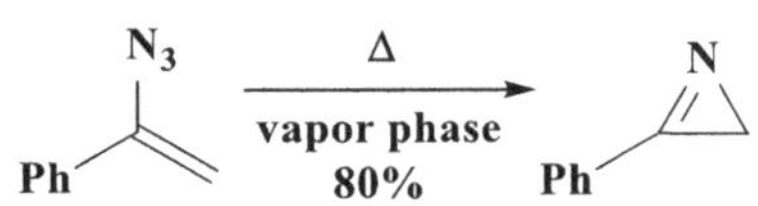

SCHEME 10

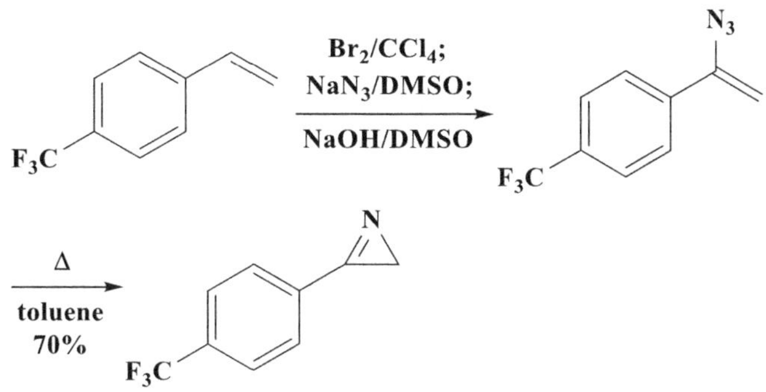

SCHEME 11

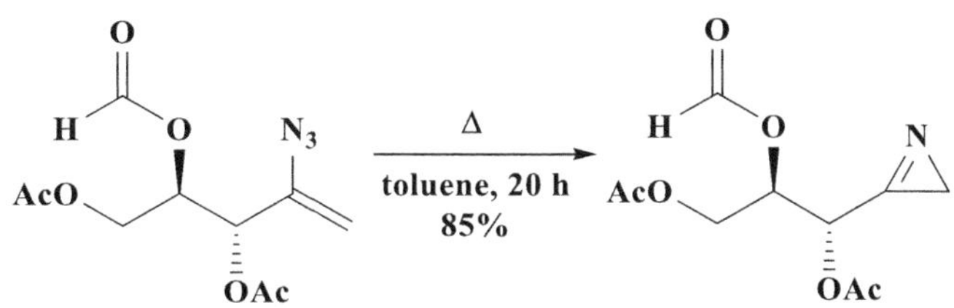

SCHEME 12

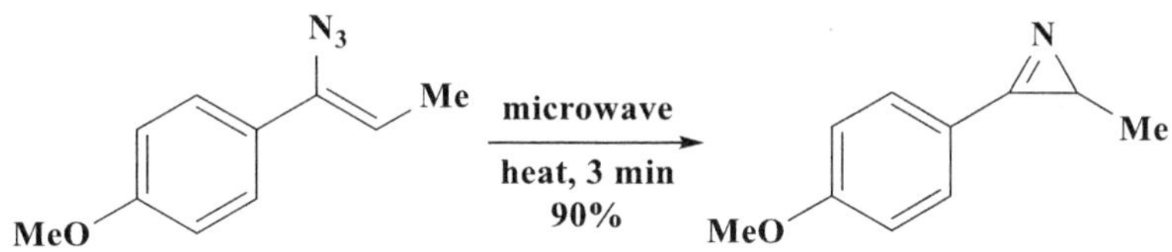

SCHEME 13

Smolinsky and Pryde [50,58] prepared azirine along with ketenimin in small amounts by the gas-phase pyrolysis of α-aryl-substituted vinyl azides (Scheme 14).

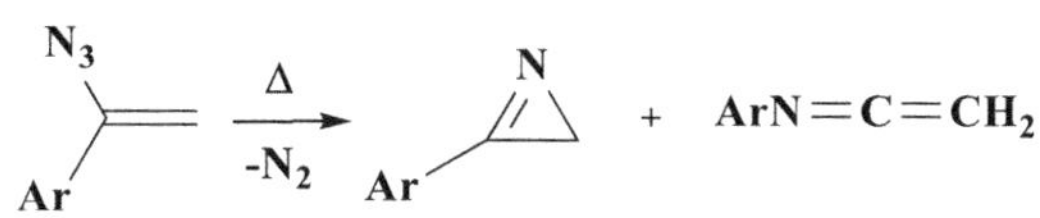

SCHEME 14

The electrocyclic ring opening of 3,4-diazidocyclobutenes synthesized 1,4-diazidobutadienes [59]. The diazidobutadienes were photolyzed to provide, among other products, the bi-2*H*-azirin-2-yls (Scheme 15), but these compounds were not isolable and were detected spectroscopically.

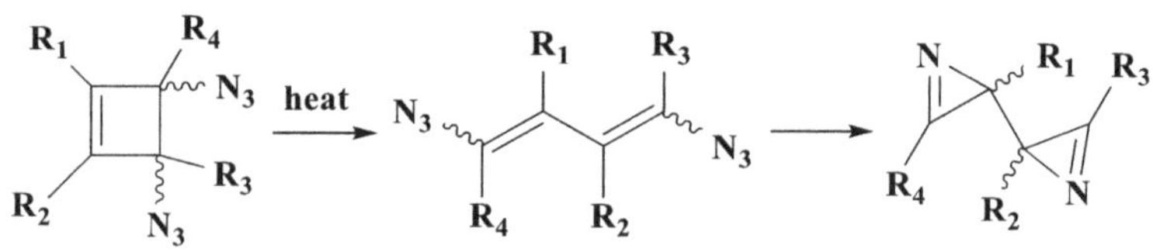

SCHEME 15

The heterocyclic products were obtained in excellent ($X = Cl$, Br) or modest ($X = I$) yields by heating solutions of halovinyl azides (Scheme 16) [40,42,60]. The azirines were synthesized by the substitution reactions using nucleophiles NuH such as $C_6H_5NH_2$ or potassium phthalimide [41], and the dehalogenation of heterocyclic products using $NaBH_4$ or tributyltin hydride was also possible [61]. Since thermolysis of β-azido-α,β-unsaturated ketones, aldehydes, or esters led to isoxazoles through the vinyl nitrenes and azirines, such products were also expected in the thermal ring expansion of heterocyclic products at first [62].

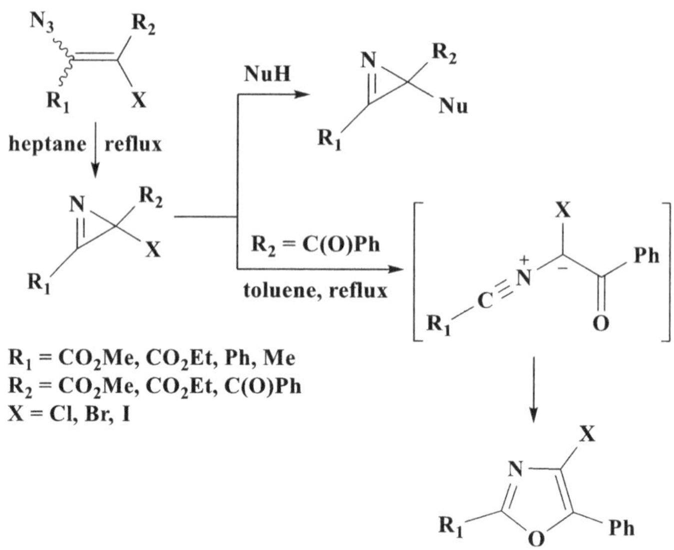

SCHEME 16

The alkene could not be isolated since the reaction of ylide with NBS and azidotrimethylsilane afforded 2-bromo-2*H*-azirine in 89% yield. The transformation of 3-azido-2-halo-3-phenylpropenoates to 2-halo-3-phenyl-2*H*-azirine-2-carboxylate was very promising. The reactions with NCS and NBS completed after 5 min. However, an excess of halogenating agent was needed and reaction completed after 4.5 h when NIS was utilized. The majority of these reactions provided high stereoselectivity affording alkenes as single isomers. However, the 2-azido-3-halobut-2-enedioates were formed as mixtures of (*Z*/*E*)-isomers. The assignment was made on the basis of a comparison of its ^{13}C nuclear magnetic resonance spectra with those of dimethyl azidomaleate and dimethyl azidofumarate in the case of 2-azido-3-chlorobut-2-enedioates [42]. The alkenes were smoothly transformed into 2-halo-2*H*-azirines on heating in heptane for 2 ± 3 h (Scheme 17). The 2-bromo-2*H*-azirines and 2-chloro-2*H*-azirines were formed in high yield (89 ± 99%), while the yields of 2-iodo-2*H*-azirine derivatives were 36%–85%.

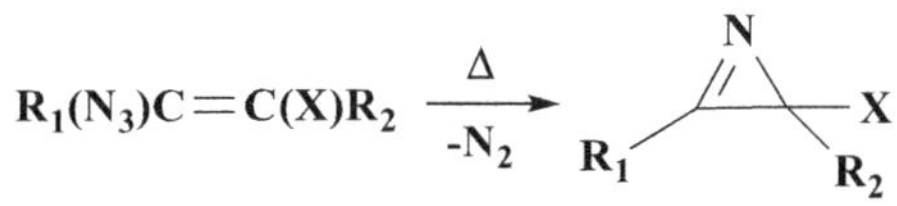

$R_1 = R_2 = CO_2Et$, X = Cl, 98%
$R_1 = R_2 = CO_2Me$, X = Cl, 99%
$R_1 = R_2 = CO_2Me$, X = Br, 98%
R_1 = Ph, $R_2 = CO_2Et$, X = Cl, 97%
R_1 = Ph, $R_2 = CO_2Et$, X = Br, 97%
R_1 = Ph, $R_2 = CO_2Et$, X = I, 36%
R_1 = Ph, $R_2 = CO_2Me$, X = Br, 89%
R_1 = Ph, $R_2 = CO_2Me$, X = I, 77%
$R_1 = CO_2Me$, R_2 = COPh, X = Cl, 97%
$R_1 = CO_2Me$, R_2 = COPh, X = Br, 98%
$R_1 = CO_2Me$, R_2 = COPh, X = I, 36%
R_1 = Me, $R_2 = CO_2Me$, X = Br, 99%
R_1 = Me, $R_2 = CO_2Me$, X = I, 85%

SCHEME 17

Smolinsky [50] reported the formation of 1-azirine ring system, who further discovered that the 1-azirines were synthesized by the pyrolysis of vinyl azides (Scheme 18). This technique was successful for the formation of three 1-azirines. This synthesis was limited only to the accessibility of the required vinyl azide, which could be a rather severe limitation [38].

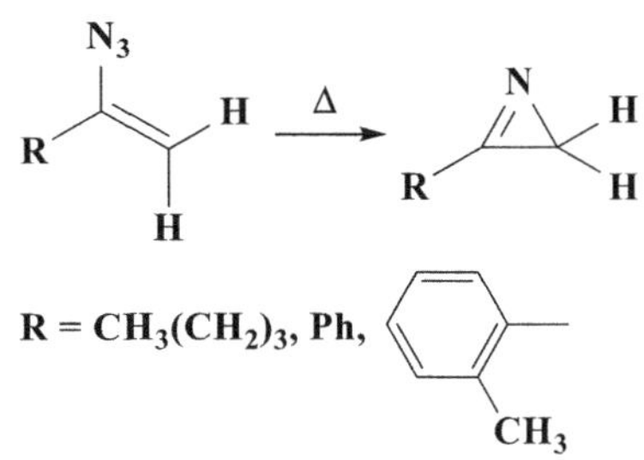

SCHEME 18

The thermolysis of vinyl azides provided azirines (Scheme 19) [63].

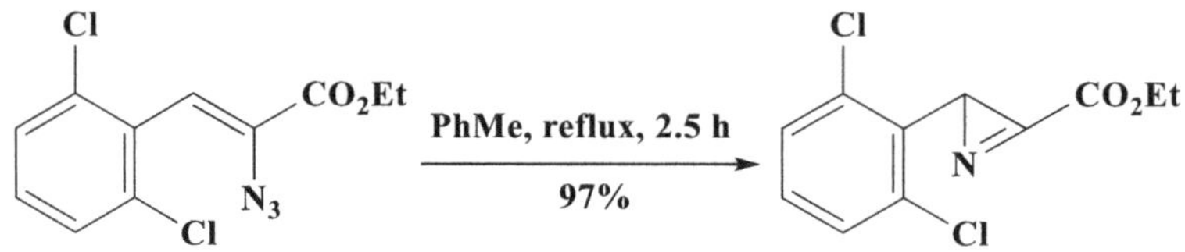

SCHEME 19

The 2-bromo-2*H*-azirines were prepared from alkenes, which were synthesized by the reaction of ylides with azidotrimethylsilane and NBS (Scheme 20). The reaction of ylides with azidotrimethylsilane and NIS provided iodoazidoalkenes. The iodoazidoalkenes were transformed to 2-iodo-2*H*-azirines. The reactivity of 2-halo-2*H*-azirine was examined utilizing potassium phthalimide as a nucleophile, which allowed the development of a procedure for the formation of new amino acids. The reaction of 2-bromo-2*H*-azirine provided ethyl 3-phenyl-2-phthalimido-2*H*-azirine-2-carboxylate in high yield (96%). The ethyl 3-phenyl-2-phthalimido-2*H*-azirine-2-carboxylate was also formed in 35% yield when 2-iodo-2*H*-azirine was reacted with potassium phthalimide. It was expected that the iodo-2*H*-azirines underwent halide displacement more smoothly than bromo derivatives. However, the lower stability of 2-iodo-2*H*-azirine provided ethyl 3-phenyl-2-phthalimido-2*H*-azirine-2-carboxylate in modest yield. The 2-iodo-2*H*-azirine was utilized for the synthesis of a new phthalimido-2*H*-azirine derivative in 28% yield. The formation of 2-substituted 2*H*-azirine derivatives in low yield was reported by the halide displacement reactions of 2-halo-2*H*-azirine [40,41].

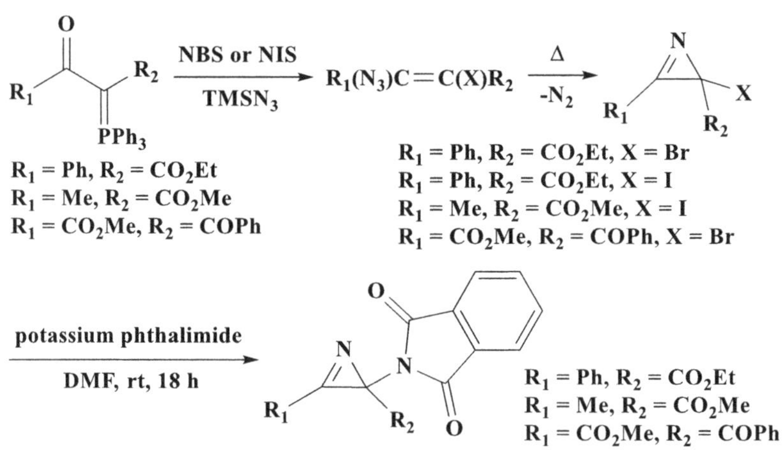

SCHEME 20

Michael addition to alkynyl esters synthesized fluorinated vinyl azides. The vinyl azide was formed as a stereoisomeric mixture, E/Z = 1:4, when azide ion addition to ester occurred at room temperature [64]. The azirinyl ester was prepared in high yield by the decomposition of azide in refluxing toluene

(Scheme 21). The dehydrohalogenation has offered additional pathway to vinyl azides. The β-iodoazide of carbohydrate origin was reacted with DBU to afford the ene-azide as a 1:3 *E*/*Z* mixture [65]. An unstable fluoroazirine in each case as a 7:3 mixture of diastereomers was obtained by the thermolysis of (*E*)-isomer occurring easily at 80°C and that of (*Z*)-isomer at 120°C (Scheme 22).

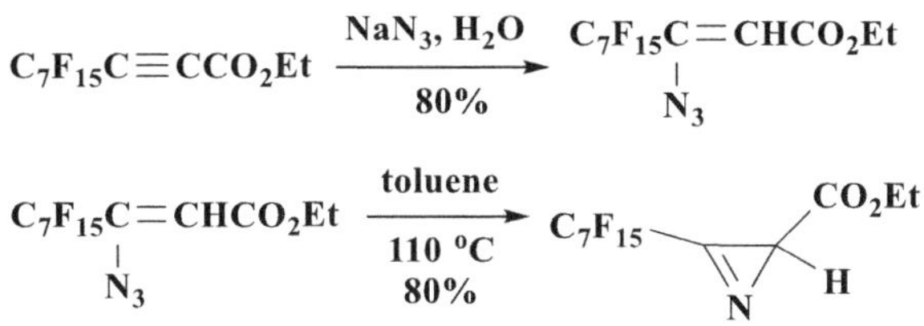

SCHEME 21

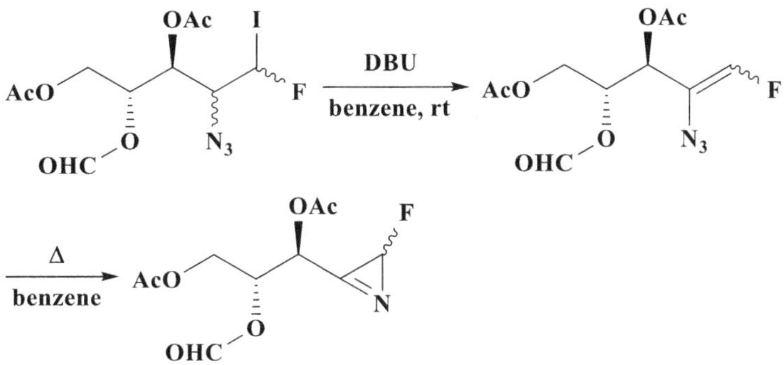

SCHEME 22

2.3 Synthesis of azirines from sodium azide

Brel [66] reported an efficient and convenient process in which phosphorylated allenes were prepared from alcohols by the Horner-Mark [2,3]-sigmatropic rearrangement of unstable phosphites synthesized in situ by reaction with diethyl chlorophosphite using Et_3N. The reaction of allenes with NaN_3 afforded 2-substituted 2-(diethylphosphono)-3-*i*-propenyl-2*H*-azirines (Scheme 23) [67].

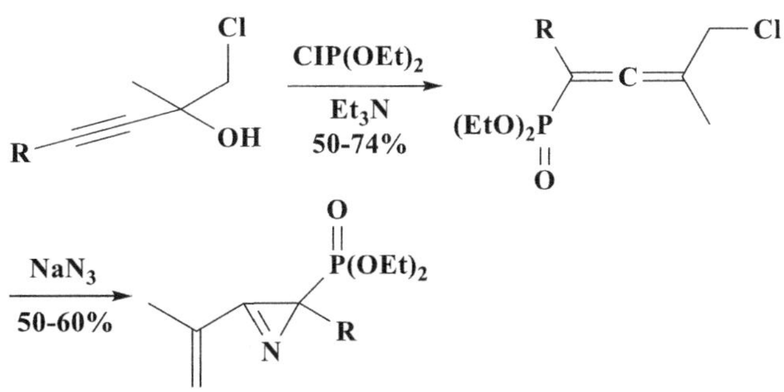

R = H, $C_2H_5O(CH_3)CHOCH_2$, C_4H_9, C_5H_{11}, $HOCH_2$

SCHEME 23

The dediazoniation of vinyl azide proceeded through the vinyl nitrene. A variant of this synthesis afforded 3-(dialkylamino)-2*H*-azirines from disubstituted acid amides (Scheme 24) [39,68–70].

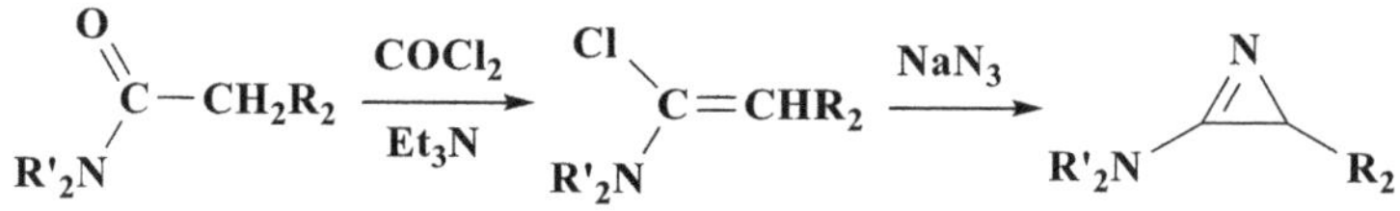

SCHEME 24

The *N*,*N*-disubstituted carboxamides were utilized for the formation of 3-(dialkylamino)-2*H*-azirines (Scheme 25) [71].

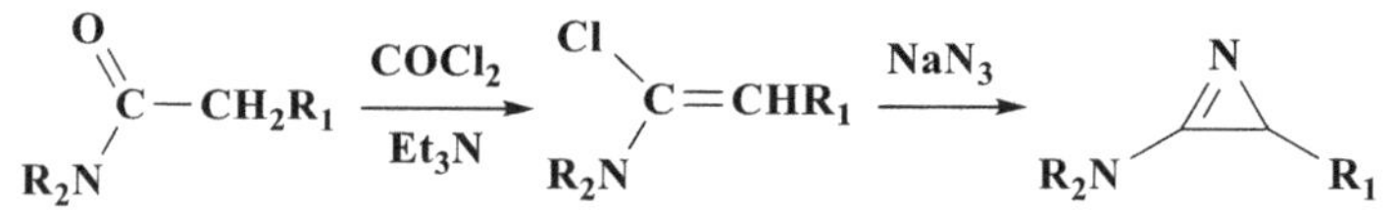

SCHEME 25

The highly toxic $COCl_2$ could be substituted by DPPCl (route A) [33]. The diphenylphosphorazidate was utilized as an alternative azide source for one-pot synthesis of azirines in very good yield (route B) (Scheme 26) [35].

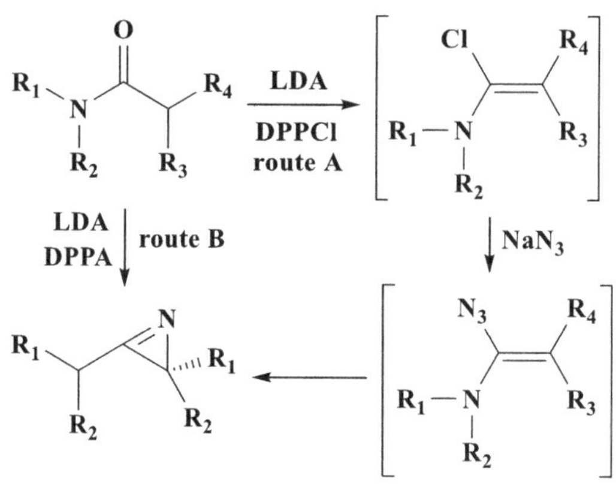

SCHEME 26

The 2,2-diphenyl-2*H*-azirine was synthesized via rearrangement of an intermediate $PhC(NCH_2CN)Ph$ to diphenylacetonitrile (Scheme 27) by an irradiation of diphenylvinyl azide [72–74]. Since 2*H*-azirines are air sensitive, once a 37% yield was reached, the solvent (deuterated chloroform) was instantly removed with a stream of N_2, and the mixture containing 2,2-diphenyl-2*H*-azirine was reacted with potassium carbonate at 150°C in dimethylformamide [75].

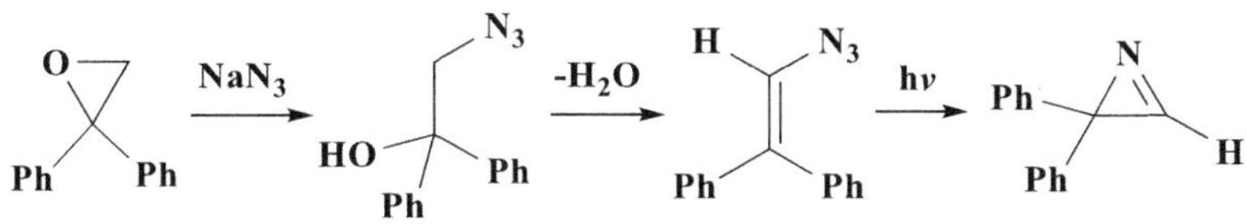

SCHEME 27

The reaction of phosphorylated allenes with NaN_3 synthesized 3-dihydro- and 3-vinylisoxazolyl-2*H*-azirines (Scheme 28) [44,66,76].

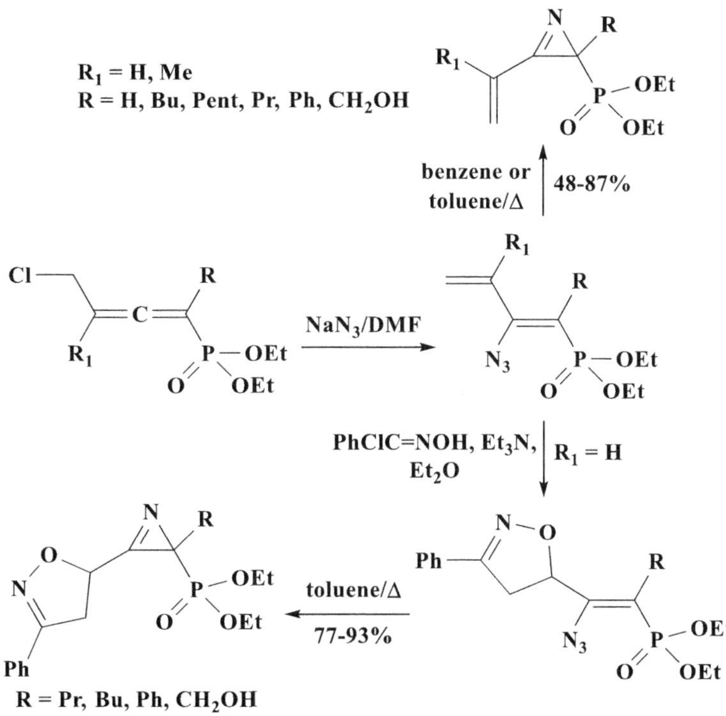

SCHEME 28

The decomposition of α-aminoalkenyl azides to aminoazirine was spontaneous and fast even at rt. Many procedures afforded vinyl azide in situ from other precursors. For instance, the keteneiminium salts, prepared by the reaction of amides with $COCl_2$, were added to NaN_3 to synthesize the transient α-aminovinyl azides that lose N_2 at rt to synthesize the azirine products (Scheme 29) [77]. The utilization of $COCl_2$ could be avoided with diphenylphosphorazidate, an aziridinating agent was reacted with amidate anions to synthesize the azirines directly, as given by the transformation of *N*-methyl-*N*-phenyl amide into aminoazirine in 94% yield (Scheme 30) [35]. This procedure tolerated various moieties [78], but some substrates must first be transformed into thioamide in the presence of

Lawesson's reagent [79,80]. Other appropriate precursors included α-chloroenamines, which underwent sequential azide displacement and quick thermal transformation into azirine under very mild conditions (Scheme 31) [31].

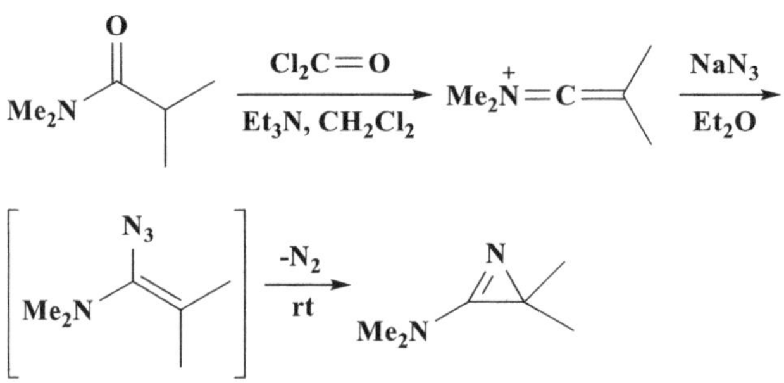

SCHEME 29

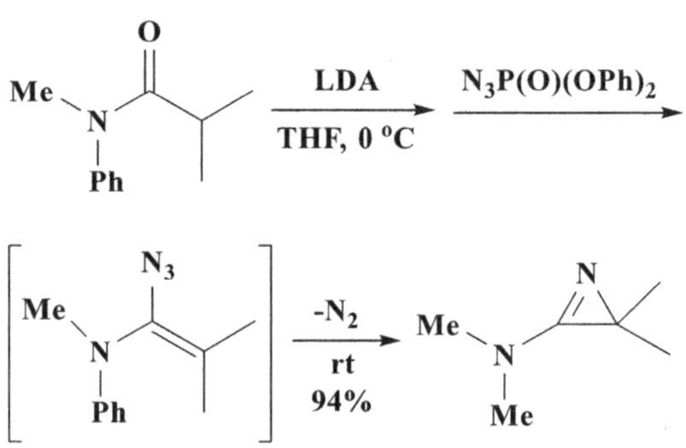

SCHEME 30

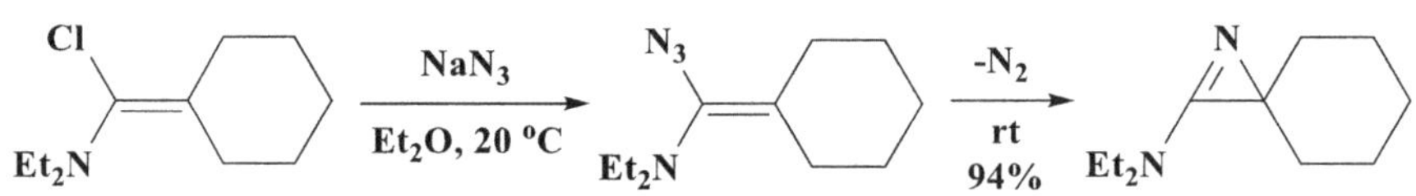

SCHEME 31

A few 2*H*-azirines were isolated from one-pot reactions in which the vinyl azide precursors were synthesized in situ. For instance, the 3-dialkylamino-2*H*-azirines were prepared from tertiary amides; the vinyl azide intermediates were generated by the consecutive chlorination and reaction with NaN_3 (Scheme 32) [32,81]. The transformation of these vinyl azides into azirines occurred at rt, and it was clear that their decomposition temperature was affected by the nature of substituents on the double bond of vinyl azides. The fluoro-substituted vinyl azides also decomposed at rt [33].

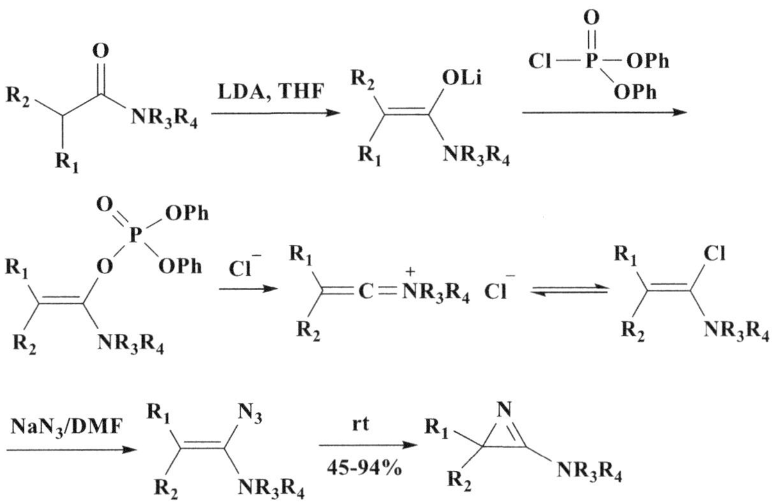

SCHEME 32

The vinyl azides were synthesized by the fragmentation of azide-substituted alkoxy radicals [56], which were easily accessible from 3-azido-2,3-didesoxyhexopyranoses [82–84] and consequently from NaN_3 and glycals. Using iodine source, the fragmentation resulted primarily in *vic*-iodoazides, which were subjected to an elimination under the mild conditions to afford the azidoalkenes. This reaction sequence was utilized for the formation of 2*H*-azirines (Scheme 33) [45].

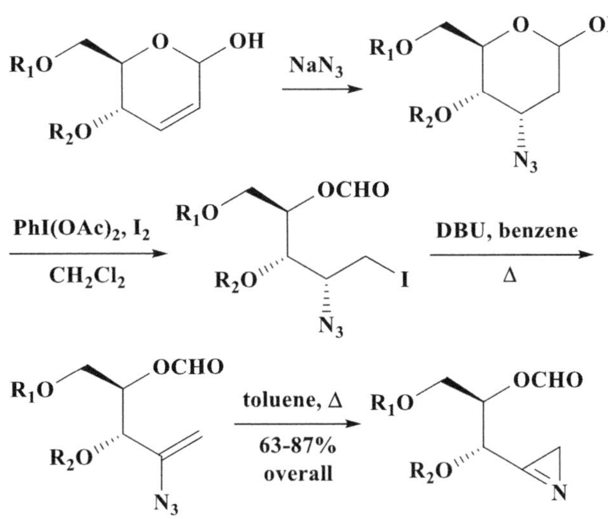

SCHEME 33

The bad-smelling azirine was prepared from styrene in 69% overall yield (Scheme 34) [52].

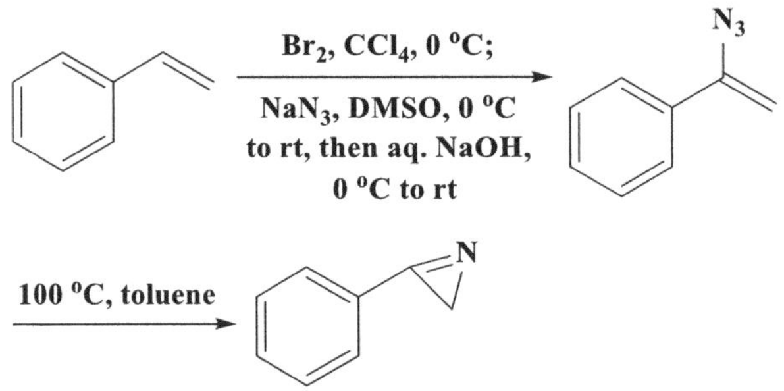

SCHEME 34

When efforts failed with Neber-type reactions, formation of 2*H*-azirines through the vinyl azides was examined and worked nicely with minor changes of published pathways (Scheme 35). The olefins were brominated to give the dibromo compound, while the olefin (R = $(CH_2)_2Ph$) was reacted with sodium azide and iodine monochloride to synthesize the iodo azide with high regioselectivity favoring azide addition to secondary position. The iodo azide and dibromo compounds were then transformed into vinyl azides in high yields [37,52,54]. These vinyl azides were subjected to ring closure on heating to synthesize the 2*H*-azirines. The choice of solvent influenced the convenience of this reaction step. The ring closure of azides into 2*H*-azirines had been achieved in refluxing toluene but losses of rather volatile 2*H*-azirines upon concentration inspired to utilize other solvents. High temperatures favored the clean transformation of 2*H*-azirines and increased the rate of reaction. The reactions were carried out in closed vessels at 125°C to avoid the need of solvents with high boiling points. The pentane worked well for ring closure of vinyl azides, but the solubility of vinyl azides (R = CO_2Bn) was poor in pentane and instead diethyl ether was utilized. Other solvents have been utilized, and good solubility appeared to be a general requirement. One way to prevent the side reactions such as polymerization and decomposition was photolytic ring closure of vinyl azides at rt [38]. The formation of side products was prevented by reducing the concentration of vinyl azide solution to 0.1 M. Other solvents, times, and temperatures could improve this cyclization. An addition of tertiary amines was found to catalyze the cyclization [85] but was not necessary in this case. The 3-(2-naphthyl)-2*H*-azirine was prepared in the similar way as for 2*H*-azirines, but the cyclization was carried out in diethyl ether.

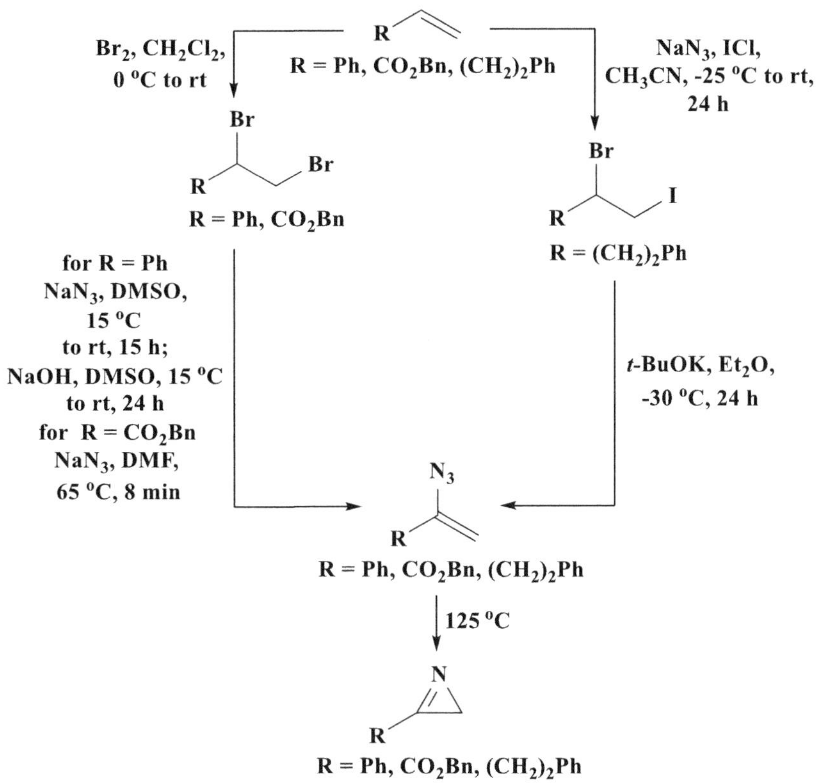

SCHEME 35

Rens and Ghosez [31] reported a reaction of α-chloroenamines with NaN_3 to synthesize the 3-amino-2*H*-azirines in excellent yields (Scheme 36). These products were obtained from vinyl azide intermediates, which were generated from α-chloroenamines by an elimination–addition mechanism. Clearly, the electron-rich vinyl azide loses N_2 at comparatively low temperature. Small ring amidines were highly beneficial synthetic equivalents for the formation of α,α-disubstituted α-amino acids and building blocks to synthesize the heterocyclic compounds and peptides as reported by Linden et al. [69,86–88]. In some specific cases, the short-lived α-azidoenamines similar to vinyl azide did not remove N_2 but cyclized to synthesize the 5-amino-4*H*-1,2,3-triazoles, which could be isomerized to 2*H*-1,2,3-triazoles [30,89].

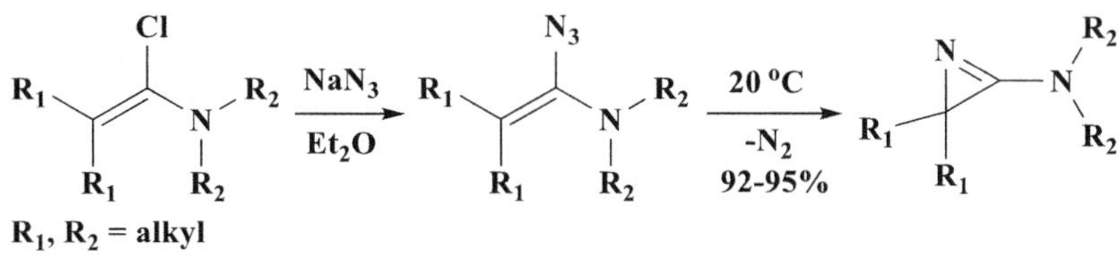

SCHEME 36

The fluoroazirines were usually prepared from vinyl azides, whether they could be reaction intermediates or stable starting compounds. The hexafluoropropene and NaN_3 in an addition–elimination reaction afforded *trans*-1-azidopentafluoropropene in 21% yield [64]. The fluoroazirine, i.e., 2,3-difluoro-2-trifluoromethyl-2*H*-azirine, was obtained as yellowish-green liquid, which being very unstable decomposed easily at 20°C (Scheme 37). It seemed that the 2,3-difluoro-2-trifluoromethyl-2*H*-azirine was formed in a concerted process in which double bond supported the removal of nitrogen. The reaction of vinyl ether with NaN_3 afforded 2,2-bis(trifluoromethyl)-3-methoxy-2*H*-azirine (Scheme 38) [65]. In this case, an intermediate vinyl azide started eliminating nitrogen even at 0°C. The vinyl azide formed en route to azirine also decomposed during the reaction at room temperature.

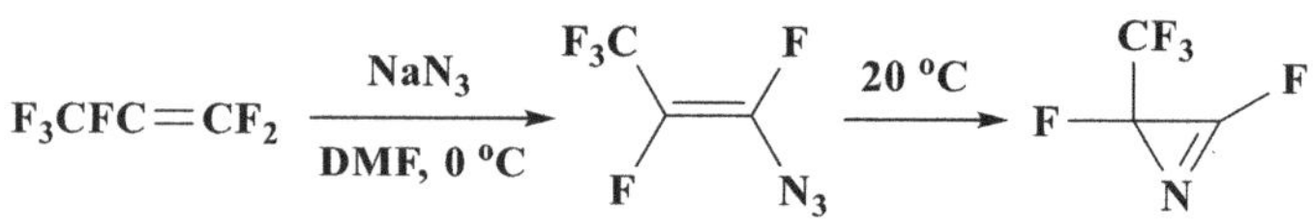

SCHEME 37

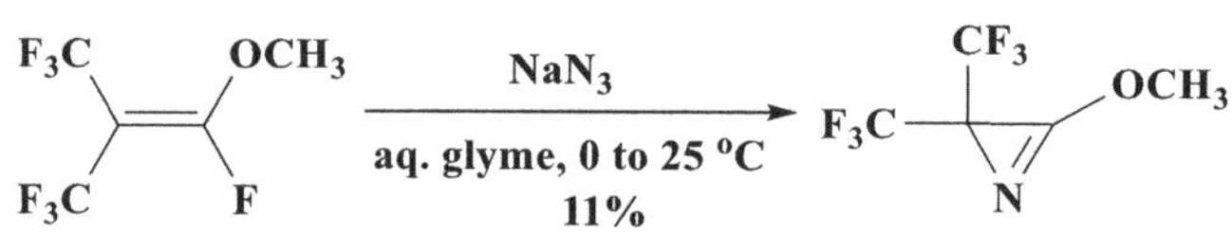

SCHEME 38

The 2*H*-azirin-3-amine, i.e., a synthon for 2-methylglutamate (Glu(2Me)), was synthesized (Scheme 39). The reaction was started from freshly distilled tetrahydro-2*H*-pyran-2-one, which is commercially accessible. The methylation of tetrahydro-2*H*-pyran-2-one in α-position to the C=O group by deprotonation with LDA followed by reaction with methyl iodide afforded methyl tetrahydro-2*H*-pyran-2-one. Instead of hexamethylphosphoramide, DMI (1,3-dimethylimidazolidin-2-one), which was of lower toxicological risk, was utilized as an additive. Although Li and coworkers [90] utilized lithium hexamethyldisilazanide as a base for similar reactions, lithium diisopropylamide in combination with 1,3-dimethylimidazolidin-2-one was preferred. Therefore, the yield (44%) was lower than reported (70%) [91]. The hydroxyamide was formed when methyl tetrahydro-2*H*-pyran-2-one was reacted with *N*-methylaniline using aluminum chloride at room temperature. Because of its cancer-causing nature, the suggested solvent 1,2-dichloroethane [92] was substituted by dichloromethane. The yield in dichloromethane (81%) was only slightly lower than in 1,2-dichloroethane (88%). In the subsequent step, the hydroxy group of hydroxyamide was oxidized with $NaIO_4$ (sodium metaperiodate) and $RuCl_3.H_2O$ (ruthenium trichloride hydrate) to synthesize the carboxylic acid. This technique, in which ruthenium tetraoxide was the active species [93], has the benefit that ruthenium trichloride hydrate could be utilized in catalytic amounts, and the stoichiometric oxidizing agent was sodium metaperiodate. The function of sodium metaperiodate was to reoxidize the reduced forms of ruthenium complex to ruthenium tetraoxide. As a solvent, a mixture of carbon tetrachloride, acetonitrile, and water in 2:2:3 ratio was utilized in a first effort [94]. As carbon tetrachloride was environmentally unwanted and toxic, it was substituted by the same amount of ethyl acetate [95]. As a consequence, the yield turned out to be a bit lower (73%) than with carbon tetrachloride (84%). The methylation of carboxylic acid with diazomethane provided ester in quantitative yield. The formation of 2*H*-azirin-3-amine reported by Villalgordo and Heimgartner [33,96] failed, even though ester was *N*-alkyl-*N*-phenyl amide. It was proposed that the deprotonation took place in α-position to the ester group instead of the α-position to the amide group. According to other well-known route, the amide was first transformed to thioamide in 93% yield with Lawesson reagent in toluene at 130°C. Finally, the 2*H*-azirin-3-amine was formed in 90% yield by the consecutive reaction of thioamide with 2 N phosgene solution in dichloromethane, deprotonation with DABCO in tetrahydrofuran, and reaction with sodium azide in tetrahydrofuran/dimethylformamide.

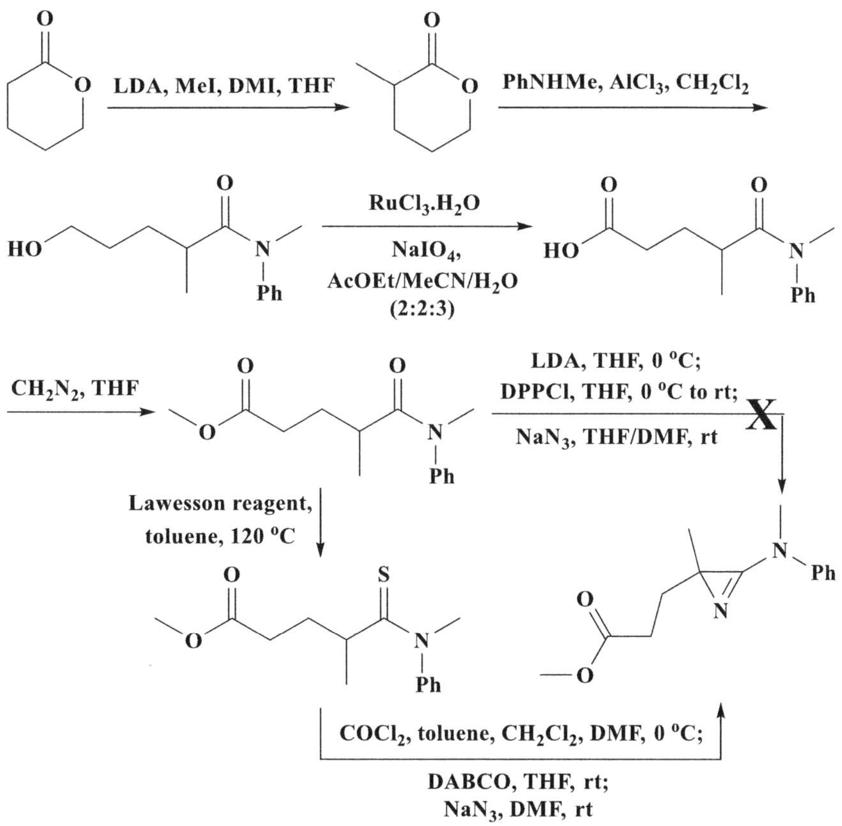

SCHEME 39

As the technique of Villalgordo and Heimgartner [32] for the formation of 2*H*-azirin-3-amines, utilized by Strässler et al. [34] for the formation of azirine (n = 1), is limited to *N*-alkyl-*N*-phenyl derivatives, some *N*-[(tetrahydro-2*H*-thiopyran-4-yl)thiocarbonyl]-L-proline methyl esters were prepared with the aim of following the procedure of aminoazirine synthesis published by Rens and Ghosez [31] and modified by Dietliker and Heimgartner [97]. The tetrahydro-2*H*-thiopyran-4-carboxylic acid chloride was reacted with L-proline methyl ester to synthesize the *N*-[(tetrahydro-2*H*-thiopyran-4-yl)carbonyl]-L-proline methyl ester in 85% yield, which was transformed to *N*-[(tetrahydro-2*H*-thiopyran-4-yl)thiocarbonyl]-L-proline methyl ester by thionation with Lawesson reagent in 67% yield (Scheme 40). In similarity to the process reported [98] and to the procedure for the formation of azirine (X = O, n = 1) [99], successive treatment of a solution of *N*-[(tetrahydro-2*H*-thiopyran-4-yl)thiocarbonyl]-L-proline methyl ester in dichloromethane and catalytic amounts of dimethylformamide with phosgene, evaporation of the solvent, dissolution of the residue in tetrahydrofuran, an addition of DABCO, filtration, and reaction with sodium azide provided azirine in 67% yield. Due to the decomposition of crude product during its purification by column chromatography (silica), the compound was utilized in the next reactions without further purification [100].

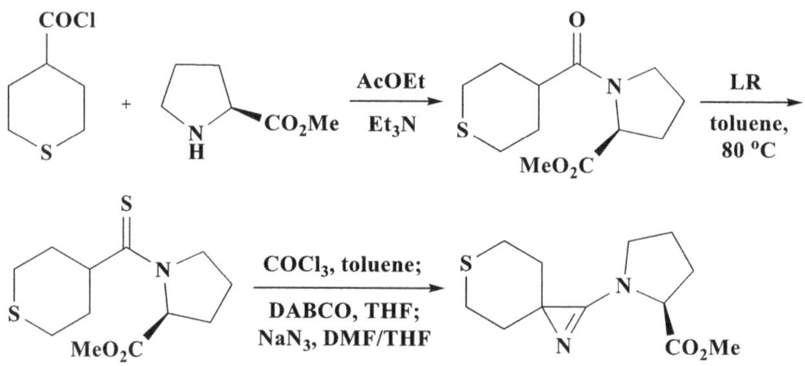

SCHEME 40

A series of chiral aminoazirines was prepared from amino acids. These azirines were utilized as reagents in peptide formation [79]. An instance was afforded by the formation of homoproline-derived azirine (Scheme 41) [101].

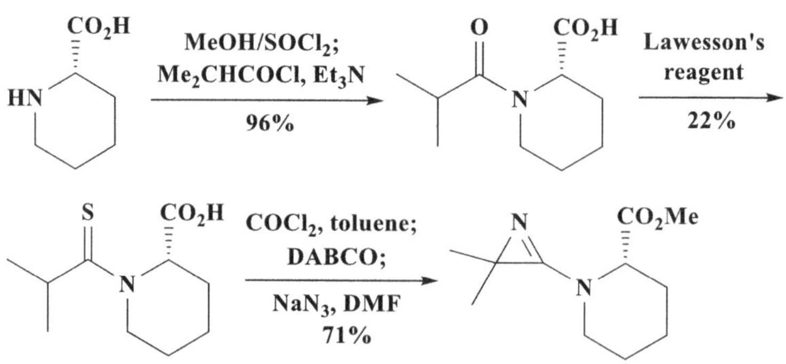

SCHEME 41

The optically active 3-amino-2*H*-azirines were prepared by the reaction of mono- or disubstituted thioamides containing a chiral substituent at the amino group with $COCl_2/Et_3N$ and NaN_3 [39,102]. This reaction was based on an earlier reported procedure that proceeded via α-chloro enamine and vinyl azide intermediates, which were not isolated [30]. The beneficial process for the formation of optically active 3-amino-2*H*-azirines used thioamide carrying a chiral substituent at the amino group (Scheme 42). The chromatographic separation of diastereomeric mixture of 3-amino-2*H*-azirines provided pure diastereoisomers, which, after electrochemical cleavage of the benzene sulfonyl group, were utilized as synthons in the formation of pentapeptides [36].

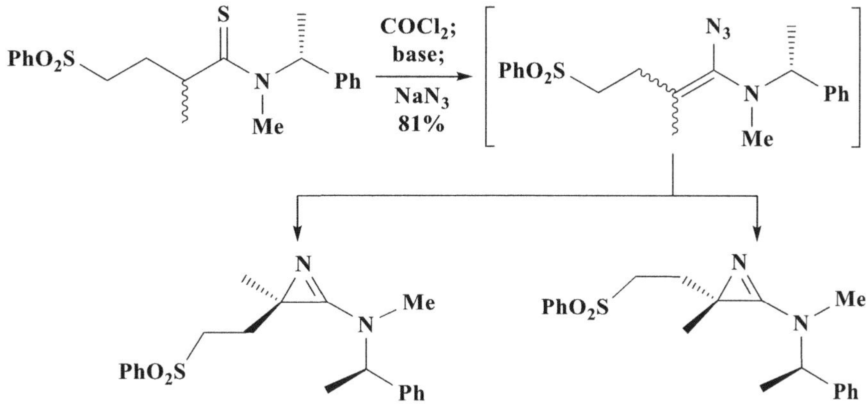

SCHEME 42

The photolysis or thermolysis of vinyl azides, prepared from alkenes, afforded 2*H*-azirines (Scheme 43) [103].

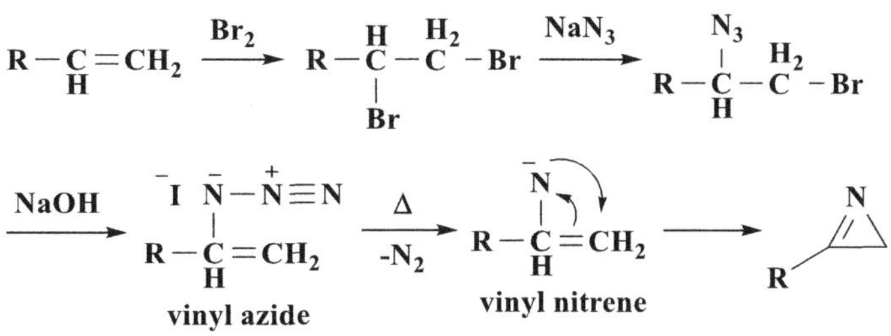

SCHEME 43

References

1. (a) N. Kaur. 2015. Role of microwaves in the synthesis of fused five-membered heterocycles with three *N*-heteroatoms. Synth. Commun. 45: 403-431. (b) N. Kaur. 2015. Synthesis of five-membered *N,N,N*- and *N,N,N,N*-heterocyclic compounds: applications of microwaves. Synth. Commun. 45: 1711-1742. (c) N.

Kaur. 2015. Six-membered heterocycles with three and four *N*-heteroatoms: microwave-assisted synthesis. Synth. Commun. 45: 151-172. (d) N. Kaur and D. Kishore. 2014. Microwave-assisted synthesis of six-membered *O*-heterocycles. Synth. Commun. 44: 3047-3081. (e) N. Kaur and D. Kishore. 2014. Microwave-assisted synthesis of six-membered *O,O*-heterocycles. Synth. Commun. 44: 3082-3111. (f) N. Kaur and D. Kishore. 2014. Synthesis of oxadiazolo, pyrimido, imidazolo and benzimidazolo containing derivatives of 1,4-benzodiazepin-5-(4'-methylpiperazinyl)-carboxamide through phenylamino spacer. Synth. Commun. 44: 2789-2796. (g) N. Kaur. 2013. Recent trends in the chemistry of privileged scaffold: 1,4-benzodiazepine. Int. J. Pharm. Biol. Sci. 4: 485-513. (h) J. Dwivedi, N. Kaur, D. Kishore, S. Kumari and S. Sharma. 2016. Synthetic and biological aspects of thiadiazoles and their condensed derivatives: an overview. Curr. Top. Med. Chem. 16: 2884-2920. (i) N. Kaur. 2019. Multiple nitrogen-containing heterocycles: metal and non-metal assisted synthesis. Synth. Commun. 49: 1633-1658. (j) N. Kaur. 2018. Synthesis of six- and seven-membered and larger heterocycles using Au and Ag catalysts. Inorg. Nano Met. Chem. 48: 541-568. (k) N. Kaur, P. Grewal, P. Bhardwaj, M. Devi and Y. Verma. 2019. Nickel-catalyzed synthesis of five-membered heterocycles. Synth. Commun. 49: 1543-1577. (l) N. Kaur. 2018. Recent developments in the synthesis of nitrogen-containing five-membered polyheterocycles using rhodium catalysts. Synth. Commun. 48: 2457-2474. (m) N. Kaur. 2017. Gold catalysts in the synthesis of five-membered *N*-heterocycles. Curr. Organocatal. 4: 122-154. (n) N. Kaur. 2019. Palladium acetate and phosphine assisted synthesis of five-membered *N*-heterocycles. Synth. Commun. 49: 483-514. (o) N. Kaur. 2019. Applications of palladium dibenzylideneacetone as catalyst in the synthesis of five-membered *N*-heterocycles. Synth. Commun. 49: 1205-1230. (p) T.L. Gilchrist. 2001. Activated 2*H*-azirines as dienophiles and electrophiles. Aldrichimica Acta 34: 51-55.

2. E.O. Stapley, D. Hendlin, M. Jackson and A.K. Miller. 1971. Azirinomycin. I. J. Antibiot. 24: 42-47.
3. H.M.I. Osborn and J. Sweeney. 1997. The asymmetric synthesis of aziridines. Tetrahedron: Asymmetry 8: 1693-1715.
4. D. Tanner. 1994. Chiral aziridines - their synthesis and use in stereoselective transformations. Angew. Chem. Int. Ed. 33: 599-619.
5. M.M.H. Verstappen, G.J.A. Ariaans and B. Zwanenburg. 1996. Asymmetric synthesis of 2*H*-azirine carboxylic esters by an alkaloid-mediated Neber reaction. J. Am. Chem. Soc. 118: 8491-8492.
6. P. Kafarski and B. Lejezak. 1991. Biological activity of aminophosphonic acids. Phosphorus, Sulfur, Silicon Relat. Elem. 63: 193-215.
7. P.W. Neber and A. Burgard. 1932. Über den reaktionsverlauf einer neuen art von umlagerung bei ketoximen. III. Justus Liebigs Ann. Chem. 493: 281-294.
8. P.W. Neber and G. Huh. 1935. Eine neue, allgemeine methode zur gewinnung von α-aminoketonen. I. Justus Liebigs Ann. Chem. 515: 283-296.
9. B. Zwanenburg and P. ten Holte. 2001. The synthetic potential of three-membered ring aza-heterocycles. Top. Curr. Chem. 216: 93-124.
10. B.A. Trofimov and A.I. Mikhaleva. 1987. Azirines in reactions involving the formation of a pyrrole ring (review). Chem. Heterocycl. Compd. (Engl. Transl.) 23: 1037-1049.
11. D.J. Anderson and A. Hassner. 1975. Synthesis of heterocycles via cycloadditions to 1-azirines. Synthesis 8: 483-495.
12. T. Wurthwein, T. Hergenrother and H. Quast. 2002. *Exo/endo* preferences of double bonds in three-membered ring compounds - the bias toward endocyclic unsaturation in 3-alkyl- and 3-amino-2*H*-azirines: a theoretical and experimental study. Eur. J. Org. Chem. 11: 1750-1755.
13. K. Banert and B. Meier. 2006. Synthesis and reactions of highly strained 2,3-bridged 2*H*-azirines. Angew. Chem. Int. Ed. 45: 4015-4019.
14. T.W. Miller, E.W. Tristram and F.J. Wolf. 1971. Azirinomycin. II. J. Antibiot. 24: 48-50.
15. F.A. Davis, G.V. Reddy and H. Liu. 1995. Asymmetric synthesis of 2*H*-azirines: first enantioselective synthesis of the cytotoxic antibiotic (*R*)-(-)-dysidazirine. J. Am. Chem. Soc. 117: 3651-3652.

16. F.A. Davis, H. Liu, C.-H. Liang, G.V. Reddy, Y. Zhang, T. Fang and D. Titus. 1999. Asymmetric synthesis of 2*H*-azirine 2-carboxylate esters. J. Org. Chem. 64: 8929-8935.
17. L. Gentilucci, Y. Grijzen, L. Thijs and B. Zwanenburg. 1995. Convenient synthesis of optically active 2*H*-azirine-2-carboxylic esters by Swern oxidation of aziridine-2-carboxylic esters. Tetrahedron Lett. 36: 4665-4668.
18. D.M. Hodgson, P.G. Humphreys, Z. Xu and J.G. Ward. 2007. Lithiation-induced migrations from nitrogen to carbon in terminal aziridines. Angew. Chem. Int. Ed. 46: 2245-2248.
19. T.M.V.D. Pinho e Melo and A.M.A.R. Gonsalves. 2004. Exploiting 2-halo-2*H*-azirine chemistry. Curr. Org. Synth. 1: 275-292.
20. T.F. Molinski and C.M. Ireland. 1988. Dysidazirine, a cytotoxic azacyclopropene from the marine sponge *Dysidea fragilis*. J. Org. Chem. 53: 2103-2105.
21. C.E. Salomon, D.H. Williams and D.J. Faulkner. 1995. New azacyclopropene derivatives from *Dysidea fragilis* collected in Pohnpei. J. Nat. Prod. 58: 1463-1466.
22. J. Lehmann, A. Linden and H. Heimgartner. 1999. Site-selective incorporation of thioamide-linkages into a growing peptide. Tetrahedron 55: 5359-5376.
23. (a) T.M.V.D. Pinho e Melo, C.S.J. Lopes, A.M.AR. Gonsalves, A.M. Beja, J.A. Paixão, M.R. Silva and L. Alte DaVeiga. 2002. Reactivity of 2-halo-2*H*-azirines. 1. Reactions with nucleophiles. J. Org. Chem. 67: 66-71. (b) F.A. Davis, C.H. Liang and H. Liu. 1997. Asymmetric synthesis of β-substituted α-amino acids using 2*H*-azirine-2-carboxylate esters. Synthesis of 3,3-disubstituted aziridine-2-carboxylate esters. J. Org. Chem. 62: 3796-3797. (c) C.A. Ray, E. Risberg and P. Somfai. 2001. Lewis acid-catalyzed hetero Diels-Alder cycloadditions of 3-alkyl, 3-phenyl and 3-carboxylated 2*H*-azirines. Tetrahedron Lett. 42: 9289-9291.
24. K.N. Koch, A. Linden and H. Heimgartner. 2001. Synthesis of conformationally restricted cyclic penta-depsipeptides via direct amide cyclization. Tetrahedron 57: 2311-2326.
25. (a) C.-Y. Huang and A.G. Doyle. 2014. The chemistry of transition metals with three-membered ring heterocycles. Chem. Rev. 114: 8153-8198. (b) A.F. Khlebnikov and M.S. Novikov. 2013. Recent advances in 2*H*-azirine chemistry. Tetrahedron 69: 3363-3401. (c) N.V. Rostovskii, M.S. Novikov, A.F. Khlebnikov, S.M. Korneev and D.S. Yufit. 2013. Cu(II)-Catalyzed domino reaction of 2*H*-azirines with diazotetramic and diazotetronic acids. Synthesis of 2-substituted 2*H*-1,2,3-triazoles. Org. Biomol. Chem. 11: 5535-5545. (d) A. Padwa. 2010. Cycloaddition and cyclization chemistry of 2*H*-azirines. Adv. Heterocycl. Chem. 99: 1-31. (e) N.V. Rostovskii, M.S. Novikov, A.F. Khlebnikov, G.L. Starova and M.S. Avdontseva. 2015. Azirinium ylides from α-diazoketones and 2*H*-azirines on the route to 2*H*-1,4-oxazines: three-membered ring opening vs 1,5-cyclization. Beilstein J. Org. Chem. 11: 302-312.
26. F. Palacios, A.M.O. Retana, E.M. Marigorta and J.M. Santos. 2001. 2*H*-Azirines as synthetic tools in organic chemistry. Eur. J. Org. Chem. 13: 2401-2414.
27. A. Padwa, M. Dharan, J. Smolanoff and S.I. Wetmore. 1973. Photochemical transformations of small ring heterocyclic compounds. XLVII. Electronic details of the photocycloaddition of arylazirines. J. Am. Chem. Soc. 95: 1954-1961.
28. A. Padwa, J.K. Rasmussen and A. Tremper. 1976. Photochemical transformations of small ring heterocyclic compounds. 75. Photochemistry of arylazirines in hydroxylic media. J. Am. Chem. Soc. 98: 2605-2614.
29. A. Padwa and T. Stengel. 2005. Grubb's and Wilkinson catalyzed reactions of 2-phenyl-3-vinyl substituted 2*H*-azirines. ARKIVOC (v): 21-32.
30. M. Henriet, M. Houtekie, B. Techy, R. Touillaux and L. Ghosez. 1980. Synthesis of 5-dialkylamino-1,2,3-triazoles and 2-amino-1-azirines from tertiary amides. Tetrahedron Lett. 21: 223-226.
31. M. Rens and L. Ghosez. 1970. Synthesis and reactions of 2-amino-1-azirines. Tetrahedron Lett. 11: 3765-3768.

32. J.M. Villalgordo and H. Heimgartner. 1992. Reaction of diphenyl phosphorochloridate with amide enolates: a new and convenient synthesis of 2-monosubstituted 3-(*N*-methyl-*N*-phenylamino)-2*H*-azirines. Helv. Chim. Acta 75: 1866-1871.
33. J.M. Villalgordo and H. Heimgartner. 1993. A novel synthesis of 2,2-disubstituted 3-amino-2*H*-azirines based on the reaction between amide enolates and diphenyl phosphorochloridate. Helv. Chim. Acta 76: 2830-2837.
34. C. Strässler, A. Linden and H. Heimgartner. 1997. Novel heterospirocyclic 3-amino-2*H*-azirines as synthons for heterocyclic α-amino acids. Helv. Chim. Acta 80: 1528-1554.
35. J.M. Villalgordo, A. Enderli, A. Linden and H. Heimgartner. 1995. Diazo-transfer reaction with diphenyl phosphorazidate. Helv. Chim. Acta 78: 1983-1998.
36. F. Palacios, M.O. de Retana, M.M. de Marigorta and J.M. de los Santos. 2002. Preparation, properties and synthetic applications of 2*H*-azirines: a review. Org. Prep. Proced. Int. 34: 219-269.
37. F.W. Fowler, A. Hassner and L.A. Levy. 1967. Stereospecific introduction of azide functions into organic molecules. J. Am. Chem. Soc. 89: 2077-2082.
38. A. Hassner and F.W. Fowler. 1968. Stereochemistry. XXXII. Synthesis and reactions of 1-azirines. J. Am. Chem. Soc. 90: 2869-2875.
39. C.B. Bucher and H. Heimgartner. 1996. Optisch aktive 3-amino-2*H*-azirine als bausteine für enantiomerenreine αα-disubstituierte α-aminosäuren: synthese von isovalin -synthonen und einbau in ein trichotoxin-A-50-segment. Helv. Chim. Acta 79: 1903-1915.
40. T.M.V.D. Pinho e Melo, A.M.A.R. Gonsalves, C.S.J. Lopes and T.L. Gilchrist. 1999. Synthesis of 2-halo-2*H*-azirines from phosphorus ylides. Tetrahedron Lett. 40: 789-792.
41. T.M.V.D. Pinho e Melo, C.S.J. Lopes and A.M.A.R. Gonsalves. 2000. Synthesis and reactivity of 2-halo-2*H*-azirines towards nucleophiles. Tetrahedron Lett. 41: 7217-7220.
42. T.M.V.D. Pinho e Melo, C.S.J. Lopes, A.L. Cardoso and A.M.R. Gonsalves. 2001. Synthesis of 2-halo-2*H*-azirines. Tetrahedron 57: 6203-6208.
43. F. Palacios, A.M. Ochoa de Retana, J.I. Gil and J.M. Ezpeleta. 2000. Simple asymmetric synthesis of 2*H*-azirines derived from phosphine oxides. J. Org. Chem. 65: 3213-3217.
44. A. Lemos. 2009. Addition and cycloaddition reactions of phosphinyl- and phosphonyl-2*H*-azirines, nitrosoalkenes and azoalkenes. Molecules 14: 4098-4119.
45. S. Brase, C. Gil, K. Knepper and V. Zimmermann. 2005. Organic azides: an exploding diversity of a unique class of compounds. Angew. Chem. Int. Ed. 44: 5188-5240.
46. K. Banert and F. K'hler. 2001. Synthesis of 1,4-diazidobuta-1,3-dienes by electrocyclic ring opening: precursors for bi-2*H*-azirin-2-yls and their valence isomerization to diazabenzenes. Angew. Chem. Int. Ed. 40: 174-177.
47. L. Garanti, A. Locatelli and G. Zecchi. 1976. Synthesis of sulphur containing fused ring heterocycles by intramolecular 1,3-dipolar-cycloadditions. J. Heterocycl. Chem. 13: 657-659.
48. M. Bertrand, J.P. Dulcere and M. Santelli. 1977. Intramolecular dipido-1,3-cycloaddition reaction of azido-1 ene-2 ynes-4: a method of synthesis of 4*H*-pyrrolo [1,2-*c*] *v*-triazoles. Tetrahedron Lett. 18: 1783-1784.
49. C.L. Branch, S.C. Finch and M.J. Pearson. 1982. Synthesis of novel fused β-lactams by intramolecular 1,3-dipolar cycloadditions. 51. Triazolocephems and triazolopenams. Tetrahedron Lett. 23: 4381-4384.
50. G. Smolinsky. 1962. Formation of azacyclopropenes by pyrolysis of vinyl azides. J. Org. Chem. 27: 3557-3559.
51. G. Smolinsky. 1961. Vinyl azene chemistry: formation of azacyclopropene. J. Am. Chem. Soc. 83: 4483-4484.
52. A.G. Hortmann, D.A. Robertson and B.K. Gillard. 1972. Convenient procedure for the preparation of 2-arylazirines. J. Org. Chem. 37: 322-324.

53. M.J. Alves and T.L. Gilchrist. 1998. Generation and Diels-Alder reactions of *t*-butyl 2*H*-azirine-3-carboxylate. Tetrahedron Lett. 39: 7579-7582.
54. T.L. Gilchrist and R. Mendonca. 2000. Addition of pyrimidine and purine bases to benzyl 2*H*-azirine-3-carboxylate. Synlett 12: 1843-1845.
55. A.S. Timen, E. Risberg and P. Somfai. 2003. Improved procedure for cyclization of vinyl azides into 3-substituted-2*H*-azirines. Tetrahedron Lett. 44: 5339-5341.
56. C.R. Alonso-Cruz, A.R. Kennedy, M.S. Rodriguez and E. Suárez. 2003. Fragmentation of carbohydrate anomeric alkoxyl radicals. A new synthesis of chiral β-iodo azides, vinyl azides, and 2*H*-azirines. Org. Lett. 5: 3729-3732.
57. P.N.D. Singh, C.L. Carter and A.D. Gudmundsdottir. 2003. A simple green procedure for the synthesis of 2*H*-azirines. Tetrahedron Lett. 44: 6763-6765.
58. G. Smolinsky and C.A. Pryde. 1968. Vinyl azide chemistry. Thermally induced reactions. J. Org. Chem. 33: 2411-2416.
59. K. Banert, S. Grimme, R. Herges, K. Heß, F. Köhler, C. Mück-Lichtenfeld and E.-U. Würthwein. 2006. Experimental and theoretical characterization of the valence isomerization of bi-2*H*-azirin-2-yls to diazabenzenes. Chem. Eur. J. 12: 7467-7481.
60. A. Gómez-Zavaglia, A. Kaczor, A.L. Cardoso, T.M.V.D. Pinho e Melo and R. Fausto. 2006. Unusual photochemical C-N bond cleavage in the novel methyl 2-chloro-3-methyl-2*H*-azirine-2-carboxylate. J. Phys. Chem. A 110: 8081-8092.
61. T.M.V.D. Pinho e Melo, A.L. Cardoso and A.M.A.R. Gonsalves. 2003. Reactivity of 2-halo-2*H*-azirines. Part 3: dehalogenation of 2-halo-2*H*-azirine-2-carboxylates. Tetrahedron 59: 2345-2351.
62. T.M.V.D. Pinho e Melo, C.S.J. Lopes, A.M.A.R. Gonsalves and R.C. Storr. 2002. Reactivity of 2-halo-2*H*-azirines. Part II. Thermal ring expansion reactions: synthesis of 4-haloisoxazoles. Synthesis 5: 605-608.
63. L. Henn, D.M.B. Hickey, C.J. Moody and C.W. Rees. 1984. Formation of indoles, isoquinolines, and other fused pyridines from azidoacrylates. J. Chem. Soc. Perkin Trans. 1 0: 2189-2196.
64. T. Eicher and S. Hauptmann. 2003. The chemistry of heterocycles: structures, reactions, and applications. John Wiley & Sons: Chichester, UK, 5-16.
65. R.E. Banks and J.C. Tatlow. 1994. Organofluorine chemistry: nomenclature and history. R.E. Banks, B.E. Smart and J.C. Tatlow (Eds.). Organofluorine chemistry: principles and commercial applications. Plenum Press: New York, 2-4.
66. V.K. Brel. 2002. New synthetic route to 2-(diethylphosphono)-2*H*-azirines. Synthesis 13: 1829-1832.
67. K. Moonen, I. Laureyn and C.V. Stevens. Synthetic methods for azaheterocyclic phosphonates and their biological activity. Chem. Rev. 104: 6177-6216.
68. P. Wipf and H. Heimgartner. 1988. Anwendung der azirin/oxazolon-methode in der peptid-chemie: synthese von modell-tripeptiden. Helv. Chim. Acta 71: 140-154.
69. M. Blagoev, A. Linden and H. Heimgartner. 1999. Lewis acid catalyzed reactions of thioketones with 1,2-epoxycyclohexane and 1,2-epoxycyclopentane. Helv. Chim. Acta 82: 2316-2335.
70. C. Alonso, J.M. de los Santos, J. Vicario and F. Palacios. 2011. Phosphorus substituted hydroxylamine and hydroxamic acid derivatives: synthesis and reactivity. ARKIVOC (iii): 221-253.
71. A.V. Eremeev and I.P. Piskunova. 1990. Synthesis and properties of substituted 3-amino-2*H*-azirines (review). Chem. Heterocycl. Compd. 26: 719-738.
72. W. Tumas, R.F. Foster and J. Brauman. 1988. Mechanistic studies on gas-phase negative ion unimolecular decompositions. Alkoxide anions. J. Am. Chem. Soc. 110: 2714-2722.
73. K. Isomura, M. Okada and H. Taniguchi. 1969. Azirine formation by decomposition of terminal vinyl azides. Tetrahedron Lett. 46: 4073-4076.
74. K. Isomura, H. Taniguchi, M. Mishima, M. Fujio and Y. Tsuno. 1977. Carbon-13 nuclear magnetic resonance spectra of some methyl and phenyl substituted 2*H*-azirines. Org. Magn. Reson. 9: 559-562.

75. A. Perosa, M. Selva and P. Tundo. 2002. The synthesis of alkyl aryl nitriles from *N*-(1-arylalkylidene) cyanomethylamines. Part 2. Mechanism. J. Chem. Soc. Perkin Trans. 2 5: 1033-1037.
76. V.K. Brel. 2007. Synthesis of 3-azido-4-(diethoxyphosphoryl)alka-1,3-dienes and their transformation to derivatives of 2*H*-azirine. Synthesis 17: 2674-2680.
77. H. Heimgartner. 1991. 3-Amino-2*H*-azirines. Synthons for α,α-disubstituted α-amino acids in heterocycle and peptide synthesis. Angew. Chem. Int. Ed. 30: 238-264.
78. M.K.G. Mekhael and H. Heimgartner. 2003. A novel synthetic approach to (±)-desoxynoreseroline. Helv. Chim. Acta 86: 2805-2813.
79. S. Stamm, A. Linden and H. Heimgartner. 2003. Chiral heterospirocyclic 2*H*-azirin-3-amines as synthons for 3-amino-2,3,4,5-tetrahydrofuran-3-carboxylic acid and their use in peptide synthesis. Helv. Chim. Acta 86: 1371-1396.
80. F.M. Hilty, K.A. Brun and H. Heimgartner. 2004. A new 2*H*-azirin-3-amine as a synthon for α-methyl glutamate. Helv. Chim. Acta 87: 2539-2548.
81. C.S. Cleaver and C.G. Krespan. 1965. Fluoroazirines. Synthesis and polymerization. J. Am. Chem. Soc. 87: 3716-3719.
82. J.-C. Florent and C. Monneret. 1987. Stereocontrolled route to 3-amino-2,3,6-trideoxy-hexopyranoses. K-10 Montmorillonite as a glycosidation reagent for acosaminide synthesis. J. Chem. Soc. Chem. Commun. 15: 1171-1172.
83. J. Wengel, J. Lau, E.B. Pedersen and C.M. Nielsen. 1991. Synthesis of L-3'-azido-3'-deoxythymidine (L-AZT) and its stereoisomers. J. Org. Chem. 56: 3591-3594.
84. B. Liberek, A. Sikorski, A. Melcer and K. Konitz. 2003. X-Ray diffraction and high-resolution NMR spectroscopy of methyl 3-azido-2,3-dideoxy-α-D-lyxo-hexopyranoside. Carbohydr. Res. 338: 795-799.
85. M. Kumatsu, S. Ichijima, Y. Ohshiro and T. Agawa. 1973. Catalysis by tertiary amines in the thermolysis of vinylazides to 1-azirines. J. Org. Chem. 38: 4341-4342.
86. B. Iliev, A. Linden and H. Heimgartner. 2003. An unexpected formation of a 14-membered cyclo-depsipeptide. Helv. Chim. Acta 86: 3215-3234.
87. P. Köttgen, A. Linden and H. Heimgartner. 2006. Synthesis of enniatin-like cyclic depsipeptides via 'direct amide cyclization'. Helv. Chim. Acta 89: 731-746.
88. K.A. Brun, A. Linden and H. Heimgartner. 2002. New optically active 2*H*-azirin-3-amines as synthons for enantiomerically pure 2,2-disubstituted glycines: synthesis of synthons for Tyr(2Me) and Dopa(2Me), and their incorporation into dipeptides. Helv. Chim. Acta 85: 3422-3443.
89. C. Bernard and L. Ghosez. 1980. Synthesis and reactivity of 4*H*-triazoles. J. Chem. Soc. Chem. Commun. 20: 940-941.
90. B. Li, R.A. Buzon and M.J. Castaldi. 2001. Case study of a γ-butyrolactone alkylation with 1,3-dimethyl-2-imidazolidinone as a promoter. Org. Process Res. Dev. 5: 609-611.
91. M. Oikawa, H. Oikawa and A. Ichihara. 1995. Reductive opening of α-methylspiroketals. Tetrahedron 51: 6237-6254.
92. P. Lesimple and D.C.H. Bigg. 1991. Aluminum chloride mediated aminolysis of lactones: a general method for the preparation of ω-hydroxyalkylamides. Synthesis 4: 306-308.
93. W.P. Griffith, S.V. Ley, G.P. Whitcombe and A.D. White. 1987. Preparation and use of tetra-*n*-butylammonium per-ruthenate (TBAP reagent) and tetra-*n*-propylammonium per-ruthenate (TPAP reagent) as new catalytic oxidants for alcohols. J. Chem. Soc. Chem. Commun. 21: 1625-1627.
94. P.H.J. Carlsen, T. Katsuki, V.S. Martin and K.B. Sharpless. 1981. A greatly improved procedure for ruthenium tetroxide catalyzed oxidations of organic compounds. J. Org. Chem. 46: 3936-3938.
95. M. Prashad, Y. Lu, H.-Y. Kim, B. Hu, O. Repic and T.J. Blacklock. 1999. An improved and practical Sharpless oxidation of primary alcohols to the carboxylic acids. Synth. Commun. 29: 2937-2942.

96. J.M. Villalgordo and H. Heimgartner. 1993. Synthesis of a novel heterospirocyclic 3-(*N*-methyl-*N*-phenylamino)-2*H*-azirine and its use as an amino acid equivalent in the preparation of a model tripeptide. Tetrahedron 49: 7215-7222.
97. K. Dietliker and H. Heimgartner. 1983. Photochemisch induzierte reaktionen von 3-amino-2*H*-azirinen. Helv. Chim. Acta 66: 262-295.
98. R. Luykx, C.B. Bucher, A. Linden and H. Heimgartner. 1996. Ein neues 3-amino-2*H*-azirin als Aib-Pro-baustein: synthese des *C*-terminalen nonapeptids von trichovirin I 1B. Helv. Chim. Acta 79: 527-540.
99. G. Suter, S.A. Stoykova, A. Linden and H. Heimgartner. 2000. Heterospirocyclic *N*-(2*H*-azirin-3-yl)-L-prolinates: new dipeptide synthons. Helv. Chim. Acta 83: 2961-2974.
100. R.A. Breitenmoser and H. Heimgartner. 2002. Synthesis and use of 2*H*-azirin-3-amines as dipeptide synthons. Helv. Chim. Acta 85: 885-912.
101. S. Stamm and H. Heimgartner. 2006. Novel *N*-(2,2-dimethyl-2*H*-azirin-3-yl)-L-prolinates as Aib-Pro synthons. Helv. Chim. Acta 89: 1841-1855.
102. C.B. Bucher, A. Linden and H. Heimgartner. 1995. Optisch aktive 3-amino-2*H*-azirine als bausteine für enantiomerenreine α,α-disubstituierte α-aminosäuren: synthese des α-methylphenylalanin-synthons and einbau in modell-peptide. Helv. Chim. Acta 78: 935-946.
103. R. Breslow, T. Eicher, A. Krebs, R.A. Peterson and J. Posner. 1965. Diphenylcyclopropenone. J. Am. Chem. Soc. 87: 1320-1325.

CHAPTER

Synthesis of diaziridines

3

Chapter outline

3.1 Introduction

Heterocycles, particularly *N*-containing heterocyclic compounds, are the most important groups of compounds in the agrochemical and pharmaceutical industries; heterocyclic compounds are present in around 60% of all drugs. The development of new approaches for the formation of heterocycles continues to be a very active area of research due to the importance of these frameworks in drug designing and in medicinal chemistry [1—3].

The diaziridines can be supposed as the smallest cyclic hydrazine derivatives. The diaziridines have also been examined as possible high-energy compounds (although not particularly promising in this respect), and there is at least one report of diaziridine derivatives showing psychotropic activity, especially antidepressant nature and monoamine oxidase inhibition [4].

Earlier, three groups of chemists synthesized diaziridines (1,2-diazacyclopropanes) by similar procedures [5,6]. During the first 20 years, Schmitz [7] made the most important contribution to this chemistry. Since then, numerous reviews have been published on the formation and chemical properties of diaziridines [8]. The diaziridines are among a few materials that possess nitrogen atoms, which are configuration stable under the trivial conditions (inversion barriers 20—27 kcal/mol); therefore, these compounds are broadly utilized to examine the stereochemistry of nitrogen [9]. Shustov et al. [10] fundamentally studied this side of diaziridine chemistry. Moreover, the diaziridine derivatives are of interest as neurotropically active compounds [11]. The diaziridines are prone to ring-enlargement reactions with electrophilic reagents (isocyanates, ketenes, acylating reagents, and isothiocyanates), which led to the development of novel easy approaches for the formation of both known and earlier unknown heterocyclic systems [12—14].

3-Membered Heterocycle Synthesis. https://doi.org/10.1016/B978-0-443-13583-5.00003-7

3.2 Synthesis of diaziridines

The diaziridines were obtained when pyrazolidinone derivatives (R_1 = aryl, R_2 = methyl or phenyl, R_3 = H or methyl) underwent a photoreversible transformation (Scheme 1) [15].

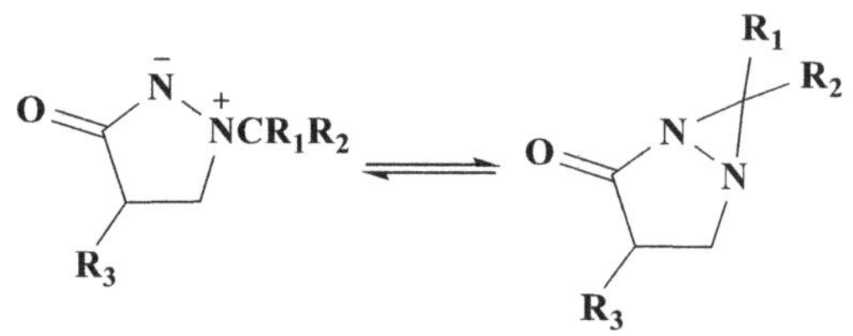

SCHEME 1

The diaziridine (R = phenyl) was formed in approximately quantitative yield by adding ethyl diazoacetate to 4-phenyl-1,2,4-triazolin-2,4-dione (R = phenyl) (Scheme 2) [16]. An array of diaziridine derivatives was obtained when 4-substituted 1,2,4-triazolin-2,4-diones were reacted with alkyl diazoacetates [17]. Due to the mild reaction conditions (0°C), the reaction occurred by a 1,3-dipolar cycloaddition of diazoacetate followed by nitrogen removal instead of an addition of a free carbene (its generation needed temperatures over 100°C).

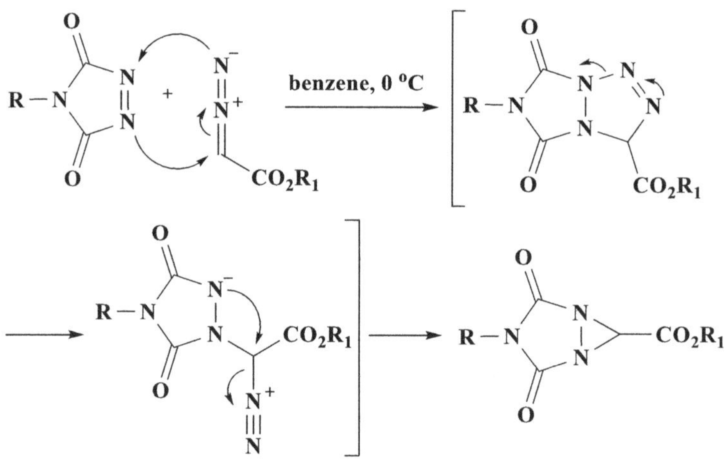

SCHEME 2

The slow reaction rate of this reaction impeded its utilization. The reaction needed many days for aliphatic *N*-chloroalkylamines, and it took many weeks for those prepared from amines with lower basicity. This variant of diaziridines synthesis could be beneficial when CO compounds were less accessible in comparison with the amines. In light of these observations, this reaction was accelerated under the high pressure. The reaction was performed in a Teflon container progressed in a special reaction block at 300, 500, and 700 MPa. The kinetics of reaction was studied for the reaction of *N*-chloroethylamine (R = methyl) and $CH_3CH_2NH_2$ and showed that these reactions were second order at all the pressures. Thus, the bimolecular reaction of *N*-chloroethylamine with $CH_3CH_2NH_2$ was the limiting step of reaction, and then, hydrogen chloride molecule was removed. It was reported that the optimum conditions for this reaction, i.e., $CH_3CH_2NH_2$ and *N*-chloroethylamine (2.5:1) with potassium carbonate (0.5 mol), H_2O (1%–2% by volume), and 500 MPa pressure at 15°C afforded 1,2-diethyl-3-methyldiaziridine in 95% yield. Under the optimized reaction conditions, many other *N*-chloroalkylamines were utilized in the similar reaction. The target 1,2,3-trialkyldiaziridines were formed in high yields in all cases (Scheme 3) [18]. Therefore, a novel technique was developed for the synthesis of 1,2,3-trialkyldiaziridines in high yields based on the reaction of *N*-chloroalkylamines with primary aliphatic amines without carbonyl compounds using same alkyl fragment, K_2CO_3, and a small amount of H_2O at high temperature [19].

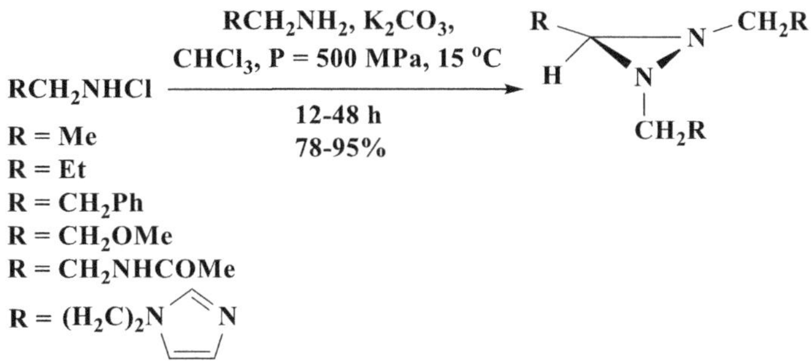

SCHEME 3

3.3 Synthesis of diaziridines from amines and carbonyl compounds

The three-membered rings with two heteroatoms generally come across only as reagents [20]. The diazirines are beneficial carbene precursors. They are more stable than the isomeric diazo compounds, though they are occasionally explosive in the pure state. They were prepared by an oxidation of diaziridines that were accessible through the condensation of aldehyde or ketone with chloramines and NH_3 [21]. The chlorodiazirines, prepared by the reaction of amidines with hypochlorite, underwent S_N2 or S_N2' displacement reactions (Scheme 4) [22].

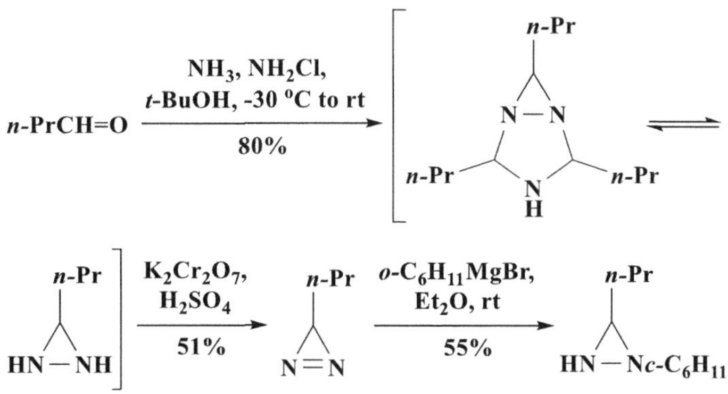

SCHEME 4

The 2,4,6-trisubstituted 1,3,5-triazabicyclo[3.1.0]hexanes were synthesized by two reactions: cyclization of 2,4,6-trialkyl(aryl)hexahydro-1,3,5-triazines using *t*-butyl hypochlorite [23] (pathway A) or reaction of *N*-chloroamine, aldehydes, and NH_3 at low temperature [24] (pathway B). Pathway B involved the synthesis of 3-monosubstituted diaziridines followed by their condensation with 1 mol of NH_3 and 2 mol of aldehyde (Scheme 5). Pathway A afforded 1,3,5-triazabicyclo[3.1.0]hexane from the equilibrium mixture of formaldehyde and ammonia condensation products in H_2O (after its reaction with potassium carbonate—"Henry solution") with sodium hypochlorite at low temperature (−20°C). As represented in that work [25], this solution consisted of a small amount of unsubstituted hexahydro-1,3,5-triazine. Pathway B in this reaction should be excluded because the original unsubstituted diaziridine (if it would be synthesized) under these conditions must be oxidized to diazirine. The 1,3,5-triazabicyclo[3.1.0]hexane was formed in only 5% yield, and the main product was urotropin. The 1,3,5-triazabicyclo[3.1.0]hexane was transformed into phenylisocyanate derivative (Scheme 6) [19,26].

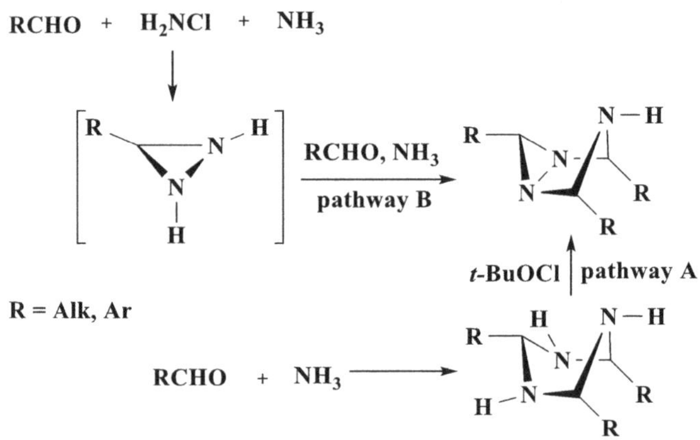

SCHEME 5

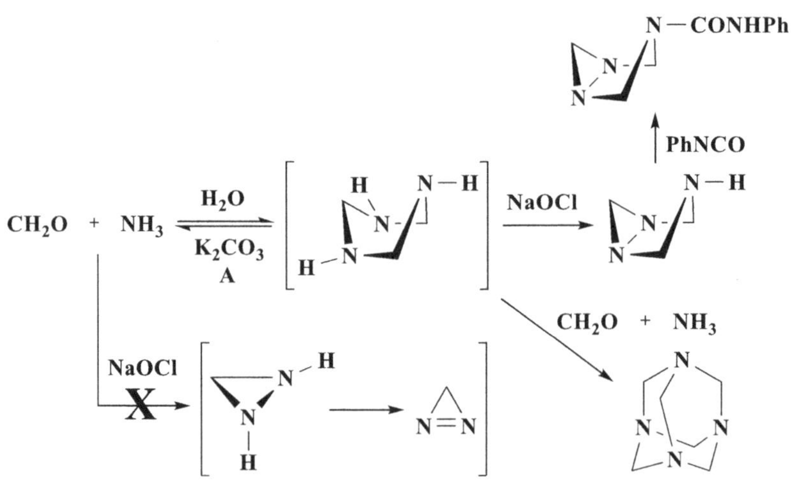

SCHEME 6

The 3*H*-diazirines were formed when diaziridines unsubstituted on the nitrogen atoms were oxidized. The reaction of ketones with NH_3 and chlorine afforded diaziridines. Mainly, the chloramine was formed (Scheme 7) [27,28].

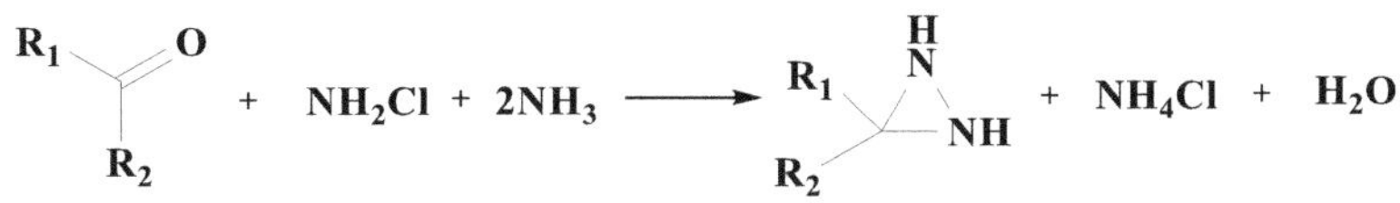

SCHEME 7

An oxidative ring closure of aminals is one of the more common pathways for the synthesis of diaziridines; in turn, aminals were synthesized in situ from amines and CO compounds. The *n*-butylamine was reacted with HCHO using aqueous NaOCl for the synthesis of *N,N*-dibutyldiaziridine (Scheme 8). The procedure was employed to diamine substrates such as propylenediamine to synthesize the fused bicyclic derivatives in excellent yield (Scheme 9). In addition, the diamines were condensed with substituted aldehydes (Scheme 10) and ketones (Scheme 11) with similar efficiency. The product distributions and yields were influenced by the pH of the reaction medium [29].

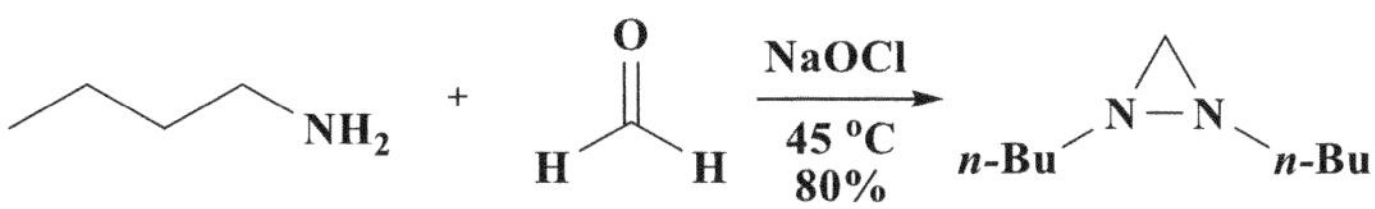

SCHEME 8

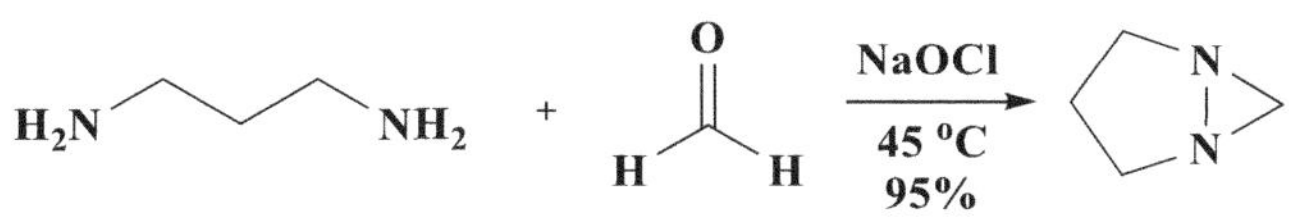

SCHEME 9

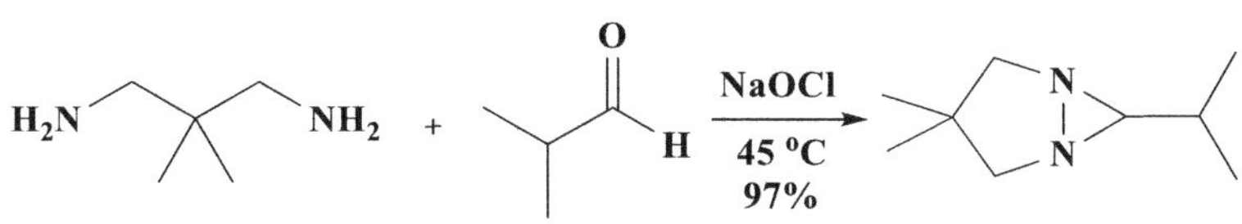

SCHEME 10

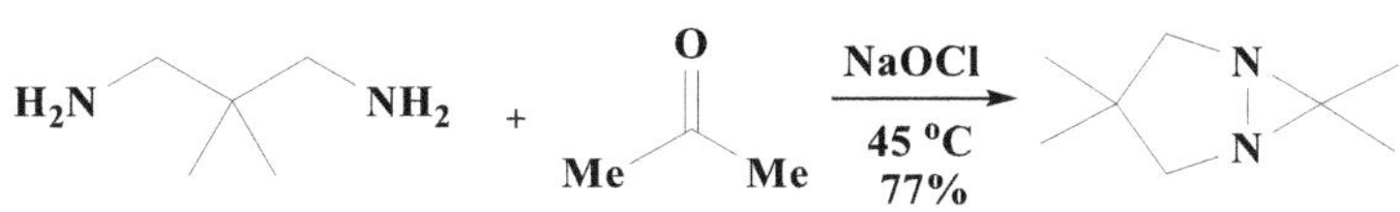

SCHEME 11

Previously [30,31], the 1,5-diazabicyclo[3.1.0]hexanes were synthesized by the halogenation of 1,3-diazacyclohexanes, prepared by the reaction of 1,3-diaminopropane with aldehydes in the presence of sodium hypochlorite in H_2O followed by an intramolecular cyclization of formed cyclic *N*-chloroaminal (Scheme 12). This modified formation of bicycles in combination with controlled pH of the medium (9.5–10.5) was utilized to synthesize the first representative of 6,6′-bis(1,5-diazabicyclo[3.1.0]hexanes) from 1,3-diaminopropane and glyoxal. Both bicyclic fragments exist in boat conformation in 6,6′-bis(1,5-diazabicyclo[3.1.0]hexanes) as indicated by X-ray diffraction data (Scheme 13) [19,32].

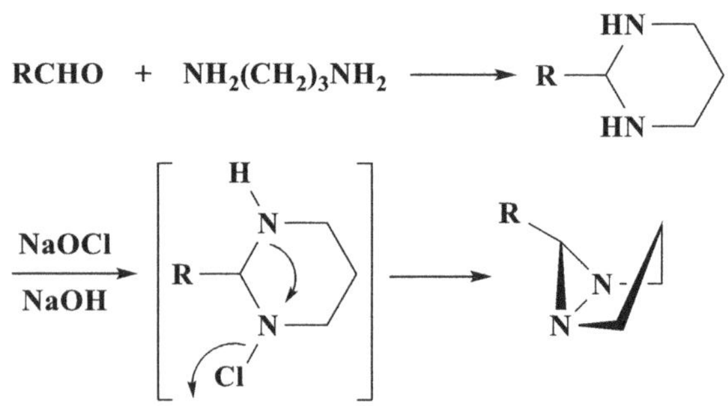

SCHEME 12

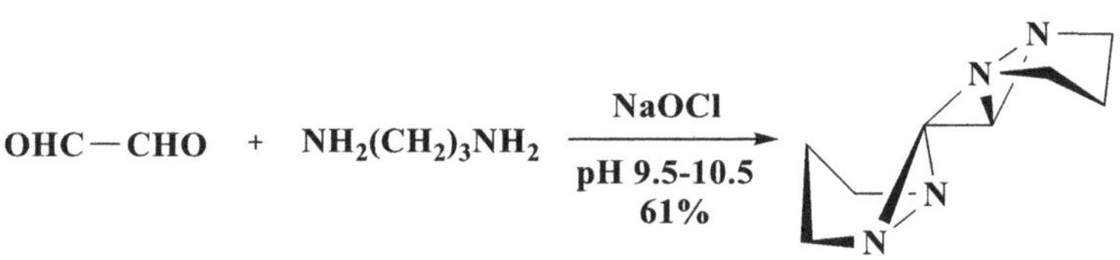

SCHEME 13

The reaction to synthesize the monocyclic diaziridines in aprotic solvents [33] was extended to the formation of 1,5-diazabicyclo[3.1.0]hexanes and 1,6-diazabicyclo[4.1.0]heptanes (a reaction of equimolar amounts of 1,3-diaminopropane or 1,4-diaminobutane, respectively, CO compounds, and *t*-butyl hypochlorite using potassium carbonate in chloroform) [34]. This variant afforded bicycle to be more smoothly isolated in increased yields (50%–52%). However, this process seemed to be inappropriate for reactions with CO compounds bearing substituents sensitive to strong halogenating reagents such as sodium hypochlorite or *t*-butyl hypochlorite. The 1,3-diaminopropane or 1,4-diaminobutane was first halogenated to their monochloro derivatives followed by condensation with carbonyl compounds to overcome aforementioned limitation. The formed 1,3-diaza- or 1,4-diaza-1-chlorocycloalkanes were converted into 6-substituted 1,5-diazabicyclo[3.1.0]hexanes and 7-substituted 1,6-diazabicyclo[4.1.0]heptanes under the basic conditions. The best outcomes were reported where starting diamines were utilized in an equimolar excess (Scheme 14) [19].

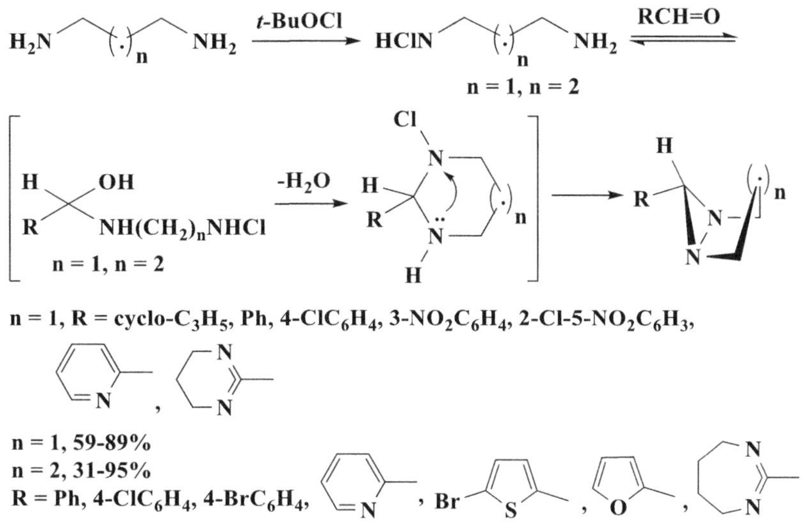

SCHEME 14

Although the generalization of procedures has not been firmly reported, two procedures merit special attention. The first was the direct electrochemical formation of diaziridines from aldehydes and amines in a bicarbonate buffered solution but avoided the requirement for chemical oxidants. Under these electrochemical conditions, the propylenediamine and HCHO afforded bicyclic diaziridine cleanly (Scheme 15). Under the optimum conditions, the efficiency reached to 85% [35]. The second technique was the photolysis of tetrazolone derivatives to diaziridinones in high purity (Scheme 16) [36].

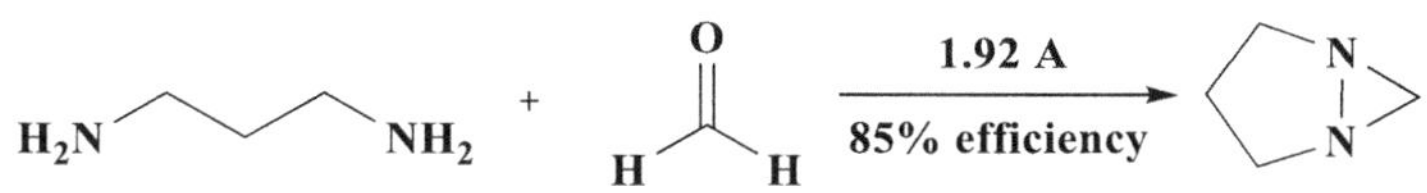

SCHEME 15

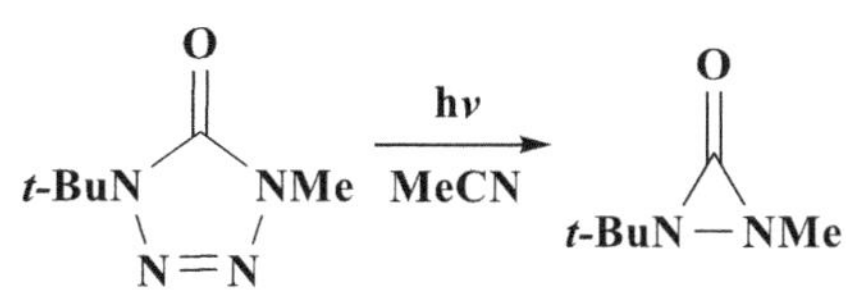

SCHEME 16

An effort was made to stabilize the diimine of glyoxal and CH_3NH_2 at the instant of its synthesis as the dihydrochloride salt, to synthesize the bidiaziridines with aliphatic radicals at nitrogen atoms. The treatment of organic layer formed after mixing of aqueous solutions of reagents with concentrated hydrochloric acid followed by two-fold treatment with potassium carbonate afforded only bis(methylamino)ethane-1,2-diol dihydrochloride whose structure was established as a mixture of two diastereomers by proton nuclear magnetic resonance, ^{13}C nuclear magnetic resonance, ^{2}D COSY-LR experimentation, and mass spectrometry data (Scheme 17). This has exhibited that the aliphatic α-amino carbinoles could be stabilized as hydrochlorides for the first time [37]. The dihydrochloride was added to a solution of *N*-chloromethylamine in an aprotic organic solvent (dichloromethane or chloroform) using different bases (triethylamine, diethylamine, or potassium carbonate) to synthesize the 1,1′,2,2′-tetramethyl-3,3′-bidiaziridine. The dihydrochloride in the presence of organic bases released free bis(methylamino)ethane-1,2-diol, which underwent a self-condensation to give the oligomers. The formation of bidiaziridine was fruitful only in heterogenous media when potassium carbonate was utilized as a base [19,38].

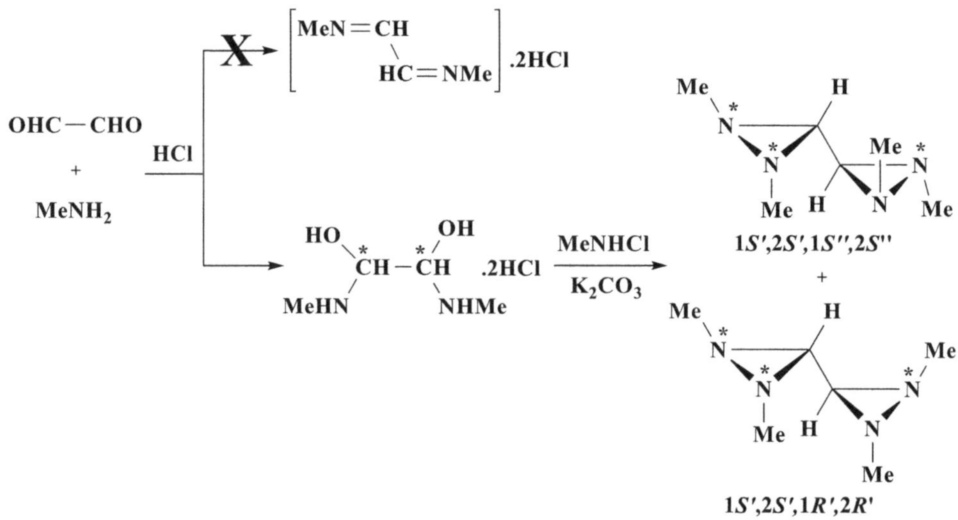

SCHEME 17

The benzaldehyde was reacted with H_2NNHCH_3 and diborane using CH_3COOH to provide the diaziridine in 60% yield (Scheme 18). The reaction occurred through the formation of carbinolamine [39]. The carbinolamine was cyclized by unimolecular nucleophilic substitution like an intramolecular substitution.

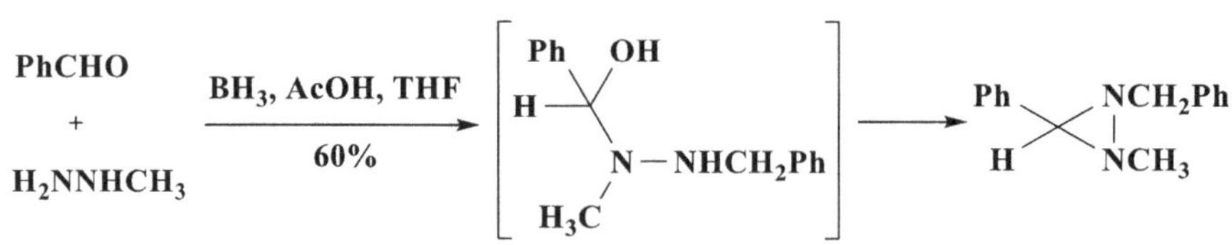

SCHEME 18

The synthetic method was developed by reacting primary aliphatic amines, *N*-chloroalkylamines, and carbonyl compounds using potassium carbonate in aprotic organic solvents for the formation of diaziridines from H_2O insoluble reagents [33]. The preliminary proton nuclear magnetic resonance analysis of a model reaction between HCHO and allyl amine has exhibited that the α-aminicarbinol

($R = R_1 = H$, $R_2 = CH_2CH = CH_2$) could survive in the aprotic organic solvent in the presence of potassium carbonate for a reasonably long time without transformation to hexahydro-1,3,5-triazine ($R = R_1 = H$) (Scheme 19). The α-aminocarbinol was stabilized with potassium carbonate, which shows both dehydration and basic properties. However, the reaction mixture was required to be stirred uninterruptedly and efficiently for interaction with potassium carbonate to carry out the reaction. The water layer appeared within a few minutes, and the hexahydro-1,3,5-triazine was obtained when stirring was stopped. Depending on these outcomes, the 1,2-disubstituted diaziridines were synthesized by an addition of *N*-chloroalkylamine to the mixture of primary aliphatic amine and carbonyl compound using potassium carbonate in chloroform or dichloromethane under the intensive stirring. The *N*-chloroalkylamines were synthesized by the reaction of analogous amines with sodium hypochlorite followed by removal of chloroform or dichloromethane solvent. The *t*-butyl hypochlorite (1 mol) was added to amine (2 mol) in the same solvents in the case of amine and *N*-chloroalkylamine with identical alkyl groups. This method was proved to be highly successful. The carbonyl compounds and primary aliphatic amines with different substituents, such as functional and sterically hindered amines, were introduced to this reaction, and a number of 1,2-di-, 1,2,3-tri-, and 1,2,3,3-tetraalkyldiaziridines were formed in 35%–70% yield. The *N*-chloroaminals, prepared by the reaction of *N*-chloroalkylamines and α-aminocarbinols under these conditions, were cyclized to diaziridines [19,39].

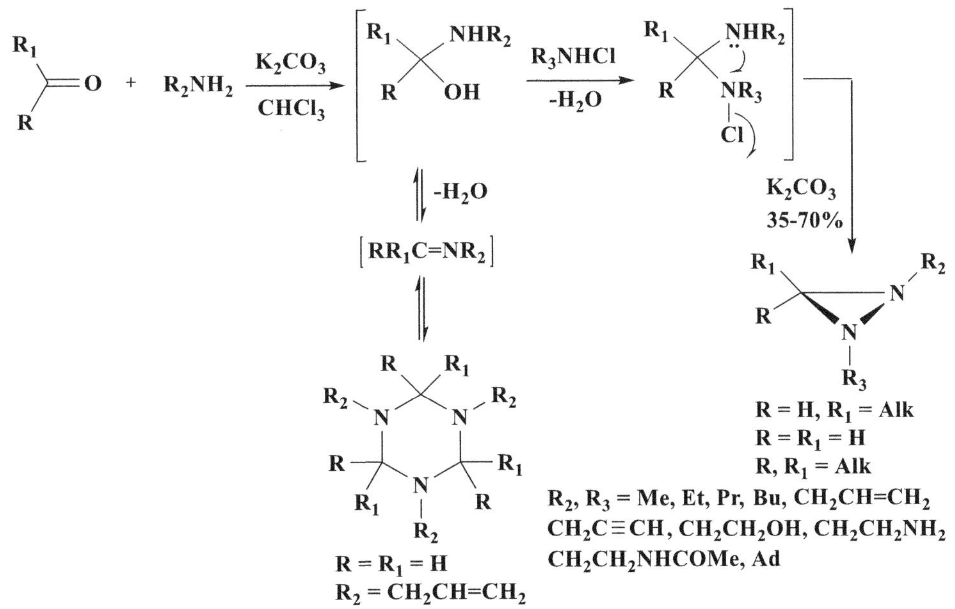

SCHEME 19

During the optimization of procedures for the formation of 1,2-di-, 1,2,3-tri-, and 1,2,3,3-tetraalkyldiaziridines, the behavior of *N*-chloroalkylamines was studied in the presence of an excess of amine in chloro-organic solvents. It was observed that these types of reactions with an amine bearing the same alkyl moiety using K_2CO_3 and small amount of H_2O at rt afforded 1,2,3-trialkyldiaziridines in which the 3-alkyl moiety possesses one methylene group less than the alkyl fragment of the starting materials as it becomes the carbon center of the three-membered ring. Given the structure of the products, it was supposed that the first stage of this reaction was the transformation of *N*-chloroalkylamines into aldimines by the E2-loss of hydrochloric acid using amine. The hydrolysis of aldimines with H_2O provided aldehydes, which were reacted with amines and *N*-chloroalkylamines to synthesize the 1,2,3-trialkyldiaziridines (Scheme 20) [19,40].

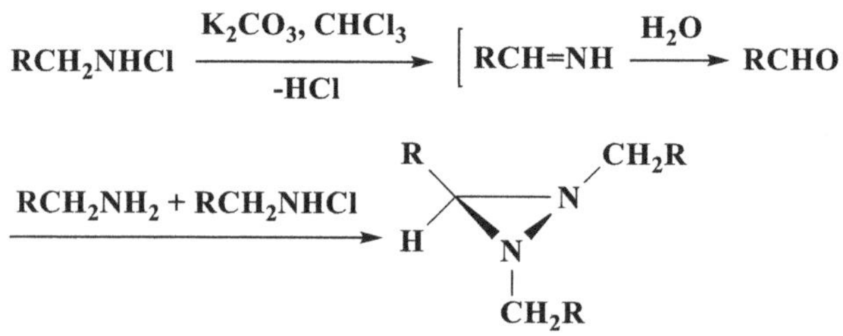

SCHEME 20

3.4 Synthesis of diaziridines from hydroxylamine-*O*-sulfonic acid

The diaziridines were synthesized by the reaction of hydroxylamine-*O*-sulfonic acid and NH_3 or primary amines with ketones. Similarly, the diaziridines were prepared by an amination of imines (azomethines) with hydroxylamine-*O*-sulfonic acid (Scheme 21) [27,28].

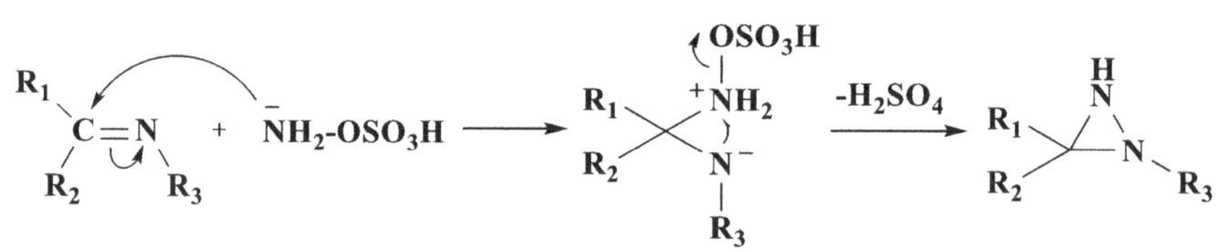

SCHEME 21

The diaziridine was accessible from *O*-sulfonated hydroxylamine derivatives, which provided the benefit of a very less oxidizing environment. As an instance, the cyclic imino ester (Scheme 22) was transformed into diaziridine on reaction with hydroxylamine *O*-sulfonic acid [41], and the similar process was efficient in synthesizing the glycosylidene-derived diaziridine (Scheme 23), which acted as a precursor for the diazirine and thus the glycosylidene carbene [42]. An operationally more direct variant of the procedure could be utilized for the formation of many 2,2-disubstituted derivatives. The ketone such as CH_3COCH_3 was reacted with hydroxylamine *O*-sulfonic acid in aqueous NH_3 to synthesize the diaziridine in one-pot protocol (Scheme 24) [43].

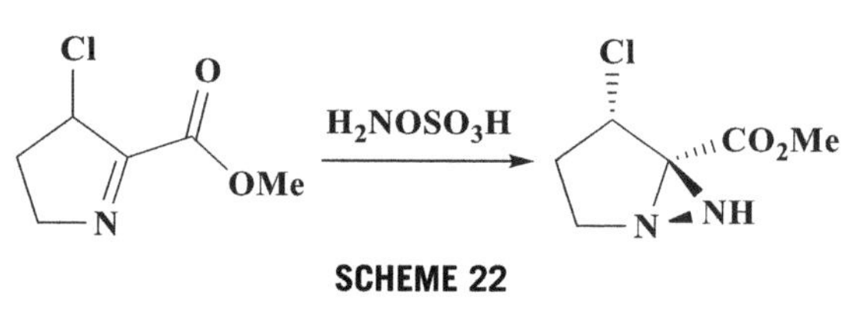

SCHEME 22

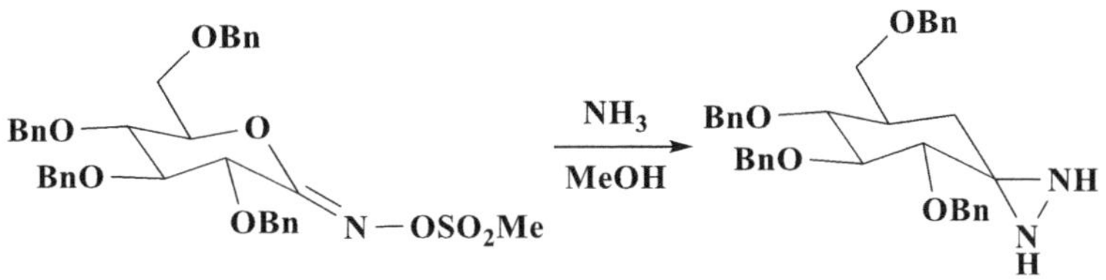

SCHEME 23

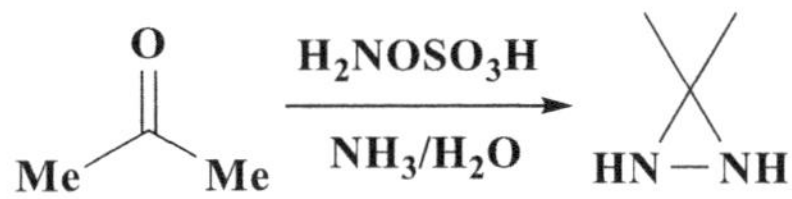

SCHEME 24

Both the oxime sulfonate and the Schmitz [44] syntheses of diaziridines occurred through an *N*,*N*-acetal derivative, which was synthesized by a nucleophilic addition of amine to oxime ester or by a nucleophilic addition of nitrenoid to imine, respectively (Scheme 25). Then, the diaziridine was synthesized by an intramolecular substitution. The cyclization of aminal intermediate through the nitrene (generated by α-elimination) could be excluded, as optically active diaziridines were obtained from oxime esters of CSA (camphor sulfonic acid), proving that the sulfonylate leaving group participated in the cyclization step [45].

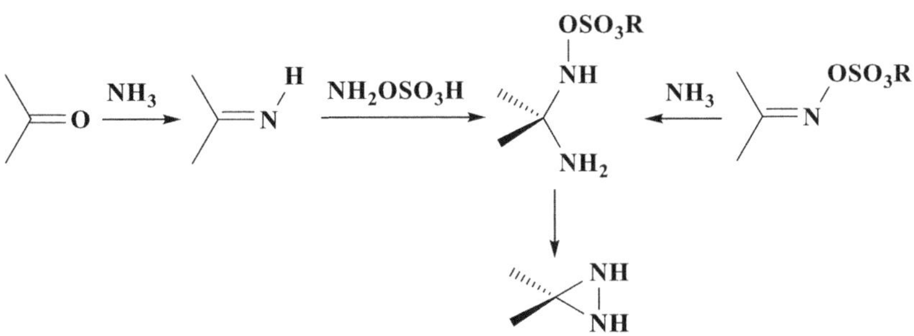

SCHEME 25

The aim of this method is to synthesize the carbasugar-derived spirodiaziridines. The validone was selected as a starting compound for the formation of spirodiaziridines and its derivatives by Schmitz method [44]. The protected derivatives of spirodiaziridines were prepared from validone derivative and were deprotected under the mild conditions to afford the spirodiaziridines. The benzylidene, tri-alkylsilyl, and *i*-propylidene groups were appropriate as protecting groups, as their removal was possible under the mild conditions without influencing the diaziridine ring [44].

The validone (Scheme 26), which was accessible by the *N*-bromosuccinimide-cleavage of validoxylamine A [46], appeared the starting compound of choice for the formation of diaziridines. The protected validone derivative was converted into diaziridines derivative by Schmitz and Ohme's [47] method for the formation of diaziridines from ketones [48], and to deprotect this diaziridine under the mild conditions.

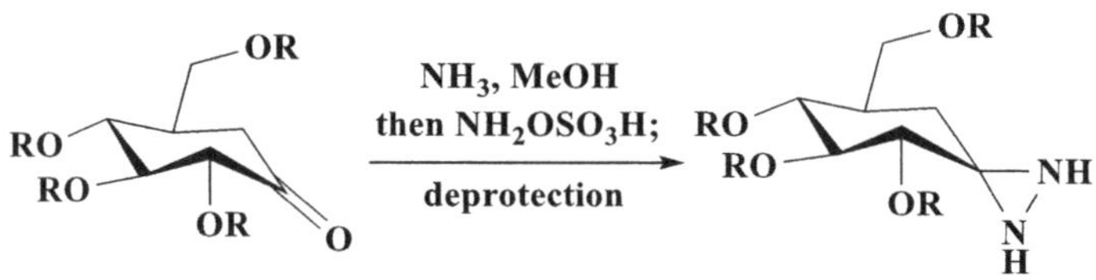

SCHEME 26

The reaction of silylated validone with benzyl amine in methanol followed by with hydroxylamine-*O*-sulfonic acid in methanol provided, after FC, a ca. 1:4 mixture of *N*-benzyl diaziridine and benzylamine, which further provided pure *N*-benzyl diaziridine in 27% yield by chromatography on a weak acid ion exchanger (Scheme 27). The configuration of *N*-benzyl diaziridine resulted from an equatorial attack of hydroxylamine-*O*-sulfonic acid on (*E*)-configured *N*-benzyl imine synthesized in situ from benzyl amine (the (*Z*)-configured *N*-benzyl imine and silylated validone were expected to be disfavored by a repulsive 1,5-interaction between BnO-C(2) and Bn-N). An equatorial attack of hydroxylamine-*O*-sulfonic acid on an intermediate imine was detected by Shustoy and coworkers [49] in the formation of 5-methyl-3,3-pentamethylenediaziridine. The diastereoselective addition of ammonia or methylamine to glyconolactone oxime resulted in mesylation that afforded alkoxydiaziridines [50].

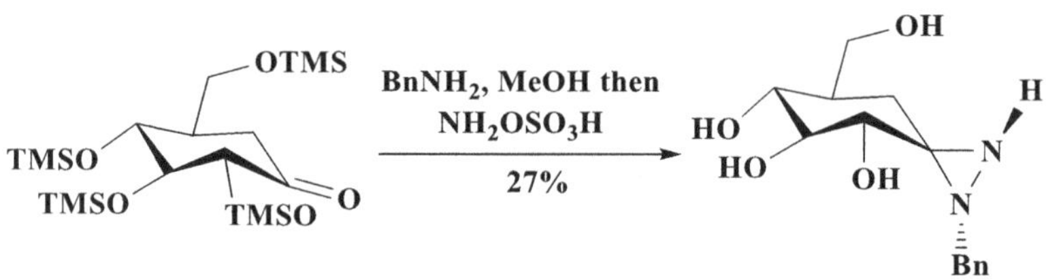

SCHEME 27

The diaziridine was prepared, and further, the tertiary nitrogen was selectively alkylated to synthesize the isoquinoline diaziridine. The chemists initially followed a known method reported by Eckenhoff and coworkers [51] that involved the reaction of imine with hydroxylaminesulfonic acid and Et_3N in C_2H_5OH at 0°C (Scheme 28). It was observed that a similar method reported by Ortega et al. [52], which utilized NH_4OH in CH_3OH instead of triethylamine in C_2H_5OH, was more reliable and much higher yielding.

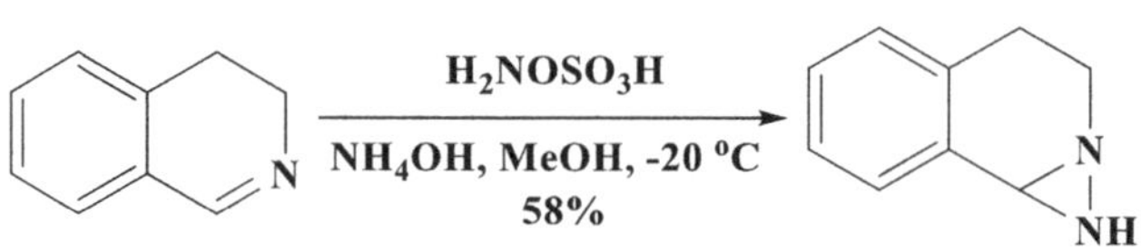

SCHEME 28

Following a methylation process for the formation of oxaziridinium ions [53], it was found that the regiospecific *N*-alkylation of tertiary nitrogen of diaziridine was performed utilizing Meerwein's salt (trimethyloxonium tetrafluoroborate). Then, the recrystallization by dissolving the crude yellow oil in a hot 1:1 $CH_2Cl_2/CH_3COOC_2H_5$ solvent mixture followed by an addition of hexanes afforded diaziridinium salt as colorless crystals in good yield (72% yield) (Scheme 29).

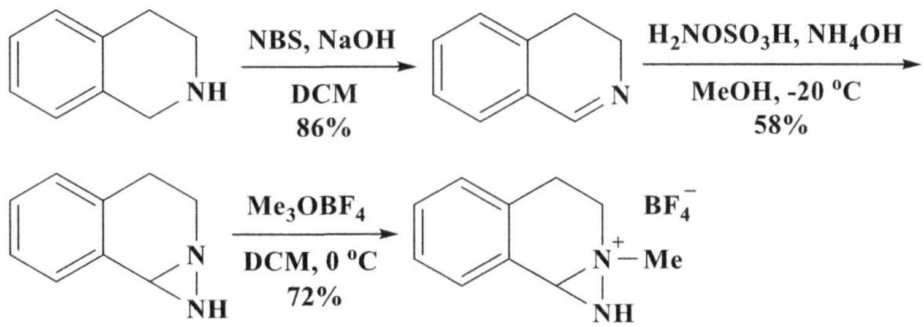

SCHEME 29

The influence of placing an alkyl substituent at the C-1 benzylic position of the diaziridine was explored. These substituted diaziridines were easily prepared from commercially accessible phenethylamine derivatives in just three steps. For instance, an initial acetylation of 2-phenethylamine provided methyl-substituted diaziridine. Then, the 1-methyl-3,4-dihydroisoquinoline was synthesized in 81% yield by the Bischler–Napieralski cyclization of *N*-phenethylethanamide. Finally, the reaction of this imine with hydroxylamine-*O*-sulfonic acid under the basic conditions afforded 1-methyl-substituted diaziridine in 63% yield (Scheme 30) [54].

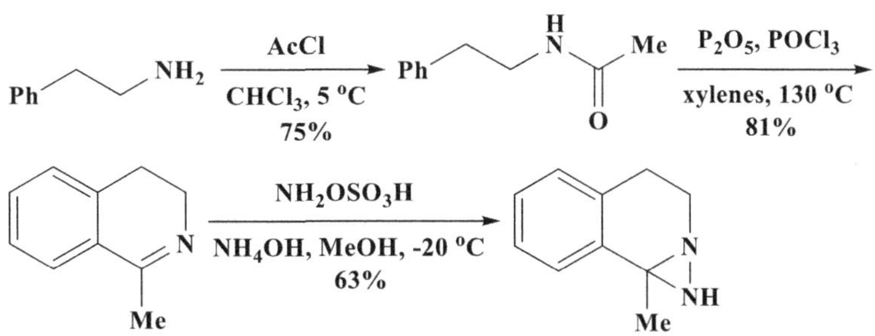

SCHEME 30

The method for the formation of monocyclic diaziridines was well established by Schmitz [55]. This method provided first representatives of 3,3′-bidiaziridines from diimines of glyoxal. It was observed that this method could be utilized only for diimines with branched alkyl substituents. The reaction of glyoxal with primary aliphatic amines bearing normal aliphatic radicals was not stopped at the stage of diimines as they underwent a polymerization or condensation. The reaction of diimines with HASA (hydroxylamine-*O*-sulfonic acid) using triethylamine in methanol at −5–20°C provided 1,1′-dialkyl-3,3′-bidiaziridines in high yields. These compounds were formed as a mixture of many diastereomers as demonstrated by the proton nuclear magnetic resonance spectra of crude products because 1,1′-dialkyl-3,3′-bidiaziridines contained besides nitrogen atoms one more chiral center, i.e., the carbon atom of diaziridine ring. However, an isolation of final products by vacuum distillation or sublimation provided a mixture of only two most thermodynamically favorable diastereomers in 2:1 to 9:1 ratio all the time. The nuclear Overhauser effect experiments and X-ray diffraction study of one of the obtained diaziridines showed that the isolated diastereomers were (1*S**,2*S**,3*R**,1′*S**,2′*S**,3′*R**) racemate and (1*S**,2*S**,3*R**,1′*R**,2′*R**,3′*S**) *meso*-form diimines (Scheme 31) [19].

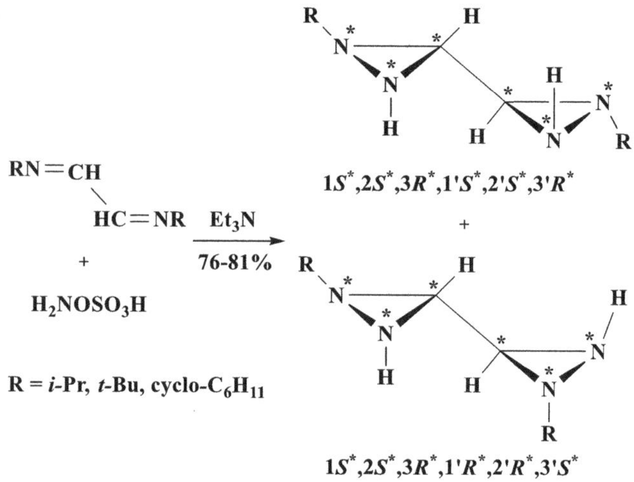

SCHEME 31

The diaziridinium ion was prepared to use the heteroatom transfer reactivity. It was observed that the imine was smoothly available from commercially accessible 1,2,3,4-tetrahydroisoqinoline through an *N*-bromosuccinamide oxidation in 77% yield [56]. Preliminary method for the synthesis of diaziridinium ion involved the reaction of methyl iminium ion with hydroxylamine-*O*-sulfonic acid (Scheme 32).

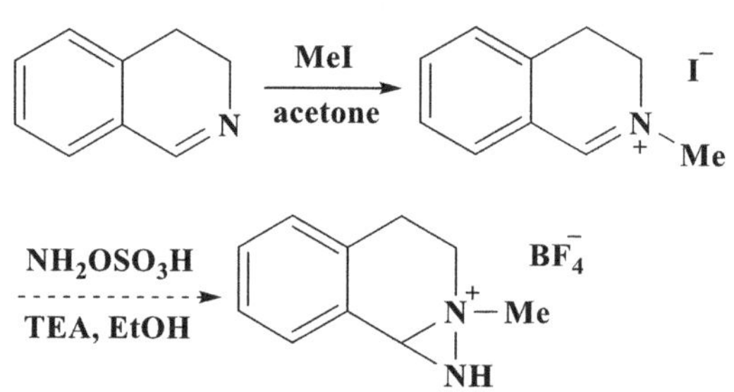

SCHEME 32

3.5 Synthesis of diaziridines from imines

The diaziridines were obtained in modest to high yields (40%–70%) when Schiff bases were reacted with ethereal solutions of chloramines for a few hours at rt (Scheme 33) [57]. This formal cycloaddition reaction had a wide range of scope and was applicable, utilizing chloramine itself (or the similar reagent hydroxylamine-*O*-sulfonic acid) along with *N*-substituted chloramines, to various Schiff bases derived from acyclic and cyclic ketones and aldehydes. In several cases, the Schiff base could be transformed in situ into diaziridine in 50%–80% yield by reacting carbonyl compound with NH_3 or primary or secondary amines using *N*-substituted hydroxylamine-*O*-sulfonic acid or hydroxylamine-*O*-sulfonic acid. The diaziridine was synthesized by an initial nucleophilic addition to the C–N double bond followed by eliminative ring closure.

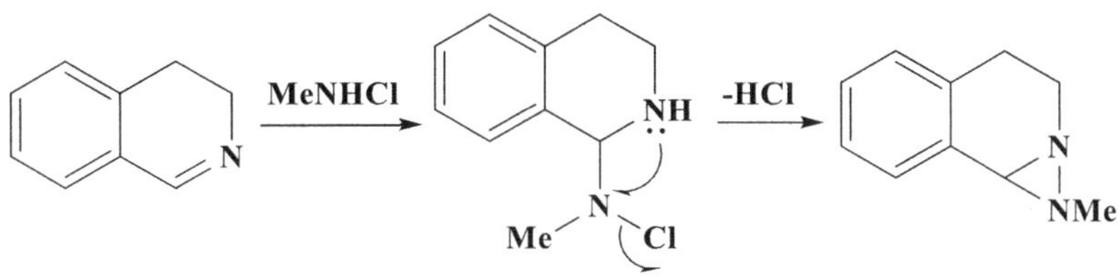

SCHEME 33

The bis-aziridination reaction of ethyl nosyloxycarbamate as aminating agent with symmetric (*E-s-trans-E*)-α-diimines occurred diastereospecifically, as the azaanion attacked on the opposite faces of conjugated system to synthesize the diaziridine (Scheme 34) [57].

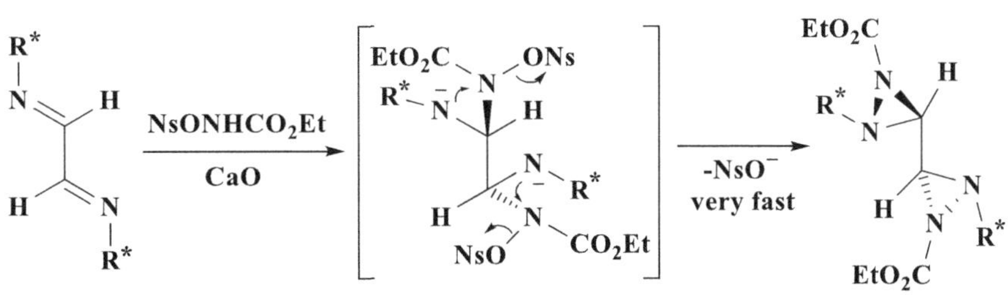

SCHEME 34

The polyfluorinated derivatives of fluorodiaziridines, including perfluorodiaziridine, were reported, but the utilized synthetic methods were specific for polyfluorinated organic compounds and could not be utilized for the formation of monofluorinated heterocycles (Scheme 35) [58]. The same was true for fluorooxaziridines [59].

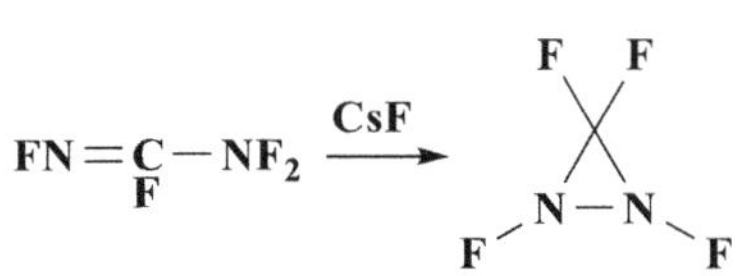

SCHEME 35

The explosive 1,2,3-trifluoro-3-difluoroaminodiaziridine and 1,2,3,3-tetrafluorodiaziridine were synthesized from pentafluoroguanidine and tetrafluoroamidine on reaction with potassium fluoride, cesium fluoride, or rubidium fluoride (Schemes 36 and 37) [58] presumably via an intramolecular nucleophilic nitrogen–nitrogen bond formation [59].

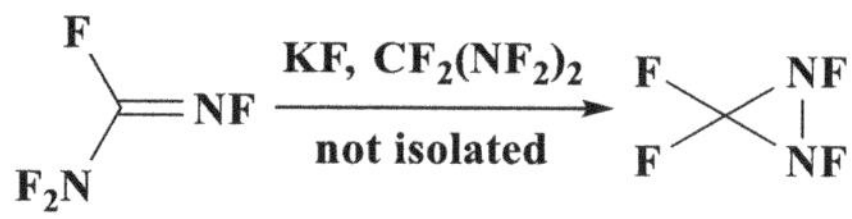

SCHEME 36

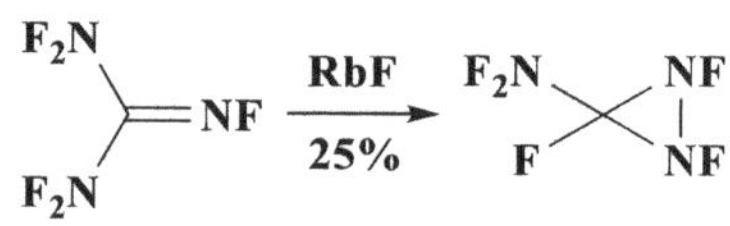

SCHEME 37

The aminals could be prepared from other precursors. The aminoalkylimine was reacted with aqueous NaOCl to afford the fused tricyclic diaziridine (Scheme 38) [60]. The benzimidamide acted as a precursor for diazirines under similar oxidizing conditions; however, an intermediate chlorodiaziridine was recognized in the reaction mixture, and it was sufficiently stable to isolate (Scheme 39) [61]. Likewise, the diaziridinimines could be obtained in good yield when tosyl guanidine was subjected to sequential reaction with *t*-BuOCl and *t*-BuOK (Scheme 40) [62].

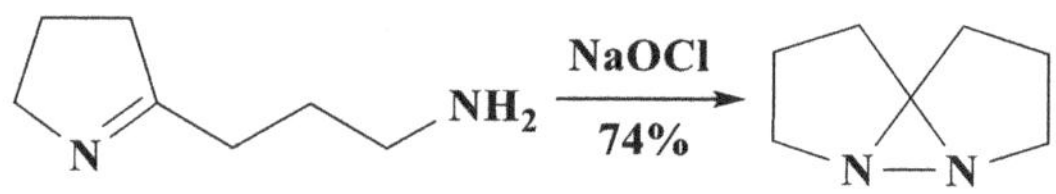

SCHEME 38

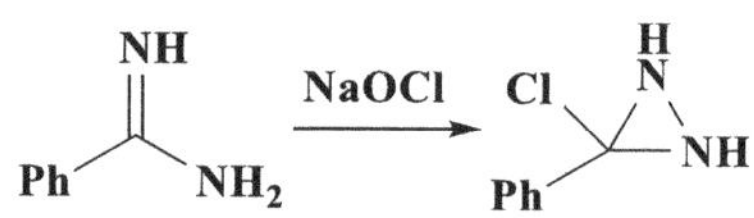

SCHEME 39

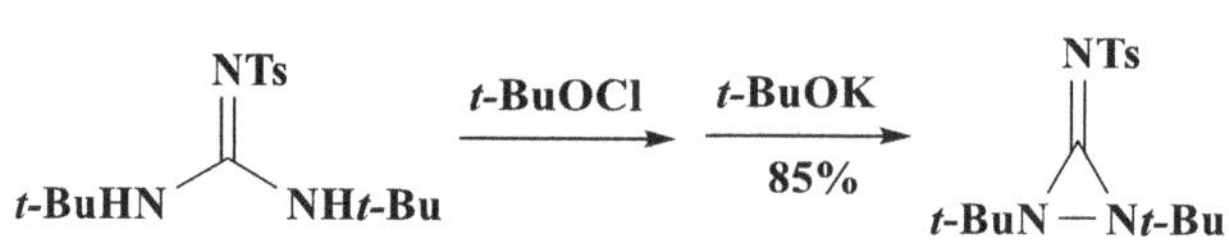

SCHEME 40

3.6 Synthesis of diaziridines from oximes

The diaziridines were obtained when acetonoxime *O*-esters were reacted with propyl amine. It was observed that only phosphates were capable to synthesize the 1-propyl-3,3-dimethyldiaziridines in modest yields (52%–67%). In this case, not very high yields were possibly associated with side alkylation or phosphorylation reactions. In the acyloximes, the carbon atom of acyl fragment seemed to be the strongest electrophilic center and afforded amides. The 1-propyl-3,3-dimethyldiaziridine was formed in 82% yield only when sterically hindered *O*-mesitylcarbonyl acetonoxime was utilized. The sulfinates and picrate did not synthesize diaziridine similar to *O*-esters (Scheme 41) [19,63].

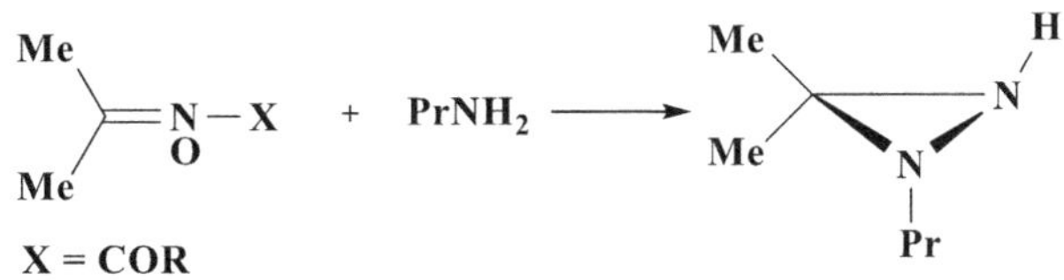

X = COR
R = Me, R = Ph, R = CH_2Cl, R = CCl_3
R = 2,4,6-$Me_3C_6H_2$
X = $P(O)R_2$
R = OPh, R = O*i*-Pr
X = S(O)R
R = OMe, R = 4-$MeOC_6H_4$
X = 2,4,6-$(NO_2)_3C_6H_2$

SCHEME 41

The utilization of ketoxime *O*-sulfonates prepared from aliphatic ketones elevated doubts in view of their higher inclination to Beckman rearrangement [64]. The 3,3-di- and 1,3,3-trialkyldiaziridines were prepared utilizing 3,3-dimethyl- and 1-alkyl-3,3-dimethyldiaziridines as model from acetonoxime *O*-sulfonates and from other acetonoxime *O*-esters (sulfinates, phosphates, picrates, and acyloximes) and NH_3 or primary aliphatic amines, respectively, in aprotic organic solvents (chloroform, dioxane, ether, and dichloromethane) [65,66]. There were at least two electrophilic centers in oxime *O*-esters, to which nucleophilic attack could be applied, the oxime carbon atom and central atom of the acidic moiety. The amine attack at the central atom of the acidic moiety in acetonoxime *O*-sulfonates was sterically hindered. Therefore, the 1-alkyl-3,3-dimethyldiaziridines were efficiently formed when all oxime *O*-esters were reacted with primary aliphatic amines at rt (Scheme 42). The yields of diaziridines, obtained by the Beckman rearrangement of esters, decreased upon elevating the temperature. The 3,3-dimethyldiaziridine was formed in an autoclave at 9 MPa [19].

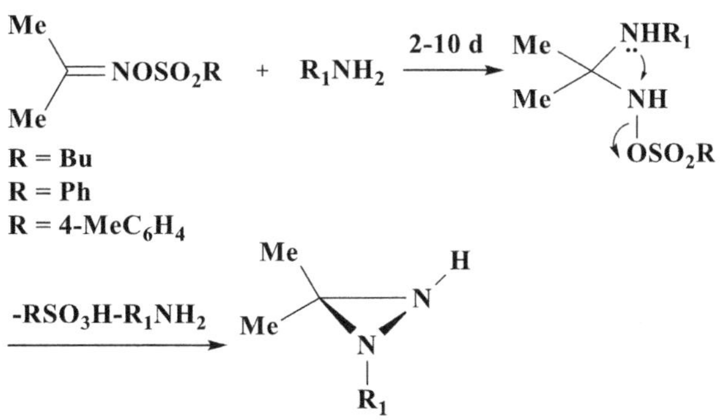

R = Bu
R = Ph
R = 4-MeC_6H_4

R_1 = H, 90.5%
R_1 = Pr
R_1 = Bu
R_1 = $(CH_2)_2OH$, 88-89.5%

SCHEME 42

The reaction of ketoxime *O*-*p*-toluenesulfonates and 1,ω-diaminoalkanes was used for the formation of α,ω-bis(3,3-dialkyldiaziridin-1-yl)alkanes. The diastereomeric racemate *meso*-(1*S**,2*S**,1′*R**,2′*R**) and (1*S**,2*S**,1′*S**,2′*S**) were isolated for α,ω-bis(3,3-dimethyldiaziridin-1-yl) ethane by fractional crystallization, and the *meso*-structure was established by X-ray diffraction study (Scheme 43) [19,67].

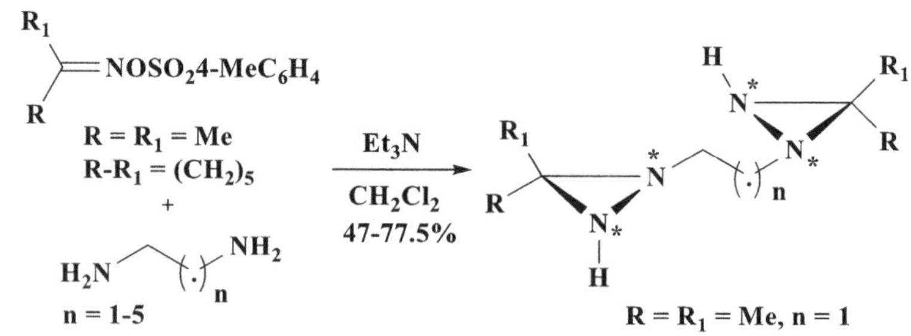

SCHEME 43

The established technique for the formation of 3,3-dimethyldiaziridine using acetonoxime *O*-sulfonates was considered inappropriate for synthesizing, practically, previously inaccessible 3-monoalkyldiaziridines from aldehydes derivatives. This was because aldoxime *O*-sulfonic acids and aldoxime *O*-sulfonates decomposed instantly after formation to provide the sulfuric or sulfonic acids and nitriles, respectively. It was demonstrated for the first time that aliphatic aldoxime *O*-sulfonic acids could be stabilized as alkyl ammonium or ammonium salts. It was observed that alkyl ammonium salts were stable at rt, but ammonium salts only survived in saturated aqueous solution of ammonia (>40%) at a lower temperature. To synthesize the ammonium salts, aqueous ammonia was saturated with gaseous ammonia under cooling, and further a freshly prepared cold solution of aldoxime *O*-sulfonic acids was added. A flow of gaseous ammonia passing through the solution of alkyl ammonium salts at a lower temperature provided 3-alkyldiaziridines in modest yields. The reaction of primary aliphatic amines with alkyl ammonium salts in water or in methanol−water mixture afforded 1,3-dialkyldiaziridines (Scheme 44) [19,68].

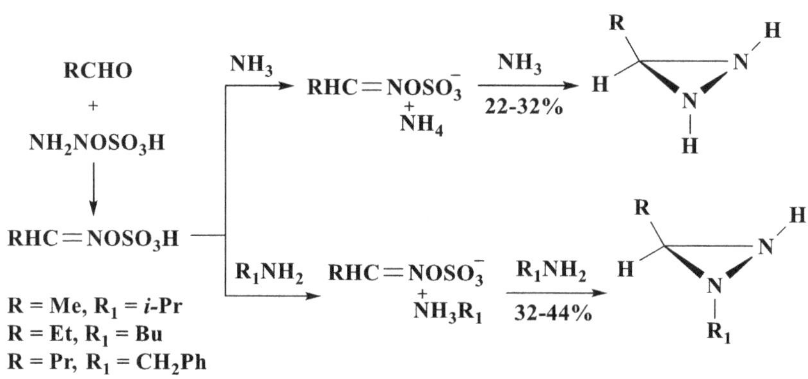

SCHEME 44

The tosyloxime was reacted in the presence of $Yb(OTf)_3$ to synthesize the diaziridine (Scheme 45) [69,70].

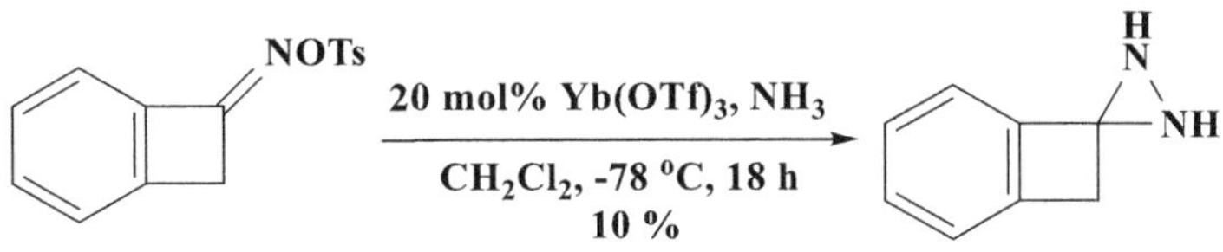

SCHEME 45

References

1. (a) N. Kaur, J. Dwivedi and D. Kishore. 2014. Solid-phase synthesis of nitrogen containing five-membered heterocycles. Synth. Commun. 44: 1671-1729. (b) N. Kaur. 2019. Ionic liquids: a versatile medium for the synthesis of six-membered two nitrogen containing heterocycles. Curr. Org. Chem. 23: 76-96. (c) N. Kaur and D. Kishore. 2014. Nitrogen-containing six-membered heterocycles: solid-phase synthesis. Synth. Commun. 44: 1173-1211. (d) N. Kaur. 2018. Photochemical reactions as key steps in five-membered *N*-heterocycles synthesis. Synth. Commun. 48: 1259-1284. (e) R. Chauhan, N. Kaur, Rajendra, S. Sharma and J. Dwivedi. 2015. Application of chalcone in synthesis of 1-(1,5-benzodiazepino) substituted analogues of indole. Rasayan J. Chem. 8: 115-122. (f) N. Kaur, P. Sharma, R. Sirohi and D. Kishore. 2012. Microwave assisted synthesis of 2-hetryl amino substituted novel analogues of 1,4-benzodiazepine-5-piperidinyl carboxamides. Arch. Appl. Sci. Res. 4: 2256-2260. (g) N. Kaur. 2013. Solid-phase synthetic approach to the synthesis of azepine heterocycles of medicinal interest. Int. J. Pharm. Biol. Sci. 4: 357-372. (h) N. Kaur. 2015. Advances in microwave-assisted synthesis for five-membered *N*-heterocycles synthesis. Synth. Commun. 45: 432-457. (i) N. Kaur. 2015. Recent impact of microwave-assisted synthesis on benzo derivatives of five-membered *N*-heterocycles. Synth. Commun. 45: 539-568.
2. (a) N. Kaur, Y. Verma, P. Grewal, N. Ahlawat, P. Bhardwaj and N.K. Jangid. 2020. Photochemical C-N bond forming reactions for the synthesis of five-membered fused *N*-heterocycles. Synth. Commun. 50: 1286-1334. (b) N. Kaur, P. Bhardwaj, M. Devi, Y. Verma and P. Grewal. 2019. Photochemical reactions in five and six-membered polyheterocycles synthesis. Synth. Commun. 49: 2281-2318. (c) N. Kaur. 2015. Microwave-assisted synthesis: fused five-membered *N*-heterocycles. Synth. Commun. 45: 789-823. (d) N. Kaur. 2015. Benign approaches for the microwave-assisted synthesis of five-membered 1,2-*N,N*-heterocycles. J. Heterocycl. Chem. 52: 953-973. (e) N. Kaur. 2015. Environmentally benign synthesis of five-membered 1,3-*N,N*-heterocycles by microwave irradiation. Synth. Commun. 45: 909-943. (f) P. Sharma, N. Kaur, R. Sirohi and D. Kishore. 2013. Microwave assisted facile one pot synthesis of novel 5-carboxamido substituted analogues of 1,4-benzodiazepin-2-one of medicinal interest. Bull. Chem. Soc. Ethiop. 27: 301-307. (g) N. Kaur. 2013. Expedient protocol for the installation of thiadiazole on 2-position of 1,4-benzodiazepin-5-carboxamide through a phenoxyl spacer. Int. J. Pharm. Biol. Sci. 4: 366-373. (h) N. Kaur. 2019. Copper catalyzed synthesis of seven and higher-membered heterocycles. Synth. Commun. 49: 879-916. (i) N. Kaur. 2019. Synthesis of seven and higher-membered heterocycles using ruthenium catalysts. Synth. Commun. 49: 617-661. (j) N. Kaur, N. Ahlawat, Y. Verma, P. Bhardwaj, P. Grewal and N.K. Jangid. 2020. Rhodium catalysis in the synthesis of fused five-membered *N*-heterocycles. Inorg. Nano Met. Chem. 50: 1260-1289.
3. (a) N. Kaur. 2017. Photochemical reactions: synthesis of six-membered *N*-heterocycles. Curr. Org. Synth. 14: 972-998. (b) N. Kaur. 2018. Synthesis of seven- and higher-membered nitrogen-containing heterocycles using photochemical irradiation. Synth. Commun. 48: 2815-2849. (c) N. Kaur. 2015. Insight into microwave-assisted synthesis of benzo derivatives of five-membered *N,N*-heterocycles. Synth. Commun. 45: 1269-1300. (d) N. Kaur. 2015. Synthesis of fused five-membered *N,N*-heterocycles using microwave irradiation. Synth. Commun. 45: 1379-1410. (e) N. Kaur and D. Kishore. 2013. Application of chalcones in heterocycles synthesis: synthesis of 2-(isoxazolo, pyrazolo and pyrimido) substituted analogues of 1,4-benzodiazepin-5-carboxamides linked through an oxyphenyl bridge. J. Chem. Sci. 125: 555-560. (f) N. Kaur. 2013. A new

approach to anti-HIV chemotherapy devised by linking the vital fragments of active RT inhibitors such as etravirine to the molecular framework of anti-HIV prone privileged nucleus of 1,4-benzodiazepine as possible substitute to 'HAART'. Int. J. Pharm. Biol. Sci. 4: 309-317. (g) N. Kaur and D. Kishore. 2014. Microwave-assisted synthesis of seven- and higher-membered *N*-heterocycles. Synth. Commun. 44: 2577-2614. (h) N. Kaur, N. Ahlawat, Y. Verma, P. Grewal, P. Bhardwaj and N.K. Jangid. 2020. Cu-Assisted C-N bond formations in six-membered *N*-heterocycle synthesis. Synth. Commun. 50: 1075-1132. (i) N. Kaur, N. Ahlawat, P. Bhardwaj, Y. Verma, P. Grewal and N.K. Jangid. 2020. Ag-Mediated synthesis of six-membered *N*-heterocycles. Synth. Commun. 50: 753-795. (j) N. Kaur. 2016. Metal catalysts for the formation of six-membered *N*-polyheterocycles. Synth. React. Inorg. Met. Org. Nano Met. Chem. 46: 983-1020.

4. (a) K.J. Gessner and D.W. Ball. 2005. Cyclic diamines as potential high energy materials. Thermochemical properties of diaziridine, 1,2-diazetidine, and 1,3-diazetidine. J. Mol. Struct. Theochem. 730: 95-103. (b) O. Trapp, V. Schurig and R.G. Kostyanovsky. 2004. The control of the nitrogen inversion in alkyl-substituted diaziridines. Chem. Eur. J. 10: 951-957. (c) R.G. Kostyanovsky, O.R. Malyshev, K.A. Lyssenko, Y.A. Strelenko, G.K. Kadorkina, V.R. Kostyanovsky, O.G. Nabiev and I.V. Fedyanin. 2004. Stereochemistry of 1,2,4,5-tetraazanorbornanes and diaziridines: exciting history and news. Mendeleev Commun. 14: 315-318. (d) R.G. Kostyanovskii, G.V. Shustov, O.G. Nabiev, S.N. Denisenko, S.A. Sukhanova and E.F. Lavretskaya. 1986. Synthesis and psychotropic activity of functionally substituted diaziridines and bisdiaziridines. Pharm. Chem. J. 20: 385-388.
5. H.J. Abendroth and G. Henrich. 1959. Über ein isomeres acetonhydrazon. Angew. Chem. 71: 283-283.
6. S.R. Paulsen. 1959. Process for the preparation of diazacyclopropanes. Belg. Patent 588352.
7. E. Schmitz. 1959. Neue synthesen des C-N-N-dreiringes. Angew. Chem. 71: 127-127.
8. S.M. Hai, A.W. Qureshi and A. Begum. 1975. Pakistan J. Sci. Ind. Res. 18: 116.
9. A. Mannschreck, R. Radeglia, E. Grundemann and R. Ohme. 1967. Protonenresonanz-untersuchungen zur inversion am dreibindigen stickstoffatom, i der diaziridin-ring als asymmetriezentrum. Chem. Ber. 100: 1778-1785.
10. G.V. Shustov, A.B. Zolotoi, N.L. Zaichenko, O.A. Dyachenko, L.O. Atovmyan and R.G. Kostyanovsky. 1984. Asymmetric nitrogen-35. Tetrahedron 40: 2151-2159.
11. C.J. Paget and C.S. Davis. 1964. Synthesis and in vitro activity of some aryl diaziridines as potential monoamine oxidase inhibitors. J. Med. Chem. 7: 626-628.
12. A.V. Shevtsov, V.Y. Petukhova, Y.A. Strelenko, K.A. Lyssenko, I.V. Fedyanin and N.N. Makhova. 2003. A new direction of ring expansion of 1,2-dialkyldiaziridines in the reactions with arylketenes. Mendeleev Commun. 13: 221-223.
13. A.V. Shevtsov, V.Y. Petukhova, Y.A. Strelenko and N.N. Makhova. 2005. An unexpected transformation of 1,2-dialkyldiaziridines into *N*-{[acetyl(alkyl)amino]methyl}-*N*-(alken-1-yl)acetamide under the action of the parent ketene. Mendeleev Commun. 15: 29-31.
14. A.V. Shevtsov, V.V. Kuznetsov, A.A. Kislukhin, V.Y. Petukhova, Y.A. Strelenko and N.N. Makhova. 2006. Ring transformation of 1,5-diazabicyclo[3.1.0]hexanes under the action of arylketenes. J. Heterocycl. Chem. 43: 881-888.
15. G. Tomaschewski, G. Geissler and G. Schauer. 1980. Azomethinimine. I. Zur photoreversibilität des systems azomethinimin/diaziridin, untersucht an pyrazolidon-azomethiniminen. J. Prakt. Chem. 322: 623-628.
16. R.A. Izydore and S. McLean. 1975. 1,2-Addition reaction of ethyl diazoacetate and 4-phenyl-1,2,4-triazoline-3,5-dione. J. Am. Chem. Soc. 97: 5611-5612.
17. I.H. Hall, K. Taylor, R.A. Izydore, D.E. Coleman, J.A. Mitchell and R. Cummings. 1998. The cytotoxic action of 1,3,5-triazabicyclo[3.1.0]hexane-2,4-diones and 1,3,5-triazine-4,6-(1*H*,5*H*)-diones in murine and human tumor cells. Pharmazie 53: 398-405.
18. V.V. Kuznetsov, Y.S. Syroeshkina, D.I. Moskvin, M.I. Struchkova, N.N. Makhova and A.A. Zharov. 2008. High pressure-assisted synthesis of 1,2,3-trialkyldiaziridines from *N*-chloroalkylamines. J. Heterocycl. Chem. 45: 497-502.
19. N.N. Makhova, V.Y. Petukhova and V.V. Kuznetsov. 2008. Synthesis of monocyclic diaziridines and their fused derivatives. ARKIVOC (i): 128-152.
20. R. Bonneau, M.T.H. Liu and R. Lapouyade. 1989. Formation of indolizines by the addition of α-chloroacrylonitrile to pyridinium ylides: regioselectivity and Hammett correlation. J. Chem. Soc. Perkin Trans. 1 8: 1547-1548.

21. E. Schmitz and R. Ohme. 1962. Cyclische diazoverbindungen, III. Cyclo-diazomethan. Chem. Ber. 95: 795-802.
22. A. Padwa and D. Eastman. 1969. Reactions of phenylchlorodiazirine with nucleophiles and substituted acetylenes. J. Org. Chem. 34: 2728-2732.
23. A.T. Nilsen, R.L. Atkins, D.W. Moore and J.M. LaBerge. 1973. Stereochemistry of 2,4,6-trialkyl-1,3,5-triazabicyclo[3.1.0]hexanes. Tetrahedron Lett. 14: 1167-1170.
24. E. Schmitz. 1962. Diaziridine, III. Die einwirkung von chloramin und ammoniak auf aldehyde. Chem. Ber. 95: 688-691.
25. H.H. Richmond, G.S. Myers and G.F. Wright. 1948. The reaction between formaldehyde and ammonia. J. Am. Chem. Soc. 70: 3659-3664.
26. N.N. Makhova, V.Y. Petukhova, T.V. Chabina and L.I. Khmel'nitskii. 1993. Synthesis of unsubstituted 1,3,5-triazabicyclo[3.1.0]hexane. Russ. Chem. Bull. (Engl. Transl.) 42: 1602-1603.
27. M.T.H. Liu. 1982. The thermolysis and photolysis of diazirines. Chem. Soc. Rev. 11: 127-140.
28. I.R. Likhotvorik, E.L. Tae, C. Ventre and M.S. Platz. 2000. A facile one-pot conversion of non-enolizable aldehydes to diazirines. Tetrahedron Lett. 41: 795-796.
29. V.V. Kuznetsov, N.N. Makhova and L.I. Khmel'nitskii. 1997. The role of pH in the synthesis of diaziridines. Russ. Chem. Bull. 46: 1354-1356.
30. R. Ohme, E. Schmitz and P. Dolge. 1966. Diaziridine, VII. Diaziridine aus aminalen des formaldehyds. Chem. Ber. 99: 2104-2109.
31. G.V. Shustov, S.N. Denisenko, I.I. Chervin, N.L. Asfandiarov and R.G. Kostianovsky. 1985. Asymmetric nitrogen - 41. Tetrahedron 41: 5719-5731.
32. V.V. Kuznetsov, N.N. Makhova and M.O. Dekaprilevich. 1999. 6,6'-Bis(1,5-diazabicyclo[3.1.0]hexane). Izv. Acad. Nauk Ser. Khim. 3: 623-625.
33. N.N. Makhova, A.N. Mikhailyuk, V.V. Kuznetsov, S.A. Kutepov and P.A. Belyakov. 2000. Effective synthesis of 1,2-di-, 1,2,3-tri-, 1,2,3,3-tetraalkyldiaziridines and 1,5-diazabicyclo[3.1.0]hexanes. Mendeleev Commun. 10: 182-184.
34. V.V. Kuznetsov, S.A. Kutepov, N.N. Makhova, K.A. Lyssenko and D.E. Dmitriev. 2003. 1,5-Diazabicyclo [3.1.0]hexanes and 1,6-diazabicyclo[4.1.0]heptanes: a new method for the synthesis, quantum-chemical calculations, and X-ray diffraction study. Russ. Chem. Bull. Int. Ed. 52: 665-673.
35. B.V. Lyalin and V.A.E. Petrosyan. 2002. Direct electrochemical synthesis of diaziridines. Russ. J. Electrochem. 38: 1220-1227.
36. L.B.H. Quast. 1981. Synthese und photolyse von 1,4-dialkyl-1,4-dihydro-5*H*-tetrazol-5-onen und -thionen: neue wege zu diaziridinonen und carbodiimiden. Chem. Ber. 114: 3253-3272.
37. V.Y. Petukhova, N.N. Makhova, V.P. Ananikov, Y.A. Strelenko and I.V. Fedyanin. 2004. 1,2-Bis(methylamino)ethane-1,2-diol dihydrochloride as a new precursor of 1,2,1",2"-tetramethyl-3,3"-bidiaziridine. Russ. Chem. Bull. Int. Ed. 53: 641-646.
38. J.A. Blair and R.J. Gardner. 1972. A novel synthesis of a diaziridine. J. Chem. Soc. Perkin Trans. 1 0: 485-485.
39. N.N. Makhova, V.V. Kuznetsov and R.G. Kostyanovsky. 1996. 1,2-Di(1-adanmantyl)diaziridine. Russ. Chem. Bull. 45: 1780-1781.
40. V.V. Kuznetsov, N.N. Makhova, D.E. Dmitriev and V.V. Seregin. 2005. Syntheses of 1,2-di- and 1,2,3-trialkyldiaziridines. Mendeleev Commun. 15: 116-118.
41. S.N. Denisenko. 1998. Stereospecific synthesis of bicyclic diaziridines: 4a-chloro-; 4e,6a- and 4a,6e-dichloro-5-methoxycarbonyl-1,6-diazabicyclo[3.1.0]hexanes. Mendeleev Commun. 8: 54-56.
42. A.V.K. Briner. 1989. Glycosylidene carbenes a new approach to glycoside synthesis. Part 1. Preparation of glycosylidene-derived diaziridines and diazirines. Helv. Chim. Acta 72: 1371-1382.
43. K. Yu, C. Hwang and I.Y. Lee. 1994. Bull. Korean Chem. Soc. 523-524.
44. E. Schmitz. 1961. Synthesen mit *N*-halogen-aminen. Angew. Chem. 73: 23-25.
45. R.G. Kostyanovskii, G.B. Shustov, A.I. Mishchenko and V.I. Markov. 1976. Asymmetric nonbridgehead nitrogen. Bull. Acad. Sci. USSR Chem. Ser. 25: 1901-1905.
46. S. Ogawa, A. Nakajima and Y. Miyamoto. 1991. Cleavage of validoxylamine A derivatives with *N*-bromosuccinimide: preparation of blocked synthons useful for the construction of carba-oligosaccharides composed of imino linkages. J. Chem. Soc. Perkin Trans. 1 12: 3287-3290.

47. E. Schmitz and R. Ohme. 1961. Cyclische diazoverbindungen, I. Herstellung und umsetzungen von diazirinen. Chem. Ber. 94: 2166-2173.
48. H.J. Abendroth. 1961. Strukturisomere hydrazone. Angew. Chem. 73: 67-67.
49. G.V. Shustov, S.V. Varlamov, I.I. Chervin, A.E. Aliev, R.G. Kostyanovsky, D. Kim and A. Rauk. 1989. Asymmetric nitrogen. 72. Geminal systems. 46. *N*-chlorooxaziridines: optical activation, absolute configuration, and chiroptical properties. J. Am. Chem. Soc. 111: 4210-4215.
50. B. Bernet, S.E. Mangholz, K. Briner and A. Vasella. 2003. Glycosylidene carbenes. Part 31. Helv. Chim. Acta 86: 1488-1521.
51. R.G. Eckenhoff, F.J. Knoll, E.P. Greenblatt and W.P. Dailey. 2002. Halogenated diazirines as photolabel mimics of the inhaled haloalkane anesthetics. J. Med. Chem. 45: 1879-1886.
52. H. Ortega, S. Ahmed and H. Alper. 2007. Facile regioselective synthesis of pyrazolo[5,1-*a*]isoquinolines via ring-opening cyclization/oxidation reactions of stable aroyldiaziridines of 3,4-tetrahydroisoquinoline with alkynes. Synthesis 23: 3683-3691.
53. G. Hanquet, X. Lusinchi and P. Milliet. 1993. Action de l'acide paranitroperbenzoique sur le tetrafluoroborate de *N*-methyl-3,4-dihydroisoquinolinium. Formation d'un sel d'oxaziridinium. Tetrahedron 49: 423-438.
54. J.M. Allen and T.H. Lambert. 2011. Tropylium ion mediated α-cyanation of amines. J. Am. Chem. Soc. 133: 1260-1262.
55. E. Schmitz. 1979. Three-membered rings with two heteroatoms. Adv. Heterocycl. Chem. 24: 63-107.
56. J.C. Pelletier and M.P. Cava. 1987. Synthesis of the marine alkaloids aaptamine and demethyloxyaaptamine and of the parent structure didemethoxyaaptamine. J. Org. Chem. 52: 616-622.
57. L. Carroccia, M. Delfini, S. Fioravanti, L. Pellacani and F. Sciubba. 2012. Stereoselective synthesis and ROESY ^{1}H NMR study of bidiaziridines. J. Org. Chem. 77: 2069-2073.
58. W.C. Firth. 1968. Reactions of pentafluoroguanidine and tetrafluoroformamidine with alkali metal fluorides. Fluorine-containing diaziridines. J. Org. Chem. 33: 3489-3492.
59. A.A. Gakh. 2012. Monofluorinated heterocycles. Top. Heterocycl. Chem. 27: 33-64.
60. S.N. Denisenko, E. Pasch and G. Kaupp. 1989. The first tricyclic diaziridines - a simple synthetic route. Angew. Chem. Int. Ed. 28: 1381-1383.
61. H.S.H. Berneth. 1980. 3-Aryl-3-chlor-1,2-diaziridine als isolierbare zwischenprodukte bei der darstellung von 3-aryl-3-chlor-3*H*-1,2-diazirinen. Chem. Ber. 113: 2040-2042.
62. G. Labbe, A. Verbruggen, T. Minami and S. Toppet. 1981. Synthesis and decomposition of functionalized diaziridinimines. J. Org. Chem. 46: 4478-4481.
63. N.N. Makhova, G.A. Karpov, A.N. Mikhailyuk and L.I. Khmel'nitskii. 1999. *N*-Alkylation of diaziridines. Mendeleev Commun. 9: 112-114.
64. L.G. Donaruma and W.Z. Heldt. 1960. Organic reactions. A.C. Cope (Ed.). Wiley and Sons, Inc. New York - London, vol 11: 1.
65. A.N. Mikhailyuk, N.N. Makhova, A.E. Bova, L.I. Khmel'nitskii and S.S. Novikov. 1978. Synthesis of diaziridines from oxime esters. Izv. AN SSSR Ser. Khim. 7: 1566-1570.
66. N.N. Makhova, V.Y. Petukhova and L.I. Khmel'nitskii. 1982. Synthesis of diaziridines from oxime esters. Izv. AN SSSR Ser. Khim. 9: 2107-2110.
67. V.Y. Petukhova, Y.A. Strelenko, K.A. Lyssenko and N.N. Makhova. 2007. Synthesis and structure of 1-[ω-(3,3-dialkyldiaziridin-1-yl)alkyl]-3,3-dialkyldiaziridines. Russ. Chem. Bull. Int. Ed. 56: 1550-1554.
68. A.N. Mikhailyuk, V.Y. Petukhova and N.N. Makhova. 1997. 3-Alkyldiaziridines and 1,3-dialkyldiaziridines from aliphatic aldoxime-*O*-sulfonic acid salts. Mendeleev Commun. 7: 60-61.
69. K.M. Broadus and S.R. Kass. 2000. Benzocyclobutadienyl anion: formation and energetics of an antiaromatic molecule. J. Org. Chem. 65: 6566-6571.
70. S. Kobayashi, M. Sugiura, H. Kitagawa and W.W.-L. Lam. 2002. Rare-earth metal triflates in organic synthesis. Chem. Rev. 102: 2227-2302.

CHAPTER

Synthesis of diazirines

4

Chapter outline

4.1 Introduction

The chemistry of small ring compounds (three- and four-membered rings) has played an important role in the development of modern organic chemistry. Leading among these reactive molecules are the small ring heterocyclic compounds. One or more heteroatoms present in these strained rings convey a considerable dipole moment to these types of molecules. It also adds a novel dimension of intrinsic difficulty concerning the formation and stability of these types of heterocyclic analogs of cyclobutanes and cyclopropanes. If one considers the compressed bond angles (approximately 90°C in four-membered rings and approximately 60°C in three-membered rings), the mere synthetic challenge, particularly for the unsaturated analogs of these heterocyclic compounds, appears vast. The small ring heterocyclic compounds have much greater reactivity toward a number of reagents than do their five- or six-membered ring analogs. The 2*H*-azirines are highly reactive three-membered unsaturated *N*-heterocyclic compounds that have been utilized for the formation of a variety of polyfunctional cyclic and acyclic *N*-containing compounds [1]. This small-membered ring can participate in several chemical reactions as it can serve as a dipolarophile or as a dienophile and can also work as nucleophile or electrophile. The ring strain undoubtedly contributes the high reactivity of 2*H*-azirine functionality and makes it an attractive intermediate for the formation of complex heterocyclic ring systems [2,3].

With a literature age of less than 50 years old [4,5], the diazirines are relative youngsters among their companion three-membered heterocyclic compounds. Although this ring system might instinctively seem relatively unstable, the strain energy is estimated at an extraordinarily low 21 kcal mol^{-1} [6]. This, along with very low basicity, is mainly responsible for the detected stability (even in vivo) of the diazirine ring at rt. The double-bond character between two nitrogen results in a short nitrogen—nitrogen bond distance, which, in turn, considerably compresses the N—C—N bond angle [7]. Similarly, the diaziridine functionality is symmetrical and planar.

3-Membered Heterocycle Synthesis. https://doi.org/10.1016/B978-0-443-13583-5.00011-6

4.2 Synthesis of diazirines

The 5-chloromercuro-2′-deoxyuridine was synthesized by the method reported by Chang and Mertes [8] utilizing deoxyuridine as starting compound and was further reacted with Pd catalyst. The final compound was formed in 10% yield. This low yield was explained by the incomplete reaction of 5-chloromercuro-2′-deoxyuridine; some amounts of unreacted 5-chloromercuro-2′-deoxyuridine were reported in the reaction mixture. The reaction of final compound with dimethoxytrityl chloride in dry Py and the purification of product on a silica gel column utilizing $CHCl_3$—MeOH mixture afforded 5′-*O*-dimethoxytrityl-5-[4-(3-(trifluoromethyl)-3*H*-diazirin-3-yl)phenyl]-2′-deoxyuridine as a chromatographically pure compound in 93% yield. The 5′-*O*-dimethoxytrityl-5-[4-(3-(trifluoromethyl)-3*H*-diazirin-3-yl)phenyl-2′-deoxyuridine-3′-*O*-(2-cyanoethyl)-*N*,*N*′-diisopropylphosphoramidine was prepared utilizing chloro-(2-cyanoethoxy)-diisopropylaminophosphine (Scheme 1). The modified oligonucleotide was formed on an automated synthesizer and purified by the reverse-phase HPLC. The modified phosphoramidite was formed in low yield (about 40%) [9].

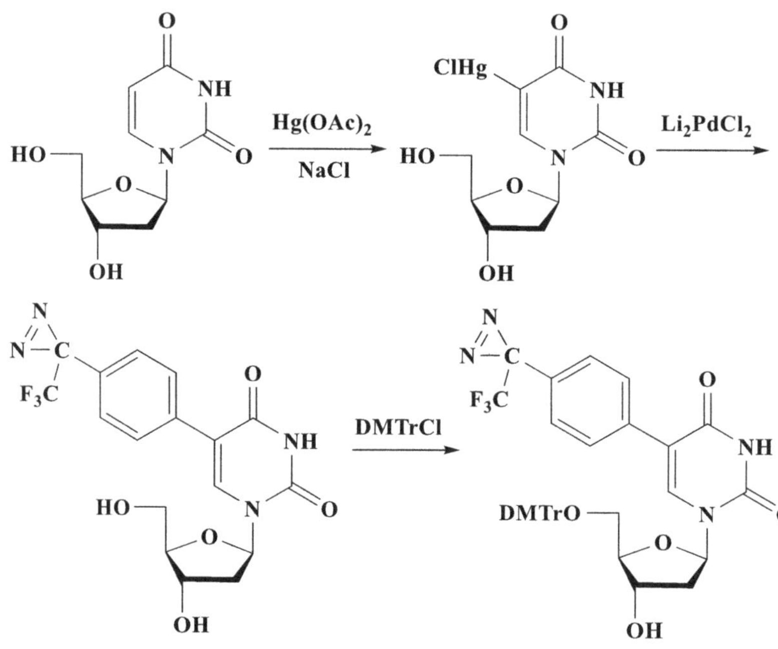

SCHEME 1

The phenothiazine diazirine was synthesized from phenothiazine acid through the ketone. However, the repeated attempts to influence the conversion of 2-chloro-*N*-(3-ketobutyl) phenothiazine, prepared from the acid, to more interesting diazirine failed. Although the reason for this failure was not known, this outcome and low yield (15%) reported in the formation of phenothiazine diazirine from ketone demonstrated the less satisfactory nature, in various examples, of the ketone to diazirine conversion (Scheme 2) [10,11].

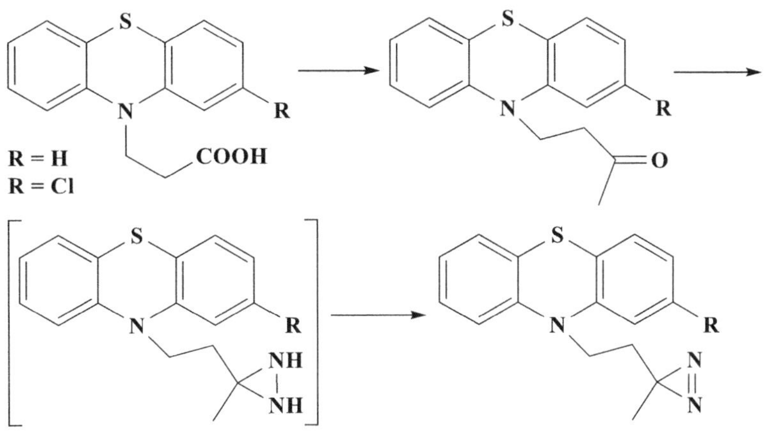

SCHEME 2

The δ-lactone was refluxed with benzylazide in the presence of trimethyl orthoformate to provide the sole product triazoline. The triazoline was suspended in water/CH_3CN mixture (8:2) with catalytic amount of BiO_3 and heated under the reflux for 2 h for the cleavage of acetal functionality. Surprisingly, the diazirine was obtained as a nearly pure product in 63% yield from the crude material after neutralization and evaporation. In the literature [12], it was observed that the diazodioxanediones isomerized under the photochemical conditions to synthesize the diazirines. Also, the diazirines gave back the diazo compounds under either the photochemical or thermal conditions. According to Scheme 3, the mechanistic suggestion for the synthesis of diazirine was based on the diazo-diazirine isomerization to produce an intermediate first and after that cyclization to diazirine.

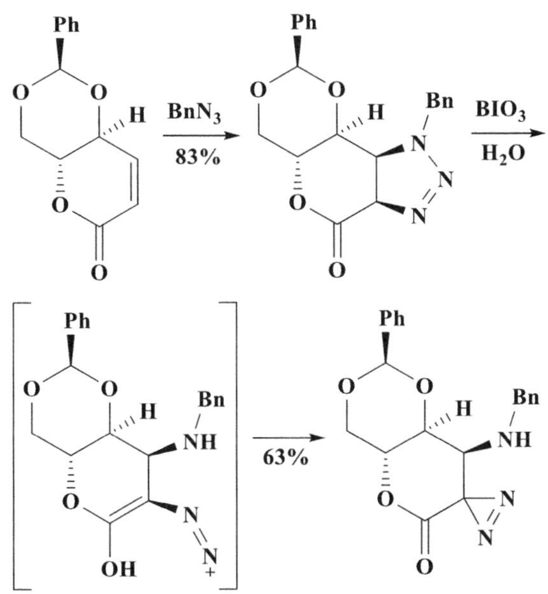

SCHEME 3

The reaction of alcohol and diazirine derivative pentanoic acid in the presence of *N,N′*-diisopropylcarbodiimide and 4-dimethylaminopyridine afforded ester. An elimination of tetrahydropyran-protecting group was achieved using pyridinium *p*-toluenesulfonate in ethanol followed by transformation of free alcohol to bromide by reacting with carbon tetrabromide and triphenylphosphine. The purification of bromide was performed utilizing a small reversed-phase cartridge to reduce the decomposition of allylic bromide on silica gel. The reversed-phase purification also removed colored impurities that interfere with the consequent peptide alkylation; although this procedure did not remove unreacted alcohol starting compound, that compound did not interfere with the consequent alkylation chemistry (Scheme 4) [13–15].

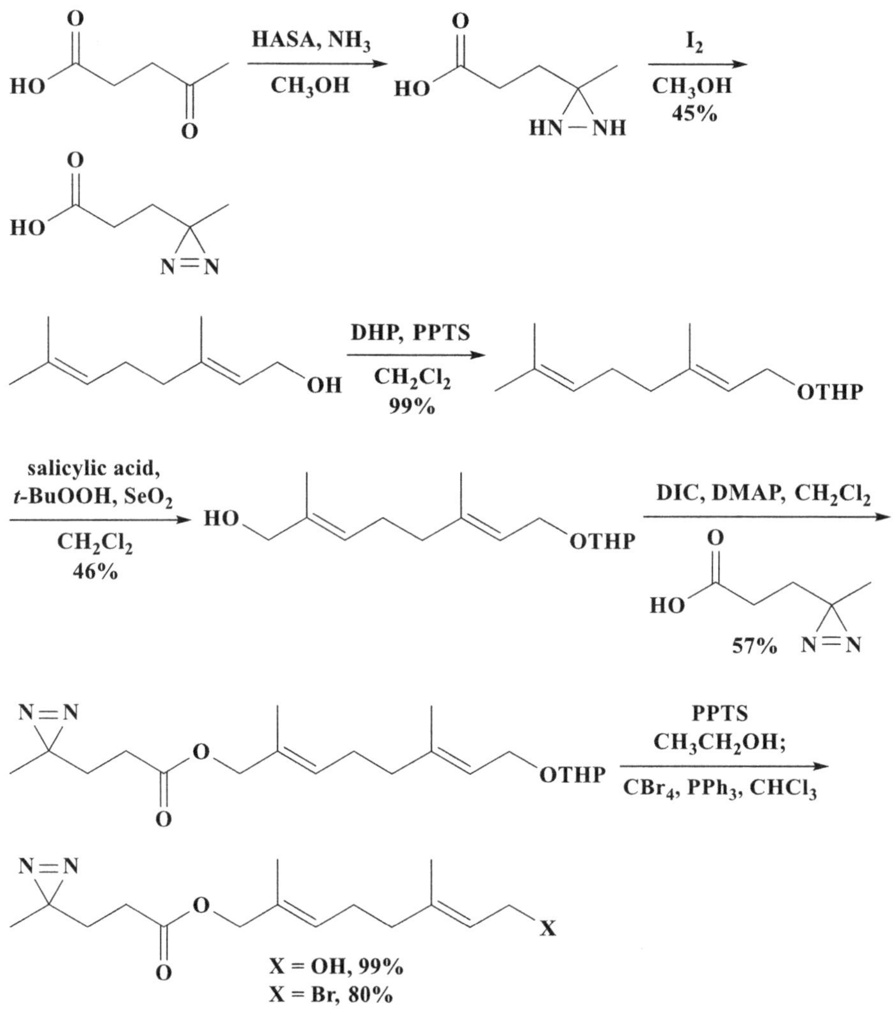

SCHEME 4

4.3 Synthesis of diazirines from amines

The aryl- or alkyl-3-halodiazirine was synthesized by the halogenation of aryl- or alkyl-amidines utilizing NaOCl in aqueous DMSO solution (Scheme 5) [16,17].

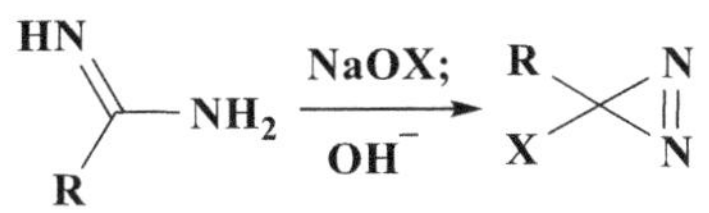

SCHEME 5

The 3-chlorodiazirines were synthesized by an oxidation of amidines in the presence of NaOCl (Scheme 6) [18,19].

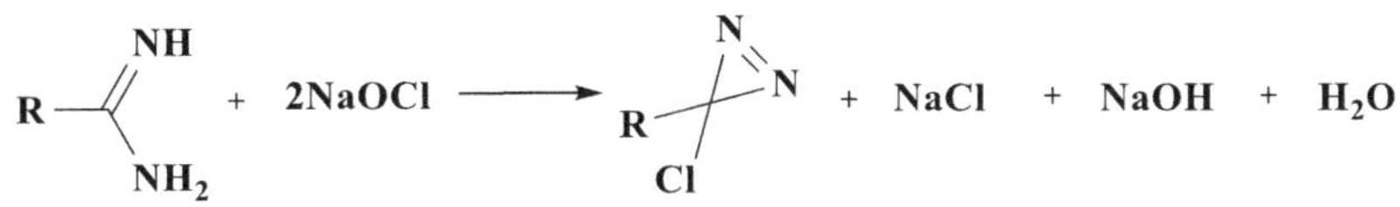

SCHEME 6

The 2,2-diaminohexafluoropropane was prepared from hexafluoroacetone. The bis(trifluoromethyl)diazirine [20,21] was synthesized by an oxidation of 2,2-diaminohexafluoropropane with NaOCl (Scheme 7).

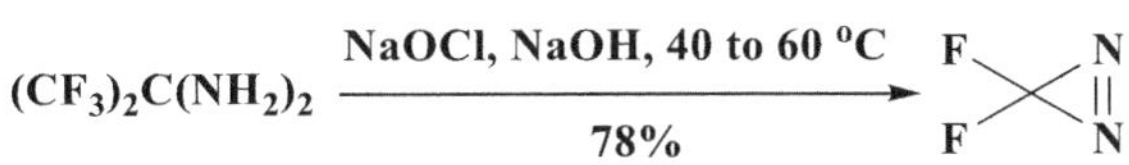

SCHEME 7

In the case of diazirines, both bis(trifluoromethyl)diazirine and difluorodiazirine were formed. The reductive defluorination of bis(difluoroamino)difluoromethane synthesized difluorodiazirine (Scheme 8) [22,23], but it could also be prepared by an interesting fluoride ion–induced rearrangement of difluorocyanamide (Scheme 9) [24].

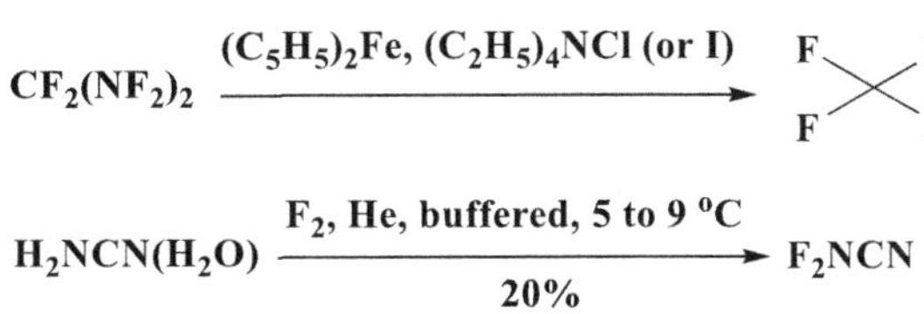

SCHEME 8

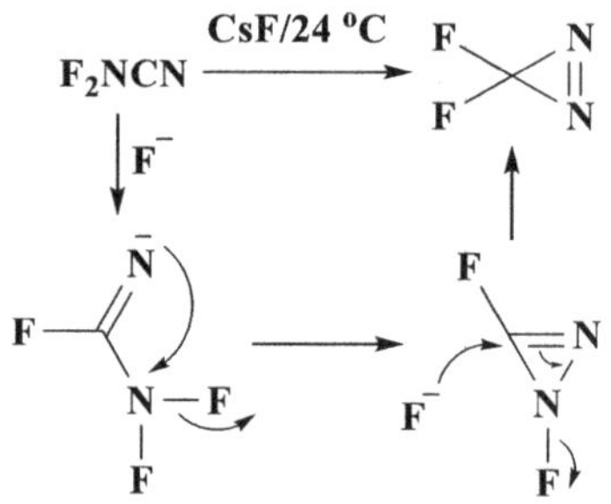

SCHEME 9

The second *t*-butyloxycarbonyl group hindered the transformation of semialdehyde to methyl ketone. Hernández et al. [25] have efficiently utilized LiBr to selectively remove one *t*-butyloxycarbonyl group from a number of diprotected compounds (Scheme 10).

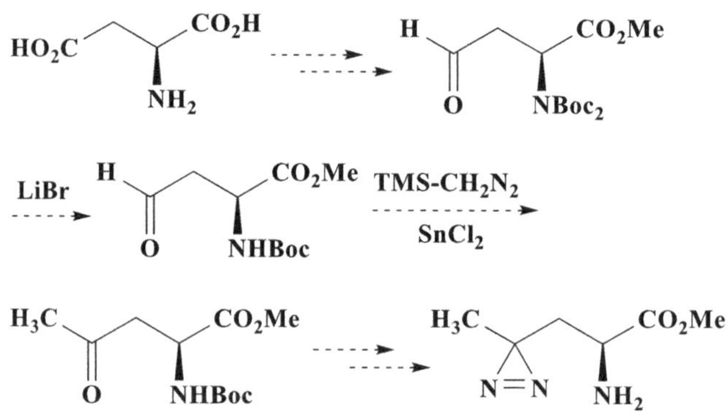

SCHEME 10

The lysine was selected to attach the LAB with dansyl group and diazirine functionality as a trifunctional linker. The reaction of 5-(dimethylamino)naphthalene-1-sulfonyl chloride with protected lysine as a starting compound under the basic conditions afforded sulfonamide compound. The deprotection of *t*-butyloxycarbonyl group of sulfonamide compound using trifluoroacetic acid/dichloromethane followed by amide reaction with diazirine acid utilizing 1-ethyl-3-(3-dimethylaminopropyl)carbodiimide and dimethylaminopyridine afforded a compound with ester group. The ester group of formed compounds was hydrolyzed utilizing aqueous base to synthesize the diazirine-linked 5-(dimethylamino)naphthalene-1-sulfonyl acid (Scheme 11) [26,27].

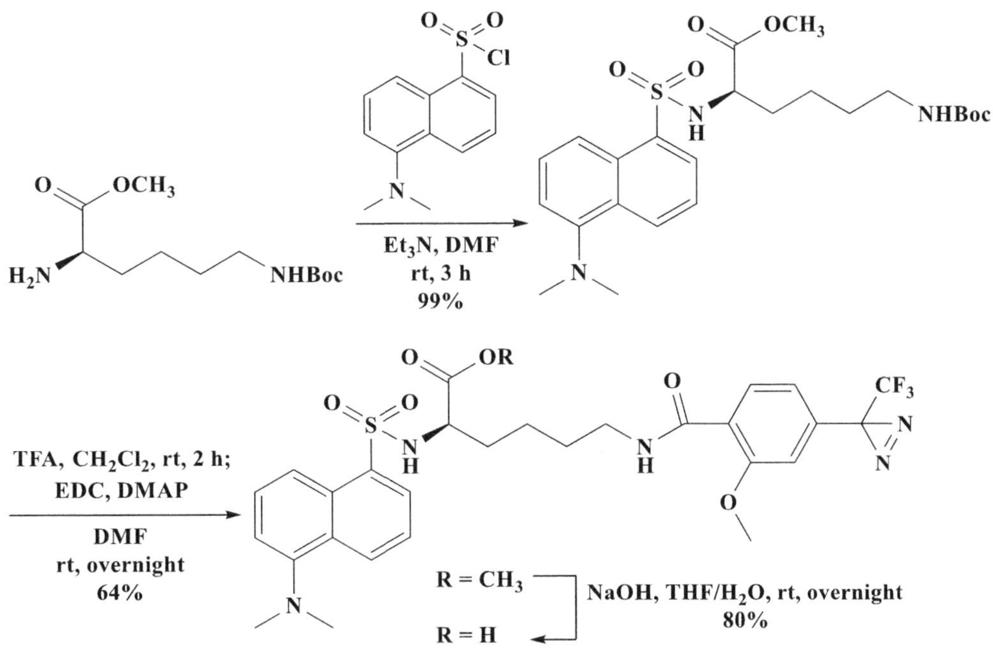

SCHEME 11

4.4 Synthesis of diazirines from oximes

The photo-cross-linking analog, 5-[4-(3-(trifluoromethyl)-3*H*-diazirin-3-yl)phenyl]-2′-deoxyuridine, was synthesized and introduced into the recognition site of MvuI and EcoRII restriction-modification enzymes. The resultant base-modified substrate was tested for cross-linking to these enzymes. The cross-linking was affected by an irradiation of enzyme–substrate complexes at 366 nm. The key compound, 4’-(3-(trifluoromethyl)-3*H*-diazirin-3-yl)iodobenzene, having photoaffinity label was prepared from 1,4-diiodobenzene in five steps. The 4’-(3-(trifluoromethyl)-3*H*-diazirin-3-yl)iodobenzene was transformed to 4′-iodo-2,2,2-trifluoroacetophenone by reaction with butyl lithium followed by reaction with $CF_3COOC_2H_5$ (Scheme 12) [9,28].

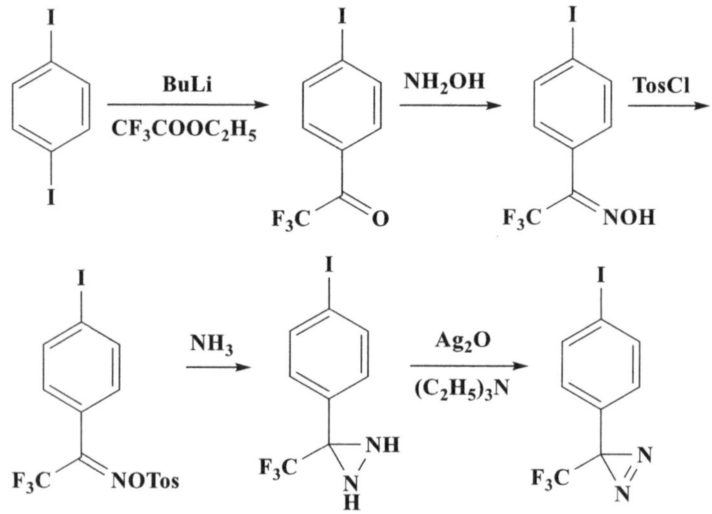

SCHEME 12

The formation of diaziridine group was attained from a protected bromo phenyl methanol, which was reacted with *N*-(trifluoroacetyl)piperidine to synthesize the ketone. The ketone was transformed to oxime followed by tosylation to synthesize the tosyl oxime. Then, the reaction with liquid NH_3 afforded diaziridine group. The diazirine was synthesized by subsequent oxidation with I_2 (Scheme 13) [29,30].

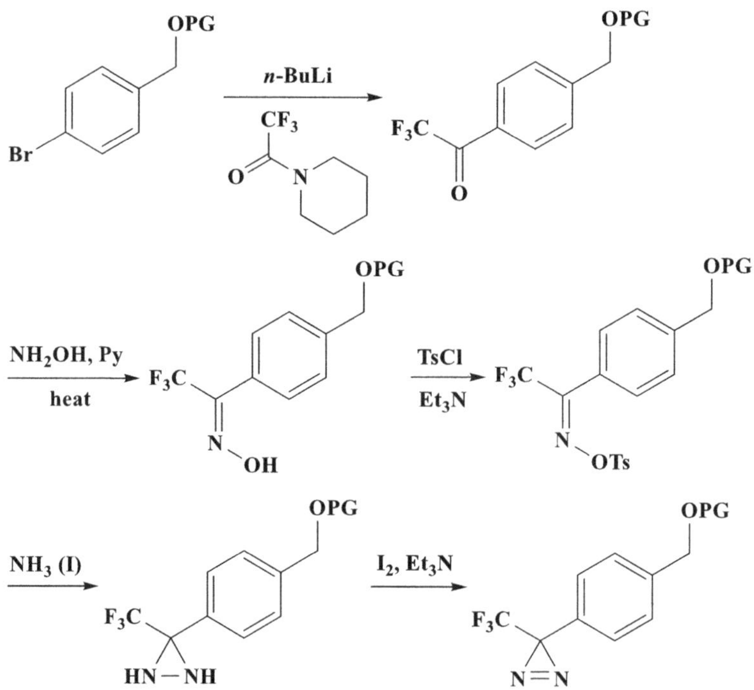

SCHEME 13

The reaction commenced with an alkylation of 4-hydroxybenzaldehyde with *t*-butyl bromoacetate to provide the *t*-butyl(4-formylphenoxy) acetate under the MWI (Scheme 14). The trifluoromethyl acetophenone was prepared in two steps: introduction of trifluoromethyl group upon reacting aldehyde with Ruppert's reagent [31] and an oxidation of formed alcohol with Dess–Martin periodinane to synthesize the ketone. Next, the oxime was synthesized with $NH_2OH.HCl$ in Py, which was then tosylated with *p*-toluenesulfonyl chloride to synthesize the tosyl oxime. Both reactions were carried out under the MWI, which enhanced the yield and shortened the reaction time. The diaziridine was synthesized by the reaction of tosylated oxime with liquid NH_3 overnight at high pressure. The trifluoromethyl aryldiazirine was formed when I_2 was utilized for an oxidation of diaziridine. Finally, the *t*-butyl group was removed with TFA (50%) in dichloromethane. This synthetic pathway afforded target compound in a reasonable 21% overall yield.

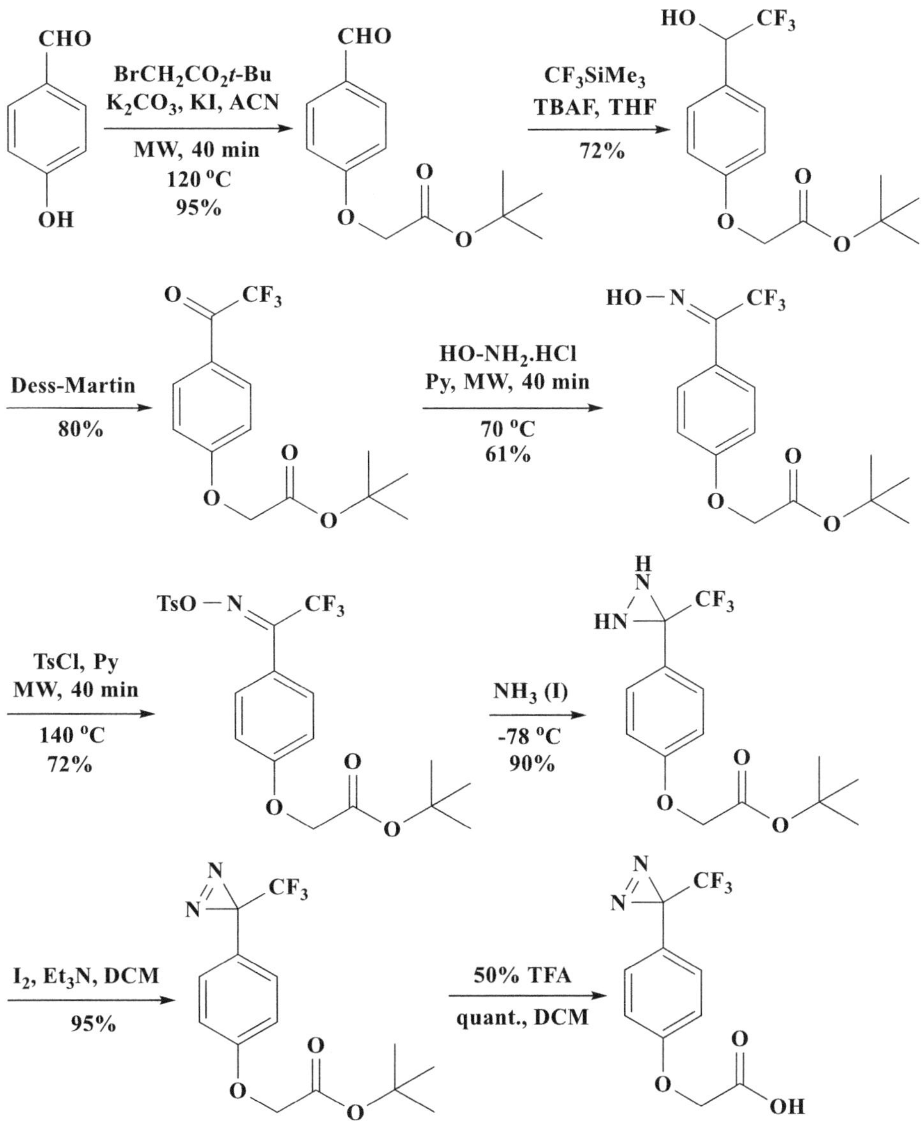

SCHEME 14

Although Ruppert's reagent and Dess–Martin oxidation provided good yields on small scale, Dess–Martin oxidation was very temperamental and lower yields were obtained when large amounts were needed. For the large-scale production, the 4-bromophenol was utilized as a starting compound, and the bromo group allowed an introduction of trifluoromethyl acetate group using *n*-butyl lithium or Grignard reagent. Both reactions were performed utilizing *N*-(trifluoromethyl)piperidine as an electrophile [32–35] under different reaction conditions. Although *n*-butyl lithium worked better than Grignard reagent, the yields were very poor. One way to improve the *n*-butyl lithium reaction yield was the protection of phenol. Finally, different electrophiles were screened, and the best yields were obtained when $CF_3COOC_2H_5$ was utilized. Therefore, the reaction was started by protecting the phenolic OH of 4-bromophenol with trifluoromethylsilane group and reaction with *n*-butyl lithium and $CF_3COOC_2H_5$ to synthesize the deprotected alcohol in 68% yield. The rest of the synthesis was achieved as explained in Scheme 15 providing target compound in total 35% yield.

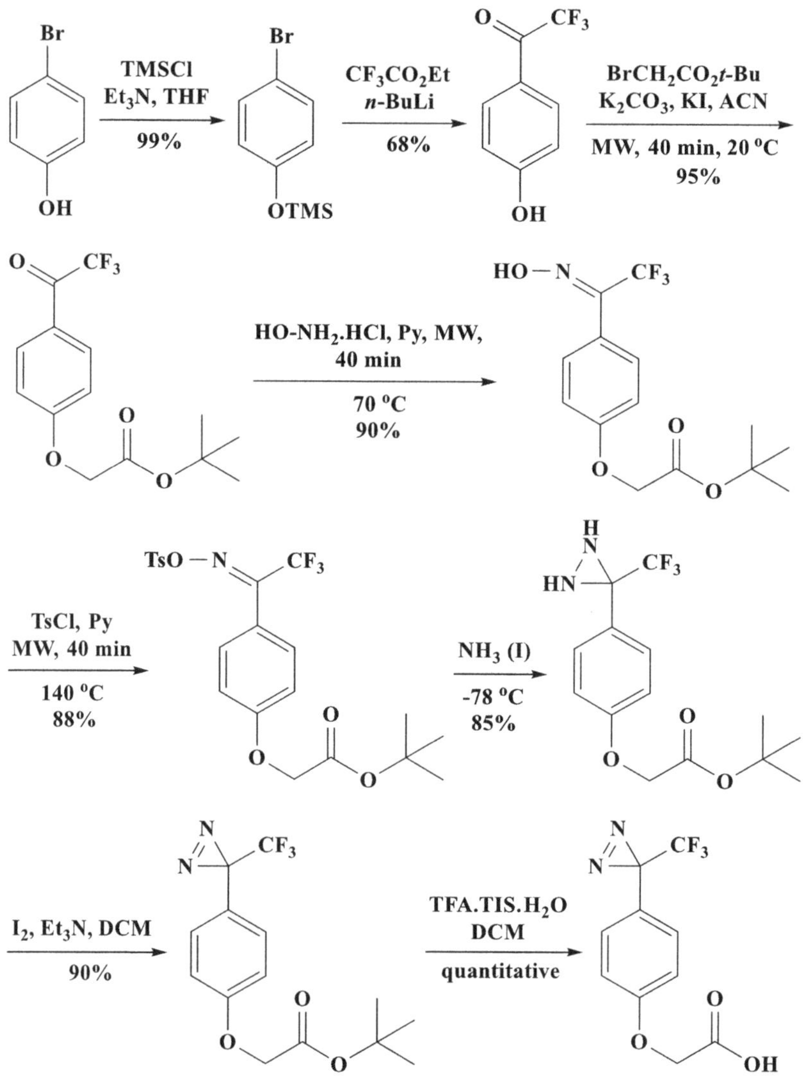

SCHEME 15

The target compound was considered because the hydrophilic part was conserved, the lipophilic side chain, however, without CH_3 groups, and the photoactive group substitutes the terminal isoprene unit. The 3-trifluoromethyl-3-phenyldiaziridine group was selected as a photoaffinity label because an irradiation (358 nm) was different from the absorption maximum of the enzyme. Further on an irradiation, carbenes were synthesized, which were more beneficial than the nitrenes [36], and because of the reactivity of carbenes regarding hydrogen abstraction (from solvents), chemists should not encounter critical difficulties with pseudo-photoaffinity labeling. The side chain of desired compound was constructed from 1,8-octanediol via bromo compound, the precursor for an attachment of essential diazirine. The lithiation of bromo compound followed by an addition of trifluoroacetylpiperidine provided trifluoroacteyl ketone, which was reacted with hydroxylamine and then with tosyl chloride to synthesize the tosyl oxime. The reaction of tosyl oxime with liquid ammonia in an autoclave afforded diaziridine, which was oxidized with silver(II) oxide, and brominated after deprotection to synthesize the bromodiazirine. The commercially accessible nitrile was reduced to tetrahydroisoquinoline, which was submitted to Pictet–Spengler conditions [37] and deprotected to synthesize the tetrahydroisoquinoline. An alkylation of tetrahydroisoquinoline with bromide provided desired compound (Scheme 16). Since most of the steps proceeded with over 90% yield, the formation of desired compound was quite efficient and was carried out with radiolabeled hydroxyisoquinoline also [8].

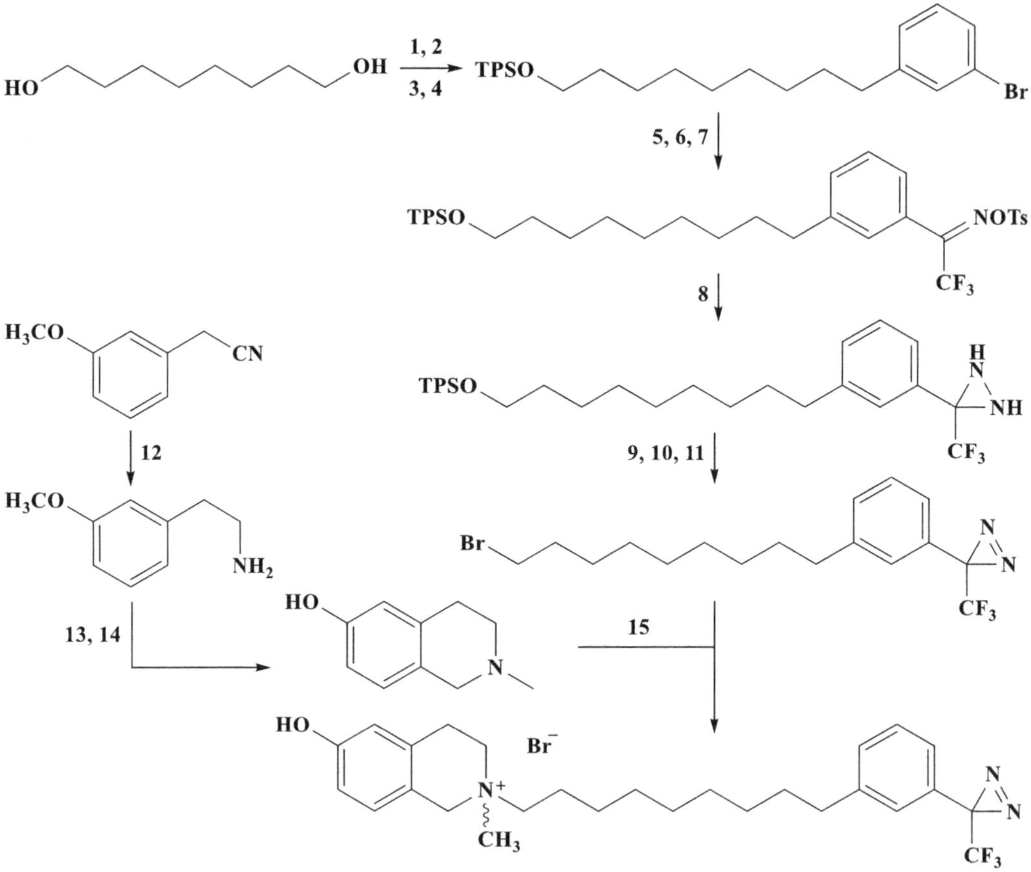

Reagents and conditions: (1) 1 eq. HBr (48%), toluene, reflux, 8 h, 77%, (2) PPh_3, EtOH, reflux, 48 h, 84%, (3) 2 eq. PhLi, 1 eq. 3-bromobenzaldehyde, THF, -40 °C, 30 min, 92.5%, (4) H_2, Pd/C, EtOH, rt, 2 h, TIPSCl, imidazole, DMF, rt, 1 h, 95%, (5) BuLi, $CF_3CONpip$, Et_2O, -50 °C, 67%, (6) NH_2OH/AcONa, EtOH, reflux, 18 h, 97%, (7) TsCl, Py, 120 °C, 3 h, 99%, (8) NH_3, Et_2O, -78 °C to rt, 18 h, 93%, (9) Ag_2O, Et_2O, rt, 1.5 h, 91%, (10) TBAF, THF, rt, 1 h, 99%, (11) Ph_3P, NBS, CH_2Cl_2, 0 °C, 45 min, 91%, (12) $LiAlH_4/AlCl_3$, Et_2O, 5 °C, 2 h, 76%, (13) HCHO, HCOOH, 95 °C, 5 h, 88%, (14) BBr_3, CH_2Cl_2, -78 °C, 3 h, 80%, (15) CH_2Cl_2, MeOH, 50 °C, 18 h, 95%.

SCHEME 16

Although the formation of TmdPhe has yet been attained (Scheme 17), the yields for the first three steps were usually high. The diazirine unit underwent photodegradation under the normal light conditions. Hence, care should be taken to perform the following steps in the dark [38].

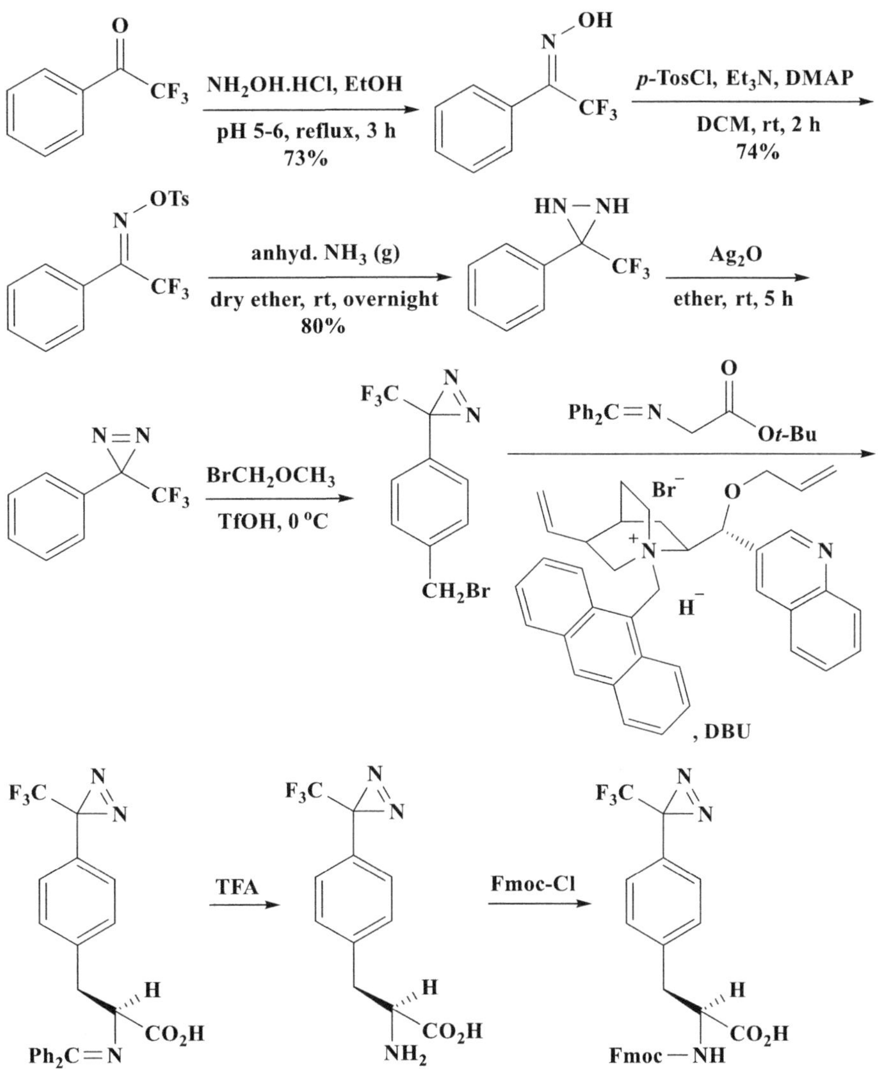

SCHEME 17

The phenyl diazirine is extensively utilized for photoaffinity labeling in chemical biology. Numerous reports are accessible about its utilization in affinity-based target fishing [39–46]. As a result, their synthesis has been studied in several applications [44,47]. By utilizing some of these efficient literature methods, the desired diazirine was prepared in high overall yield. In the starting, the ketone was synthesized by a Grignard reaction utilizing 4-bromotoluene-derived Grignard reagent and the amide as an electrophile. The ketone on reaction with hydroxylamine hydrochloride in a mixture of pyridine and C_2H_5OH followed by subsequent modification utilizing tosyl chloride afforded tosyl oxime in 83% yield over two steps. The tosylated oxime was reacted with liquid ammonia in a closed flask to synthesize the diaziridine in 93% yield. The diaziridine was easily transformed to diazirine in excellent yield using freshly prepared silver(I) oxide. In the last step, the CH_3 group of toluene derivative was oxidized with potassium permanganate in aqueous Py at rt to afford the desired acid in 56% yield ready for the reaction with TEG-derived linker (Scheme 18).

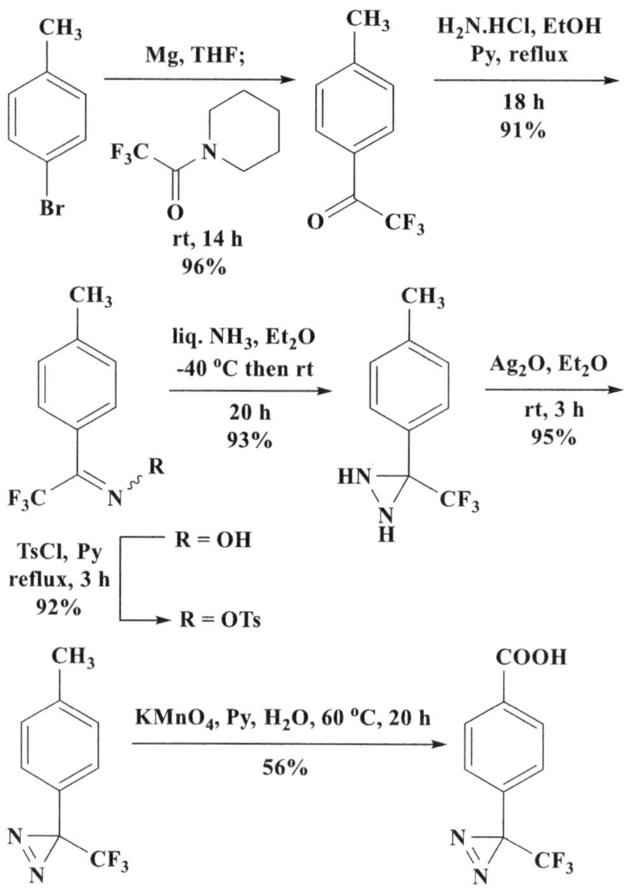

SCHEME 18

Since end-application properties are required, the 3-trifluoromethyldiazirines are of particular attention, and these are prepared in a similar manner. For instance, the diazirinyl angiotensin II analog (Scheme 19) was obtained in very good yield by the Ag_2O-assisted oxidation of diaziridine [48]. An equally efficient diazirine formation was attained starting from *O*-tosyl oxime, which was transformed into diaziridine by reaction with liquid NH_3. After the fluoride-assisted desilylation of alcohol moiety, the pyridinium dichromate oxidation afforded the desired diazirine in remarkably good overall yield (Scheme 20). The target was utilized as a centerpiece for the preparation of a vitamin D analog, an evidence to the robustness of diazirinyl functionality toward several experimental conditions [49a].

Finally, the CF_3 group was usually introduced by an activated TFA acid derivative. For instance, the aryl Grignard derived from *m*-bromoanisole was easily added to *N*-trifluoroacetylpiperidine to synthesize the trifluoroacetophenone derivative, which was transformed into oxime in excellent yield by traditional approaches. The consequent *O*-tosylation and reaction with NH_3 gave diaziridine, which was oxidized to diazirine with *t*-BuOCl (Scheme 21) [49b].

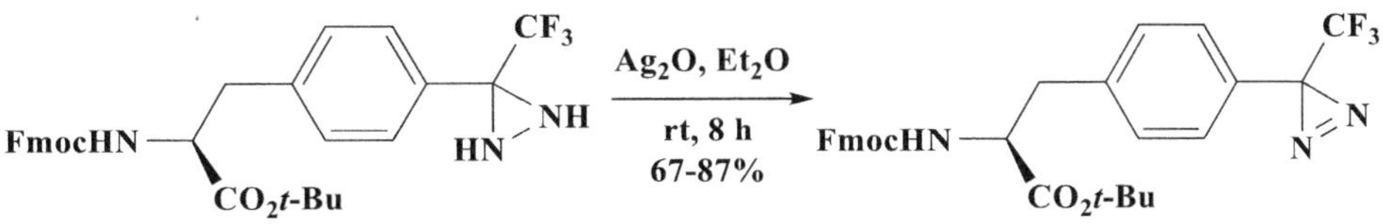

SCHEME 19

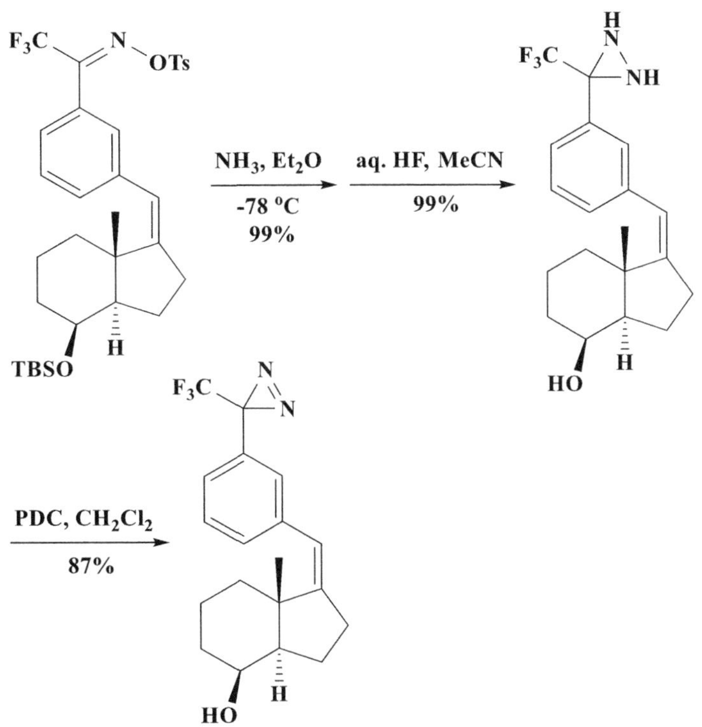

SCHEME 20

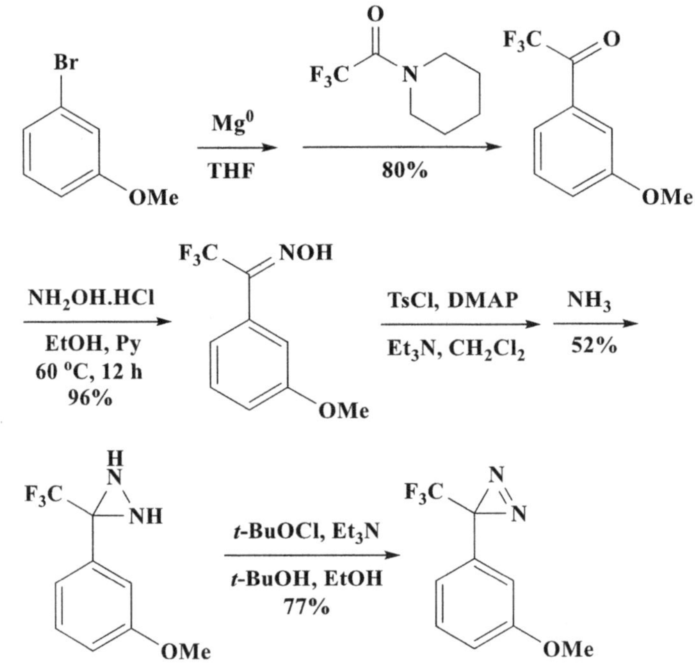

SCHEME 21

The diazirine was synthesized [50] by a five-step sequence from bromide. An initial metalation of bromide with *n*-BuLi followed by reaction with CH_3COOCF_3 provided ketone. The oxime was prepared via reaction with $NH_2OH.HCl$ using base and then protected with TosCl to synthesize the *p*-tolylsulfonyloxime. The reaction of diaziridine with NH_3 and then an oxidation with Ag_2O provided desired probe diazirine (Scheme 22).

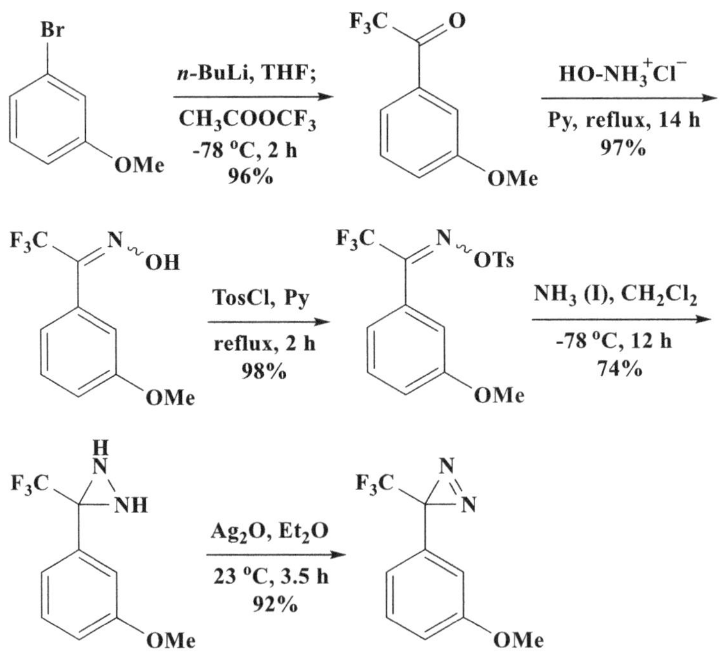

SCHEME 22

The aldehyde was prepared in eight-step sequence as shown in Scheme 23 [51,52]. This pathway was selected even though it needed three additional steps. The bromobenzaldehyde was transformed into its acetal in 91% yield with 1,3-propanediol and catalytic Amberlyst 15 with removal of H_2O through a Dean–Stark trap. The acetal was transmetalated with *n*-butyl lithium to synthesize the intermediate aryl lithium species, which was further reacted with CH_3COOCF_3 to synthesize the ketone in 96% yield. The reaction of ketone with $NH_2OH.HCl$ in a 2:1 mixture of Py-C_2H_5OH at reflux for 14 h gave oxime in 99% overall yield. The transformation of *N*-hydroxy group into *O*-mesylate, a good leaving group, occurred by reaction with MsCl in dichloromethane in the presence of Et_3N and 4-dimethylaminopyridine forming *O*-mesyl ketoxime. The mesyl ketoxime was converted into diaziridine by reaction with liquid NH_3 followed by an oxidation to diazirine with Ag_2O in 66% yield over three steps. Finally, the cleavage of protecting group provided aldehyde in 57% overall yield or quantitative yield for the entire eight-step sequence.

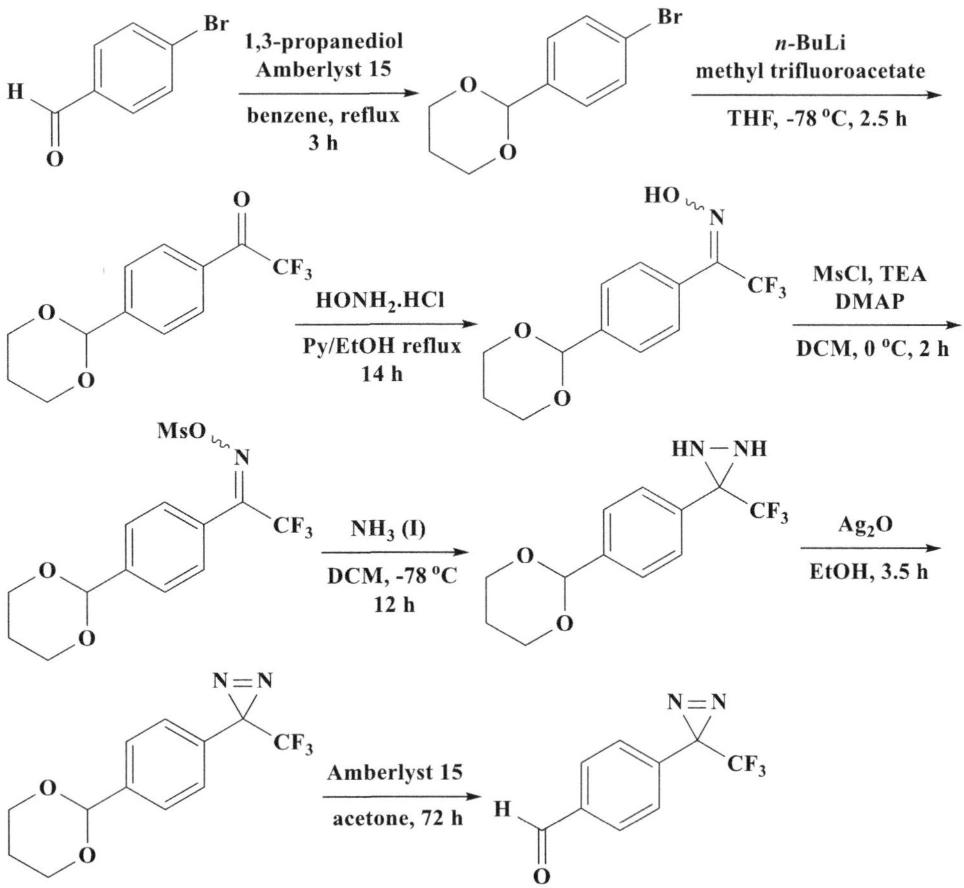

SCHEME 23

The formation of diazirine was proved to be direct (Scheme 24) [44c]. The trifluoroacetophenone was heated at reflux in the presence of $NH_2OH.HCl$ in a mixture of Py and C_2H_5OH (2:1) to synthesize the ketoxime. The *N*-hydroxy group of ketoxime was transformed to tosylate, an efficient leaving group, by reaction with tosyl chloride. The reaction of tosyl ketoxime with liquid NH_3 provided diaziridine, which was then oxidized with Ag_2O to synthesize the diazirine in 92% overall yield over four steps.

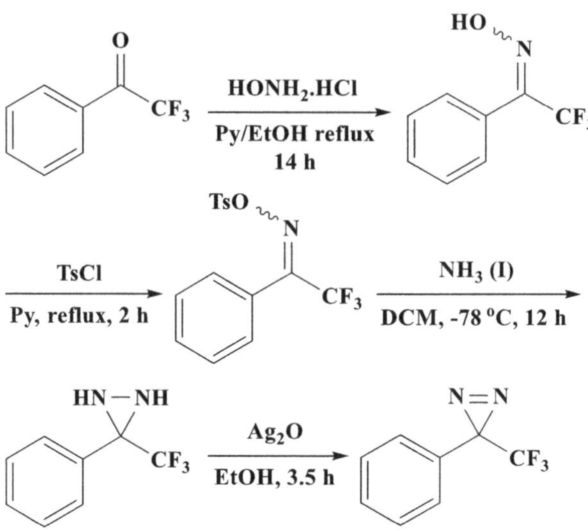

SCHEME 24

Weber [53] reported the formation of free diaziridine and diazirine (Scheme 25). The TES-protected lactone oxime mesylate (prepared from lactone oxime in three steps and 59% yield) did not react with ammonia in MeOH, 2-methoxyethanol, or 2,2,2-trifluoroethanol at temperatures of up to 45°C, while the triflate (prepared from lactone oxime in three steps and 55% yield) decomposed quickly. The reaction of unprotected lactone oxime mesylate (formed in 76% yield from TES-protected lactone oxime mesylate) with ammonia in methanol at 25°C provided crude diaziridine in up to 30% yield, but purification was not straightforward. Finally, the reaction of acetyl-protected lactone oxime triflate with ammonia in dichloromethane/methanol provided acetyl-protected diaziridine in 71% yield. The hydrolysis of diaziridine (ammonia, methanol, and water) and the crystallization of crude product from methanol provided diaziridine in 72% yield as a white solid. This diaziridine was rather unstable; in aqueous solution, it decomposed within hours. A mixture of D-glucose and methyl α- and β-D-glucosides was obtained in an oxidation of acetyl-protected diaziridine, most probably through the diazirine.

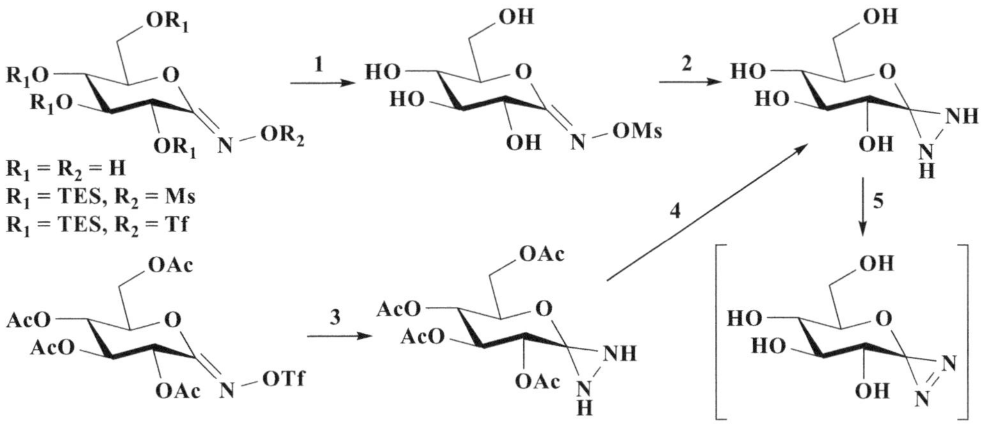

Reagents and conditions: (1) HF, Py, THF, 76%, (2) NH_3, MeOH, 25 °C, 31% of crude, (3) NH_3, MeOH, CH_2Cl_2, -20 °C, 71%, (4) NH_3, MeOH, H_2O, 25 °C, 72%, (5) I_2, Et_3N, H_2O, MeOH, mixture of D-glucose and methyl α- and β-D-glucopyranoside formed.

SCHEME 25

The endohedral metallofullerene glycoconjugates were synthesized under the mild conditions by carbene addition utilizing suitable glycosylidene-derived diazirine. The nuclear magnetic resonance spectroscopic analysis revealed that the glycoconjugate contains two diastereomers of [6,6]-open monoadducts. The glycosylidene carbene was beneficial to introduce carbohydrate functionalities onto endohedral metallofullerene surfaces. As given in Scheme 26, the glycosylidene-derived diazirine was prepared according a literature procedure reported by Vasella and coworkers [54]. The *O*-benzyl derivative was formed when commercially accessible methyl-D-glucopyranoside was reacted with benzyl chloride and NaH [55]. The 2,3,4,6-tetra-*O*-benzyl-D-glucopyranose was obtained by the deprotection of pyranoside anomeric hydroxyl group with H_2SO_4 [56]. The 2,3,4,6-tetra-*O*-benzyl-D-glucopyranose was condensed with $NH_2OH.HCl$ using Na to synthesize the open-chain oxime as a mixture of stereoisomers [57]. An oxidative cyclization of oxime with sodium metaperiodate afforded desired ring-closed material hydroximinolactone. The methanesulfonate was synthesized by the reaction of hydroximinolactone with MsCl under the basic conditions. The reaction of methanesulfonate with NH_3 afforded diaziridine, which was then oxidized with I_2 to synthesize the diazirine [58].

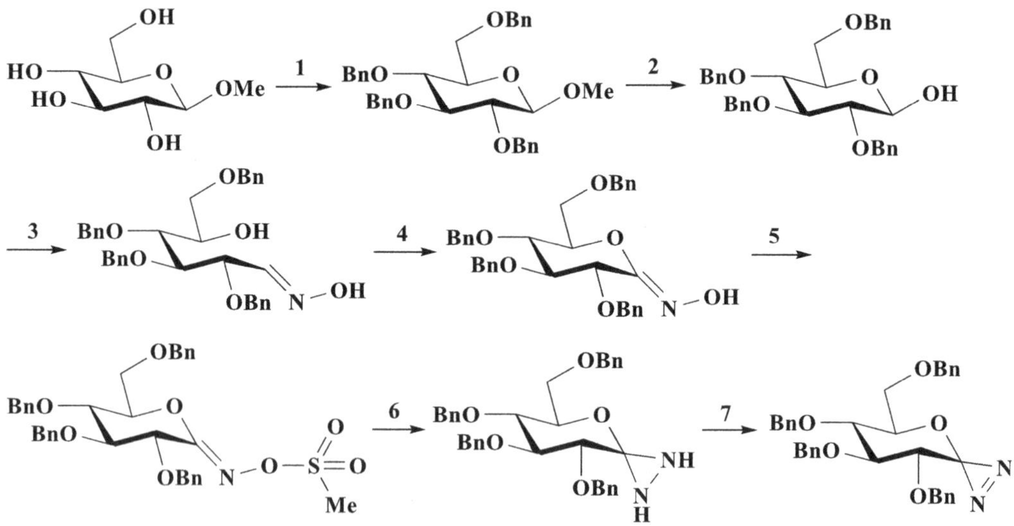

Reagents and conditions: (1) benzyl chloride, NaH, 120 °C, (2) AcOH, H_2SO_4, 90 °C, 15%, 2 steps, (3) hydroxylamine hydrochloride, Na, EtOH, 74 °C, 38%, (4) sodium metaperiodate, sodium acetate, EtOH, H_2O, 60 °C, 50%, (5) methanesulfonyl chloride, Et_3N, CH_2Cl_2, 0 °C, 67%, (6) NH_3/MeOH, rt, 76%, (7) I_2, Et_3N, MeOH, -20 °C, 50%.

SCHEME 26

The transformation of (*E*)-tosyl oxime to diaziridine was achieved quantitatively with liquid ammonia in methyl *t*-butyl ether (Scheme 27). An oxidation of diaziridine with iodine/triethylamine to diazirine and the acidic deprotection gave diazirine diol. The monosilylation of primary hydroxy group of diazirine diol provided the building block. The reaction with Boc-protected dolaproine (*N,N′*-dicyclohexylcarbodiimide/4-dimethylaminopyridine/camphorsulfonic acid) afforded ester as a mixture of two diastereomers. After deprotection of the pyrrolidine, coupling with tripeptide gave *O*-silylated diazirinyl-substituted malevamide D (68%). The desilylation (tetra-*n*-butylammonium fluoride, tetrahydrofuran/water 19:1) afforded a product mixture, the separation of which needed semipreparative high-performance liquid chromatography (RP-18, methanol/water 9:1). The diazirinyl-substituted malevamide was obtained in 33% isolated yield as a mixture of two epimers differing at the carbinol center. In the nuclear magnetic resonance spectra, four sets of signals were recorded due to the additional *cis*/*trans*-isomerism of acyl pyrrolidine functionality [59].

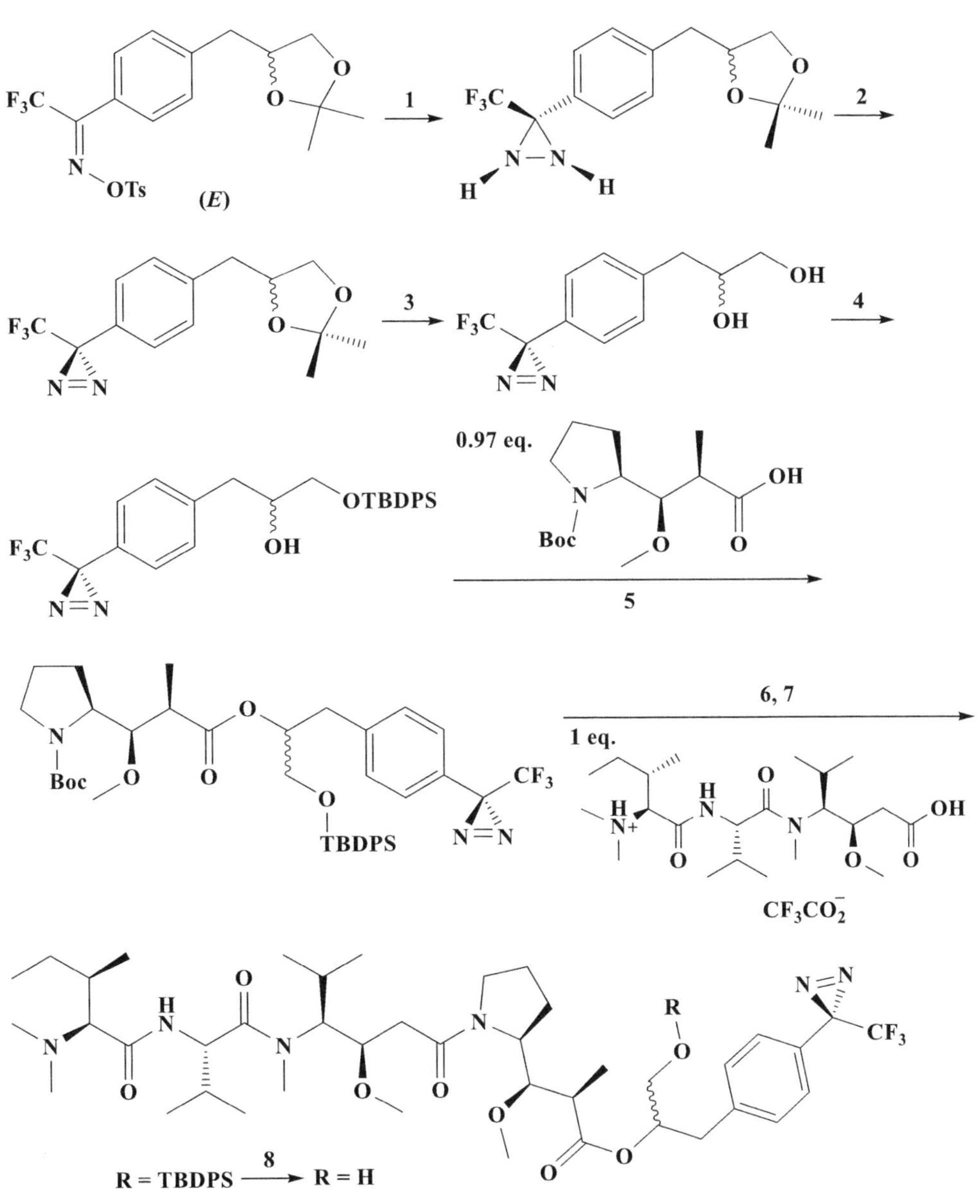

Reagents and conditions: (1) NH_3 (l), *t*-BuOMe, -40 °C, 2 h, rt, 16 h, quantitative, (2) 1.2 eq. I_2, 2.2 eq. Et_3N, Et_2O, 0 °C, 15 min, rt, 1 h, 99%, (3) THF, 2 N HCl, rt, 3 h, 97%, (4) 1.1 eq. TBDPSCl, 2.2 eq. imidazole, DMF, 0 °C, 1 h, 80%, (5) 0.41 eq. DMAP, 1 eq. DCC, 0.23 eq. CSA, 0 °C to rt, 16 h, 69%, (6) TFA, DCM, 0 °C, 90 min, (7) 6 eq. Et_3N, 2.14 eq. DEPC, DCM, 0 °C, 150 min, 68%, (8) 5 eq. TBAF·$3H_2O$, THF/H_2O (19:1), 0 °C, 2 h, 33%.

SCHEME 27

References

1. (a) T.L. Gilchrist. 2001. Activated 2*H*-azirines as dienophiles and electrophiles. Aldrichimica Acta 34: 51-55. (b) N. Kaur. 2018. Green synthesis of three to five-membered *O*-heterocycles using ionic liquids. Synth. Commun. 48: 1588-1613. (c) N. Kaur. 2019. Ionic liquid assisted synthesis of *S*-heterocycles. Phosphorus, Sulfur, Silicon Relat. Elem. 194: 165-185.
2. N. Kaur. 2017. Ionic liquids: promising but challenging solvents for the synthesis of *N*-heterocycles. Mini Rev. Org. Chem. 14: 3-23.
3. N. Kaur. 2019. Synthesis of three-membered and four-membered heterocycles with the assistance of photochemical reactions. J. Heterocycl. Chem. 56: 1141-1167.
4. S.R. Paulsen. 1960. 3,3-Dialkyldiazacyclopropen. Angew. Chem. 72: 781-782.
5. E. Schmitz. 1964. Three-membered rings containing two heteroatoms. Angew. Chem. Int. Ed. 3: 333-341.
6. A. Skancke and J.F. Liebman. 1999. The energetics of cyclopropene, 1,4-cyclohexadiene, and some of their hetero- and/or exocyclic derivatives. J. Org. Chem. 64: 6361-6365.
7. C. Puzzarini, A. Gambi and G. Cazzoli. 2004. An ab initio study of diazirine: equilibrium structure and molecular properties. J. Mol. Struct. 695-696: 203-210.
8. G. Chang and M.P. Mertes. 1987. Linear free energy relationship studies of 5-substituted 2,4-dioxopyrimidine nucleosides. J. Org. Chem. 52: 3625-3631.
9. A.N. Topin, O.M. Gritsenko, M.G. Brevnov, E.S. Gromova and G.A. Korshunova. 1998. Synthesis of a new photo-cross-linking nucleoside analogue containing an aryl(trifluoromethyl)diazirine group: application for EcoRII and MvaI restriction-modification enzymes. Nucleos. Nucleot. Nucl. Acids 17: 1163-1175.
10. E.F. Godefroi and E.L. Wittle. 1956. The preparation of some derivatives of β-(10-phenothiazinyl)propionic acid and β-(2-chloro-10-phenothiazinyl)propionic acid. J. Org. Chem. 21: 1163-1168.
11. R.F.R. Church, R.R. Maleike and M.J. Weiss. 1972. Diazirines. 3. Synthesis of a series of diazirine-containing molecules and their pharmacological evaluation. J. Med. Chem. 15: 514-518.
12. A. Bogdanova and V.V. Popik. 2003. Wavelength-dependent photochemistry of diazo Meldrum's acid and its spirocyclic isomer, diazirino Meldrum's acid: Wolff rearrangement versus isomerization. J. Am. Chem. Soc. 125: 1456-1457.
13. A.J. DeGraw, Z. Zhao, C.L. Strickland, A.H. Taban, J. Hsieh, M. Jefferies, W. Xie, D.K. Shintani, C.M. McMahan, K. Cornish and M.D. Distefano. 2007. A photoactive isoprenoid diphosphate analogue containing a stable phosphonate linkage: synthesis and biochemical studies with prenyltransferases. J. Org. Chem. 72: 4587-4595.
14. M.R. Bond, H. Zhang, P.D. Vu and J.J. Kohler. 2009. Photo crosslinking of glycoconjugates using metabolically incorporated diazirine-containing sugars. Nat. Protoc. 4: 1044-1063.
15. J.S. Vervacke, A.L. Funk, Y.-C. Wang, M. Strom, C.A. Hrycyna and M.D. Distefano. 2014. Diazirine-containing photoactivatable isoprenoid: synthesis and application in studies with isoprenylcysteine carboxyl methyltransferase. J. Org. Chem. 79: 1971-1978.
16. W.H. Graham. 1965. The halogenation of amidines. I. Synthesis of 3-halo- and other negatively substituted diazirines. J. Am. Chem. Soc. 87: 4396-4397.
17. A.A. Aly and A.M. Nour-El-Din. 2008. Functionality of amidines and amidrazones. ARKIVOC (i): 153-194.
18. M.T.H. Liu. 1982. The thermolysis and photolysis of diazirines. Chem. Soc. Rev. 11: 127-140.
19. I.R. Likhotvorik, E.L. Tae, C. Ventre and M.S. Platz. 2000. A facile one-pot conversion of non-enolizable aldehydes to diazirines. Tetrahedron Lett. 41: 795-796.
20. D.M. Gale, W.J. Middleton and C.G. Krespan. 1966. Perfluorodiazo compounds. J. Am. Chem. Soc. 88: 3617-3623.

21. R.B. Minasyan, E.M. Rokhlin, N.P. Gambaryan, Y.V. Zeifman and I.L. Knunyants. 1965. Bis(trifluoromethyl)cyclodiazomethane. Bull. Acad. Sci. USSR 14: 746-746.
22. R.A. Mitsch. 1966. Synthesis of difluorodiazirine. J. Heterocycl. Chem. 3: 245-245.
23. R.A. Mitsch. 1968. Organic fluoronitrogens. X. Reductive defluorination-cyclization. J. Org. Chem. 33: 1847-1849.
24. M.D. Meyers and S. Frank. 1966. Difluorocyanamide. Inorg. Chem. 5: 1455-1457.
25. J.N. Hernández, M.A. Ramírez and V.S. Martín. 2003. A new selective cleavage of *N,N*-dicarbamoyl-protected amines using lithium bromide. J. Org. Chem. 68: 743-746.
26. Y. Ambroise, F. Pillon, C. Mioskowski, A. Valleix and B. Rousseau. 2001. Synthesis and tritium labeling of new aromatic diazirine building blocks for photoaffinity labeling and cross-linking. Eur. J. Org. Chem. 20: 3961-3964.
27. E. Lim, J. Ricci and M. Jung. 2011. Synthesis of a dual-labeled probe of dimethyl lithospermate B with photochemical and fluorescent properties. Molecules 16: 9886-9899.
28. B.J. Jensen and J. Hergenrother. 1993. Poly(arylene ether)s with pendant ethynyl groups. Pure Appl. Chem. A30 (6-7): 449-458.
29. J. Brunner. 1993. New photolabeling and crosslinking methods. Annu. Rev. Biochem. 62: 483-514.
30. Y. Tanaka, M.R. Bond and J.J. Kohler. 2008. Photo crosslinkers illuminate interactions in living cells. Mol. Biosys. 4: 473-480.
31. G.K.S. Prakash, R. Krishnamurti and G.A. Olah. 1989. Synthetic methods and reactions. 141. Fluoride-induced trifluoromethylation of carbonyl compounds with trifluoromethyltrimethylsilane (TMS-CF_3). A trifluoromethide equivalent. J. Am. Chem. Soc. 111: 393-395.
32. M. Nassal. 1983. 4-(1-Azi-2,2,2-trifluoroethyl)benzoic acid, a highly photolabile carbene generating label readily fixable to biochemical agents. Liebigs Ann. Chem. 9: 1510-1523.
33. K. Stromgaard, D.R. Saito, H. Shindou, S. Ishii, T. Shimizu and K. Nakanishi. 2002. Ginkgolide derivatives for photolabeling studies: preparation and pharmacological evaluation. J. Med. Chem. 45: 4038-4046.
34. C. Sridar, Y. Kobayashi, H. Brevig, U.M. Kent, S.G. Puppali, J.M. Rimoldi and P.F. Hollenberg. 2006. Synthesis of substituted phenyl diaziridines and characterization as mechanism-based inactivators of human cytochrome P450 2B6. Drug Metab. Dispos. 34: 1849-1855.
35. A. Ruhmann and C. Wentrup. 1994. Synthesis of a photoactivatable 9-*Z*-oleic acid for protein kinase C labeling. Tetrahedron 50: 3785-3796.
36. H. Bayley and J.R. Knowles. 1977. Photoaffinity labeling. Meth. Enzym. 46: 91-114.
37. S. Ruchirawat and E. Davis. 1984. A convenient synthesis of simple tetrahydroisoquinolines. Synth. Commun. 14: 1221-1228.
38. H. Nakashima, M. Hashimoto, Y. Sadakane, T. Tomohiro and Y. Hatanaka. 2006. Simple and versatile method for tagging phenyldiazirine photophores. J. Am. Chem. Soc. 128: 15092-15093.
39. (a) B.J. Leslie and P.J. Hergenrother. 2008. Identification of the cellular targets of bioactive small organic molecules using affinity reagents. Chem. Soc. Rev. 37: 1347-1360. (b) R.A.G. Smith and J.R. Knowles. 1975. The preparation and photolysis of 3-aryl-3*H*-diazirines. J. Chem. Soc. Perkin Trans. 2 7: 686-694.
40. N. Kanoh, K. Honda, S. Simizu, M. Muroi and H. Osada. 2005. Photo-cross-linked small-molecule affinity matrix for facilitating forward and reverse chemical genetics. Angew. Chem. Int. Ed. 44: 3559-3562.
41. N. Kanoh, K. Honda, S. Simizu, M. Muroi and H. Osada. 2005. Photo-cross-linked small-molecule affinity matrix for facilitating forward and reverse chemical genetics. Angew. Chem. Int. Ed. 44: 4282-4282.
42. H. Bayley and J.R. Knowles. 1978. Photogenerated reagents for membrane labeling. 2. Phenylcarbene and adamantylidene formed within the lipid bilayer. Biochemistry 17: 2420-2423.
43. N. Kanoh, S. Kumashiro, S. Simizu, Y. Kondoh, S. Hatakeyama, H. Tashiro and H. Osada. 2003. Immobilization of natural products on glass slides by using a photoaffinity reaction and the detection of protein-small-molecule interactions. Angew. Chem. Int. Ed. 42: 5584-5587.

44. (a) M. Hashimoto and Y. Hatanaka. 2005. Post-biotinylation of photo crosslinking by Staudinger-Bertozzi ligation of preinstalled alkylazide tag. Chem. Pharm. Bull. 53: 1510-1512. (b) M. Hashimoto and Y. Hatanaka. 2008. Recent progress in diazirine-based photoaffinity labeling. Eur. J. Org. Chem. 15: 2513-2523. (c) T. Mayer and M.E. Maier. 2007. Design and synthesis of a tag-free chemical probe for photoaffinity labeling. Eur. J. Org. Chem. 28: 4711-4720.
45. W.-W. Qiu, J. Xu, J.-Y. Li, J. Li and F.-J. Nan. 2007. Activity-based protein profiling for Type I methionine aminopeptidase by using photo-affinity trimodular probes. ChemBioChem 8: 1351-1358.
46. A.B. Smith, K. Sugasawa, O. Atasoylu, C.-P.H. Yang and S.B. Horwitz. 2011. Design and synthesis of (+)-discodermolide-paclitaxel hybrids leading to enhanced biological activity. J. Med. Chem. 54: 6319-6327.
47. T. Bender, M. Huss, H. Wieczorek, S. Grond and P. von Zezschwitz. 2007. Convenient synthesis of a [1-^{14}C] diazirinylbenzoic acid as a photoaffinity label for binding studies of V-ATPase inhibitors. Eur. J. Org. Chem. 23: 3870-3878.
48. D. Pillion, M. Deraet, B.J. Holleran and E. Escher. 2006. Stereospecific synthesis of a carbene-generating angiotensin II analogue for comparative photoaffinity labeling: improved incorporation and absence of methionine selectivity. J. Med. Chem. 49: 2200-2209.
49. (a) A.M. Fernandez-Gacio. 2002. Studies on the introduction of a photoreactive aryldiazirine group into the vitamin D skeleton. Eur. J. Org. Chem. 15: 2529-2534. (b) Y. Hatanaka, M. Hashimoto, H. Kurihara, H. Nakayama and Y. Kanaoka. 1994. A novel family of aromatic diazirines for photoaffinity labeling. J. Org. Chem. 59: 383-387.
50. J.E. Baldwin, A.J. Pratt and M.G. Moloney. 1987. The synthesis of aryl substituted analogues of phenoxyacetyl-L-cysteinyl-D-valine and phenylacetyl-L-cysteinyl-D-valine. Tetrahedron Lett. 43: 2565-2575.
51. J.M. Delfino, S.L. Schreiber and F.M. Richards. 1993. Design, synthesis, and properties of a photoactivatable membrane-spanning phospholipidic probe. J. Am. Chem. Soc. 115: 3458-3474.
52. L. Odell, N. Chau, A. Mariana, M. Graham, P. Robinson and A. McCluskey. 2009. Azido and diazarinyl analogues of bis-tyrphostin as asymmetrical inhibitors of dynamin GTPase. ChemMedChem 4: 1182-1188.
53. M. Weber. 1999. Glycosyliden-carbene anwendungen neuer und bekannter glycosyliden-diazirine in der synthese und zur modifikation von titandioxid-oberflächen. PhD Thesis, ETH Zürich.
54. (a) D. Beer and A. Vasella. 1985. Aldonhydroximo-lactones. Preparation and determination of configuration. Helv. Chim. Acta 68: 2254-2274. (b) K. Briner and A. Vasella. 1989. Glycosylidene carbenes a new approach to glycoside synthesis. Part 1. Preparation of glycosylidene-derived diaziridines and diazirines. Helv. Chim. Acta 72: 1371-1382.
55. M.E. Tate and C.T. Bishop. 1963. Preparation, thin layer chromatography, and ultraviolet spectra of some *o*-benzyl derivatives of D-glucose. Can. J. Chem. 41: 1801-1806.
56. S. Koto, N. Morishima, Y. Miyata and S. Zen. 1976. Preparation of 2,3,4,6-tetra-*O*-benzyl-D-mannose. Bull. Chem. Soc. Jpn. 49: 2639-2640.
57. B.M. Aebischer, H.W. Hanssen, A.T. Vasella and W.B. Schweizer. 1982. Oxidation of aldose oximes. Formation and structure of hydroxydiazene oxide acetals and preparation of hydroximolactones. X-Ray crystal structure of 2,3:5,6-di-*O*-isopropylidene-α-D-mannofuranosyl-ONN-azoxy 2,3:5,6-di-*O*-isopropylidene-α-D-mannofuranoside. J. Chem. Soc. Perkin Trans. 1 0: 2139-2147.
58. M. Yamada, C.I. Someya, T. Nakahodo, Y. Maeda, T. Tsuchiya and T. Akasaka. 2011. Synthesis of endohedral metallofullerene glycoconjugates by carbene addition. Molecules 16: 9495-9504.
59. W. Telle, G. Kelter, H.-H. Fiebig, P.G. Jones and T. Lindel. 2014. Total synthesis and cytotoxicity of the marine natural product malevamide D and a photoreactive analog. Beilstein J. Org. Chem. 10: 316-322.

CHAPTER

5 Synthesis of diazirines from hydroxylamine

Chapter outline

5.1 Introduction

Small heterocyclic rings constitute systems of central importance in medicinal, theoretical, synthetic organic, and bioorganic chemistry, and specifically aziridines and azirines occur in various natural and pharmacologically active substances, and also, they are important building blocks and versatile synthetic intermediates; thus, they are very important and interesting systems. The chemistry of three-membered heterocyclic ring has played an important role in the history of organic synthesis. Uncountable classic and modern procedures exist for their synthesis from simple starting compounds. The intrinsic ring strain of these compounds primes them for facile ring opening. Therefore, much attention has been paid to the discovery of reactions that employ the reactivity of three-membered heterocyclic rings as entry points for the formation of complex heteroatom containing products [1,2].

The azirine is the smallest *N*-unsaturated heterocyclic system possessing two carbon atoms with one double bond in a three-membered ring. An interest in these heterocycles stems from the usual effect of ring strain on chemical reactivity and to the potential of their derivatives to serve as starting compounds to more elaborate heterocycles. The 2*H*-azirines are versatile and reactive substrates due to certain inherent features in their structure including an electron-rich π bond, ring strain, and a nitrogen lone pair. The lone pair of electrons on the nitrogen atom can interact with electrophiles; the π bond, associated with the inherent ring strain, makes 2*H*-azirines good dienophiles and dipolarophiles; and the trigonal carbon atom is an electrophilic center. Besides, the ring can be cleaved by photochemical and thermal excitation [3–6].

The chemistry of highly reactive 2*H*-azirines has been discovered widely for several synthetic objectives. These heterocyclic compounds undergo reactions in which they can act either as an electrophile or as a nucleophile. The reactions with nucleophiles always involve an initial addition to the imine bond [7].

3-Membered Heterocycle Synthesis. https://doi.org/10.1016/B978-0-443-13583-5.00010-4

5.2 Synthesis of diazirines from hydroxylamine

The formation of 3-phenyl-3-(trifluoromethyl)diazirine three-membered ring needed at least five steps starting from aryl halide derivatives. The tether should be incorporated onto the benzene ring to connect to the ligand skeleton. Earlier syntheses of 3-phenyl-3-(trifluoromethyl)diazirine derivatives had pre-installed appropriate tethers before synthesis of the diazirine functionality, but it appeared that the three-membered diazirinyl ring was not stable toward various organic reactions. The repeated synthesis of a diazirine functionality for each derivative was a disadvantage for application of the photophore for photoaffinity labeling. The "postfunctional" adaptation of diazirinyl compounds [8,9] exposed that the CF_3-substituted three-membered ring was stable under various organic reaction conditions. Although the 3-(trifluoromethyl)diazirinyl functionality was considered as an alkyl substituent, polarization means that the quaternary carbon atom was slightly positively charged; therefore, the functionality was less activated toward an electrophilic aromatic substitution than its unsubstituted counterpart [10]. Earlier *m*-methoxy-substituted 3-phenyl-3-(trifluoromethyl)diazirine was selected as a mother skeleton, because (1) the OCH_3 group strongly activated for an electrophilic aromatic substitution, (2) the orientation of substitution favored the *ortho*-position against the OCH_3 group, because the *para*-position was sterically hindered by 3-(trifluoromethyl)diazirinyl functionality, and (3) demethylation of *m*-methoxy-3-phenyl-3-(trifluoromethyl)diazirine was easier than for *p*-methoxy-3-phenyl-3-(trifluoromethyl)diazirine, and realkylation of demethoxylated group was used for the introduction of a tag (Scheme 1) [11].

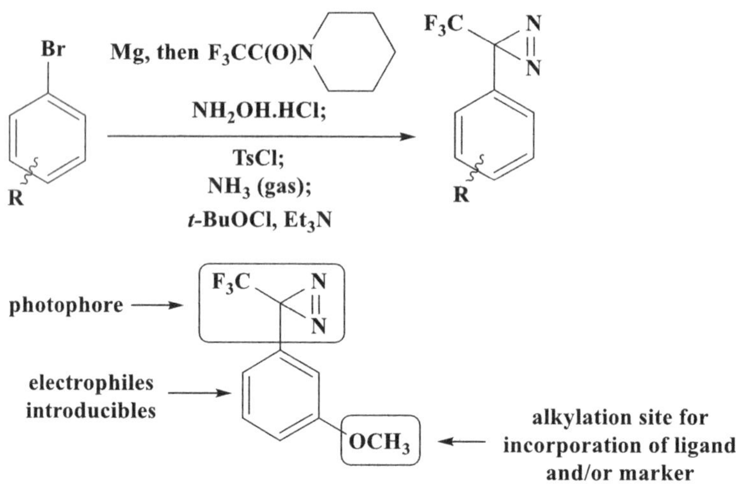

SCHEME 1

The molecules bearing azirine functional group are well-known precursors for the synthesis of carbene. Particularly, the 3-(trifluoromethyl)-3-phenylazirine is quite stable at rt, may be stored for prolonged periods of time (away from light), and as a result has seen a great deal of use in biological analysis as a photoaffinity-labeling agent [12–14]. By way of demonstration of the universal use of molecules possessing 3-(trifluoromethyl)-3-phenylazirine functionality, a substructure search was conducted utilizing SciFinder software located 828 molecules containing this substructure [15,16]. Given the popularity of 3-(trifluoromethyl)-3-phenylazirine in biology, it is no surprise that the formation of molecule (Scheme 2) must be reasonably direct and that some commercial sources of further functionalized molecules are also commercially accessible [17]. While highly reactive carbenes such as those from diazo methane will add across C—C double bonds or insert into C—H bonds, the 3-(trifluoromethyl)-3-phenylazirine carbene was known to insert best into nitrogen—hydrogen or oxygen—hydrogen bonds. Because of the resemblance between the amide moiety found in polypeptides and that of the urethane functional group usually utilized in ultraviolet cure urethane acrylate coatings, the possible use of this molecule as a cross-linking agent appeared apparent.

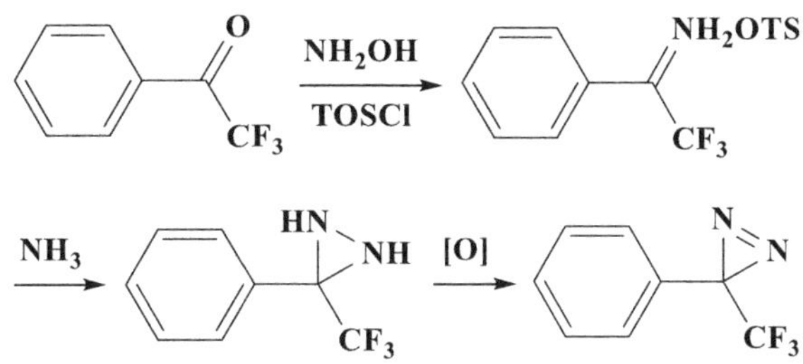

SCHEME 2

The formation of first photoprobe began with the loading of amino methyl polystyrene resin with the Rink Amide linker under the MWI for 20 min at 60°C followed by Fmoc deprotection with 20% piperidine in dimethylformamide. The Fmoc-Lys(Dde)-OH was further coupled to the resin under same reaction conditions, and deprotection was performed with $NH_2OH.HCl$ and imidazole. The 3-(trifluoromethyl)-3-phenyldiazirine was coupled to free amino group present on the side chain (Scheme 3) [18].

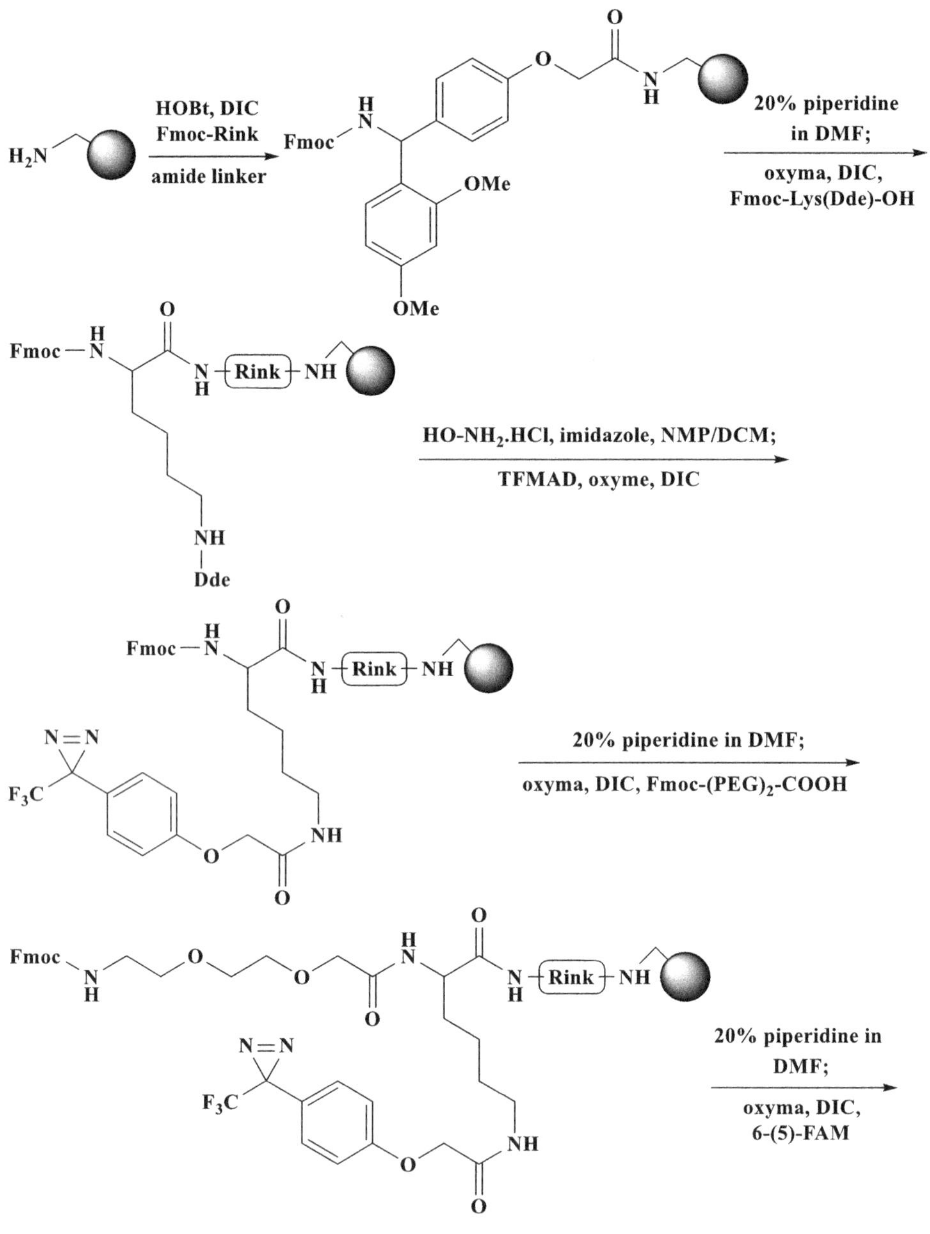
H2N
HOBt, DIC
Fmoc-Rink
amide linker
Fmoc
OMe
OMe
20% piperidine
in DMF;
oxyma, DIC,
Fmoc-Lys(Dde)-OH
Fmoc
Rink
NH
Dde
HO-NH2.HCl, imidazole, NMP/DCM;
TFMAD, oxyme, DIC
N=N
F3C
20% piperidine in DMF;
oxyma, DIC, Fmoc-(PEG)2-COOH
20% piperidine in
DMF;
oxyma, DIC,
6-(5)-FAM

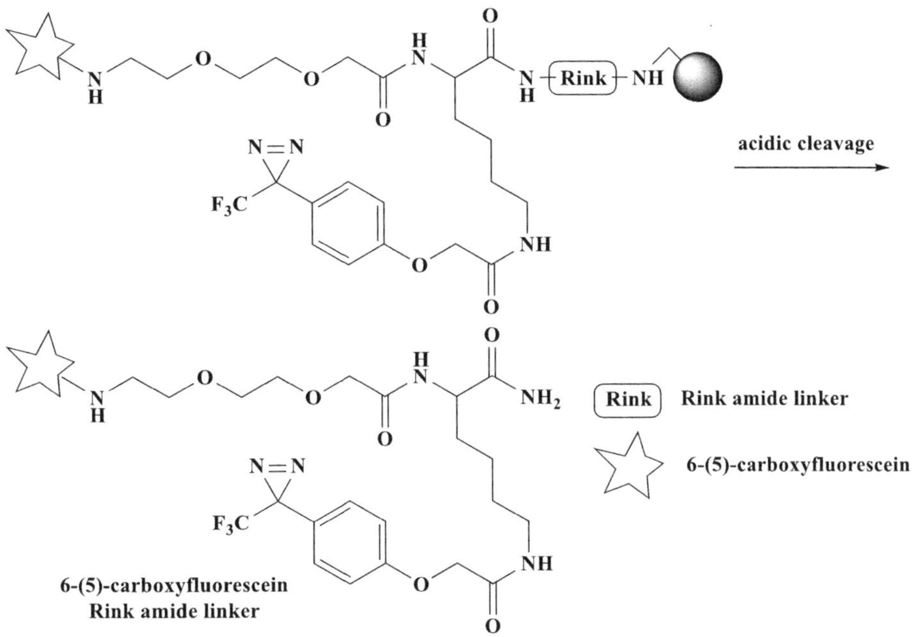

SCHEME 3

The amino-functionalized polystyrene microspheres were derivatized with Fmoc-8-amino-3,6-dioxaoctanoic acid spacer unit. The step was repeated under the same conditions to provide the microspheres to have a longer spacer. The microspheres were derivatized first with adipic acid and then cystamine to introduce the disulfide bond into the structure (Scheme 4). Finally, the deprotection of Fmoc group was performed if essential, and the 3-(trifluoromethyl)-3-phenyldiazirine photophore was added [18].

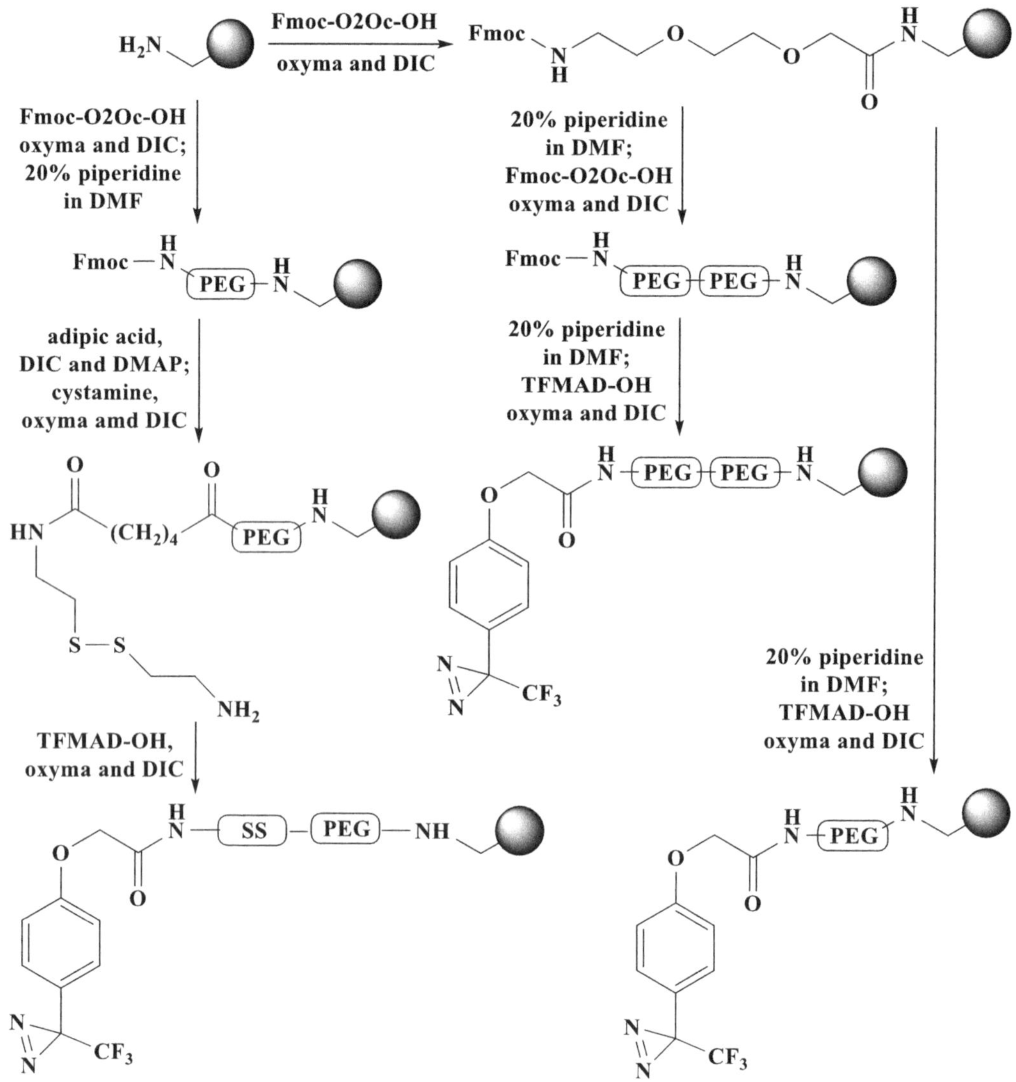

SCHEME 4

The Fmoc-Lys(Dde)-OH amino acid was utilized to add branches for derivatization during the formation of positively charged microspheres. Initially, the free amino group of lysine was coupled to 3-(trifluoromethyl)-3-phenyldiazirine, and then, the amino group was utilized to introduce the Fmoc-Arg(Boc)$_2$-OH (Scheme 5). The coupling and deprotection steps were checked utilizing colorimetric ninhydrin test and zeta potential values to determine if they were successful. The ninhydrin test afforded information about the presence of primary amino groups on the microspheres, whereas the zeta potential value was an important particle characteristic, which monitored various functional groups present on the surface [19].

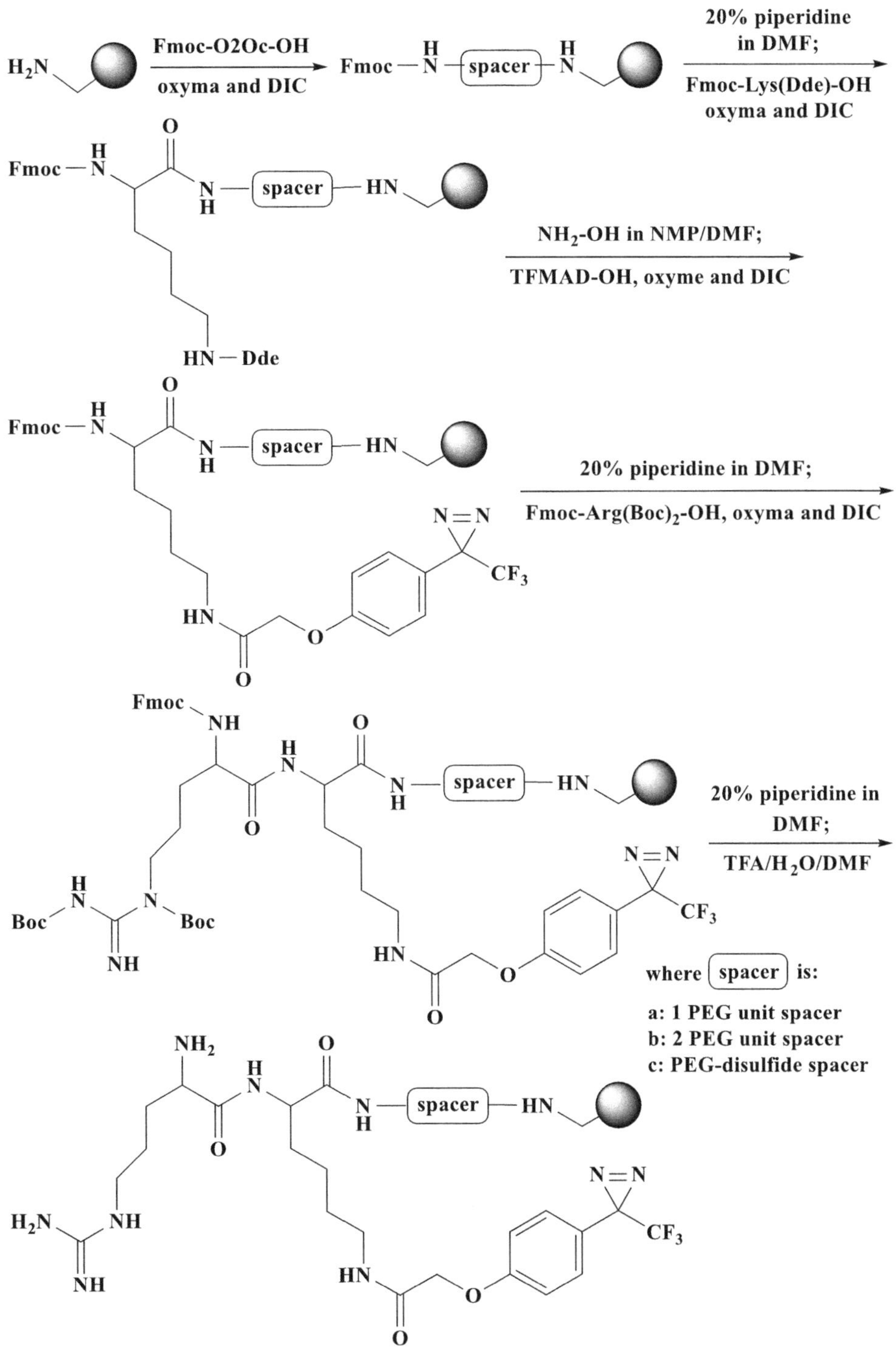

Fmoc-O2Oc-OH
oxyma and DIC
20% piperidine in DMF;
Fmoc-Lys(Dde)-OH oxyma and DIC
spacer
HN—Dde
NH2-OH in NMP/DMF;
TFMAD-OH, oxyme and DIC
CF3
20% piperidine in DMF;
Fmoc-Arg(Boc)2-OH, oxyma and DIC
Boc
20% piperidine in DMF;
TFA/H2O/DMF
where spacer is:
a: 1 PEG unit spacer
b: 2 PEG unit spacer
c: PEG-disulfide spacer

SCHEME 5

5.3 Synthesis of diazirines from hydroxylamine-*O*-sulfonic acid

The diazirine was prepared from 3-ketobutanol under the Church and Weiss's conditions [20] (Scheme 6). The alcohol was then dissolved in methanol and cooled to −78°C. The liquid ammonia was condensed into the flask, and this was further allowed to stir at −78°C for 4 h. The hydroxylamine-*O*-sulfonic acid was added, and the mixture was warmed to rt for 1 h with a dry ice condenser. The obtained diaziridine was redissolved in methanol, and then TEA and solid I_2 were added to oxidize the diaziridine to diazirine. The diazirine was formed in 67% yield after purification.

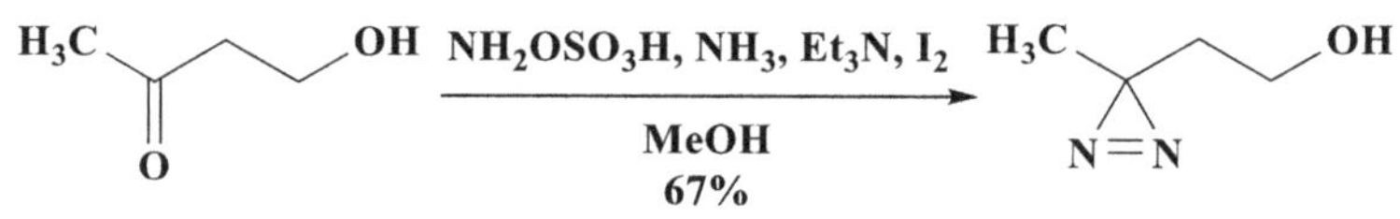

SCHEME 6

This pathway utilized commercially accessible Schöllkopf bis-lactam ether ((2*R*)-(−)-2,5-dihydro-3,6-dimethoxy-2-*i*-propylpyrazine) as a chiral auxiliary to insert the amino acid moiety, with the essential chirality, at the desired position on the compound. This reaction focused on the formation of photomethionine because of the commercial accessibility of the starting alcohol, 3-ketobutanol. The process of Church and Weiss [20] was utilized for the transformation of alcohol into diazirine. The alcohol was mesylated utilizing mesyl chloride and TEA. The mesylate will be transformed to iodide utilizing NaI to make it more labile during the reaction with chiral auxiliary. Following an alkylation of the auxiliary, hydrolysis with 6 M hydrochloric acid provided desired photoacid. This reaction could then be operated for the formation of other two photoacids, photoisoleucine and photoleucine, simply by changing the starting alcohol (Scheme 7).

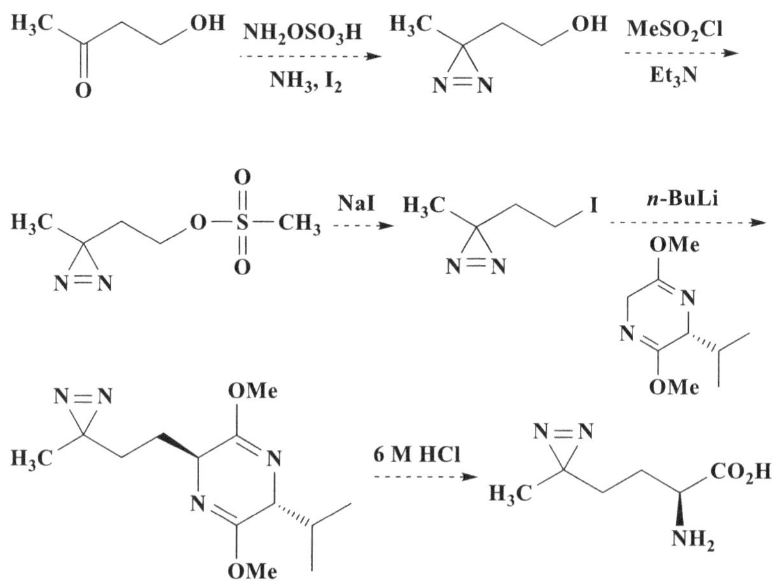

SCHEME 7

All reagents were prepared from commercial sources and utilized as received unless otherwise noted. The inert nitrogen atmosphere was used for all reactions, and all solvents were distilled. The forced-flow chromatography on EM Science Geduran silica gel 60 (35–75 μm) was utilized to purify the products. The TLC was carried out on EM Science silica gel 60 F254 plates (250 μm). The developed chromatogram was visualized by UV lamp or/and by staining with aqueous $KMnO_4$/potassium carbonate solutions. NMR spectra for ^1H were attained with deuterated chloroform on Bruker DRX-500/400 operating at 500/400 MHz; tetramethylsilane was utilized as internal standard for calibration purpose (Scheme 8) [21].

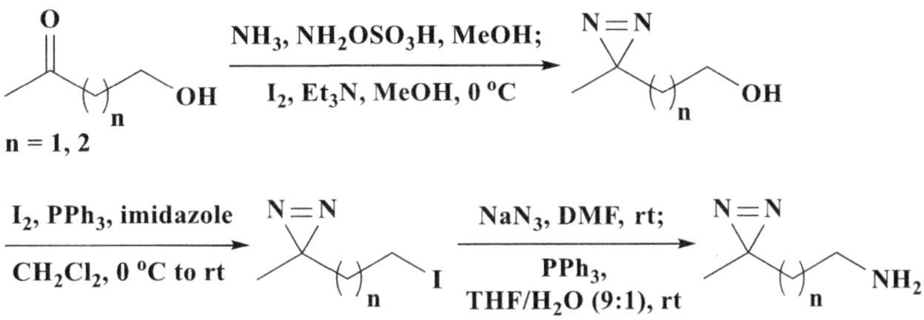

SCHEME 8

The diazirine-lysine was prepared from commercial 4-hydroxybutan-2-one and was transformed into diazirine in two steps (Scheme 9). Then, an activation of alcohol with DSC (disuccinimide carbonate) using TEA provided carbonate in 86% yield. The reaction of carbonate with Boc-lysine in dimethylformamide afforded diazirine in 88% yield. The acid-sensitive nature of diazirine group posed some synthetic challenges to cleave the *t*-butyloxycarbonyl-protecting group, which happened to be more problematic than expected. Various common procedures (for example, trifluoroacetic acid/dichloromethane (1:4 or 1:1); hydrochloric acid in MeOH (1 or 0.5 M); hydrochloric acid in dioxane/dichloromethane (1 M); all at 0°C or rt; with or without a dimethyl sulfide as a cation scavenger) were used, and the reaction progress was checked at different time intervals. All aforementioned conditions led to the decomposition of starting compounds. Further screening of deprotection conditions [22] resulted in a clean deprotection attained with a 1:20 mixture of trifluoroacetic acid in dichloroethane in the presence of Et_3SiH as a cation scavenger. Finally, the desired amino acid was formed in 76% yield.

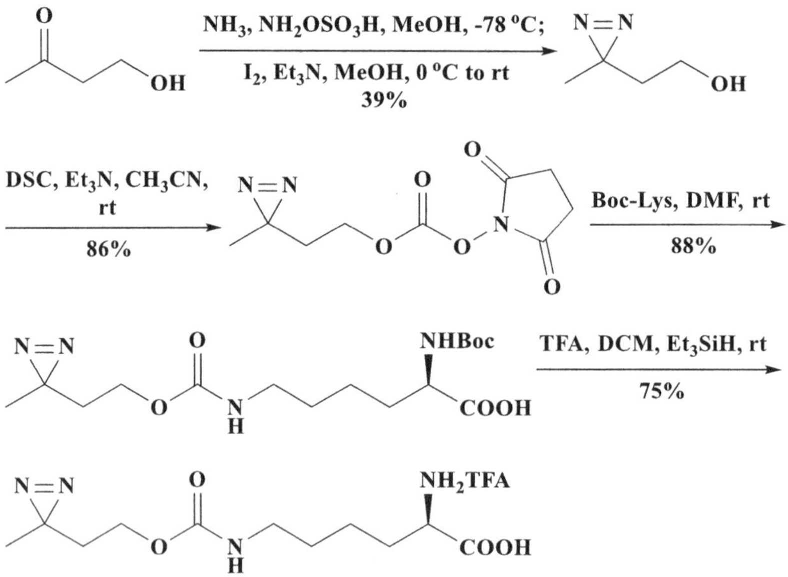

SCHEME 9

The diazirine group was selected as the most suitable photoactivatable agent. Thus, the 3-diazirine decanoic acid was synthesized according to a method reported by Husain and coworkers [23]. The β-keto decanoic acid was reacted with hydroxylamine-*O*-sulfonic acid in liquid NH_3 to synthesize the aziridine intermediate. An oxidation with I_2 provided diazirine in 30% yield (Scheme 10) [24].

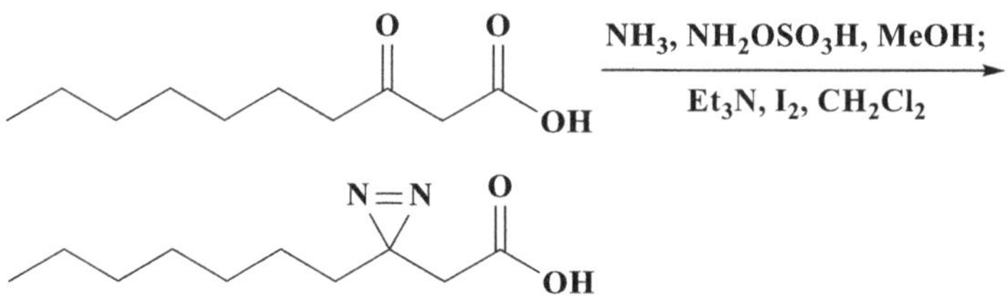

SCHEME 10

The formation of octanoic acid derivative with diazirine at C-7 was first attained by a one-pot, two-step sequence starting from 7-oxo-octanoic acid. The reaction of acid with 7 N NH_3 in MeOH followed by hydroxylamine-*O*-sulfonic acid synthesized diaziridine intermediate, which was oxidized using I_2 to synthesize the diazirine side chain. Because an irradiation at 350 nm will induce (*E*)- to (*Z*)-isomerization of 4,4′-diaminostilbene-2,2′-disulfonate in addition to the photolysis of diazirine, alkene was hydrogenated. This reaction afforded *i*-propyl-protected 4,4′-diaminostilbene-2,2′-disulfonic acid, which was utilized to synthesize the photoaffinity-labeled inhibitor. The octanoyl chloride (1 eq.) was added to *i*-propyl-protected 4,4′-diaminostilbene-2,2′-disulfonic acid forming a 1:2:1 statistical mixture of *i*-propyl-protected 4,4′-diaminostilbene-2,2′-disulfonic acid and the mono- and bis-*N*-

acylated products. The diazirine derivative was prepared from mono-*N*-acylated product, then *N*-acylation with acid chloride of diazirine side-chain, and deprotection with NaI. The isolated material was stable at 5 mm under the ambient conditions but decomposed when irradiated with 365 nm light (Scheme 11) [25].

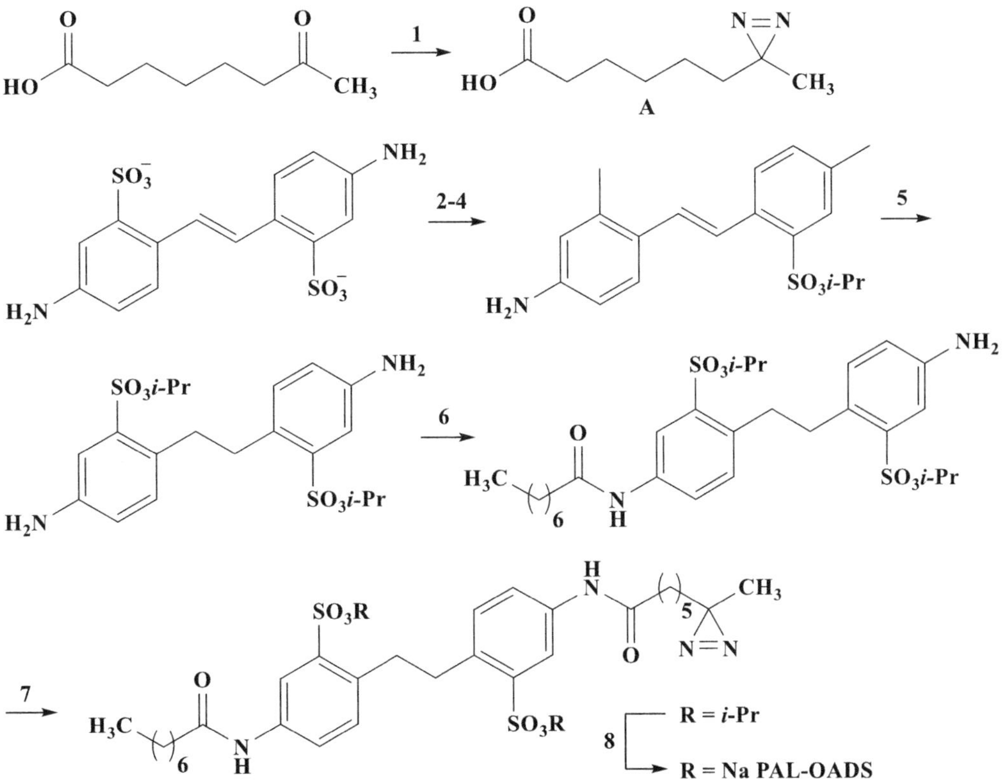

Reagents and conditions: (1) 7 N NH_3, MeOH; then NH_2OSO_2OH; then I_2, (2) FmocCl, Na_2CO_3, water, dioxane, (3) triisopropylorthoformate, dioxane, reflux, (4) pyrrolidine, THF, (5) H_2, Pd/C, THF, (6) 1 eq. octanoyl chloride, K_2CO_3, THF, (7) a) A, $(COCl)_2$, cat. DMF, CH_2Cl_2, b) THF, Et_3N, (8) NaI, acetone, water.

SCHEME 11

Haarp et al. [26] synthesized photoleucine by the α-bromination of azicarboxylic acid (Scheme 12). This was followed by an aminolysis of azibromocarboxylic acid. An acetylation followed by enzymatic kinetic resolution utilizing porcine kidney acylase I provided L-photoleucine.

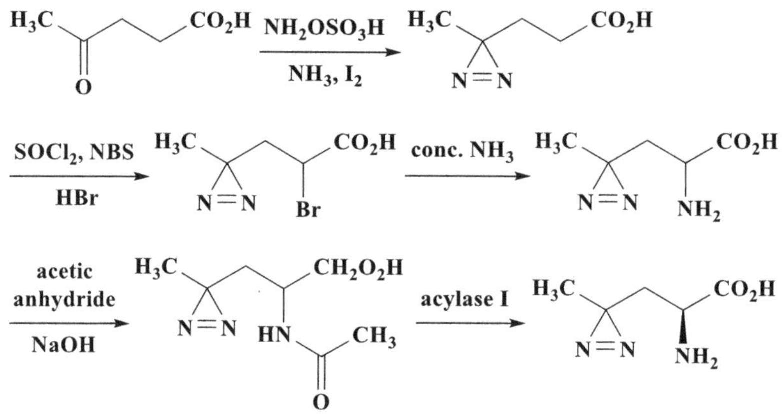

SCHEME 12

Work is concurrently being performed to synthesize the photoleucine. As this proved more successful and because of limited time, focus is shifted to the formation of photoleucine. In the case of photoleucine, the L-aspartic acid was protected as the diester, and the amino moiety was protected with two *t*-butyloxycarbonyl groups. As before, this afforded the necessary selectivity during the reduction of β-ester to aldehyde. This aldehyde was transformed to methyl ketone with TMS-CH_2N_2 and Me_3Al. Following this, one of the *t*-butyloxycarbonyl groups was eliminated with LiBr, and the ester was hydrolyzed to acid, to limit any interference at an aziridination step. An aziridination was then performed to synthesize the diazirine, which was deprotected to afford the photoleucine (Scheme 13) [27].

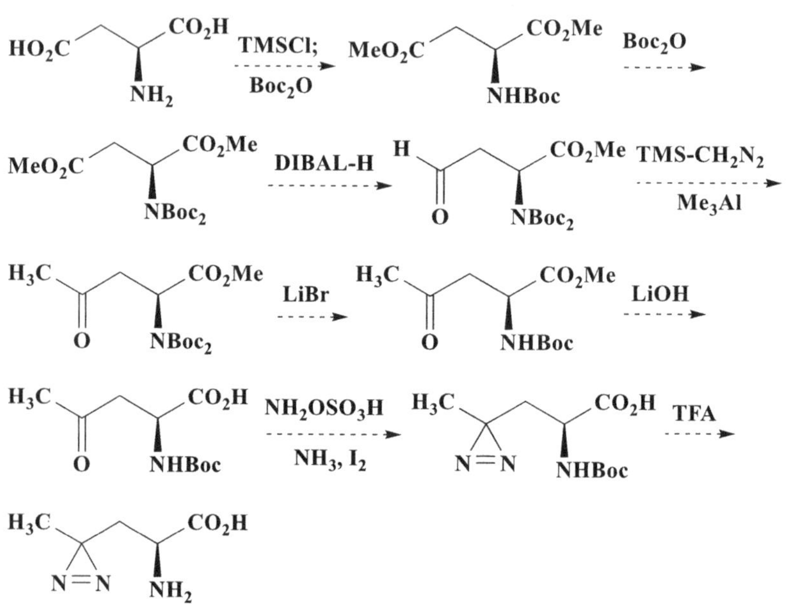

SCHEME 13

The commercially accessible substrate was smoothly substituted to create all three photoacids. The diazirine ring was incorporated late in the synthesis to minimize any difficulties that this group may cause. A new pathway was planned for photomethionine and is given in the following Scheme 14. The pathway was started with commercially accessible L-glutamic acid, which was protected as a diester dibutyloxycarbonyl analog. This was selectively reduced to aldehyde [27–30]. This aldehyde was transformed to methyl ketone utilizing TMS-CH_2N_2. Following this, one of the butyloxycarbonyl groups was selectively removed, and the ester was hydrolyzed to acid to limit any interference at an aziridination step. The formed acid was aziridinated to synthesize the diazirine, which was deprotected to provide the photomethionine.

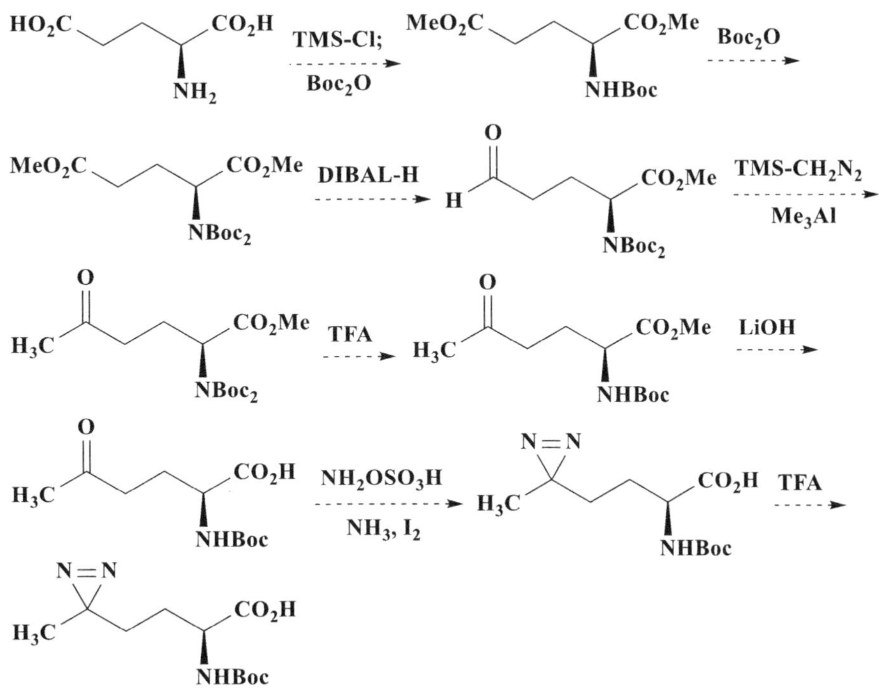

SCHEME 14

In the synthesis of a cyclic five-coordinate phosphorus intermediate, the stereochemistry in the product was dictated by the steric interactions between two aryl groups. The diazirines were prepared from nonenolizable aldehydes in a one-pot reaction (Scheme 15) [31].

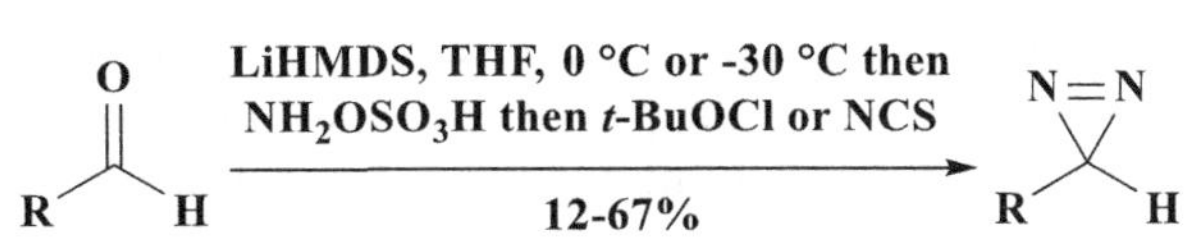

SCHEME 15

The diazirine group was introduced by an oxidation of diaziridine prepared from the ketone by reaction with NH_3 and hydroxylamine-*O*-sulfonic acid or chloramine (Scheme 16) [26]. Earlier, an oxidation was performed utilizing Ag_2O. However, certain functional groups such as amines could interfere with the reagent, and the reaction was slow because of the competing side reactions. It also creates Ag mirror with potential explosive character on the inside of the reaction vessel as an oxidation proceeded. Church and Weiss [20] reported that I_2 as an oxidant (in basic media) resulted in a cleaner, faster, and safer reaction.

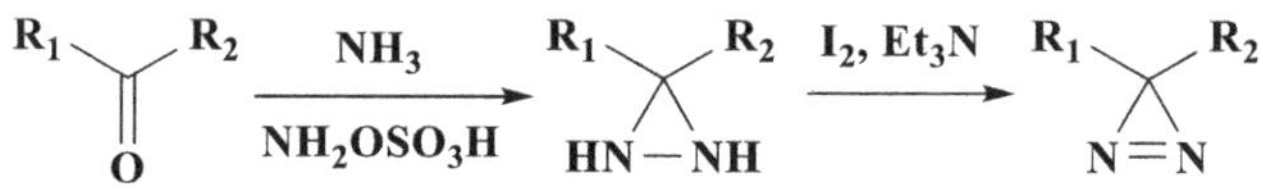

SCHEME 16

The cyclohexanone was prepared from octa-*O*-benzyl validoxylamine in two steps (benzylation followed by *N*-bromosuccinimide cleavage) in 50% yield (Scheme 17) [32]. The spirodiaziridine (35%) was synthesized by the saturation of a solution of cyclohexanone in methanol at −20°C with ammonia and then reaction with hydroxylamine-*O*-sulfonic acid [33–38]. The spirodiaziridine was oxidized to spirodiazirine (85%) with iodine using triethylamine [20]. The flash chromatography was performed for the purification of both spirodiaziridine and spirodiazirine. The diaziridine oxidized iodide to iodine in acidic medium, a reaction characteristic for diaziridines [39–43]. The vicinal coupling constants for the ring hydrogen of spirodiaziridine and spirodiazirine proved a $^{6}C3$ conformation.

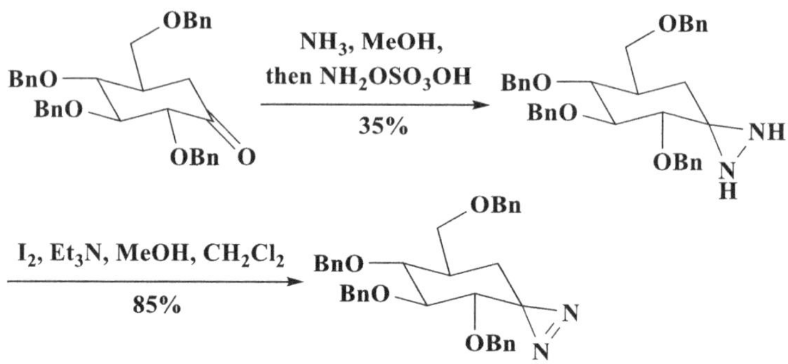

SCHEME 17

The trimethylsilylation [44] of validone provided a crude trimethylsilyl derivative (94%) that was isolated by chromatography (27%). By applying Kugelrohr distillation (130°C, 0.5 bar), an effort to purify trimethylsilyl derivative resulted in the partial decomposition. Good results in the subsequent step were provided by chromatographically purified samples of trimethylsilyl derivative only. The reaction of a solution of trimethylsilyl derivative in methanol, saturated at −20°C using ammonia, with hydroxylamine-*O*-sulfonic acid provided crude diaziridine (69%), which oxidized iodide in acidic solution. The flash chromatography provided a solution of diaziridine, purified by thin-layer chromatography, but partial decomposition occurred at even below 10°C by the complete evaporation of solvent (ethyl acetate, *i*-propanol, and water). The pure diaziridine was obtained in 50% yield by the partial evaporation of column eluate, application of the residual colorless solution of pure diaziridine to a strong acid ion exchanger, elution with 2% aqueous NH_3 solution, and lyophilization. The stability of diaziridine on the strong acid ion exchanger was in contrast to its sensitivity to aqueous hydrogen chloride, although thin-layer chromatography experiments showed that diaziridine was stable in aqueous buffers at pH 4.2, pH 5, and in pure H_2O, an addition of 1 M hydrochloric acid to a MeOH solution of diaziridine (to pH 1) resulted in rapid decomposition to a complex mixture. The pure diaziridine hydrolyzed partially to validone (according to thin layer chromatography), when it was subjected to Sephadex G-10 chromatography (eluent water). The diaziridine was also prepared by the reaction of validone with ammonia and hydroxylamine-*O*-sulfonic acid in methanol. Though in agreement with the results of Kurz and coworkers [38], the yield of diaziridine was lower (28% crude diaziridine) and the reaction was less clean, so that diaziridine cannot be obtained pure. A mixture of triethylamine and diazirine (1:2.5), obtained when diaziridine was oxidized with iodine in methanol using triethylamine (Scheme 18), could not be separated even by repeated flash chromatography. Replacing triethylamine with more volatile trimethylamine, which was successfully utilized in the formation of pyranosylidene diazirines, provided a mixture of diazirine and many unidentified side products. An acetylation of mixture of triethylamine and diazirine provided tetraacetate (84% from diaziridine). The pure diazirine (89%) was obtained from diazirine and triethylamine by chromatography on a strong acid ion exchanger, resulting in the decomposition to a complex mixture. This was in agreement with the known sensitivity of β-hydroxydiazirines and more general of diazirines having a (partial) positive charge at a carbon atom adjacent to the three-membered ring [45,46].

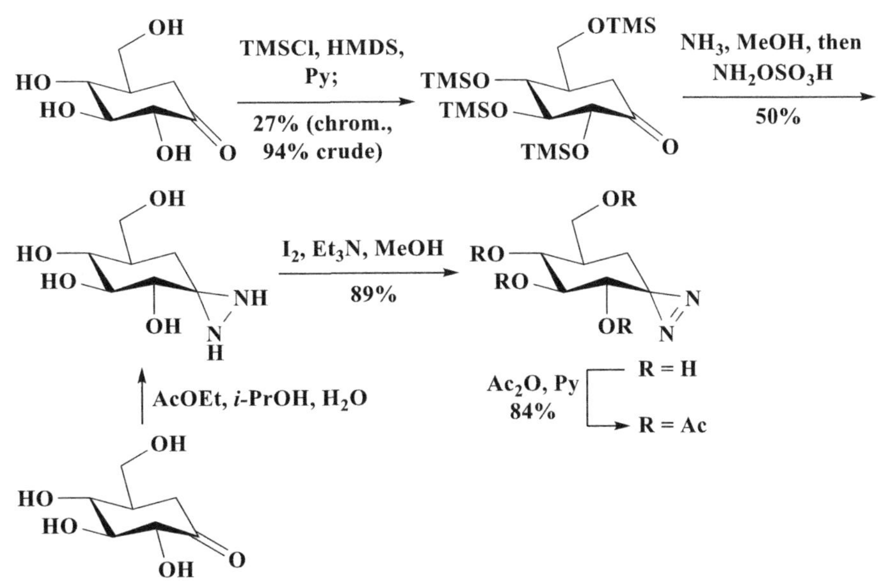

SCHEME 18

Lehmann et al. [47–54] prepared a series of pyranoside-derived diazirines (Scheme 19) with the azi substituent on C-6, on the aglycon of oxygen-, sulfur-, and carbon-glycosides, and in one case with the azi substituent spiro-linked to C-4. These diazirines were prepared from ketones, which were converted into diaziridines by Schmitz technique. The crude diaziridines were oxidized with iodine and triethylamine. The formed diazirines were utilized to photolabel glycosidases. The spirodiazirine was prepared from trimethylsilylated uloside through the spirodiaziridine in 13% yield [38]. The diaziridine was not purified. The trimethylsilyloxy groups were cleaved in parallel to the diaziridine formation, but they were critical for the synthesis of diaziridine. The reaction of free uloside with ammonia and hydroxylamine-*O*-sulfonic acid in methanol afforded a number of compounds, none of which could be oxidized to diazirine. It was assumed that the basic reaction conditions led to the decomposition of unprotected uloside, and the role of trimethylsilyloxy groups of trimethylsilylated uloside was to protect the uloside against decomposition.

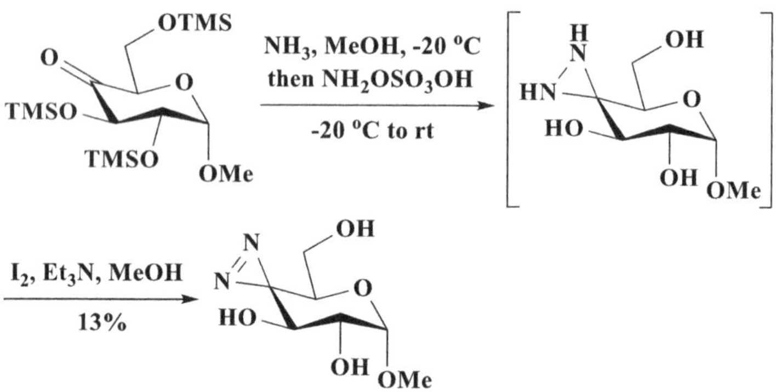

SCHEME 19

The 2-azicamphane, prepared from 2-hydrazicamphane intermediate, is of interest as a carbene precursor (Scheme 20). The diaziridine could not be prepared by standard Schmitz method. However, it was synthesized from iminium salt in a satisfactory yield (48%) [55]. The 3-alkyldiaziridines were synthesized by the reaction of *N*-trimethylsilylimines of nonenolizable aldehydes [31b], which were difficult to prepare from aldehydes themselves by standard process. However, the diaziridine precursors could be utilized as appropriate substrates for several one-pot processes. The camphor-derived iminium salt was transformed into sterically hindered chiral diaziridine 2-azicamphane by reaction with hydroxylamine-*O*-sulfonic acid in methanolic NH_3 followed by an oxidation with I_2 using Et_3N.

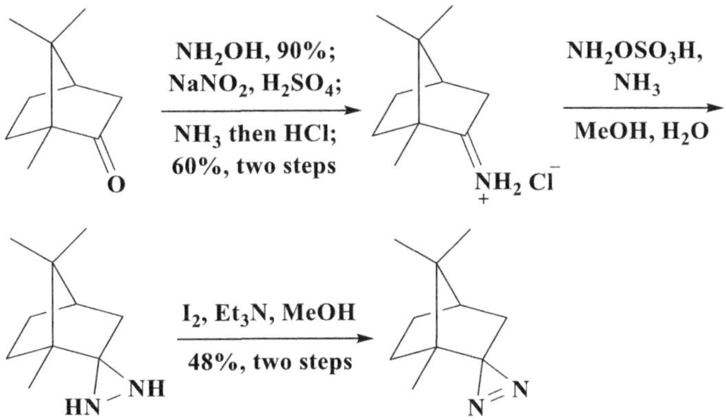

SCHEME 20

The L-proline derivative known as photoproline, possessing a diazirine group at position four of the pyrrolidine ring, was prepared from L-4-hydroxyproline. The use of Fmoc-L-photoproline in the formation of a cyclic peptidomimetic antibiotic showed that this photoprobe could be introduced into synthetic peptides by solid-phase Fmoc chemistry. The photoproline may be of high value in the formation of diverse peptide-based photoaffinity probes. The relatively inexpensive L-*trans*-4-hydroxyproline was utilized for the formation of Fmoc-protected L-photoproline (photoPro) (Scheme 21). The formation of *N*-Boc-4-oxoproline followed a simple two-step process: (1) butyloxycarbonyl protection and (2) Jones oxidation [56]. The technique utilized to install the diazirine group finds precedence in previously published pathways to photomethionine and photoleucine [57,58]. By utilizing standard solid-phase Fmoc peptide chemistry, the Fmoc derivative was needed for an incorporation of photoPro into synthetic peptides. The diazirine group was stable to basic and acidic conditions required in Fmoc solid-phase peptide preparation. It should be observed that the enantiomer (Fmoc-D-photoPro) could be obtained similarly, after an epimerization of L-*trans*-4-hydroxyproline to D-*cis*-4-hydroxyproline, which was convenient to perform on a mole scale following a published process [59,60].

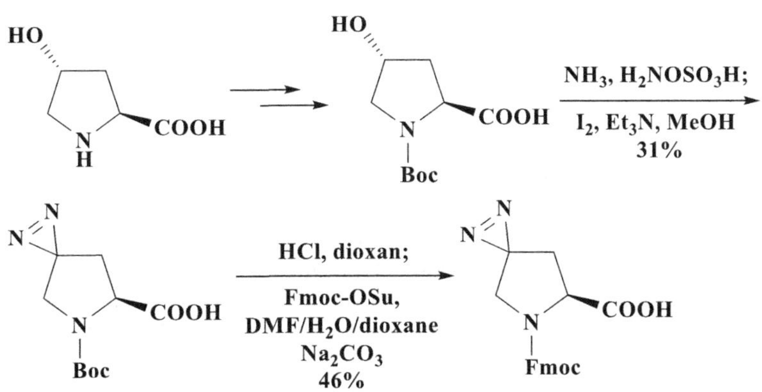

SCHEME 21

The utilization of ketones was beneficial as shown by the formation of diaziridinyl cluster mannoside from ketone [61]. The alkoxy-substituted diazirines were synthesized from alcohols (Scheme 22).

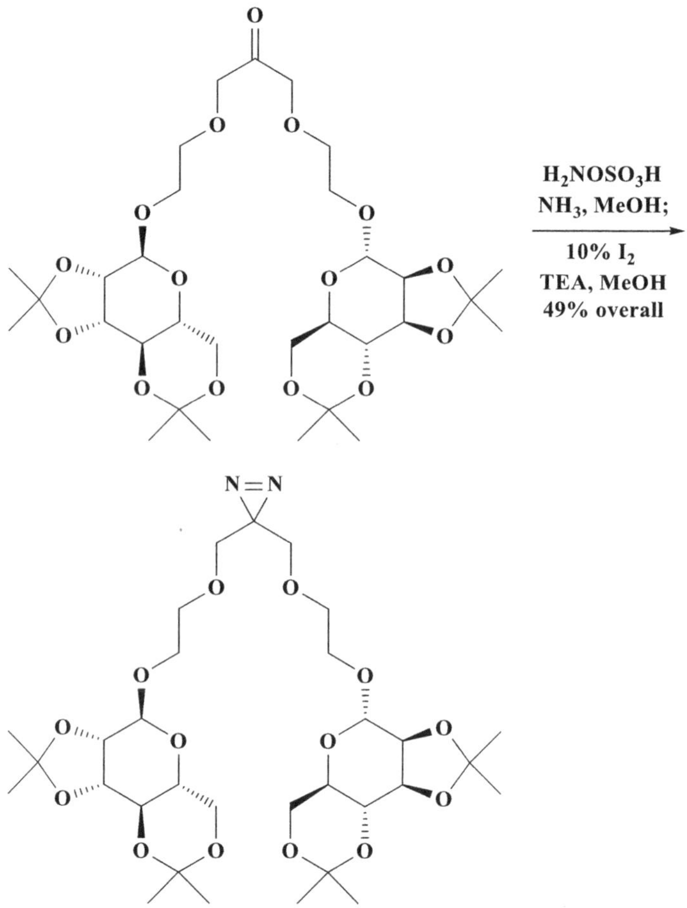

SCHEME 22

Various approaches were utilized for the formation of diazirines from aliphatic ketones [20,33,62], aromatic aldehydes [63], and hexafluoroacetone. Due to the marked change in the reactivity of the ketone because of the fluorine groups, a combination of approaches from different sources was required. The tetrafluorodiazirine was prepared efficiently as given in Scheme 23. The tetrafluoroketone imine was synthesized by a Wittig type of reaction from triphenylphosphinimine and ketone [64]. It was purified by chromatography on silicic acid and by distillation and was formed in 60% yield. The imine was surprisingly inert toward nucleophilic additions to the carbon atom. An effort to synthesize the ammonia adduct from imine hydrochloride and NH_3 was unsuccessful. The product was prepared, which could not be characterized and did not provide diazirine when reacted with NaOCl. Instead, the original ketone hydrate, obtained possibly by the rapid exchange of NH_3 with H_2O, was formed. The diaziridine was prepared when the imine was reacted with hydroxylamine-*O*-sulfonic acid in anhydrous diglyme. The diaziridine could be oxidized with $Pb(OAc)_4$. The formed diazirine was purified by chromatography on silicic acid; it was separated easily from untreated tetrafluoroketone hydrate. No efforts were made to enhance the yield, which was 17% relative to the tetrafluoroketone imine, although possibilities for doing so can easily be conceived. After purification on silicic acid, the tetrafluorodiazirine was 95% pure. The impurities were parent tetrafluoroketone and small amounts of compounds, which were later formed during photolysis. The pure samples were obtained by preparative gas chromatography [15].

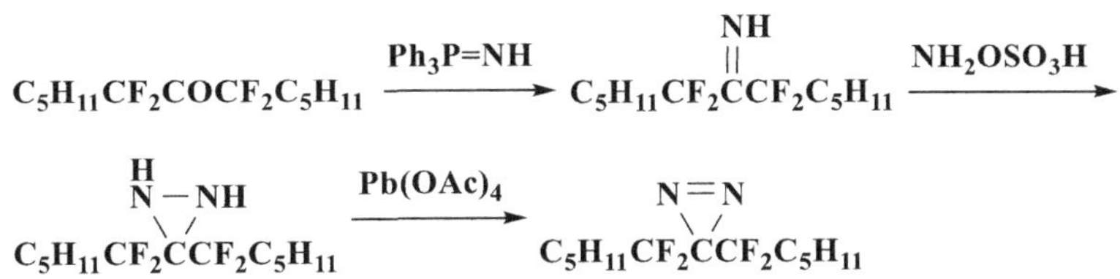

SCHEME 23

The difluorodiazirine was prepared in 45% yield from aldehyde by a technique utilized for the formation of aromatic diazirine. The aldehyde was first rendered anhydrous azeotropically by refluxing in benzene. The imine was formed spontaneously on adding *t*-butylamine, and the imine on reaction with hydroxylamine-*O*-sulfonic acid in C_2H_5OH and Et_3N provided *N*-*t*-butyldiaziridine. The *N*-*t*-butyldiaziridine was oxidized to diazirine with *t*-butyl hypochlorite in C_2H_5OH using Et_3N (Scheme 24) [15,65,66].

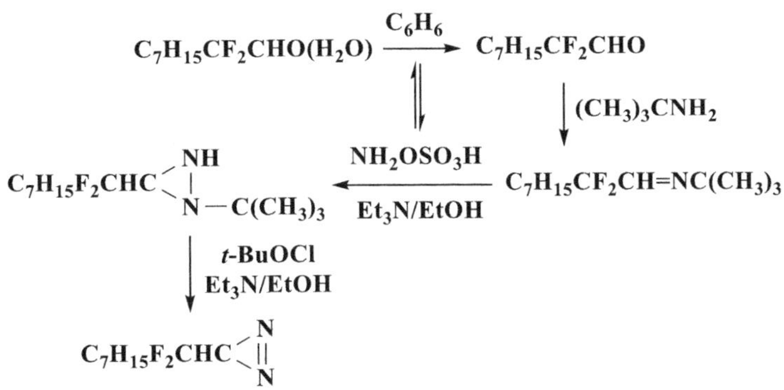

SCHEME 24

The key intermediate diazirine aldehyde was synthesized in seven steps from 3-bromoanisole as explained for an introduction of 3-(trifluoromethyl)-diazirin-3-yl group [27–29]. The final oxidation of aldehyde to acid was achieved by a combination of sodium chlorite with sulfamic acid in tetrahydrofuran and H_2O (Scheme 25) [67].

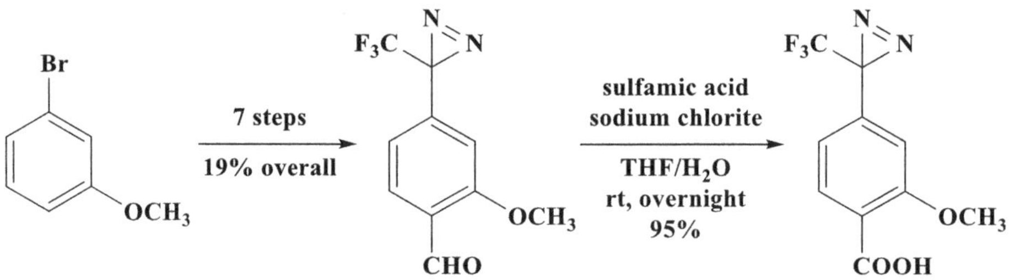

SCHEME 25

References

1. (a) B. Zwanenburg and P. ten Holte. 2001. The synthetic potential of three-membered ring aza-heterocycles. Top. Curr. Chem. 216: 93-124. (b) N. Kaur. 2019. Synthesis of three-membered and four-membered heterocycles with the assistance of photochemical reactions. J. Heterocycl. Chem. 56: 1141-1167. (c) N. Kaur. 2017. Ionic liquids: promising but challenging solvents for the synthesis of *N*-heterocycles. Mini Rev. Org. Chem. 14: 3-23. (d) N. Kaur. 2019. Ionic liquid assisted synthesis of *S*-heterocycles. Phosphorus, Sulfur, Silicon Relat. Elem. 194: 165-185. (e) N. Kaur. 2018. Green synthesis of three to five-membered *O*-heterocycles using ionic liquids. Synth. Commun. 48: 1588-1613.
2. E. Schmitz. 1964. Three-membered rings containing two heteroatoms. Angew. Chem. Int. Ed. 3: 333-341.
3. H. Inui and S. Murata. 2005. Photochemistry of 2-(1-naphthyl)-2*H*-azirines in matrixes and in solutions: wavelength-dependent C-C and C-N bond cleavage of the azirine ring. J. Am. Chem. Soc. 127: 2628-2636.
4. T.L. Gilchrist. 2001. Activated 2*H*-azirines as dienophiles and electrophiles. Aldrichimica Acta 34: 51-55.
5. C. Puzzarini, A. Gambi and G. Cazzoli. 2004. An ab initio study of diazirine: equilibrium structure and molecular properties. J. Mol. Struct. 695-696: 203-210.
6. R.F.R. Church, R.R. Maleike and M.J. Weiss. 1972. Diazirines. 3. Synthesis of a series of diazirine-containing molecules and their pharmacological evaluation. J. Med. Chem. 15: 514-518.
7. M.J. Alves, T.L. Gilchrist and J.H. Sousa. 1999. Reactions of methyl 2-aryl-2*H*-azirine-3-carboxylates with nucleophiles. J. Chem. Soc. Perkin Trans. 1 10: 1305-1310.
8. Y. Hatanaka, M. Hashimoto, H. Kurihara, H. Nakayama and Y. Kanaoka. 1994. A novel family of aromatic diazirines for photoaffinity labeling. J. Org. Chem. 59: 383-387.
9. Y. Hatanaka, M. Hashimoto, H. Nakayama and Y. Kanaoka. 1994. Syntheses of nitro-substituted aryl diazirines. An entry to chromogenic carbene precursors for photoaffinity labeling. Chem. Pharm. Bull. 42: 826-831.
10. M. Hashimoto, Y. Kato and Y. Hatanaka. 2006. Simple method for the introduction of iodo-label on (3-trifluoromethyl) phenyldiazirine for photoaffinity labeling. Tetrahedron Lett. 47: 3391-3394.
11. M. Hashimoto and Y. Hatanaka. 2008. Recent progress in diazirine-based photoaffinity labeling. Eur. J. Org. Chem. 15: 2513-2523.
12. J.E. Baldwin, C.D. Jesudason, M.G. Moloney, D.R. Morgan and A.J. Pratt. 1991. The synthesis of substituted aryl diazirines. A bifunctional reagent suitable for application to photoaffinity labelling studies. Tetrahedron 47: 5603-5614.
13. A.A. Kogon, D.E. Bochkariov, B.P. Baskunov and A.V. Cheprakov. 1992. 2,3-Dihydroxy-3-{3-[3-(trifluoromethyl)diazirin-3-yl]phenyl}propionic acid. A cleavable carbene-generating reagent used for photo crosslinking. Liebigs Ann. Chem. 8: 879-881.

14. M. Nassal. 1984. 4'-(1-Azi-2,2,2-trifluoroethyl)phenylalanine, a photolabile carbene-generating analog of phenylalanine. J. Am. Chem. Soc. 106: 7540-7545.
15. B. Ernil and H.C. Khorana. 1980. Fatty acids containing photoactivable carbene precursors. Synthesis and photochemical properties of 3,3-bis(1,1-difluorohexyl)diazirine and 3-(1,1-difluorooctyl)-3*H*-diazirine. J. Am. Chem. Soc. 102: 3888-3896.
16. Y. Hatanaka, M. Hashimoto, S. Nishihara, H. Narimatsu and Y. Kanaoka. 1996. Synthesis and characterization of a carbene-generating biotinylated *N*-acetylglucosamine for photoaffinity labeling of β-(1→4)-galactosyltransferase. Carbohydr. Res. 294: 95-108.
17. J.F. Brunner, H. Senn and F.M. Richards. 1980. 3-Trifluoromethyl-3-phenyldiazirine. A new carbene generating group for photolabeling reagents. J. Biol. Chem. 255: 3313-3318.
18. M.V. Miret. 2011. PhD Thesis, The University of Edinburgh.
19. J.J. Diaz-Mochon, L. Bialy and M. Bradley. 2004. Full orthogonality between Dde and Fmoc: the direct synthesis of PNA-peptide conjugates. Org. Lett. 6: 1127-1129.
20. R.F.R. Church and M.J. Weiss. 1970. Diazirines. II. Synthesis and properties of small functionalized diazirine molecules. Observations on the reaction of a diaziridine with the iodine-iodide ion system. J. Org. Chem. 35: 2465-2471.
21. W. Kirmse. 1971. Rearrangements. Carbene chemistry. Academic Press: New York, 2nd ed., 457-503.
22. A. Metha, R. Jaouhari, T. Benson and K. Douglas. 1992. Improved efficiency and selectivity in peptide synthesis: use of triethylsilane as a carbocation scavenger in deprotection of *t*-butyl esters and *t*-butoxycarbonyl-protected sites. Tetrahedron Lett. 33: 5441-5444.
23. S.S. Husain, S.A. Forman, M.A. Kloczewiak, G.H. Addona, R.W. Olsen, M.B. Pratt, J.B. Cohen and K.W. Miller. 1999. Synthesis and properties of 3-(2-hydroxyethyl)-3-*n*-pentyldiazirine, a photoactivable general anesthetic. J. Med. Chem. 42: 3300-3307.
24. L. Dubinsky, L.M. Jarosz, N. Amara, P. Krief, V.V. Kravchenko, B.P. Krom and M.M. Meijler. 2009. Synthesis and validation of a probe to identify quorum sensing receptors. Chem. Commun. 47: 7378-7380.
25. J. Wrobel, D. Green, J. Jetter, W. Kao, J. Rogers, M.C. Perez, J. Hardenburg, D.C. Deecher, F.J. Lopez, B.J. Arey and E.S. Shen. 2002. Synthesis of (bis)sulfonic acid, (bis)benzamides as follicle-stimulating hormone (FSH) antagonists. Bioorg. Med. Chem. 10: 639-656.
26. D.N. Haarp, L.Q. Bao, C.J. Black, J.G. Gleason and R.A. Smith. 1975. Efficient alpha-halogenation of acyl chlorides by *N*-bromosuccinimide, *N*-chlorosuccinimide, and molecular iodine. J. Org. Chem. 40: 3420-3427.
27. (a) J.M. Padrón, G. Kokotos, T. Martín, T. Markidis, W.A. Gibbons and V.S. Martín. 1998. Enantiospecific synthesis of α-amino acid semialdehydes: a key step for the synthesis of unnatural unsaturated and saturated α-amino acids. Tetrahedron: Asymmetry 9: 3381-3394. (b) J.N. Hernández, M.A. Ramírez and V.S. Martín. 2003. A new selective cleavage of *N,N*-dicarbamoyl-protected amines using lithium bromide. J. Org. Chem. 68: 743-746.
28. F.A. Luzzio, D.Y. Duveau and W.D. Figg. 2006. A chiral pool approach toward the synthesis of thalidomide metabolites. Heterocycles 70: 321-334.
29. D.J. Hamilton and A. Sutherland. 2004. A flexible approach for the synthesis of selectively labelled L-arginine. Tetrahedron 45: 5739-5741.
30. R.J. Cox, J.S. Gibson and M.B.M. Martin. 2002. Aspartyl phosphonates and phosphoramidates: the first synthetic inhibitors of bacterial aspartate-semialdehyde dehydrogenase. ChemBioChem 3: 874-886.
31. (a) M.F. Ward. 2001. Heterocyclic chemistry. Annu. Rep. Prog. Chem. Sect. B 97: 143-176. (b) I.R. Likhotvorik, E.L. Tae, C. Ventre and M.S. Platz. 2000. A facile one-pot conversion of non-enolizable aldehydes to diazirines. Tetrahedron Lett. 41: 795-796.

32. S. Ogawa, A. Nakajima and Y. Miyamoto. 1991. Cleavage of validoxylamine A derivatives with *N*-bromosuccinimide: preparation of blocked synthons useful for the construction of carba-oligosaccharides composed of imino linkages. J. Chem. Soc. Perkin Trans. 1 12: 3287-3290.
33. E. Schmitz and R. Ohme. 1961. Cyclische diazoverbindungen, I. Herstellung und umsetzungen von diazirinen. Chem. Ber. 94: 2166-2173.
34. H.J. Abendroth. 1961. Strukturisomere hydrazone. Angew. Chem. 73: 67-67.
35. H.J. Abendroth and G. Henrich. 1959. Über ein isomeres acetonhydrazon. Angew. Chem. 71: 283-283.
36. E. Schmitz. 1961. Synthesen mit *N*-halogen-aminen. Angew. Chem. 73: 23-25.
37. R.F.R. Church, A.S. Kende and M.J. Weiss. 1965. Diazirines. I. Some observations on the scope of the ammonia-hydroxylamine-*O*-sulfonic acid diaziridine synthesis. The preparation of certain steroid diaziridines and diazirines. J. Am. Chem. Soc. 87: 2665-2671.
38. G. Kurz, J. Lehmann and R. Thieme. 1985. Preparation and properties of some photolabile sugar derivatives for affinity labelling. Carbohydr. Res. 136: 125-133.
39. E. Schmitz. 1964. Dreiringe mit zwei heteroatomen. Angew. Chem. 76: 197-206.
40. V.V. Kuznetsov, N.N. Makhova, Y.A. Strelenko and L.I. Khmel'nitskii. 1991. The role of pH in the synthesis of diaziridines. Bull. Acad. Sci. USSR Chem. Sci. 40: 2496-2505.
41. J.J. Uebel and J.C. Martin. 1964. Nuclear magnetic resonance studies of diazirines. An application to conformational analysis in six-membered carbocyclic rings. J. Am. Chem. Soc. 86: 4618-4623.
42. J. Lehmann, A. Linden and H. Heimgartner. 1999. Site-selective incorporation of thioamide-linkages into a growing peptide. Tetrahedron 55: 5359-5376.
43. K. Briner and A. Vasella. 1989. Glycosylidene carbenes a new approach to glycoside synthesis. Part 1. Preparation of glycosylidene-derived diaziridines and diazirines. Helv. Chim. Acta 72: 1371-1382.
44. C.C. Sweeley, R. Bentley, M. Makita and W.W. Wells. 1963. Gas-liquid chromatography of trimethylsilyl derivatives of sugars and related substances. J. Am. Chem. Soc. 85: 2497-2507.
45. E. Schmitz. 1979. Three-membered rings with two heteroatoms. Adv. Heterocycl. Chem. 24: 63-107.
46. E. Schmitz, A. Stark and C. Hörig. 1965. Cyclische diazoverbindungen, V: ein diazoketon mit dreiringstruktur. Chem. Ber. 98: 2509-2515.
47. J. Lehmann and R. Thieme. 1986. Synthese von 6-*C*-azi-6-desoxy-D-glucose und -D-galactose für die photoaffinitätsmarkierung von kohlenhydratbindenden proteinen. Liebigs Ann. Chem. 3: 525-532.
48. J. Lehmann, J. Steck and W. Weiser. 1988. Malto-oligosaccharide homologues of 3,7-anhydro-2-azi-1,2-dideoxy-D-glycero-D-gulo-octitol: improved photoaffinity reagents for labelling the malto-oligosaccharide-binding protein of *Escherichia coli*. Carbohydr. Res. 184: 113-120.
49. J. Lehmann and M. Scheuring. 1990. Synthese von undecose-derivaten als D-galactose-analoga zur spezifischen modifikation galactose-bindender proteine. Liebigs Ann. Chem. 3: 271-275.
50. S.-C. Ats, E. Lausberg, J. Lehmann and K. Sandhoff. 1990. Synthese von 4-acetamido-3,7-anhydro-2-azi-1,2,4-tridesoxy-D-glycero-D-gulo-octitol, einem potentiellen photoaffinitätsreagens mit guter affinität zu menschlicher β-hexosaminidase. Liebigs Ann. Chem. 12: 1261-1264.
51. J. Lehmann and S. Petry. 1993. Photolysis of 2-azi-2-deoxy-D-arabino-hexitol and analogous hexitols; the ineffectiveness of certain carbohydrate diazirines as photoaffinity labels. Carbohydr. Res. 239: 133-142.
52. C.P.J. Glaudemans, J. Lehmann and M. Scheuring. 1989. Ein "spacer"-modifiziertes trisaccharid als photolabiler ligand für die affinitätsmarkierung von antigalactan IgA X 24. Angew. Chem. 101: 1730-1731.
53. C.-S. Kuhn, J. Lehmann and J. Steck. 1990. Syntheses and properties of some photolabile β-thioglycosides. Potential photoaffinity reagents for β-glycoside hydrolases. Tetrahedron 46: 3129-3134.
54. C.-S. Kuhn, J. Lehmann, G. Jung and S. Stevanovic. 1992. Investigation of the active site of *Escherichia coli* β-D-galactosidase by photoaffinity labelling. Carbohydr. Res. 232: 227-233.
55. D. Krois and U.H. Brinker. 2001. Synthesis of a sterically hindered diazirine with a satisfactory yield: optically active 2-azicamphane. Synthesis 3: 379-381.

56. J.E. Baldwin, R.A. Field, C.C. Lawrence, K.D. Merritt and C.J. Schofield. 1993. Proline 4-hydroxylase: stereochemical course of the reaction. Tetrahedron Lett. 34: 7489-7492.
57. Y. Ikeda and E.J. Behrman. 2008. Improved synthesis of photo-leucine. Synth. Commun. 38: 2276-2284.
58. M. Suchanek, A. Radzikowska and C. Thiele. 2005. Photo-leucine and photo-methionine allow identification of protein-protein interactions in living cells. Nat. Methods 2: 261-268.
59. D.S. Robinson and J.P. Greenstein. 1952. Stereoisomers of hydroxyproline. J. Biol. Chem. 195: 383-388.
60. B. van der Meijden and J.A. Robinson. 2011. Synthesis and application of photoproline - a photoactivatable derivative of proline. ARKIVOC (vi): 130-136.
61. M. Walter, M. Wiegand and T.K. Lindhorst. 2006. Synthesis of cluster mannosides carrying a photolabile diazirine group. Eur. J. Org. Chem. 3: 719-728.
62. W.H. Graham. 1965. Reactions of dichloramine. I. A convenient method of preparation of diazirine. J. Org. Chem. 30: 2108-2108.
63. R.A.G. Smith and J.R. Knowles. 1975. The preparation and photolysis of 3-aryl-3*H*-diazirines. J. Chem. Soc. Perkin Trans. 2 7: 686-694.
64. R.A. Bekker, G.G. Melikyan, B.L. Dyatkin and I.L. Knunyants. 1975. Polyfluorinated enols and their derivatives. 3. Halogenation, sulfonation and nitrosation of perfluoropropen-2-ol. Zh. Org. Khim. 17: 1604-1607.
65. R.A. Mitsch. 1965. Difluorodiazirine. III. Synthesis of difluorocyclopropanes. J. Am. Chem. Soc. 87: 758-761.
66. S.A. Fugua, W.G. Duncan and R.M. Silverstein. 1965. A one-step synthesis of 1,1-difluoro olefins from aldehydes. J. Org. Chem. 30: 1027-1029.
67. T. Mayer and M.E. Maier. 2007. Design and synthesis of a tag-free chemical probe for photoaffinity labeling. Eur. J. Org. Chem. 28: 4711-4720.

CHAPTER

Synthesis of epoxides and oxaziridines

6

Chapter outline

6.1 Introduction

The major components of synthetic molecules and natural products are heterocyclic compounds and are having huge importance industrially and biologically. Many aspects show their high importance, but the major ones are their biological activities [1a–j].

The chemistry of three-membered heterocyclic ring has played an important role in the history of organic synthesis. Uncountable classic and modern procedures exist for their synthesis from simple starting compounds. The intrinsic ring strain of these compounds primes them for facile ring opening. Therefore, much attention has been paid to the discovery of reactions that employ the reactivity of three-membered heterocyclic rings as entry points for the formation of complex heteroatom containing products. Three-membered ring systems such as aziridines, cyclopropanes, and epoxides are versatile and reactive organic molecules that have received significant focus in the scientific community over a long time. The epoxides are important functional groups in the synthesis because they undergo stereospecific nucleophilic ring opening to synthesize the bifunctional compounds [1k].

In organic synthesis, the epoxides serve as useful intermediates [2,3]. They are important functional groups of natural [4] and medicinal products [5], polymer [6], and supramolecular compounds [7]. An epoxidation of terminal alkenes to epoxides is especially interesting because epoxides are appreciable intermediates for the formation of fine chemicals such as paints, epoxy resins [8–11], surfactants, vital intermediates in organic synthesis, agrochemicals, and pharmaceuticals [12,13]. The Schiff bases and their stable complexes with most of transition metals play an important role in coordination chemistry [14–17].

The epoxides are strained heterocycles that came across throughout chemistry. The epoxides are reasonably stable and simple to synthesize. They react easily, frequently via reactions that release ring strain, such as an addition of nucleophile to ring carbon. It is reasonable to think that the reactivity of epoxide is related to its strain, which is likely to be based on the nature and number of substituents on the epoxide ring. For the formation of epoxides, an epoxidation of alkenes has been an important key

3-Membered Heterocycle Synthesis. https://doi.org/10.1016/B978-0-443-13583-5.00014-1

pathway, and a number of researchers are paying attention to it. Generally, the chlorine-utilizing noncatalytic procedure, coepoxidation, and catalytic procedures utilizing organic peroxides and peracids are traditional and common technologies in industrial procedures [18,19]. However, these procedures are not capital intensive and economic. Additionally, environmental drawbacks occur as well. For example, a huge output of chloride-laden sewage will be caused during the chlorine-utilizing noncatalytic procedure, and a great deal of acid waste will also be formed using peracids, in addition to the problem in separation of homogenous catalysts [20–22]. As to the coepoxidation processes, the products from reaction may be consistently commercially desired [23]. The metal-catalyzed epoxidation of alkenes to epoxides with milder oxidants such as oxygen, hydrogen peroxide, and alkyl hydroperoxides has received a lot of interest and resulted in the improvement of highly efficient catalysts [24–26].

6.2 Synthesis of epoxides

The diepoxides and epoxy lactones could be synthesized as major products by the thermal rearrangement of alkyl- and some aryl-substituted furan *endo*-peroxides (Scheme 1). The products were formed through an initial oxygen–oxygen bond homolysis to oxygen diradicals [27]. The 1,2-hydrogen shift in intermediate ($R_2 = H$) afforded epoxy lactone. The 2,5-dimethylfuran *endo*-peroxides (R, R_2 = methyl, R_1 = H or CO_2CH_3) were transformed into diepoxides when heated in carbon tetrachloride or benzene above 60°C [28–30]. The lack of epoxy lactone as a product was in accordance with lower migratory ability of CH_3 in comparison with hydrogen substituent R_2 in intermediate. The unsubstituted and 2-alkyl-substituted furan *endo*-peroxides were rearranged to mixtures of diepoxides and epoxy lactones at rt in carbon tetrachloride, dichloromethane, or chloroform. The ratio of diepoxides and epoxy lactones strongly depend on the solvent utilized. Analogous results were reported for 3-alkyl-substituted furan *endo*-peroxides [31–33]. Generally, the alkyl substituents in 2- and 5-position of the furan *endo*-peroxides, along with apolar solvents, favored the synthesis of epoxides [34].

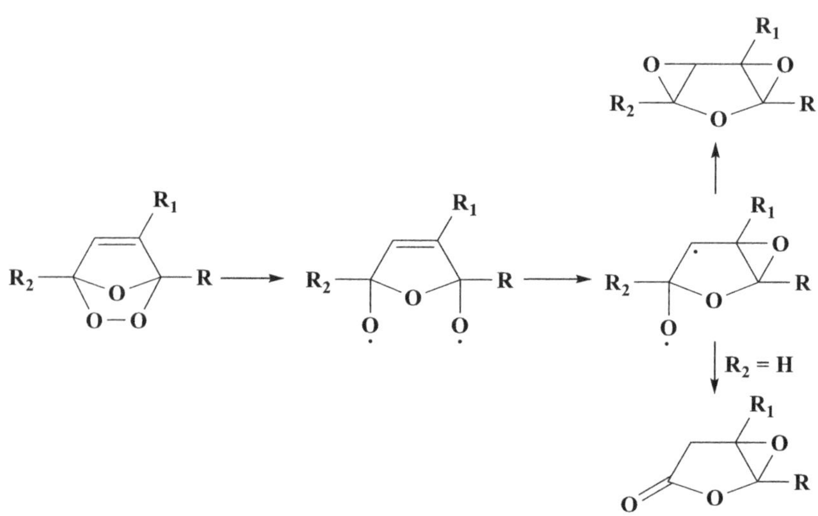

SCHEME 1

An analysis of anion-induced decomposition of 2-ethoxycarbonyl prop-2-enyl peroxides has established that the epoxides were synthesized (Schemes 2 and 3) by (1) adding nucleophile to the acrylic unsaturated bond and (2) an intramolecular anionic substitution on the peroxidic bond [35]. The synthesis of two peroxy epoxides on addition of potassium *t*-butoxide to peroxide was consistent with competing internal nucleophilic substitution reactions of carbanion intermediate in this two-step process.

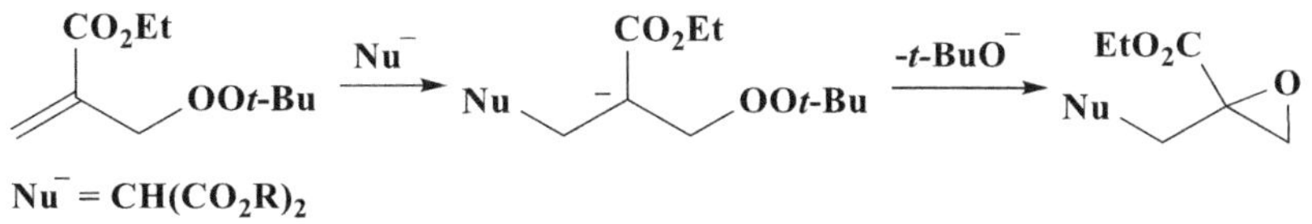

SCHEME 2

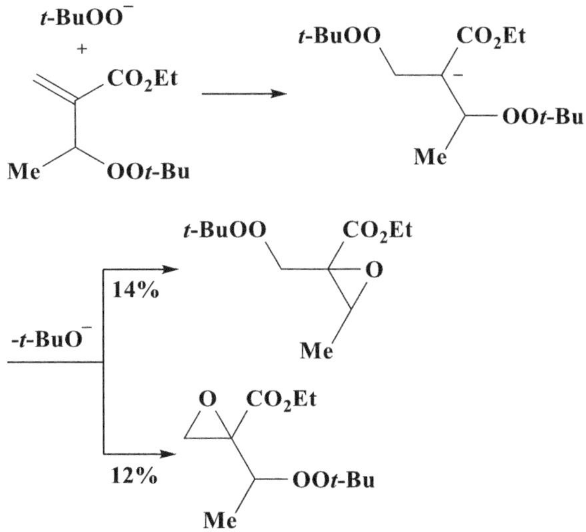

SCHEME 3

An interesting instance of aziridine formation by adding nitrogen to a double bond was reported by Barros and coworkers [36]. The Gabriel–Cromwell reaction, i.e., an addition–elimination pathway to aziridines afforded precursors in a short overall formation of (+)-bromoxone [37]. The aziridine formation given in Scheme 4 led into second and more attractive field of aziridine formation, which concerned with the cyclization of amines having a leaving group. These approaches depend on a 1,2-arrangement of the amine to leaving group resulting in a 3-*exo-tet* cyclization. Typically, the substrates for this type of aziridine formation included 1,2-azido alcohols and 1,2-amino halides.

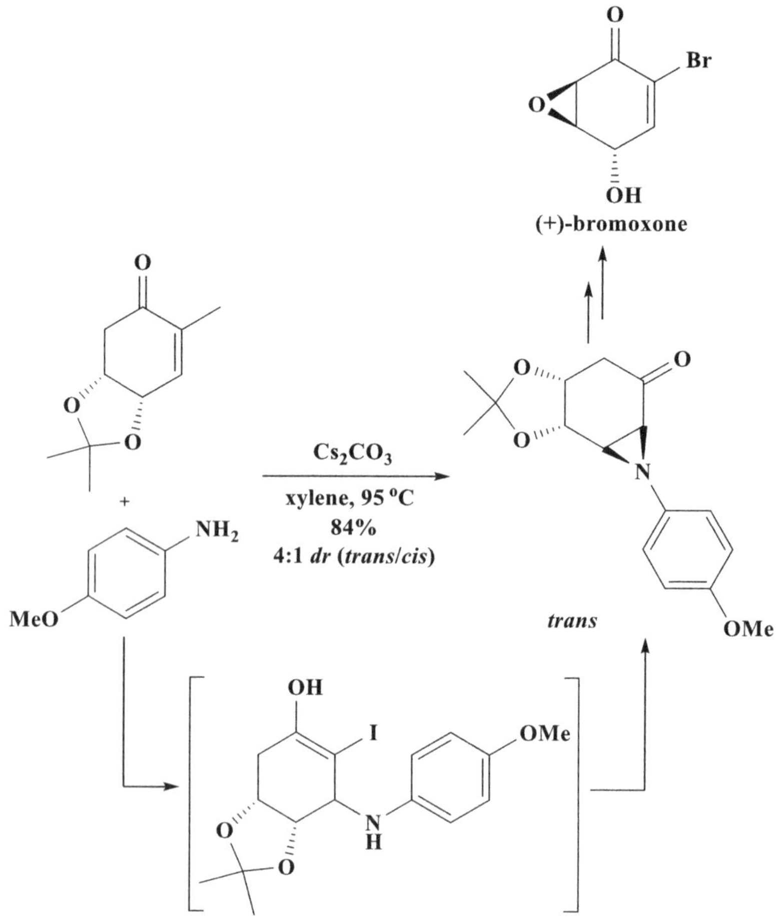

SCHEME 4

The cyclic acetals were utilized as protecting groups for 1,2- and 1,3-diols in the formation of an array of natural products. Wu et al. [38,39] modified acetonides in the overall formation of hepoxillin and trioxillin. They observed that two diastereoisomers of each natural product could be prepared from chiral aldehyde derived from gluconolactone [40]. The aldehyde was synthesized by organozinc addition to gluconolactone followed by a selective hydrolysis and an oxidative cleavage of less hindered terminal acetonide, which was then elaborated to synthesize the 10-(*R*)-trioxillin and 10-(*R*)-hepoxillin, respectively (Scheme 5).

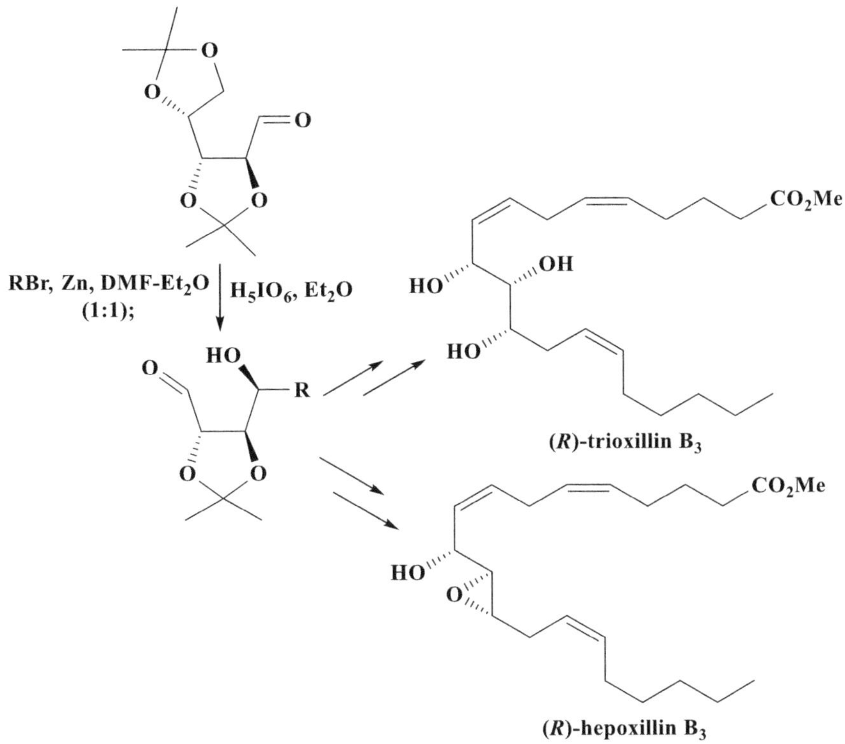

SCHEME 5

An instance of an unexpectedly facile and stereoselective formation of eight-membered lactone from an acyclic precursor diene ester was detected during the overall formation of antitumor agent octalactin A (Scheme 6) [41]. The dense substitution pattern in cyclization substrate imposed conformational constraints in a way that resulted in a conformation favorably disposed for the ring closure. The cyclization substrate was subjected to 10–20 mol% catalyst A to prepare the cyclization product within 24 h in 86% yield. On the other hand, the diene ester with epimeric *p*-methoxybenzyl ether group was subjected to ring closure under the similar conditions only with difficulty affording 3-*epi*-lactone in 20% yield after 7 d in boiling CH_2Cl_2 [42].

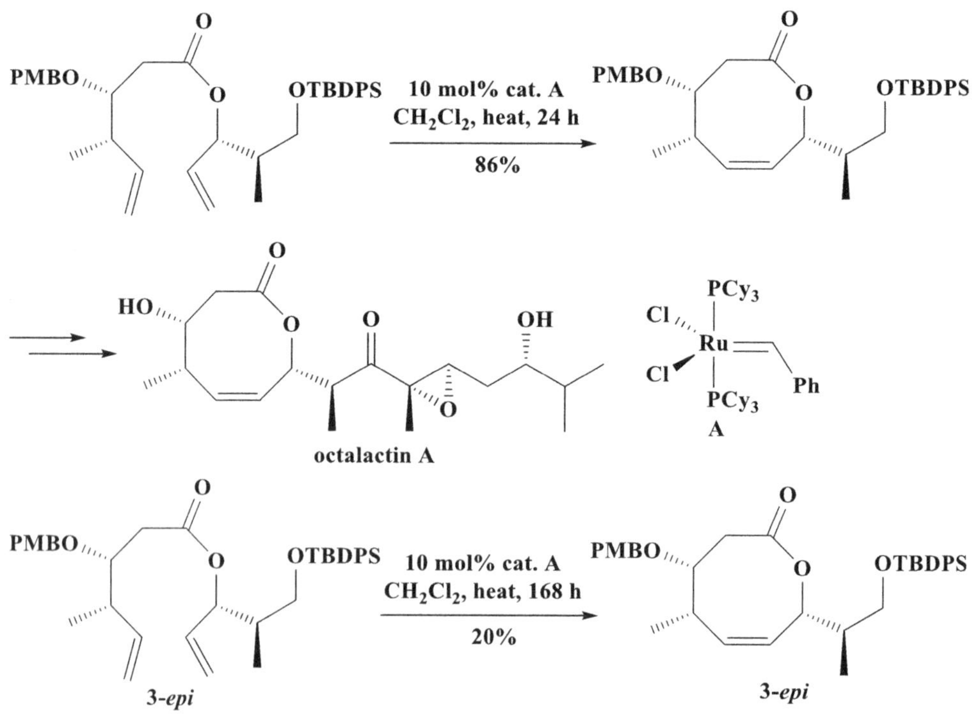

SCHEME 6

While reactions of α,β-unsaturated carbonyl compounds with singlet oxygen have been the subject of numerous analyses, the reactions of their enolic tautomers have attracted little attention. The reaction of β-hydroxy-α,β-unsaturated ketones is shown in Scheme 7 [43]. The photooxidation included an ene reaction with singlet oxygen, where oxygen extracted hydrogen either from the 2-methyl group or the enolic OH group. The decomposition of enedione by the removal of H_2O_2 afforded the major enedione product followed by reaction with hydrogen peroxide to synthesize the epoxide. In alcoholic solvents, the enol form of compounds predominated and was synthesized by the conjugate addition of alcohol on the enedione. An epoxidation to provide the epoxide was reduced by bulky substituents at the terminal CO and in the ring *ortho*-positions, and favored when all substituents were CH_3. The phenyl ring of styrene substrates dictated the *anti/syn*-stereochemistry in ene reactions with singlet oxygen and triazolinediones via a favorable interaction of enophile with phenyl ring [44].

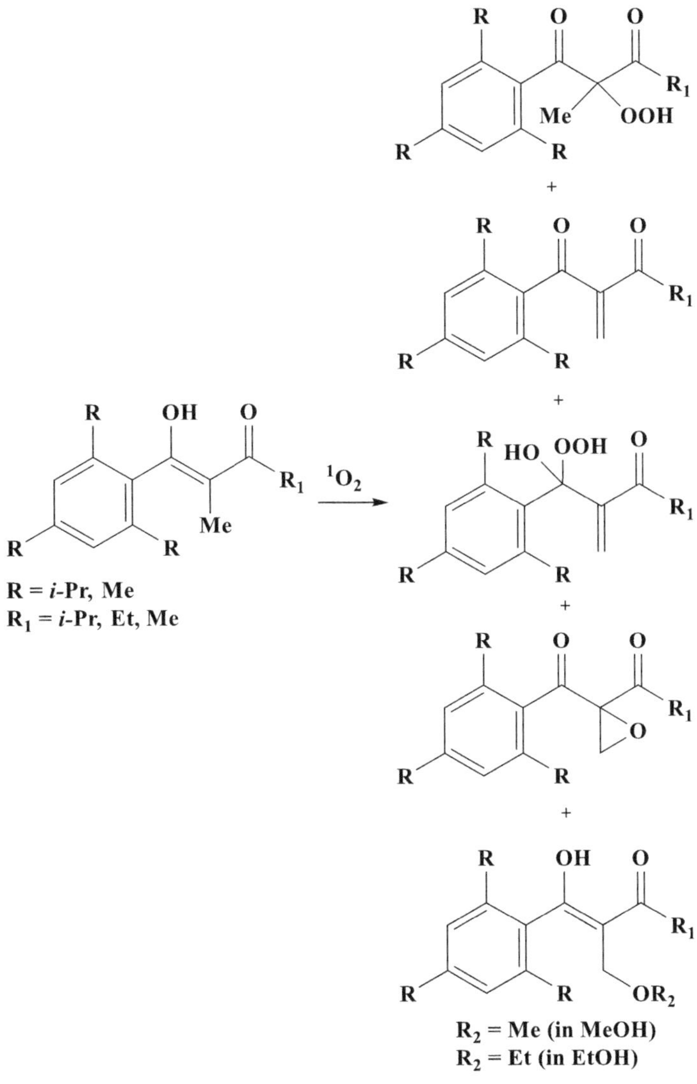

SCHEME 7

Under the mild conditions, the triethylsilyl ethers were utilized as alcohol-protecting groups in cases where deprotection of alcohol required to be conducted in the presence of sensitive moieties. They were incorporated by reacting an alcohol with triethylamine and triethylsilyl chloride in dimethylformamide. The triethylsilyl ethers could withstand these conditions: (1) hydrogenolysis with palladium–carbon in $CH_3COOC_2H_5$, although hydrogenation in CH_3OH will cleave triethylsilyl ethers; (2) ozonolysis of double bonds in CH_2Cl_2 at −78°C; (3) acetylation with acid chlorides and acid anhydrides in pyridine; and (4) reduction with sodium borohydride in CH_3OH or tetrahydrofuran at −20°C. The triethylsilyl ethers could be cleaved in high yields by mild acid hydrolysis in

CH_3COOH/tetrahydrofuran/water at rt. The deprotection could also be achieved with tetra-*n*-butyl ammonium fluoride in tetrahydrofuran; however, these conditions could not be utilized for substrates that were vulnerable to nucleophilic attack or base. Many triethylsilyl ethers could be hydrolyzed with mild acid without influencing the epoxide, the spiroketal, and ester functionality (Scheme 8) [45]. The triethylsilyl ethers are much more acid labile than *t*-butyldimethylsilyl or *t*-butyldiphenylsilyl ethers and often could be cleaved without influencing these Si ethers.

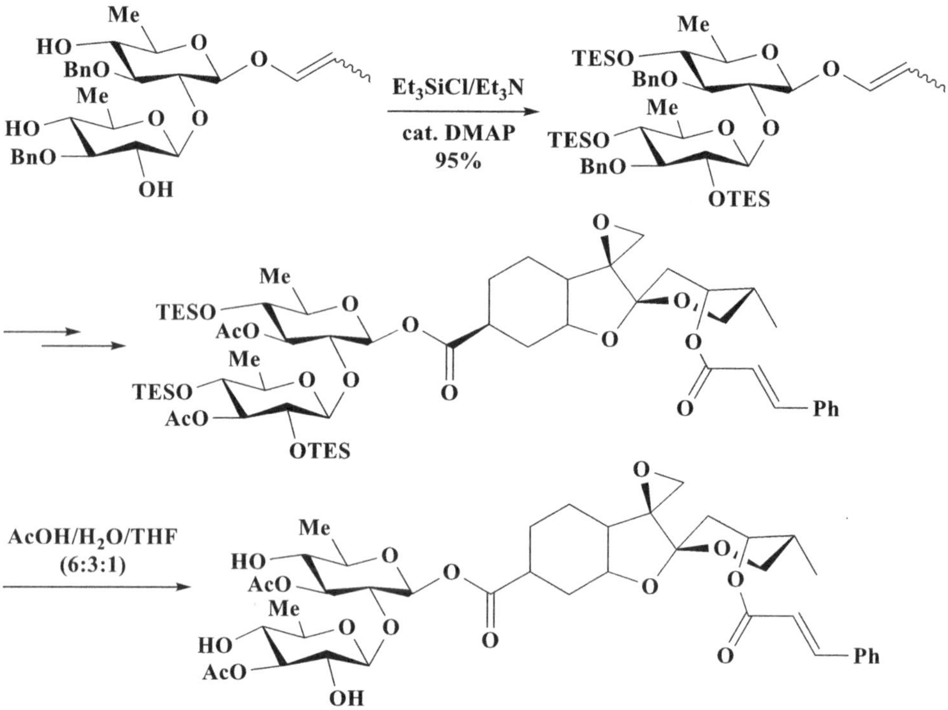

SCHEME 8

The trisubstituted butyrolactones acted as powerful synthetic substrates and have been utilized in several natural product syntheses [46,47]. One such instance is in the formation of cryptophycin A (Scheme 9) [48]. The methyl lactone was esterified, and acetal was protected utilizing 2,2-dimethoxypropane, CH_3OH, and Amberlyst 15 in a one-pot process. The resultant open-chain methyl ester–acetonide was reduced to aldehyde derivative with concomitant Mukaiyama aldol addition/dehydration providing cryptophycin unit A precursor in 42% yield over two steps.

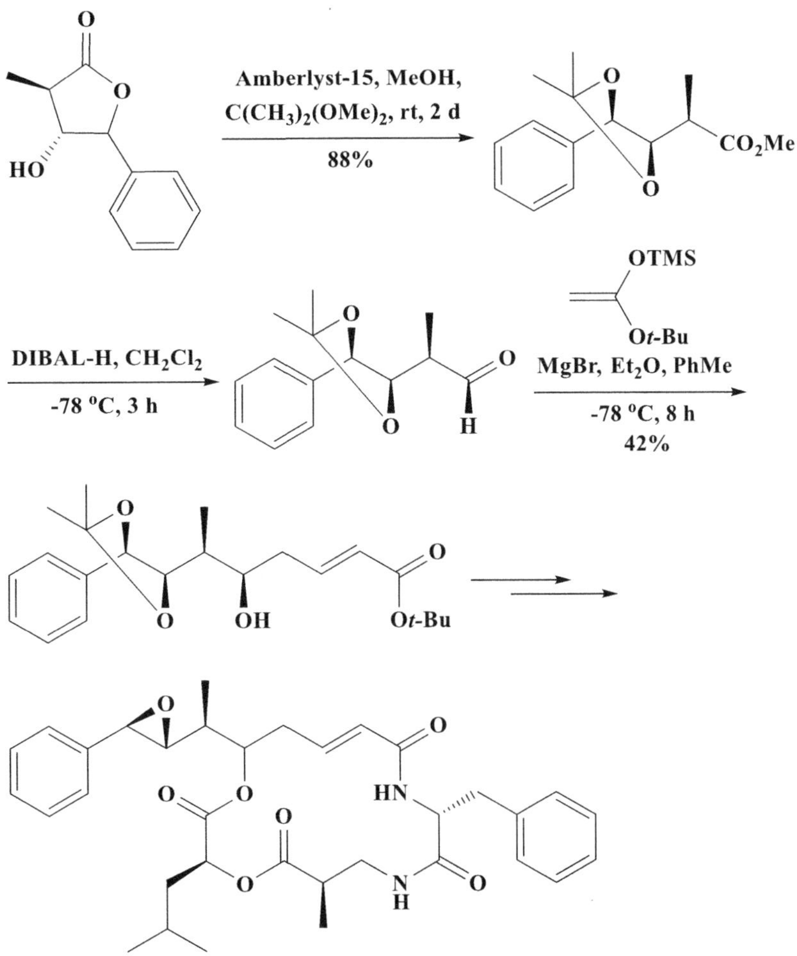

SCHEME 9

The starting compound was subjected to flash vacuum pyrolysis. Once flash vacuum pyrolysis was carried out at relatively low temperature of 500°C and subsequent study of the proton nuclear magnetic resonance spectrum of the cold trap, there seemed to be a large amount of starting compound present with other constituents. An attempted separation by preparative thin-layer chromatography provided 2,2-diphenylcyclopropanone along with an unidentified by-product possessing only phenyl groups and a compound, which seemed to be 3-methyleneoxiran-2-one (Scheme 10) [49].

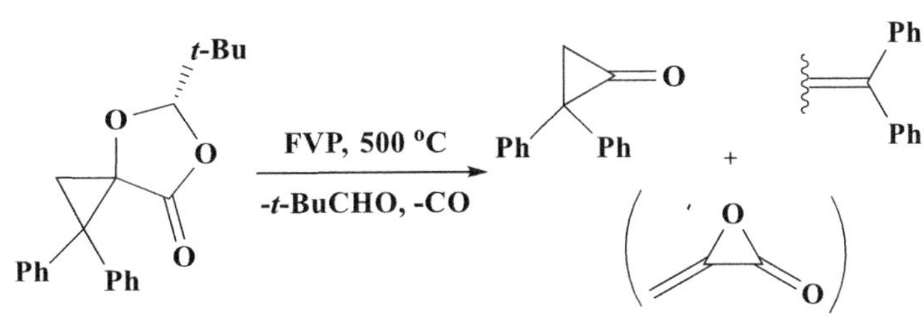

SCHEME 10

The 3-methyleneoxiran-2-one was synthesized after the removal of trimethylacetaldehyde from starting material, with the lone pair of oxygen attacking in and removing the diphenyl carbene, which was then free to react with another carbene prepared by this process or react to synthesize other products such as fluorine (Scheme 11) [49].

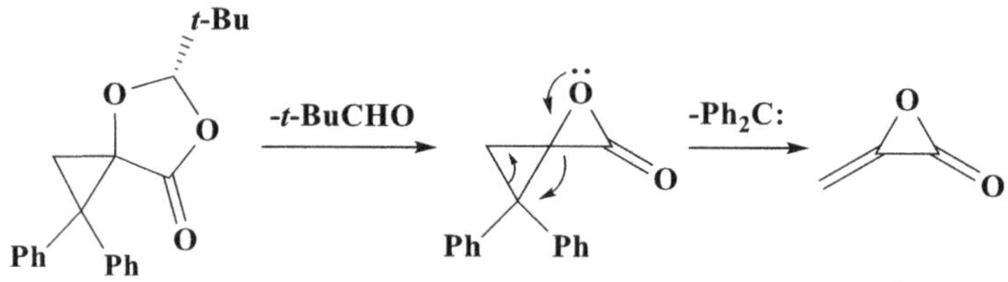

SCHEME 11

The reaction of 4,5-disubstituted oxepines [50] with azo compounds provided adducts, whereas an adduct formed using 2,7-disubstituted oxepines from the oxepine valence isomer was not stable but underwent a Claisen rearrangement to afford the cyclopropyl ketone (Scheme 12) [51].

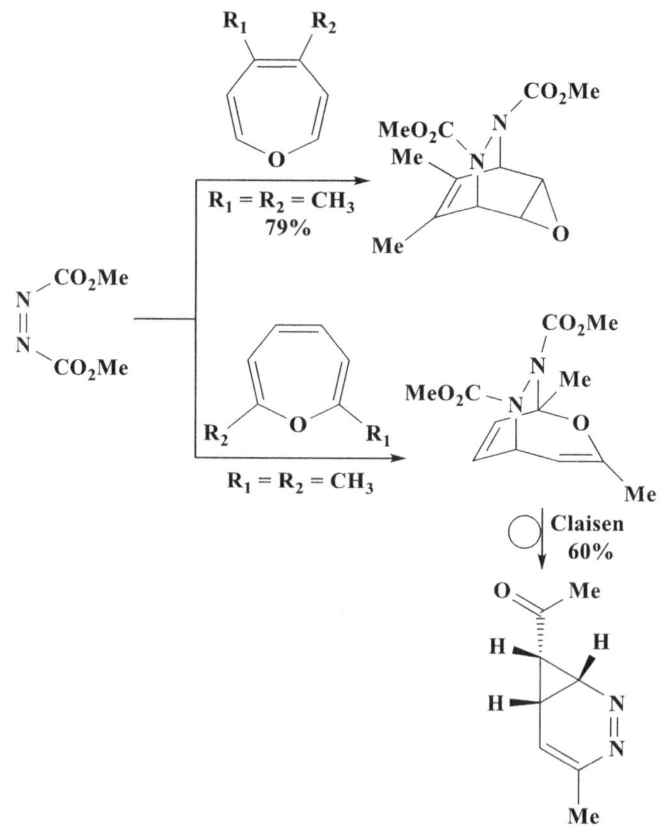

SCHEME 12

Villeneuve and Tam [52] selected Cp*Ru(COD)Cl as a catalyst to interrupt the phenol synthesis. Thus, the 1,4-epoxy-1,4-dihydronaphthalene was reacted in the presence of a Ru catalyst in ethylene dichloride at 60°C to synthesize the 1,2-naphthalene oxide (Scheme 13).

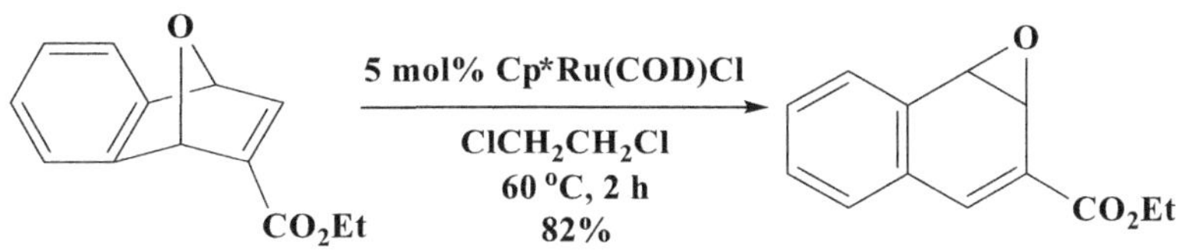

SCHEME 13

The ADA adduct was subjected to Tam's ruthenium-catalyzed isomerization conditions for carbon–oxygen bond proximal to the electron-withdrawing sulfonamide functionality as a main product (Scheme 14). The catalyst was able to coordinate to the sulfonamide allowing it to serve as a directing group, overcoming the normal reactivity preference of substrate. The optically pure 1,2-naphthalene oxides are attractive and heavily functionalized synthetic building blocks [53].

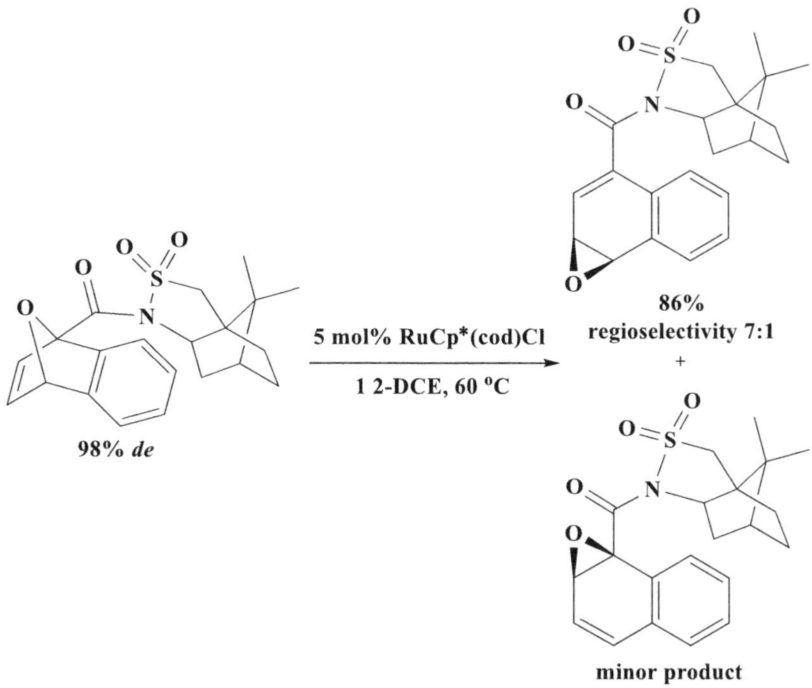

SCHEME 14

The starting compound underwent dehydroidination to afford the α-methylidene dihydropyran in very good yield. As stated in the literature [54], the reaction of hemiketal with 1,8-diazabicyclo[5.4.0]undec-7-ene provided bicyclic ketal in a nonoptimized 53% yield. The reaction with an excess of 1,8-diazabicyclo[5.4.0]undec-7-ene resulted in the formation of a 1:2 mixture of epoxide and bicyclic ketal in 80% overall yield (Scheme 15) [55].

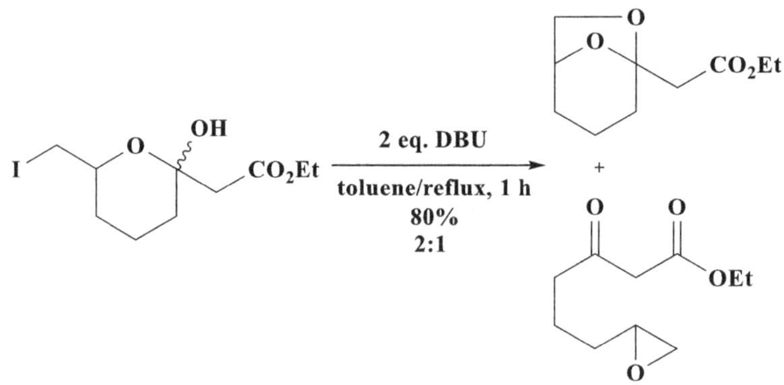

SCHEME 15

The transformation of acetonide alcohol to [^{3}H]trapoxin and trapoxin was achieved (Scheme 16). After the formation of primary tosylate in 78% yield, the acetonide was eliminated with aqueous hydrogen chloride. The epoxide formation with 1,8-diazabicyclo[5.4.0]undec-7-ene/methanol occurred rapidly at 0°C to synthesize the epoxy alcohol (two steps, 62% yield). Depending on the efficiency of the Moffatt oxidation in the last step of Schmidt's chlamydocin synthesis [56], the epoxy alcohol was reacted with a reagent combination of dimethylsulfoxide, *N,N'*-dicyclohexylcarbodiimide, and catalytic $CHCl_2COOH$. These exceptionally mild conditions provided trapoxin B in 80% yield. Given that primary motivation for pursuing the overall formation of trapoxin lay in determining the molecular basis for its detransformation activity, it was crucial that this path allowed a radioisotope incorporation at a late stage. A major benefit of this method as compared with earlier syntheses of related Aoe ((2*S*,9*S*)-2-amino-8-oxo-9,10-epoxydecanoic acid) possessing natural products rests in the intermediacy of acetonide alcohol. The tritiated variant was achieved by oxidizing acetonide alcohol with Dess-Martin periodinane [57] and reducing the aldehyde with [^{3}H]sodium borohydride. The tritiated variant was transformed to [^{3}H]trapoxin utilizing a reaction sequence analogous to that utilized in the formation of trapoxin, which acted as an excellent model system for the formation of tritiated compound. Starting with the sodium borohydride reduction, a mock preparation of [^{3}H]trapoxin was carried out under the similar reaction conditions and concentrations as in the radioactive formation [58,59].

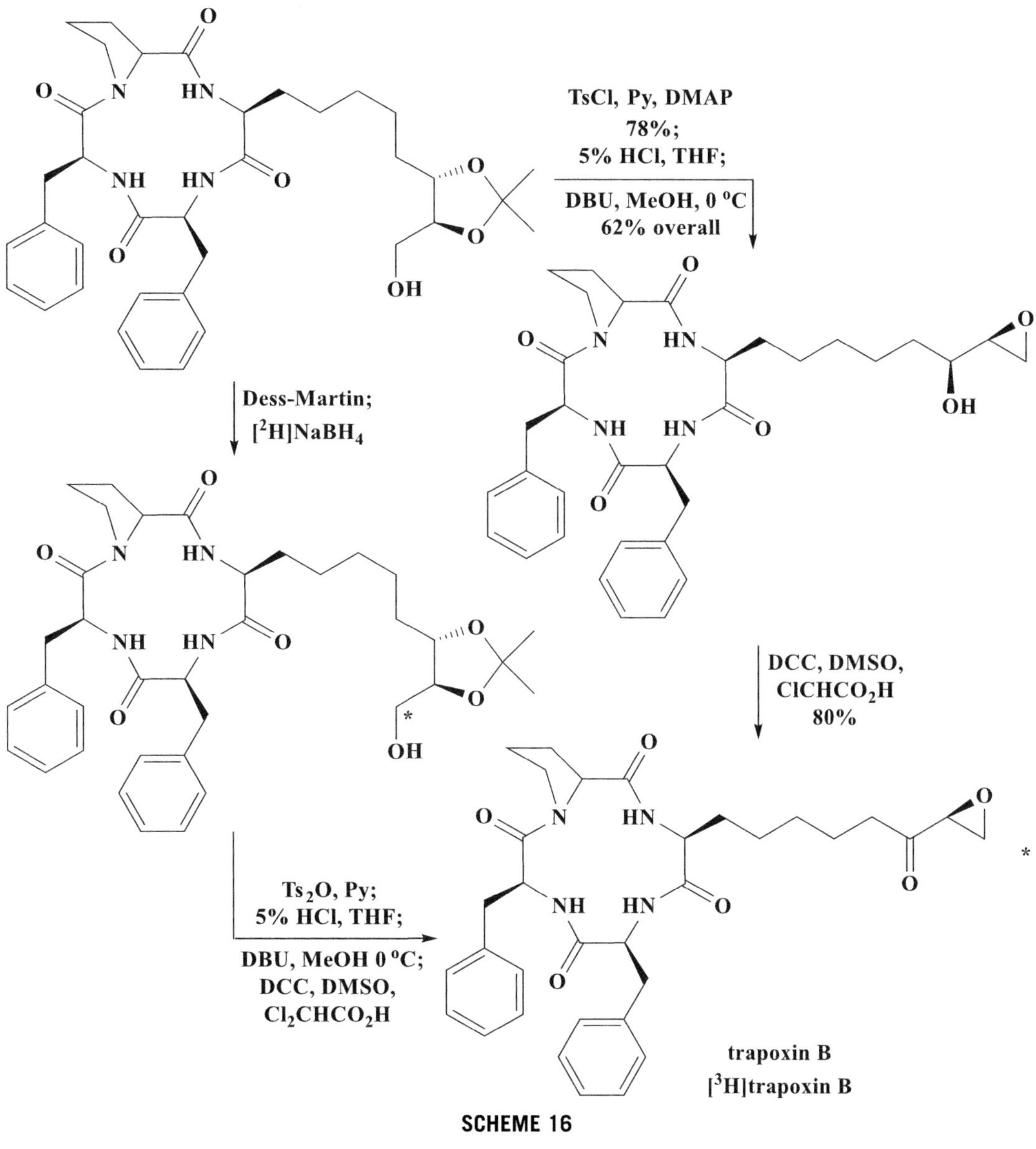

SCHEME 16

The K-trap synthesis proceeded similar to trapoxin pathway (Scheme 17), except that H_2O-soluble 1-ethyl-3-(3-dimethylaminopropyl)carbodiimide was utilized in place of *N,N′*-dicyclohexylcarbodiimide in the final Moffatt oxidation. This sequence of reactions afforded tens of milligrams of pure K-trap. The K-trap and trapoxin displayed nearly superimposable ^{1}H and ^{13}C nuclear magnetic resonance spectral characteristics corresponding to their shared structural elements, and substitution of Alloc-lysine for phenylalanine had little effect on the ability of K-trap to prevent histone deacetylase activity in vitro.

However, the K-trap was around 10 times less powerful than trapoxin in an in vivo cell growth assay (inhibition of [^{3}H]thymidine incorporation in MG-63 human osteosarcoma cells). This loss of potency in vivo may reflect K-trap's increased susceptibility or/and decreased cell permeability to reductive inactivation in vivo. The K-trap was deblocked with dimedone (5,5-dimethyl-1,3-cyclohexanedione) and catalytic Pd(0) [60]. The crude amine was purified by reverse-phase high-performance liquid chromatography, lyophilized, and coupled immediately to Affi-Gel 10 (Biorad), a cross-linked agarose matrix showing *N*-hydroxysuccinimide ester functional groups. A simple experiment was carried out to show that reaction took place without loss of biological activity. The nuclear extracts were incubated with either an ethanolamine-capped control matrix or the K-trap matrix for 12 h at 4°C. The matrix beads were removed by centrifugation, and the supernatants were assayed for both histone deacetylase and [^{3}H]trapoxin-binding activity [61]. Both activities were entirely exhausted by reaction with the K-trap matrix and yet were not influenced by the control matrix [59].

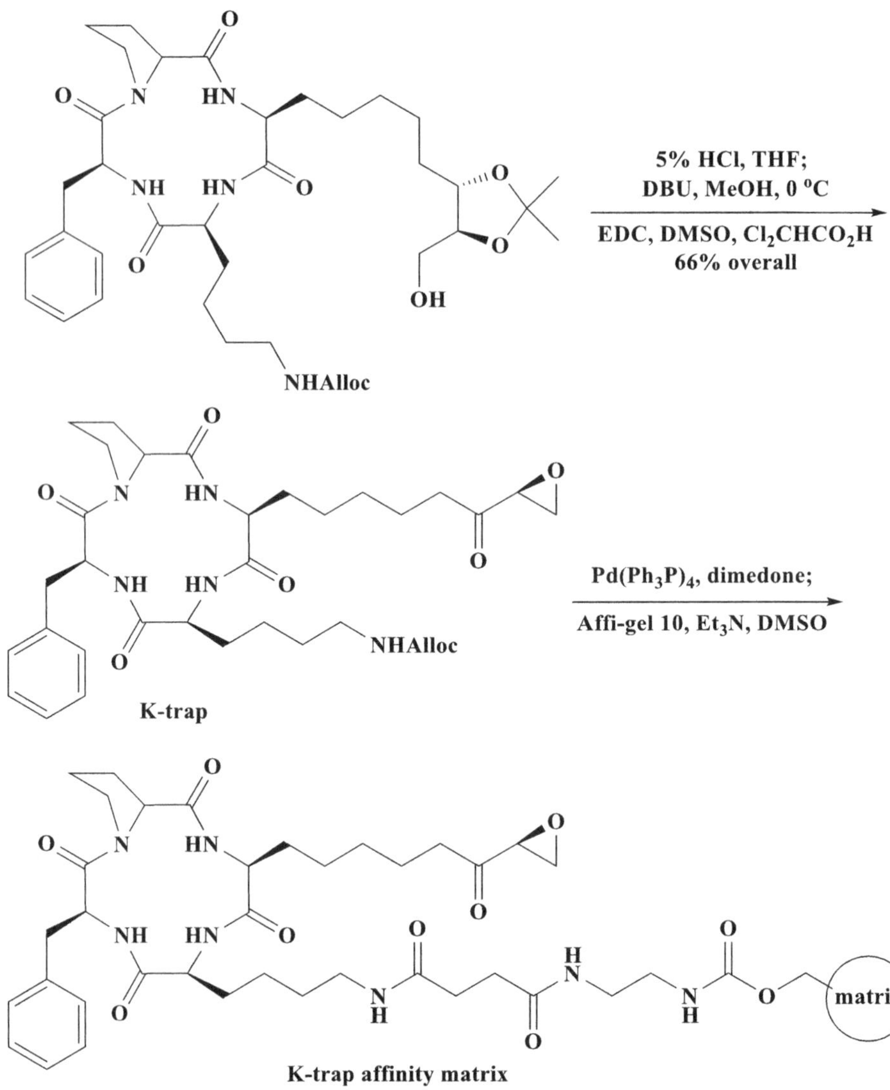

SCHEME 17

A molar-scale formation of enantiomerically pure stilbene oxide from diol was reported by Chang and Sharpless [62]. The reaction included ring opening of C2-symmetric cyclic carbonate by chloride ion followed by decarboxylative ring closure (Scheme 18).

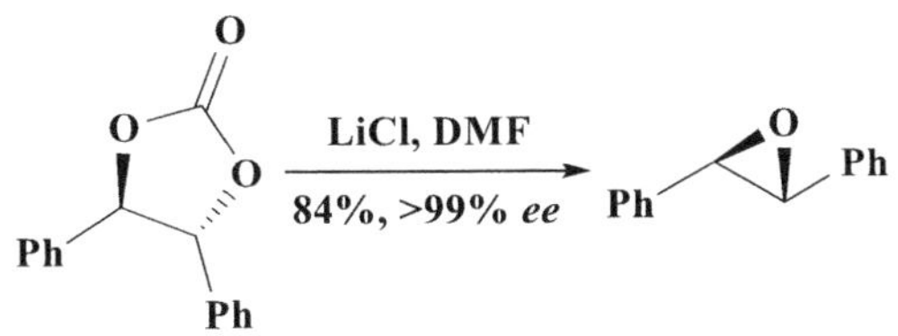

solvent	time (d)	yield (%)	*de* (*trans*) (%)	*ee* (*S,S*) (%)
9:1 CH_3CN/H_2O	1	92	88	84
9:1 *t*-BuOH/H_2O	2	92	86	88
9:1 *i*-PrOH/H_2O	8	59	86	90
9:1 EtOH/H_2O	3	15	84	94
H_2O	4	90	74	86

SCHEME 18

The epoxide was synthesized by reacting azetidine with enolate of acetophenone (Scheme 19). In larger or open-chained heterocycles, the synthesis of epoxide by displacement of an ammonium leaving group was not very successful because of the high energetic barrier for ring closure. With the utilization of a strained azetidinium ylide, however, the energy gain associated with breaking the four-membered ring compensated for that energy barrier and increased the feasibility of such method [63].

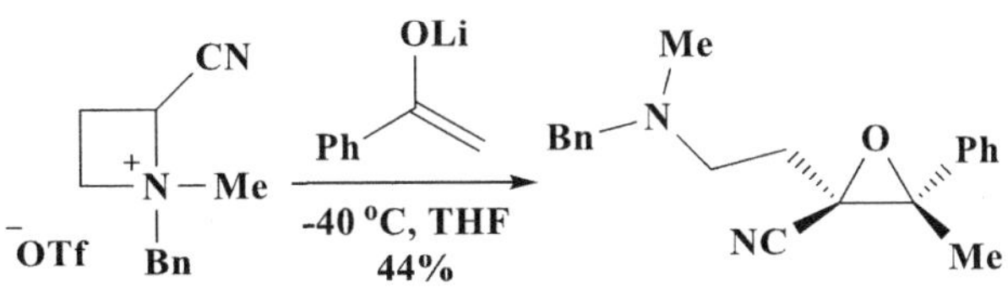

SCHEME 19

A longer alkenyl electrophile such as 5-iodo-2-methylpent-2-ene was used in the formation of difuran sesquiterpene. The dioxolanone adduct was used in the synthesis of one of the starting compounds, the oxirane (Scheme 20) [64].

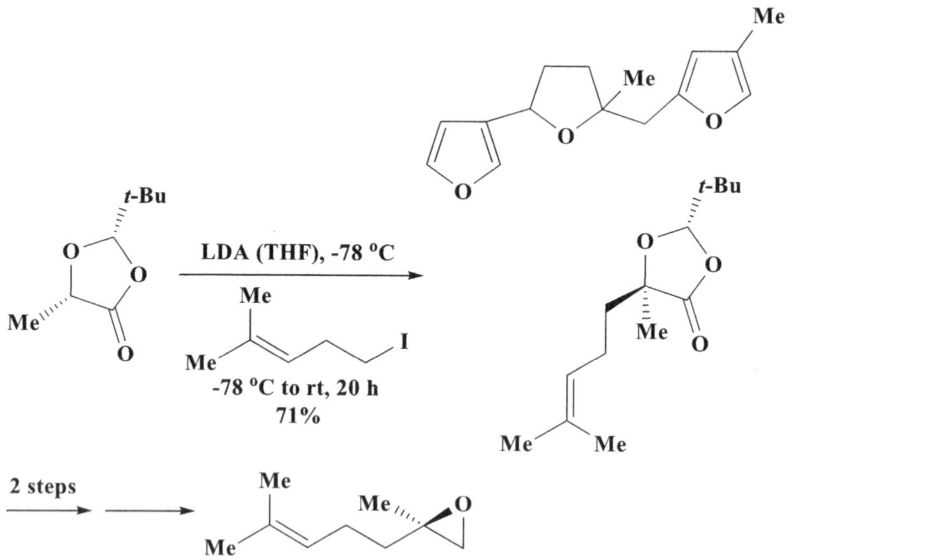

SCHEME 20

The diepoxy allylic bromide was transformed to acylsilane by an alkylation utilizing metalloenamine prepared from LDA (lithium diisopropylamine) deprotonation of imine [65–67]. An addition of vinylmagnesium bromide to acylsilane in the presence of diepoxide was proved to be a sensitive conversion. However, on screening reaction conditions, the α-silyl allylic alcohol could be prepared in quantitative yield by carrying out the reaction at −40°C and controlling the rate of addition of Grignard reagent via syringe pump. The Brook rearrangement was triggered to homoenolate by reacting α-silyl allylic alcohol with *n*-butyl lithium. The enolsilane with complete (*Z*)-selectivity was formed when an intermediate homoenolate was alkylated with geranyl bromide (Scheme 21). The C14–C15 enolsilane was placed to took the benefit of the presence of more reactive terminating nucleophile at C-14 relative to the likely placement of CH_3 substituent in the putative biosynthetic substrate.

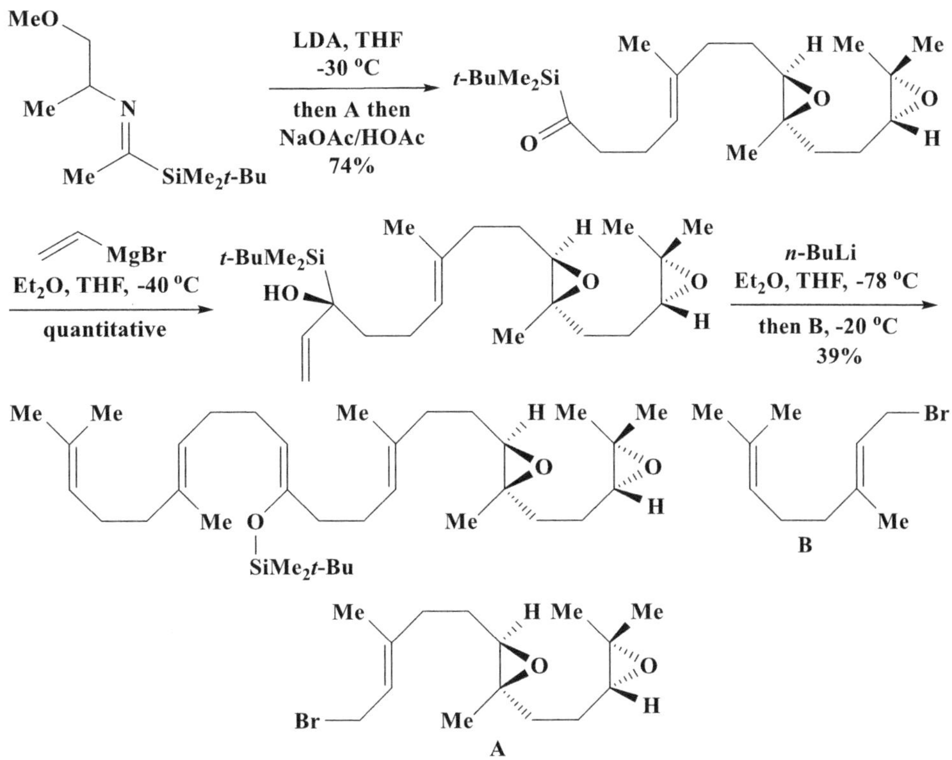

SCHEME 21

The methyl-2,3-anhydro-4,6-benzylidene-D-allopyranoside was synthesized by reacting cyclic sulfate of methyl-4,6-*O*-benzylidene-D-glucopyranoside with sodium hydroxide (Scheme 22). The result has shown that the oxy anion was created at C-2 position exclusively [68].

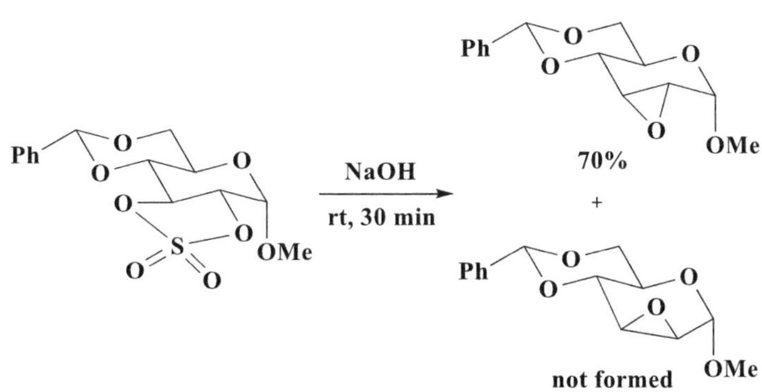

SCHEME 22

The unsaturated lactones acted as electrophile for this conjugate addition as shown in Scheme 23 [69].

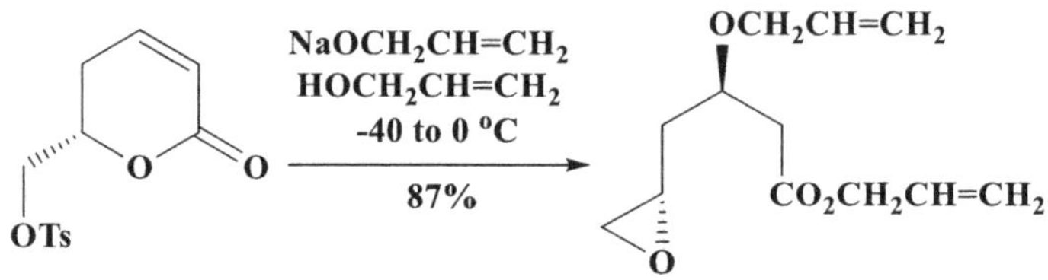

SCHEME 23

The opening of lactone in the presence of Na_2CO_3 in methanol gave episulfides, which were then transformed into free amino acids in reasonable total yields. An analogous procedure was followed by same group to reach at epoxides [70]. In this case, the lactones were directly cyclized utilizing sodium carbonate in methanol to synthesize the epoxides in diastereomerically pure form, which were easily deprotected to afford the free epoxy amino acids (Scheme 24).

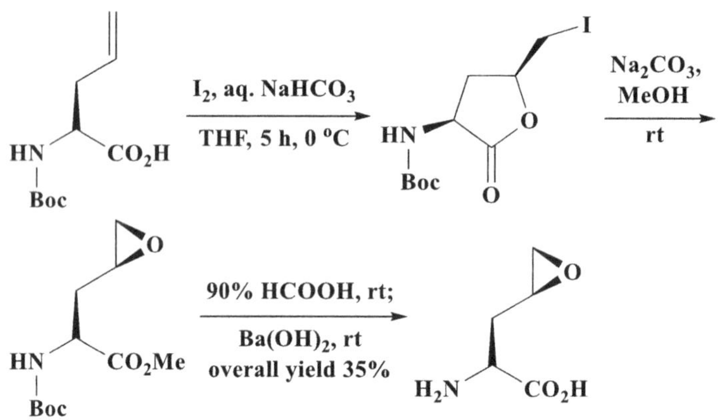

SCHEME 24

The *cis*-fused acetonide was formed when aldehyde was epimerized under the basic conditions (Scheme 25). This *cis*-fused acetonide intermediate was converted to opposite diastereoisomers 10-(*S*)-trioxillin and 10-(*S*)-hepoxillin, respectively, showing that all four isomers of natural products could be prepared from the same chiral building block [71,72].

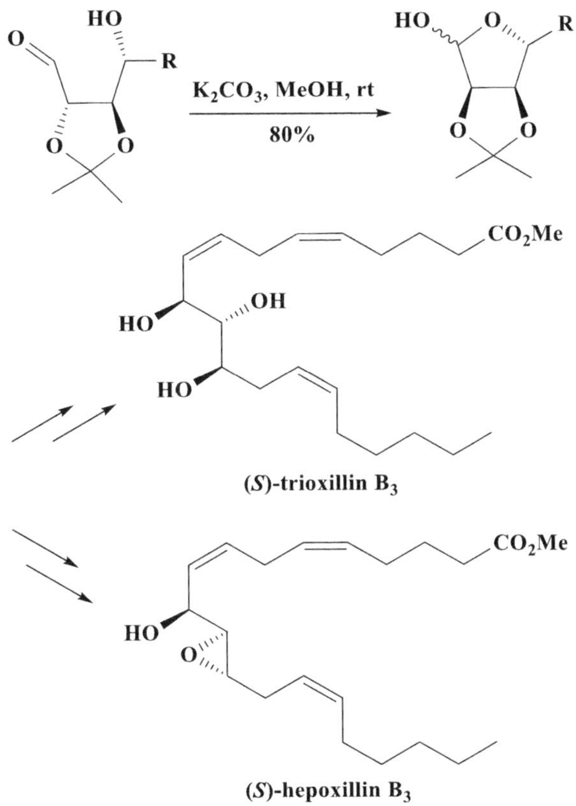

SCHEME 25

A sequential procedure was utilized as an early-stage important step in the overall formation of ovalicin, in which the doubly protected cyclic triol was subjected to mesylation, deprotection, and ring closure to synthesize the spirocyclic epoxide [73–75]. An analogous process assembled the epoxide ring late in the overall formation of neocarzinostatin chromophore aglycone [76]. The development of biocatalytic procedures for synthesizing chiral nonracemic chlorohydrins from chloroketones has opened a potentially beneficial route toward chiral epoxides by this pathway [77]. The ring closure process could be utilized as a nice complement to other concerted oxygenations. For instance, in their overall formation of (−)-kendomycin, Smith et al. [78] performed a *cis*-dihydroxylation on the macrocyclic alkene to synthesize the diol. The mesylation of secondary alcohol provided hydroxy mesylate, which underwent a ring closure with inversion under the phase-transfer conditions. In this route, the *cis*-alkene was controllably and efficiently transformed to *trans*-epoxide (Schemes 26 and 27).

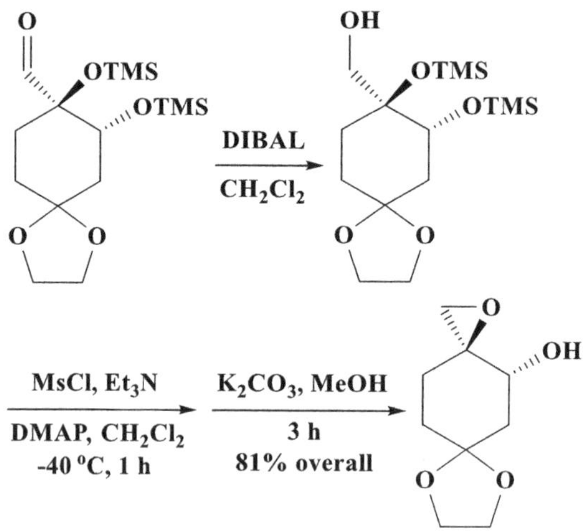

SCHEME 26

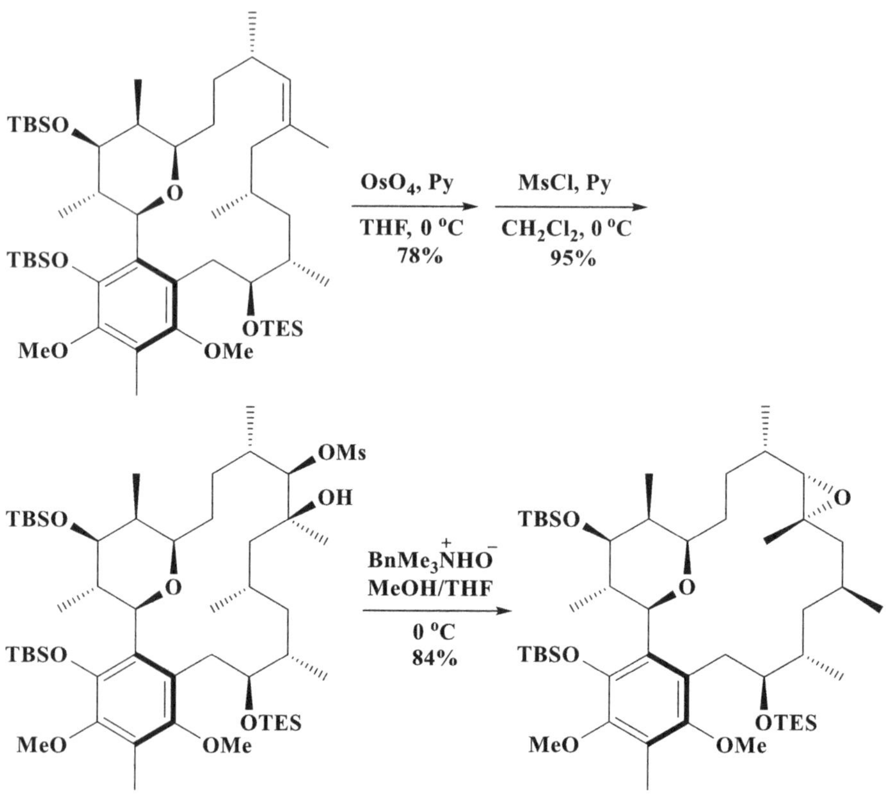

SCHEME 27

Tolstikov and Tolstikov [79] used Perlin aldehydes, prepared from glycals, as chiral templates in many natural products syntheses. The stereospecific formation of optically pure sex pheromone of gypsy moth *Porthetria dispar* L-(+)-*cis*-disparlure was achieved from Perlin aldehyde, which was prepared from 3,4,6-tri-*O*-acetyl-D-galactal (Scheme 28) [80,81]. The triacetoxy hexenal, prepared by an acetylation of aldehyde, on Wittig olefination provided triacetoxydiene. Its conversion to monotosylate was performed in four steps through the triol. The diol, obtained after the deprotection of acetonide group, was transformed to epoxide. Its oxidation with Collins reagent followed by Wittig olefination of formed aldehyde with suitable phosphonium salt provided epoxide, which on further catalytic hydrogenation gave optically pure (+)-*cis*-disparlure [82].

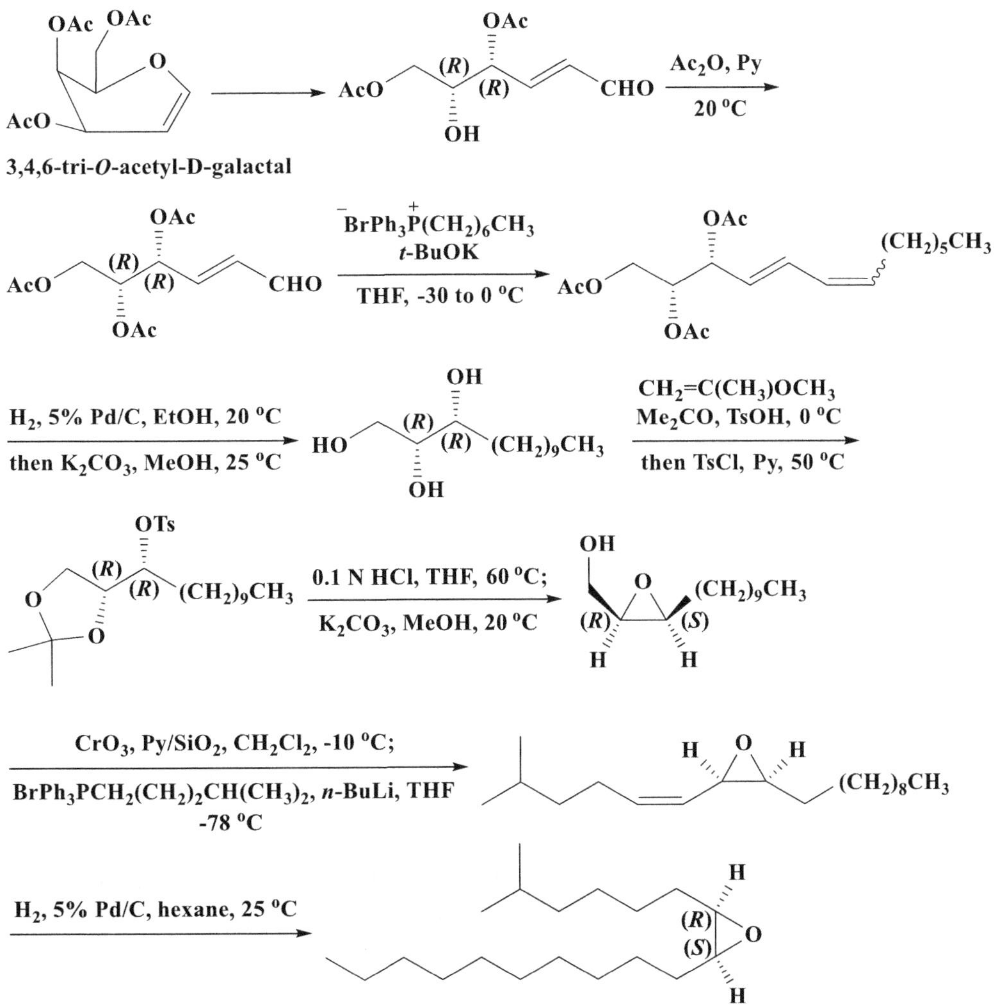

SCHEME 28

The chirality of building block was introduced into product with retention of configuration at C-5 and inversion of configuration at C-4 of the enal. The optical antipode, (−)-*cis*-disparlure, (7*S*,8*R*)-*cis*-2-methyl-7,8-epoxyoctadecane, was synthesized from galactose-derived aldehyde (Scheme 29) [83,84]. The triene was synthesized by the Wittig olefination of enal, prepared by the hydrolysis of galactal derivative, with *n*-heptyltriphenyl phosphonium bromide. The triene was further transformed to desired epoxide through the diacetate by hydrogenation followed by tosylation and then reaction with potassium carbonate in MeOH [82].

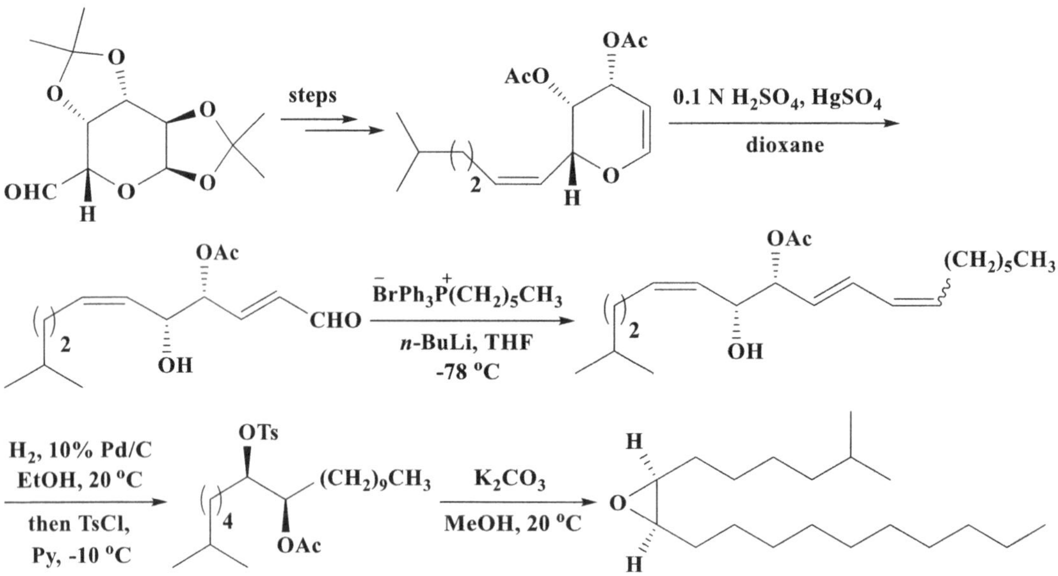

SCHEME 29

6.3 Synthesis of oxaziridines

The oxaziridinium ions are the derivatives of oxaziridine with positively charged nitrogen. The oxaziridinium salts, related electrophilic oxygen sources, were first reported by Lusinchi and Hanquet [85] as reactive intermediates for oxygen transfer to nucleophilic substrates including alkenes and sulfides. The oxaziridinium species was able to transfer oxygen to alkene substrates with the regeneration of iminium salts under the reaction conditions, providing overall a catalytic procedure. Furthermore, the propensity of iminium ions to react with Oxone triple salt to produce the oxaziridinium species renders the development of catalytic procedures possible. The oxaziridinium salts have also been prepared by either an oxidation of iminium salt with peracid or a methylation of oxaziridine with FSO_3Me (Scheme 30) [86,87]. The steroidal pyrrolinic immonium salt played an important role in an oxidation of peracid.

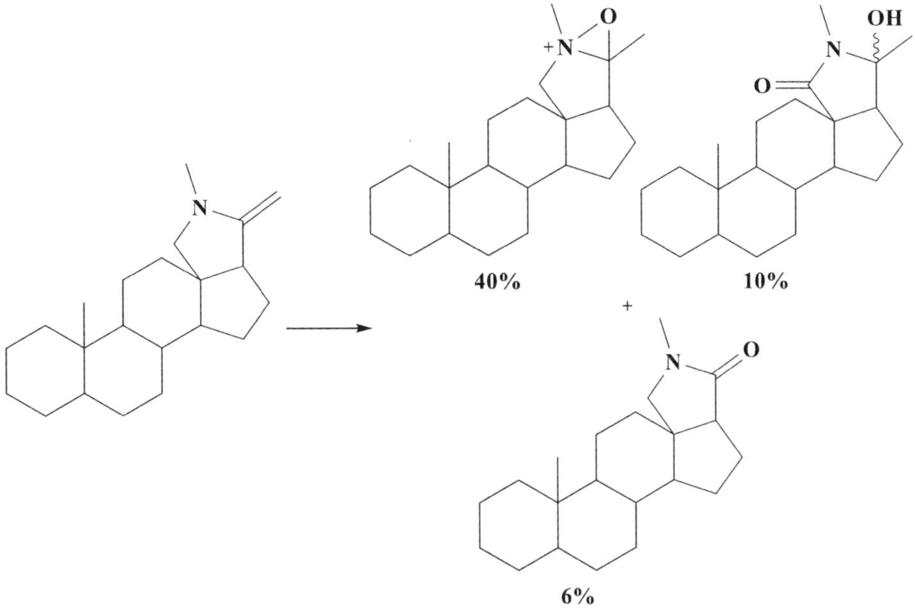

SCHEME 30

The reaction of 3-bromo-1,1,1-trifluoropropane-2-oxime and triethyl phosphite synthesized [*N*-hydroxy-*N*-(1-trifluoromethylethenyl)]amido diethylphosphate. The [*N*-hydroxy-*N*-(1-trifluoromethyl-1-hydroxyethyl)]amido diethyl phosphate was obtained upon adding H_2O to [*N*-hydroxy-*N*-(1-trifluoromethylethenyl)]amido diethylphosphate in a Markovnikov fashion across the carbon−carbon double bond [88,89]. Then, H_2O was removed utilizing DCC to synthesize the *N*-phosphorylated 3-methyl-3-trifluomethyloxaziridine (Scheme 31) [90].

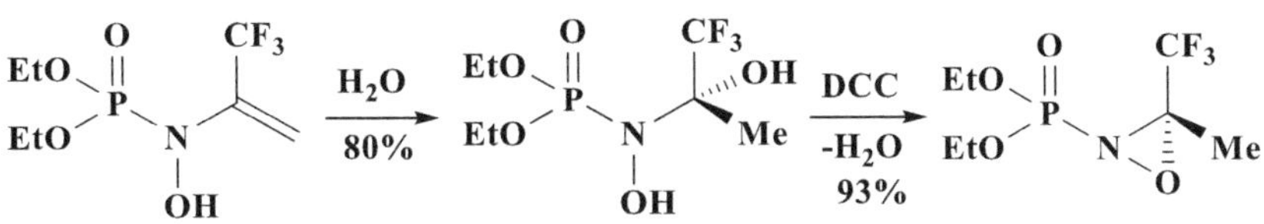

SCHEME 31

References

1. (a) N. Kaur. 2019. Gold and silver assisted synthesis of five-membered oxygen and nitrogen containing heterocycles. Synth. Commun. 49: 1459-1485. (b) N. Kaur, P. Bhardwaj, M. Devi, Y. Verma and P. Grewal. 2019. Synthesis of five-membered *O,N*-heterocycles using metal and non-metal. Synth. Commun. 49: 1345-1384. (c) R. Tyagi, N. Kaur, B. Singh and D. Kishore. 2013. A noteworthy mechanistic precedence in the exclusive formation of one regioisomer in the Beckmann rearrangement of ketoximes of 4-piperidones annulated to pyrazolo-indole nucleus by organocatalyst derived from TCT and DMF. Synth. Commun. 43:

16-25. (d) R. Tyagi, N. Kaur, B. Singh and D. Kishore. 2014. A novel synthetic protocol for the hetero-annulation of oxocarbazole and oxoazacarbazole derivatives through corresponding oxoketene dithioacetals. J. Heterocycl. Chem. 51: 18-23. (e) N. Kaur. 2014. Microwave-assisted synthesis of five-membered *O,N*-heterocycles. Synth. Commun. 44: 3509-3537. (f) N. Kaur. 2014. Microwave-assisted synthesis of five-membered *O,N,N*-heterocycles. Synth. Commun. 44: 3229-3247. (g) N. Kaur. 2018. Perspectives of ionic liquids applications for the synthesis of five and six-membered *O,N*-heterocycles. Synth. Commun. 48: 473-495. (h) N. Kaur. 2018. Photochemical-mediated reactions in five-membered *O*-heterocycles synthesis. Synth. Commun. 48: 2119-2149. (i) N. Kaur, M. Devi, Y. Verma, P. Grewal, P. Bhardwaj, N. Ahlawat and N.K. Jangid. 2019. Photochemical synthesis of fused five-membered *O*-heterocycles. Curr. Green Chem. 6: 155-183. (j) N. Kaur. 2018. Photochemical reactions for the synthesis of six-membered *O*-heterocycles. Curr. Org. Synth. 15: 298-320.(k) A.-H. Li, L.-X. Dai and V.K. Aggarwal. 1997. Asymmetric ylide reactions: epoxidation, cyclopropanation, aziridination, olefination, and rearrangement. Chem. Rev. 97: 2341-2372.

2. J.G. Smith. 1984. Synthetically useful reactions of epoxides. Synthesis 8: 629-656.
3. R.E. Parker and N.S. Isaacs. 1959. Mechanisms of epoxide reactions. Chem. Rev. 59: 737-799.
4. A.K. Ghosh and Y. Wang. 2000. Total synthesis of (-)-laulimalide. J. Am. Chem. Soc. 122: 11027-11028.
5. K.E. James, J.L. Asgian, Z.Z. Li, O.D. Ekici, J.R. Rubin, J. Mikolajczyk, G.S. Salvesen and J.C. Powers. 2004. Design, synthesis, and evaluation of aza-peptide epoxides as selective and potent inhibitors of caspases-1, -3, -6, and -8. J. Med. Chem. 47: 1553-1574.
6. Z.P. Cheng, X.L. Zhu, G.D. Fu, E.T. Kang and K.G. Neoh. 2005. Dual-Brush-type amphiphilic triblock copolymer with intact epoxide functional groups from consecutive RAFT polymerizations and ATRP. Macromolecules 38: 7187-7192.
7. P.R. Carlier. 2004. Threading the needle: mimicking natural toroidal catalysts. Angew. Chem. Int. Ed. 43: 2602-2605.
8. K. Yuan, T. Song, D. Wang, Y. Zou, J. Li, X. Zhang, Z. Tang and W. Hu. 2018. Bimetal-organic frameworks for functionality optimization: MnFe-MOF-74 as a stable and efficient catalyst for the epoxidation of alkenes with H_2O_2. Nanoscale 10: 1591-1597.
9. X. Chen, B. Gao, Y. Su and H. Huang. 2017. Enantioselective epoxidation of electron-deficient alkenes catalyzed by manganese complexes with chiral N4 ligands derived from rigid chiral diamines. Adv. Synth. Catal. 359: 2535-2541.
10. T. Zhang, X. Chen, G. Chen, M. Chen, R. Bai, M. Jia and J. Yu. 2018. Synthesis of anatase-free nano-sized hierarchical TS-1 zeolites and their excellent catalytic performance in alkene epoxidation. J. Mater. Chem. A 6: 9473-9479.
11. N.A. Grosso-Giordano, C. Schroeder, A. Okrut, A. Solovyov, C. Schottle, W. Chasse, N. Marinkovic, H. Koller, S.I. Zones and A. Katz. 2018. Outer-sphere control of catalysis on surfaces: a comparative study of Ti(IV) single-sites grafted on amorphous versus crystalline silicates for alkene epoxidation. J. Am. Chem. Soc. 140: 4956-4960.
12. Q.H. Xia, H.Q. Ge, C.P. Ye, Z.M. Liu and K.X. Su. 2005. Advances in homogeneous and heterogeneous catalytic asymmetric epoxidation. Chem. Rev. 105: 1603-1662.
13. B.S. Lane and K. Burgess. 2003. Metal-catalyzed epoxidations of alkenes with hydrogen peroxide. Chem. Rev. 103: 2457-2474.
14. B. Dede, F. Karipcin and M. Cengiz. 2009. Novel homo- and hetero-nuclear copper(II) complexes of tetradentate Schiff bases: synthesis, characterization, solvent-extraction and catalase-like activity studies. J. Hazard. Mater. 163: 1148-1156.
15. T. Ghosh and S. Pal. 2015. Synthesis and characterization of dodecahedral cerium(IV) and gadolinium(III) complexes with a tetradentate Schiff base. J. Chem. Sci. 127: 1201-1209.
16. M. Gupta, P. Sharma, M. Gupta and R. Gupta. 2015. Silica functionalized Cu(II) catalysed selective oxidation of benzyl alcohols using TEMPO and molecular oxygen as an oxidant. J. Chem. Sci. 127: 1485-1489.

17. R. Sandaroos, M.T. Goldani, S. Damavandi and A. Mohammadi. 2012. Efficient asymmetric Baeyer-Villiger oxidation of prochiral cyclobutanones using new polymer-supported and unsupported chiral Co(salen) complexes. J. Chem. Sci. 124: 871-876.
18. F. Ballistreri, R. Toscano, M. Amato, A. Pappalardo, C. Gangemi, S. Spidalieri, R. Puglisi and G.T. Sfrazzetto. 2018. A new Mn-salen micellar nanoreactor for enantioselective epoxidation of alkenes in water. Catalysts 8: 129-138.
19. J. Lu, L. Liang and M. Weck. 2016. Micelle-based nanoreactors containing Ru-porphyrin for the epoxidation of terminal olefins in water. J. Mol. Catal. A: Chem. 417: 122-125.
20. S. Tian, Q. Fu, W. Chen, Q. Feng, Z. Chen, J. Zhang, W.C. Cheong, R. Yu, L. Gu, J. Dong, J. Luo, C. Chen, Q. Peng, C. Draxl, D. Wang and Y. Li. 2018. Carbon nitride supported Fe_2 cluster catalysts with superior performance for alkene epoxidation. Nat. Commun. 9: 2353-2359.
21. E.V. Starokon, S.E. Malykhin, M.V. Parfenov, G.M. Zhidomirov and A.S. Kharitonov. 2017. Oxidation of lower alkenes by α-oxygen $(Fe^{III}{-}O^{\bullet -})_{\alpha}$ on the FeZSM-5 surface: the epoxidation or the allylic oxidation? Mol. Catal. 443: 43-51.
22. S. Huber, M. Cokoja and F.E. Kühn. 2014. Historical landmarks of the application of molecular transition metal catalysts for olefin epoxidation. J. Organomet. Chem. 751: 25-32.
23. N. Mizuno, K. Yamaguchi and K. Kamata. 2005. Epoxidation of olefins with hydrogen peroxide catalyzed by polyoxometalates. Coord. Chem. Rev. 249: 1944-1956.
24. F.P. Ballistreri, C.M. Gangemi, A. Pappalardo, G.A. Tomaselli, R.M. Toscano and G.T. Sfrazzetto. 2016. (Salen)Mn(III) catalyzed asymmetric epoxidation reactions by hydrogen peroxide in water: a green protocol. Int. J. Mol. Sci. 17: 1112-1120.
25. S.H. Kashani, M. Moghadam, S. Tangestaninejad, V. Mirkhani and I. Mohammadpoor-Baltork. 2018. Ruthenium nanoparticles immobilized on nano-silica functionalized with thiol-based dendrimer: a nanocomposite material for oxidation of alcohols and epoxidation of alkenes. Catal. Lett. 148: 1110-1123.
26. Y. Zhu, Q. Wang, R.G. Cornwall and Y. Shi. 2014. Organocatalytic asymmetric epoxidation and aziridination of olefins and their synthetic applications. Chem. Rev. 114: 8199-8256.
27. W. Adam. 1979. Cyclic peroxides. 100. Oxygen diradicals derived from cyclic peroxides. Acc. Chem. Res. 12: 390-396.
28. K. Gollnick and A. Griesbeck. 1985. Singlet oxygen photooxygenation of furans. Tetrahedron 41: 2057-2068.
29. M.L. Graziano, M.R. Iesce and R. Scarpati. 1982. Photosensitized oxidation of furans. Part 4. Influence of the substituents on the behaviour of the *endo*-peroxides of furans. J. Chem. Soc. Perkin Trans. 1 0: 2007-2012.
30. M.L. Graziano, M.R. Iesce and R. Scarpati. 1981. Unusual thermal rearrangement of the *endo*-peroxides of 2,5-dimethylfurans. J. Chem. Soc. Chem. Commun. 14: 720-721.
31. W. Grimminger and W. Kraus. 1979. Photooxidation von 3-substituierten furanen in aprotischen lösungsmitteln. Liebigs Ann. Chem. 10: 1571-1576.
32. F. Fariña, M.V. Martín and M.C. Paredes. 1973. Pseudoesters and derivatives; VI1. Synthesis of 5-hydroxy- and 5-methoxy-3-pyrrolin-2-ones. Synthesis 3: 167-168.
33. J. R. van der Merre and C.F. Garbers. 1964. J. S. Afr. Chem. Inst. 17: 149.
34. B.L. Feringa. 1987. Photo-oxidation of furans. Recl. Trav. Chim. Pays-Bas 106: 469-488.
35. C. Cramay, R. Ferdinando, M. Degueil-Castaing and B. Maillard. 1998. Anionically induced decomposition of 2-ethoxycarbonylprop-2-enyl peroxides. J. Chem. Soc. Chem. Commun. 17: 1855-1856.
36. T. Barros, P. Matias, C. Maycock and M.R. Ventura. 2003. Aziridines as a protecting and directing group. Stereoselective synthesis of (+)-bromoxone. Org. Lett. 5: 4321-4323.
37. G. Cardillo, L. Gentilucci, C. Tomasini and M. Castejon-Bordas. 1996. Synthesis of enantiomerically pure *trans* aziridine-2-carboxylates by diastereoselective Gabriel-Cromwell reaction. Tetrahedron: Asymmetry 7: 755-762.

38. Y.L. Wu, W.L. Wu, Y.L. Li, X.L. Sun and Z.H. Peng. 1996. Stereoselective synthesis of acyclic natural products containing chiral vicinal diol. Pure Appl. Chem. 68: 727-734.
39. W.L. Wu and Y.L. Wu. 1993. A concise synthesis of the methyl esters of (10*S*)-hepoxilin B3 and (10*S*)-trioxilin B3. J. Org. Chem. 58: 2760-2762.
40. T.D. Inch, R.V. Ley and P. Rich. 1968. Asymmetric synthesis. Part II. Stereospecific synthesis of benzylic centres. Some 6-deoxy-5-*C*-phenylhexose derivatives. J. Chem. Soc. C 13: 1683-1692.
41. K.R. Buszek, N. Sato and Y. Jeong. 2002. Total synthesis of octalactin A via ring-closing metathesis reaction. Tetrahedron Lett. 43: 181-184.
42. J. Mulzer and E. Ohler. 2004. Diene, enyne, and diyne metathesis in natural product synthesis. Top. Organomet. Chem. 13: 269-366.
43. M. Yoshioka, K. Hashimoto, T. Fukuhara and T. Hasegawa. 1998. Reaction of singlet oxygen with enolic tautomers of 1-aryl-2-methyl 1,3-diketones. J. Chem. Soc. Perkin Trans. 1 2: 283-288.
44. M. Stratakis, M. Orfanopoulos and C.S. Foote. 1998. Reactions of singlet oxygen and *N*-methyltriazolinediones with β,β-dimethylstyrene. Exceptional *syn* selectivity in the ene products. J. Org. Chem. 63: 1315-1318.
45. A.B. Smith, R.A. Rivero, K.J. Hale and H.A. Vaccaro. 1991. Phyllanthoside-phyllanthostatin synthetic studies. 8. Total synthesis of (+)-phyllanthoside. Development of the Mitsunobu glycosyl ester protocol. J. Am. Chem. Soc. 113: 2092-2112.
46. M. He, A.W. Lei and X.M. Zhang. 2005. Enantioselective synthesis of 3,4,5-trisubstituted γ-lactones: formal synthesis of (-)-blastmycinolactol. Tetrahedron Lett. 46: 1823-1826.
47. K. Karisalmi and A.M.P. Koskinen. 2004. Studies towards the synthesis of the C(9)-C(20) lactone-dipropionate fragment of calyculin C. Synthesis-Stuttgart 9: 1331-1342.
48. S. Eissler, M. Nahrwold, B. Neumann, H.G. Stammler and N. Sewald. 2007. Short and efficient synthesis of cryptophycin unit A. Org. Lett. 9: 817-819.
49. L.A. Power. 2008. New synthetic uses for chiral 1,3-dioxolan-4-ones. PhD Thesis, University of St. Andrews.
50. W.H. Rastetter and T.J. Richard. 1978. Generation and Cope rearrangement of two methylated oxepin oxides. Tetrahedron Lett. 19: 2995-2998.
51. W.H. Rastetter and T.J. Richard. 1978. The behavior of 2,7-dimethyloxepin toward azodicarboxylate esters: the fortuitous synthesis of a cycloheptatriene from an oxepin. Tetrahedron Lett. 19: 2999-3002.
52. K. Villeneuve and W. Tam. 2006. Ruthenium-catalyzed isomerization of oxa/azabicyclic alkenes: an expedient route for the synthesis of 1,2-naphthalene oxides and imines. J. Am. Chem. Soc. 128: 3514-3515.
53. R.A. Webster. 2011. Synthesis and reactions of unsymmetrical oxabicyclic alkenes: studies toward the total synthesis of phorbol and prostratin. PhD Thesis, University of Toronto.
54. H.M.C. Ferraz, M.K. Sano, M.R.S. Nunes and G.G. Bianco. 2002. Synthesis of cyclic enol ethers from alkenyl-β-dicarbonyl compounds. J. Org. Chem. 67: 4122-4126.
55. H.M.C. Ferraz, M.K. Sano, F.I. Bombonato and M.R.S. Nunes. 2005. Studies on the dehydroiodination of cyclic beta-iodo-enol ethers. J. Braz. Chem. Soc. 16: 884-887.
56. U. Schmidt, A. Lieberknecht, H. Griesser and F. Bartkowiak. 1984. Stereoselective total synthesis of chlamydocin and dihydrochlamydocin. Angew. Chem. Int. Ed. 23: 318-320.
57. D.B. Dess and J.C. Martin. 1991. A useful 12-I-5 triacetoxyperiodinane (the Dess-Martin periodinane) for the selective oxidation of primary or secondary alcohols and a variety of related 12-I-5 species. J. Am. Chem. Soc. 113: 7277-7287.
58. H. Itazaki, K. Nagashima, K. Sugita, H. Yoshida, Y. Kawamura, Y. Yasuda, K. matsumoto, K. Ishii, N. Uotani, H. Nakai, A. Terui, S. Yoshimatsu, Y. Ikenishi and Y. Nakagawa. Isolation and structural elucidation of new cyclotetrapeptides, trapoxins A and B, having detransformation activities as antitumor agents. J. Antibiot. 43: 1524-1532.

59. J. Taunton, J.L. Collins and S.L. Schreiber. 1996. Synthesis of natural and modified trapoxins, useful reagents for exploring histone deacetylase function. J. Am. Chem. Soc. 118: 10412-10422.
60. H. Kunz and C. Unverzagt. 1984. The allyloxycarbonyl (aloc) moiety - conversion of an unsuitable into a valuable amino protecting group for peptide synthesis. Angew. Chem. Int. Ed. 23: 436-437.
61. J. Taunton, C.A. Hassig and S.L. Schreiber. 1996. A mammalian histone deacetylase related to the yeast transcriptional regulator Rpd3p. Science 272: 408-411.
62. H.T. Chang and K.B. Sharpless. 1996. Molar scale synthesis of enantiopure stilbene oxide. J. Org. Chem. 61: 6456-6457.
63. T.M. Bott. 2011. The construction of diverse nitrogen-containing heterocycles via the reaction of amines of azides with metallocarbenes. PhD Thesis, University of Alberta.
64. G. Bojack and H. Bornowski. 1991. First synthesis of the difuransesquiterpene athanasin and the elucidation of its relative and absolute configuration. Tetrahedron 47: 9179-9186.
65. J.S. Nowick and R.L. Danheiser. 1998. Chemistry of cyclopropylacylsilanes I. α-Functionalized acylsilane reagents for the cyclopropanation of electrophilic alkenes. Tetrahedron 44: 4133-4134.
66. E.J. Corey and S. Lin. 1996. A short enantioselective total synthesis of dammarenediol II. J. Am. Chem. Soc. 118: 8765-8766.
67. E.J. Corey, G. Luo and L.S. Lin. 1997. A simple enantioselective synthesis of the biologically active tetracyclic marine sesterterpene scalarenedial. J. Am. Chem. Soc. 119: 9927-9928.
68. D.O. Jang, Y.H. Joo and D.H. Cho. 2000. Efficient synthesis of epoxides from vicinal diols via cyclic sulfates. Synth. Commun. 30: 4489-4494.
69. B. Roth, D.P. Baccanari, C.W. Sigel, J.P. Hubbell, J. Eaddy, J.C. Kao, M.E. Grace and B.S. Rauckman. 1988. 2,4-Diamino-5-benzylpyrimidines and analogs as antibacterial agents. 9. Lipophilic trimethoprim analogs as antigonococcal agents. J. Med. Chem. 31: 122-129.
70. D. Guillerm, K. Lavrador and G. Guillerm. 1995. Enantioselective synthesis of epoxy amino acids. Synth. Commun. 25: 877-882.
71. W.L. Wu and Y.L. Wu. 1993. Formal syntheses of hepoxilin B3, trioxilin B3 and substances against rice blast disease from D-mannitol. J. Chem. Soc. Chem. Commun. 10: 821-822.
72. W.L. Wu and Y.L. Wu. 1993. Formal syntheses of hepoxilin B3, trioxilin B3 and substances effective against rice blast disease and total syntheses of 11(*R*),12(*S*),13(*S*)-trihydroxyoctadeca-9(*Z*),15(*Z*)-dienoic acid. J. Chem. Soc. Perkin. Trans. 1 24: 3081-3086.
73. A. Wurtz. 1859. Ueber das aethylenoxyd. Annal. Chem. Pharm. 110: 125-128.
74. B.D. Feske, I.A. Kaluzna and J.D. Stewart. 2005. Enantiodivergent, biocatalytic routes to both taxol side chain antipodes. J. Org. Chem. 70: 9654-9657.
75. J. Yamaguchi, M. Toyoshima, M. Shoji, H. Kakeya, H. Osada and Y. Hayashi. 2006. Concise enantio- and diastereoselective total syntheses of fumagillol, RK-805, FR65814, ovalicin, and 5-demethylovalicin. Angew. Chem. Int. Ed. 45: 789-793.
76. S. Kobayashi, M. Hori, G.X. Wang and M. Hirama. 2006. Formal total synthesis of neocarzinostatin chromophore. J. Org. Chem. 71: 636-644.
77. T.M. Poessl, B. Kosjek, B. Ellmer, C.C. Gruber, K. Edegger, K. Faber, P. Hildebrandt, U.T. Bornscheuer and W. Kroutil. 2005. Non-racemic halohydrins via biocatalytic hydrogen-transfer reduction of halo-ketones and one-pot cascade reaction to enantiopure epoxides. Adv. Synth. Catal. 347: 1827-1834.
78. A.B. Smith, E.F. Mesaros and E.A. Meyer. 2006. Evolution of a total synthesis of (-)-kendomycin exploiting a Petasis-Ferrier rearrangement/ring-closing olefin metathesis strategy. J. Am. Chem. Soc. 128: 5292-5299.
79. A.G. Tolstikov and G.A. Tolstikov. 2007. Unsaturated sugars in enantiospecific synthesis of natural low-molecular bioregulators and their structural analogues. Russ. J. Bioorg. Chem. 33: 3-23.
80. A.G. Tolstikov, N.V. Khakhalina and V.N. Odinokov. 1987. A chiral synthon from triacetoxy-D-galactinal for (+)-*cis*-disparlure, the pheromone of *Porthetria dispar* L. Zh. Org. Khim. 23: 2469-2470.

81. A.G. Tolstikov, N.V. Khakhalina, V.N. Odinokov, V.M. Khalilov, L.V. Spirikhin and G.A. Tolstikov. 1989. Glycals in stereospecific synthesis. Part 1. Synthesis of (+)-*cis*-disparlure, the sex pheromone of the unpaired silk moth (*Porthetria dispar* L.). Zh. Org. Khim. 25: 296-299.
82. L.V.R. Reddy, V. Kumar, R. Sagar and A.K. Shaw. 2013. Glycal-derived δ-hydroxy α,β-unsaturated aldehydes (Perlin aldehydes): versatile building blocks in organic synthesis. Chem. Rev. 113: 3605-3631.
83. A.G. Tolstikov, O.F. Prokopenko, L.M. Khalilov, V.N. Odinokov and G.A. Tolstikov. 1991. Glycals in stereospecific synthesis. Part 5. Chiral synthons for (13*R*,15*R*)-lipoxine B (IX) and (-)-*cis*-disparlure (VI) from 6-substituted 6-deoxy-D-galactose. Zh. Org. Khim. 27: 798-802.
84. A.G. Tolstikov, O.F. Prokopenko, L.M. Khalilov, L.V. Spirikhin and G.A. Tolstikov. 1991. Synthesis and acid-induced ring opening of modified glycals. Synthons for (14*R*,15*R*)-lipoxin B and (7*S*,8*R*)-(-)-disparlure. Mendeleev Commun. 1: 51-51.
85. X. Lusinchi and G. Hanquet. 1997. Oxygen transfer reactions from an oxaziridinium tetrafluoroborate salt to olefins. Tetrahedron 53: 13727-13738.
86. P. Milliet, A. Picot and X. Lusinchi. 1976. Formation d'un sel d'oxaziridinium quaternaire par methylation d'un oxaziranne - mise en evidence de ses proprieties oxydantes. Tetrahedron Lett. 17: 1573-1576.
87. A. Picot, P. Milliet and X. Lusinchi. 1976. Action de l'acide *p*-nitroperbenzoïque et de l'eau oxygénée sur un sel d'immonium hétérocyclique stéroïdique et sur l'énamine correspondante. Tetrahedron Lett. 17: 1577-1580.
88. H. Hund and G.-V. Röschenthaler. 1996. 3-Bromo-1,1,1-trifluoro-propane-2-oxime and triethylphosphite. Phosphorus, Sulfur, Silicon Relat. Elem. 119: 87-91.
89. H. Hund and G.-V. Röschenthaler. 1993. New phosphorus derivatives of nitrosoalkanes and oximes. Phosphorus, Sulfur, Silicon Relat. Elem. 75: 209-212.
90. C. Alonso, J.M. de los Santos, J. Vicario and F. Palacios. 2011. Phosphorus substituted hydroxylamine and hydroxamic acid derivatives: synthesis and reactivity. ARKIVOC (iii): 221-253.

CHAPTER

7 Synthesis of dioxiranes

Chapter outline

7.1 Introduction

The heterocyclic chemistry comprises a half of all organic chemistry research globally. A large number of biologically active molecules that consist of heterocyclic rings have made synthetic studies of new heterocyclic rings very attractive; predominantly, polyfunctionalized heterocyclic compounds play important roles in the drug discovery procedure; and investigation of drugs in late development or on the market displays that 68% of them are heterocyclic compounds. It is not astonishing that the formation of polyfunctionalized heterocyclic compounds has attracted an attention. In current years, attention has been paid on the formation of bis-heterocyclic compounds, which show several biological activities such as tuberculostatic, antibacterial, antiamebic, fungicidal, and plant growth–regulative properties. The heterocyclic frameworks symbolize the central core of various biologically active compounds [1–5].

Given the value of epoxides as synthons, among the myriad of beneficial dioxirane oxidations, the most common reaction certainly remains an oxidation of C–C double bonds of alkenes. Actually, the utilization of dioxiranes today rivals that of peracids in epoxidation reactions due to their performance under the neutral conditions and the enhanced reaction rates giving access to even highly sensitive epoxides. The alkenes of several types, either electron-poor or electron-rich, have been efficiently epoxidized using dioxiranes. An efficient oxy-functionalization of unactivated carbon–hydrogen bonds of alkanes under the tremendously mild conditions certainly ranks at present as the highlight of dioxirane chemistry [6]. For this noteworthy conversion, the high stereoselectivity along with kinetic evidence all points to an oxenoid (electrophilic) mechanism of insertion [7]. In the past many years, the dioxiranes have quickly grownup into one of the more beneficial oxidation procedures in the arsenal of synthetic chemists. Their exclusive reactivity ranges from the transfer of oxygen atom to "unactivated" carbon–hydrogen bonds of saturated hydrocarbons, to epoxidations, to an oxidation of heteroatoms having a lone pair of electrons such as sulfides and amines. In fact, applications in the formation of

3-Membered Heterocycle Synthesis. https://doi.org/10.1016/B978-0-443-13583-5.00017-7

popular DDO (dimethyldioxirane) and TFDO (methyl(trifluoromethyl)dioxirane) in isolated form have proliferated an access to key products that are beneficial in organic synthesis [8].

The dioxiranes became outstanding electrophilic oxidants used for a number of oxidative conversions. Particularly, these highly efficient reagents are successfully utilized for the stereoselective epoxidation of alkenes. An additional use is an oxygen insertion into a carbon–hydrogen bond of saturated hydrocarbons. Several aromatic hydrocarbons have also been reacted with dioxiranes to give diverse oxidized products. As far as the dioxirane oxidation of heteroatoms is concerned, the N, P, Si, and S atoms should be highlighted. As a consequence of the wide research activity in this area, the utility of dioxiranes as versatile oxidizing agents has been reported in many review articles [9–20].

Though considerable development has been made, the asymmetric epoxidation of olefins and allylic alcohols, chiral auxiliary-based asymmetric epoxidation of α,β-unsaturated acid derivatives is also an important approach in organic synthesis. For an epoxidation of alkenes, peracids [21–25] and dioxiranes [26,27] such as DMD are the popular oxidizing agents. Adam and coworkers [28] reported the opposite π-face selective epoxidation of chiral tiglic amides utilizing *m*-CPBA and DMD (dimethyl dioxirane). To prevent the formation and storing of dimethyl dioxirane [29,30], it is more appropriate to epoxidize the alkenes by in situ–synthesized dioxirane. As dioxiranes are electrophilic in nature, epoxidation of electron-deficient alkenes utilizing in situ–produced dioxiranes is generally very slow, when normal ketones such as CH_3COCH_3, hexanones, and 2-butanone are utilized [31,32].

The dioxiranes are generally prepared from ketones and Oxone (potassium peroxomonosulfate) [33,34]. An epoxidation can be carried out utilizing either dioxiranes generated in situ or isolated dioxiranes. When the dioxirane is synthesized in situ, the ketone can be utilized as a catalyst since it can be regenerated upon epoxidation.

7.2 Synthesis of dioxiranes

The polyalkylene terephthalate containing a pendant dioxirane functionality was synthesized. The polyalkylene terephthalate can either be obtained commercially like from BASF (sold as ULTRA DUR) or synthesized by known procedures from terephthaloyl chloride and 1,4-butanediol. At least one of the carbonyl groups of polyalkylene terephthalate was oxidized by general method shown in Scheme 1 where both CO groups were transformed to dioxiranes; however, it is to be understood that only one of the groups may be transformed and not every monomer unit in the polymer is essentially transformed [35].

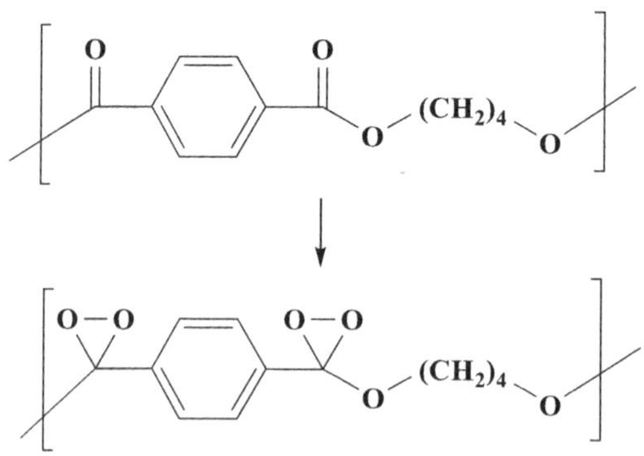

SCHEME 1

Scheme 2 shows the formation of poly(methyl)acrylate containing a pendant dioxirane functionality. The poly(methyl)acrylate is obtained commercially or prepared from methyl acrylate. The poly(methyl)acrylate was oxidized. It should be noted that not every monomeric unit of the polymer may contain a dioxirane functionality [35].

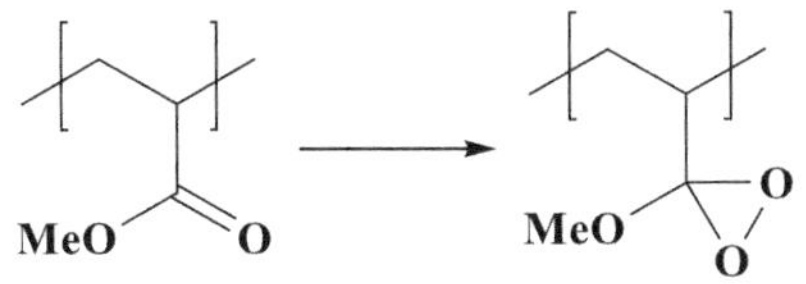

SCHEME 2

The polylactate with a pendant dioxirane functionality was prepared as given in Scheme 3. The polylactic acid, an illustrative polylactate, was synthesized or obtained commercially. The polylactic acid was oxidized [35].

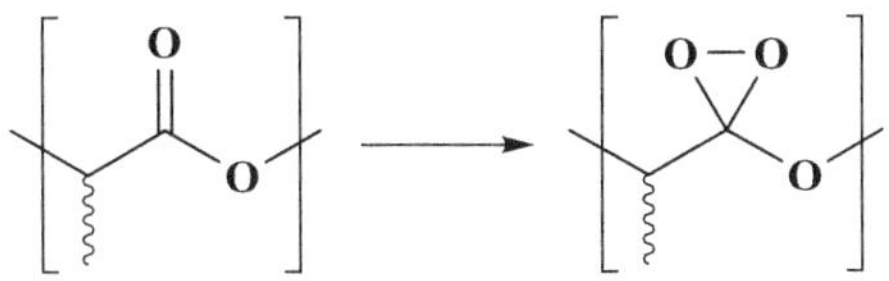

SCHEME 3

The polycarbonate, poly(bis-phenol carbonate), was synthesized by known approaches or obtained commercially. The poly(bis-phenol carbonate) was oxidized. This type of procedure is shown in Scheme 4 where both CO groups were transformed to dioxiranes; however, only one of the groups may be transformed, and not every monomer unit in the polymer was essentially transformed [35].

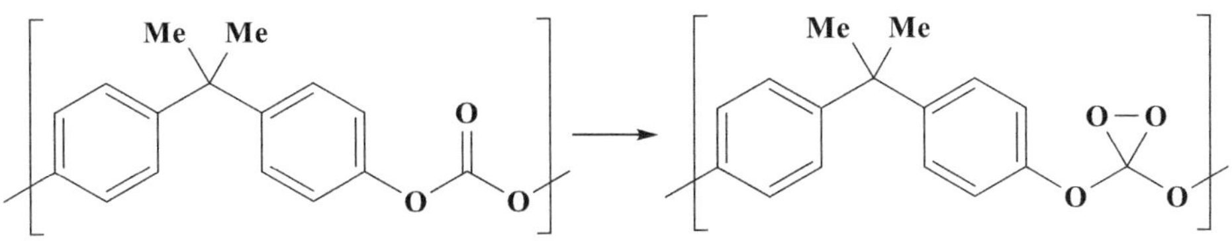

SCHEME 4

The polysaccharide, chitin, was obtained commercially or synthesized by known approaches. The chitin was oxidized according to a general process. This type of procedure is given in Scheme 5 where both CO groups were transformed to dioxiranes; however, it was assumed that only one of the groups may be transformed, and not every monomer unit in the polymer was essentially transformed [35].

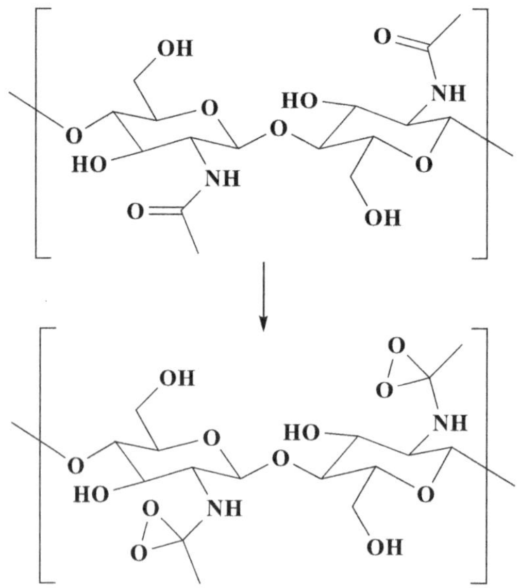

SCHEME 5

Scheme 6 shows the formation of polyacetate containing a pendant dioxirane functionality. The polyvinylacetate is obtained commercially or synthesized by known procedures. The polyvinylacetate was oxidized by a general process [35].

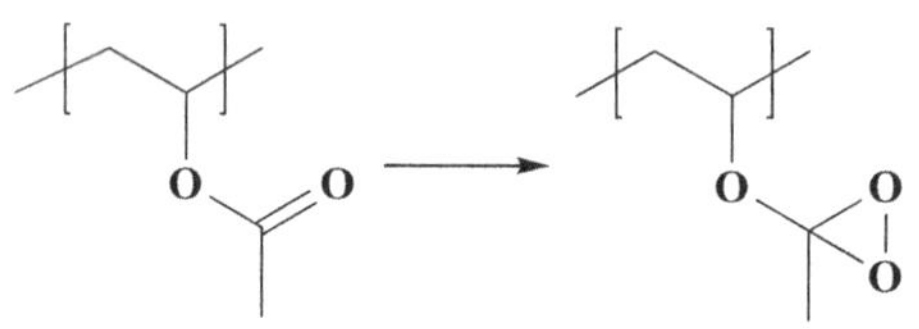

SCHEME 6

The polystyrene maleic anhydride with a pendant dioxirane functionality was prepared. The polystyrene maleic anhydride was easily accessible by known approaches or obtained commercially. The polystyrene maleic anhydride was oxidized by a general process. This type of procedure is given in Scheme 7 where both CO groups were transformed to dioxiranes; however, it was assumed that only one of the groups may be transformed, and not every monomer unit in the polymer was essentially transformed [35].

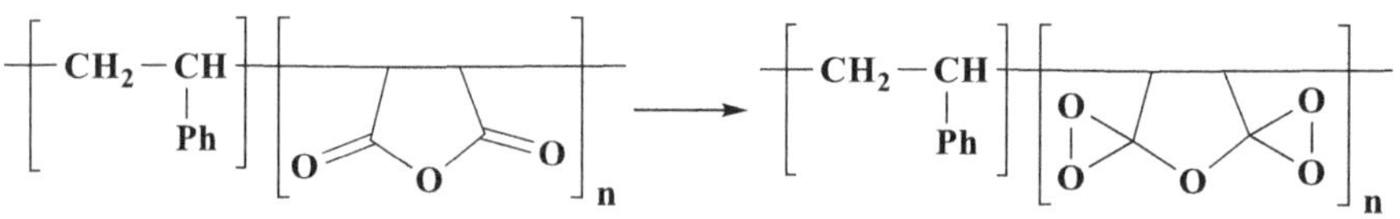

SCHEME 7

The polyurethane, i.e., poly(4,4′-diphenylmethanediisocyanate-1,2-ethanediol) and related members were synthesized. The poly(4,4′-diphenylmethanediisocyanate-1,2-ethanediol) was oxidized. This type of procedure is given in Scheme 8 in which both CO groups were transformed to dioxiranes; however, it was assumed that only one of the groups may be transformed, and not every monomer unit in the polymer was essentially transformed [35].

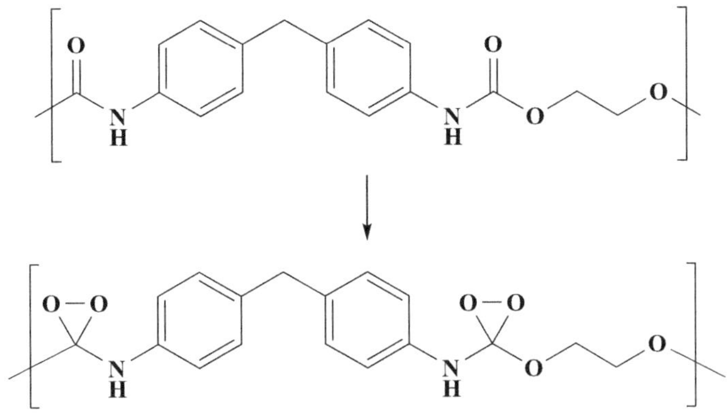

SCHEME 8

According to Scheme 9, the polyamide containing a pendant dioxirane functionality and quaternary amine were prepared. The poly(caprolactam) (nylon 6) was synthesized by known procedures or obtained commercially. The poly(caprolactam) was exhaustively methylated by adapting known procedures to give the polyamide [35].

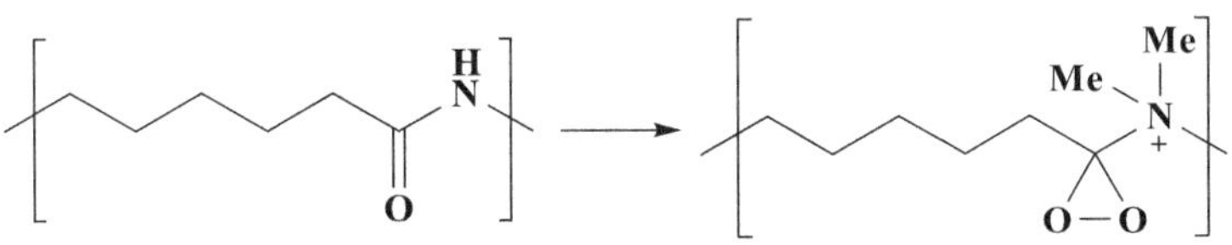

SCHEME 9

According to Scheme 10, the polyacrylamide containing a pendant dioxirane functionality was prepared. The polyacrylamide with a molecular weight range of 5,000,000–6,000,000 is accessible via known procedures and commercial sources such as Sigma–Aldrich and Acros Organics. The polyacrylamide was exhaustively methylated by published procedures to give the polyacrylamide. Then, 100 g of modified polyacrylamide was added to a mixture of H_2O (1 L), $NaHCO_3$ (2 M), and $2KHSO_5.KHSO_4.K_2SO_4$ (2 M w/r $KHSO_5$)) and stirred by magnetic stir bar for mechanical 1 h. The polymer was further purified by dialysis against distilled H_2O utilizing a dialysis membrane (molecular weight cutoff: 1000). Freeze drying afforded modified polyacrylamide containing a pendant dioxirane functionality [35].

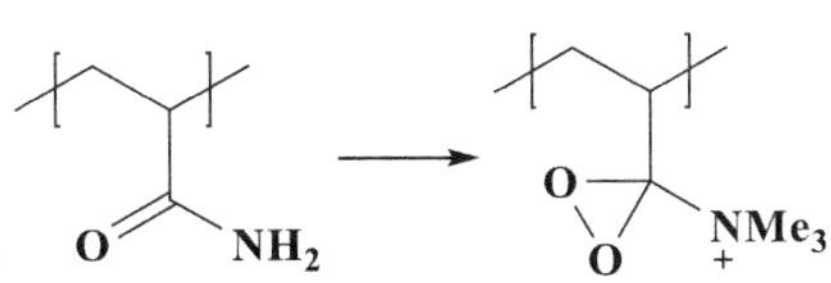

SCHEME 10

According to Scheme 11, the modified polyacrylamide containing a pendant dioxirane functionality was prepared. The polyacrylamide, polydimethylacrylamide, was prepared. The polydimethylacrylamide was exhaustively methylated by known procedures. The methylated polydimethylacrylamide (100 g) was added to a mixture of H_2O (1 L), $NaHCO_3$ (2 M), and $2KHSO_5.KHSO_4.K_2SO_4$ (2 M w/r $KHSO_5$)) and stirred through magnetic stir bar for mechanical 1 h. The polymer was purified by dialysis against distilled H_2O utilizing a dialysis membrane (molecular weight cutoff: 1000). Freeze drying delivered modified polyacrylamide containing a pendant dioxirane functionality [35].

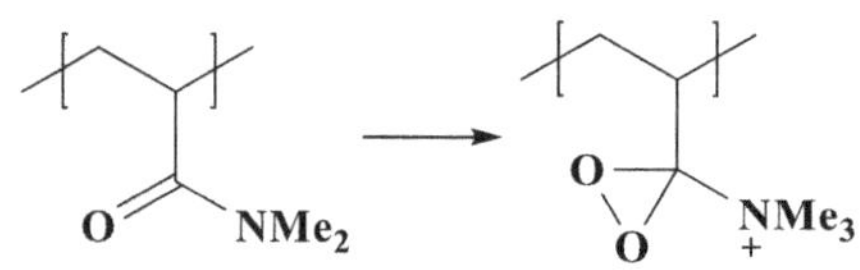

SCHEME 11

According to Scheme 12, the poly(methyl)methacrylate with a pendant dioxirane functionality was prepared. The poly(methyl)methacrylate, a methacrylate polymer, was accessible by known procedures and commercial sources such as Sigma–Aldrich. The poly(methyl)methacrylate was oxidized [35].

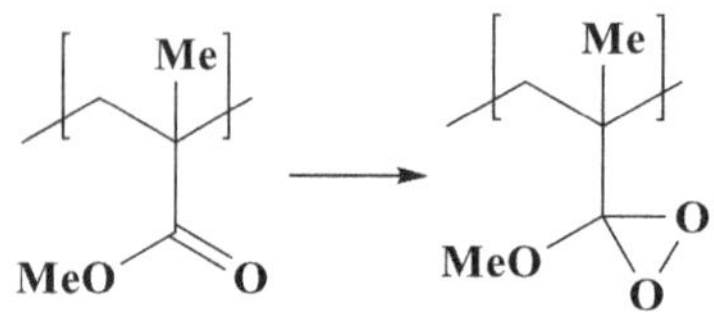

SCHEME 12

An aramid, polyparaphenylene terephthalamide, was accessible by known approaches such as the combination of terephthaloyl dichloride and *p*-phenylenediamine, and commercial sources such as DuPont (sold as KEV LAR). The polyparaphenylene terephthalamide was oxidized. This type of process is shown in Scheme 13 in which both CO groups were transformed to dioxiranes; however, it was to be understood that only one of the groups may be transformed, and not every monomer unit in the polymer was essentially transformed [35].

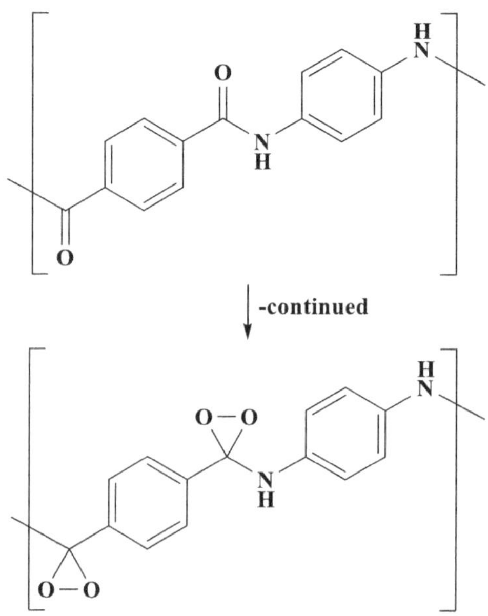

SCHEME 13

The spandex was accessible by published approaches and from commercial sources such as Invista. Spandex, the polyurea, was oxidized. This type of procedure is shown in Scheme 14 in which both CO groups were transformed to dioxiranes; however, only one of the groups may be transformed, and not every monomer unit in the polymer was essentially transformed [35].

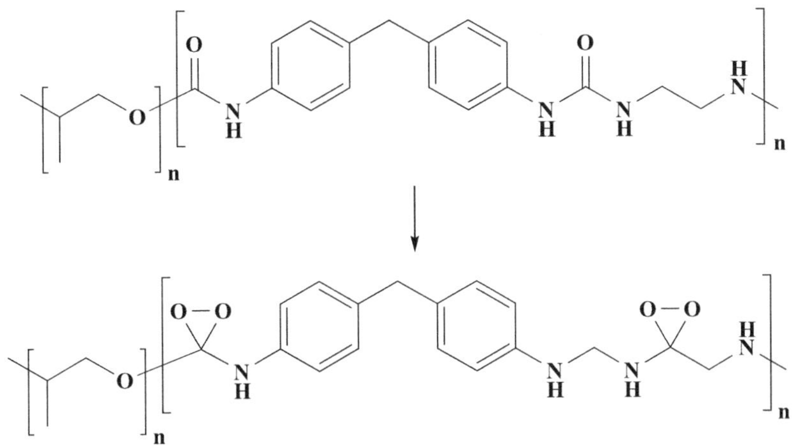

SCHEME 14

The spandex was exhaustively methylated by known approaches to give the modified polyurea. The methylated spandex (100 g) was added to a mixture of 1 L H_2O, 2 M $NaHCO_3$, and $2KHSO_5.KHSO_4.K_2SO_4$ (2 M w/r $KHSO_5$) and stirred using magnetic stir bar for mechanical 1 h. The polymer was purified by dialysis against distilled H_2O utilizing a dialysis membrane (molecular weight cutoff: 1000). Freeze drying delivered modified spandex containing a pendant dioxirane functionality. This type of procedure is shown in Scheme 15 in which both CO groups were transformed to dioxiranes; however, only one of the groups may be transformed and not every monomer unit in the polymer was essentially transformed [35].

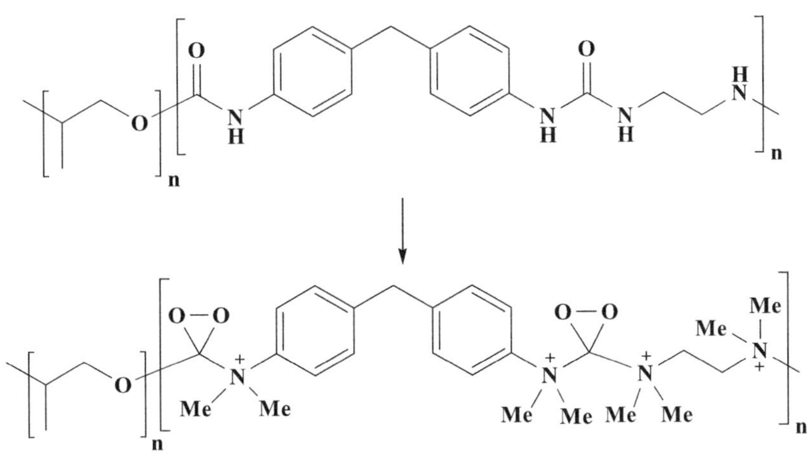

SCHEME 15

Scheme 16 shows the formation of a polymer containing a perfluoro-stabilized pendant dioxirane functionality. The 1,1,1-trifluoro-3,3,4-trimethylpent-4-en-2-one, a polymer along with a pendant perfluoro group, was synthesized according to known procedures. The published approaches were utilized for the polymerization of 1,1,1-trifluoro-3,3,4-trimethylpent-4-en-2-one [35].

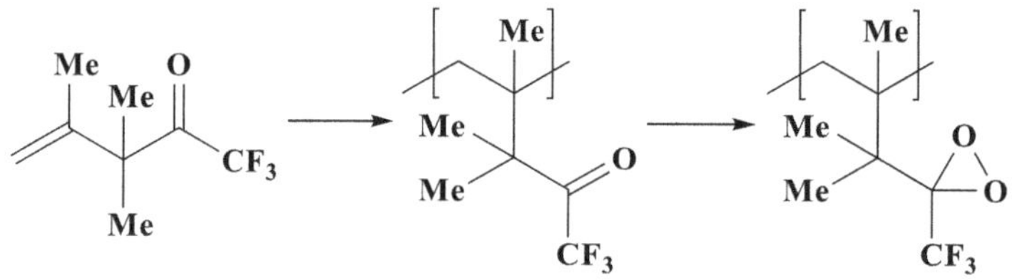

SCHEME 16

Scheme 17 shows the formation of a branched backbone polymer containing a perfluoro-stabilized pendant dioxirane functionality. The branched polymer, 4-ethoxy-1,1,1-trifluorobut-3-en-2-one, was accessible by known approaches and commercially from Sigma–Aldrich. The 4-ethoxy-1,1,1-trifluorobut-3-en-2-one was polymerized by known procedures [35].

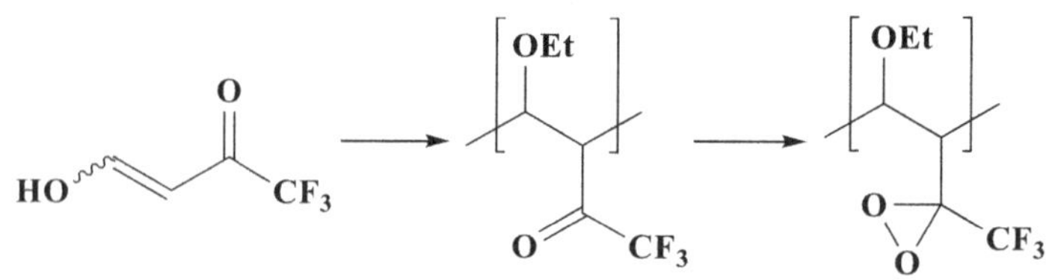

SCHEME 17

The poly(methyl-2-keto-3-butenoate) containing a pendant dioxirane functionality was prepared as shown in Scheme 18. The methyl-2-keto-3-butenoate was prepared from commercially accessible 2-keto-3-butenoic acid by known approaches. The methyl-2-keto-3-butenoate was polymerized by known procedures [35].

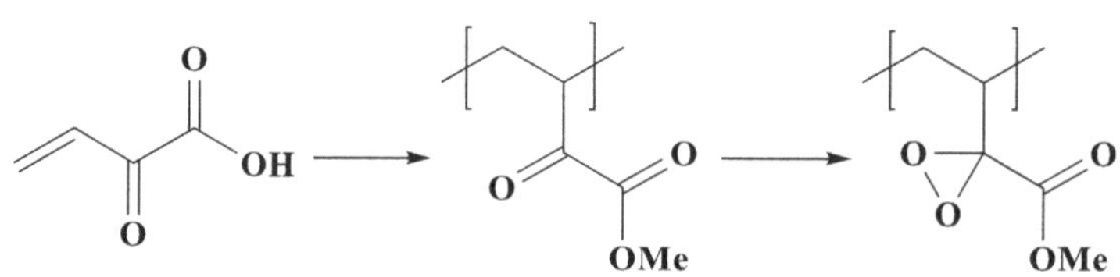

SCHEME 18

Scheme 19 shows the formation of poly(methyl-2-keto-4-pentenoate) bearing a pendant dioxirane functionality. The methyl-2-keto-4-pentenoate was synthesized from commercially accessible 2-keto-4-pentenoic acid by available approaches. The known procedures were utilized for the polymerization of methyl-2-keto-4-pentenoate [35].

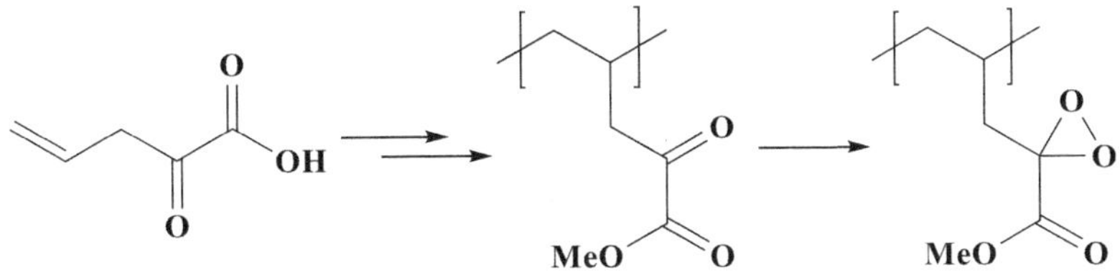

SCHEME 19

Depending on its ultraviolet absorption, the bis(trifluoromethyl)dioxirane was obtained when mono- or dilithium salt of hexafluoroacetone hydrate was reacted with elemental fluorine (Scheme 20) [36,37].

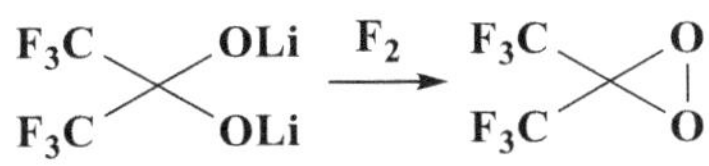

SCHEME 20

This was the first dioxirane that was stable in the gas phase at room temperature. It was first suggested as an intermediate in the reaction with oxygen of triplet CF_2 produced by an attack of arc-produced carbon atoms on CF_4. The alkenes were stereospecifically epoxidized, and dioxirane was proposed as the oxygen atom transfer reagent. Many years later, the cesium fluoride–catalyzed isomerization of fluoroformyl hypofluorite in the presence of halogen afforded dioxirane (Scheme 21). The process was improved using potassium bifluoride as a catalyst, and the similar conversion was accomplished by the IR photolysis of fluoroformyl hypofluorite utilizing a pulsed carbon dioxide laser. An unusual stability of dioxirane among dioxiranes was attributed to π donor ability of fluorine, particularly to lone pair donation into the antisymmetric Walsh orbital of the ring [36,37].

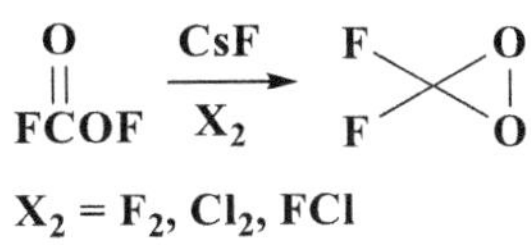

SCHEME 21

7.3 Synthesis of dioxiranes using hydrogen peroxysulfate

If the persulfate like Oxone was used as a terminal oxidant instead of H_2O_2, the dioxirane was obtained from the ketone [38]. This strained heterocycle was capable of a broad range of oxidation chemistry involving carbon–hydrogen hydroxylation. As with the perhydrate chemistry, only a catalytic amount of ketone mediator was needed in certain cases. The dioxiranes have been accessible only since the 1980s [39]. They were synthesized by an oxidation of ketones in the presence of potassium hydrogen peroxysulfate (Oxone) (Scheme 22) [40].

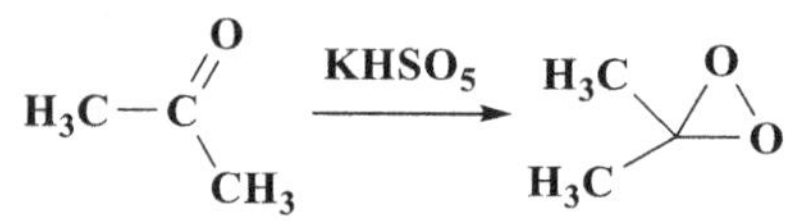

SCHEME 22

The ketone-catalyzed epoxidation of olefins is one of the most analyzed organocatalytic methods. Curci et al. [41,42] performed isotopic-labeling studies, which revealed that utilizing Oxone triple salt ($KHSO_5.KHSO_4.K_2SO_4$) as a stoichiometric oxidant, dioxirane species were prepared in situ from ketones via Criegee intermediate and that these intermediaries were responsible for the decomposition of Oxone (Scheme 23).

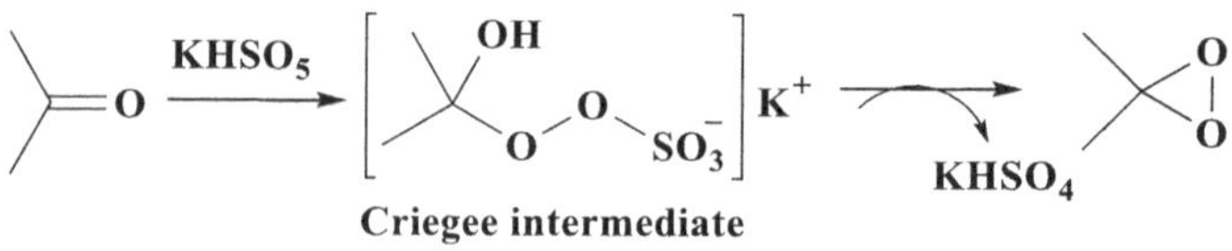

SCHEME 23

A milestone for the use of dioxiranes in synthetic organic chemistry was the formation of isolated DMD in CH_3COCH_3 solution by Murray and Jeyaraman [43]. This oxidant became tremendously popular for a couple of years. A novel easy process was worked out in the laboratory of Adam et al. [44] for the formation of DMD (Scheme 24). This technique delivered 0.09–0.10 M CH_3COCH_3 solution of DMD utilizing commercially accessible CH_3COCH_3, $NaHCO_3$, and potassium monoperoxysulfate (the triple salt $2KHSO_5.KHSO_4.K_2SO_4$, trade names Caroat, Curox, or Oxone) at a strict controlled temperature below 15°C and rigorous stirring. After the discovery of this process, DMD became "the dioxirane" one of the very extensively utilized electrophilic oxidants in synthetic organic chemistry. In case a more concentrated and acetone-free DMD was required, it could be prepared by the technique operated in the research group of Ferrer et al. [39,45]. They managed to prepare 0.4 M DMD solution in dichloromethane. Additional chlorinated solvents such as chloroform or carbon tetrachloride could also be utilized for this purpose. The DMD solution could be stored in the deep cooler of a normal refrigerator for weeks. However, it was suggested to determine the peroxide content of the solution before utilization [46].

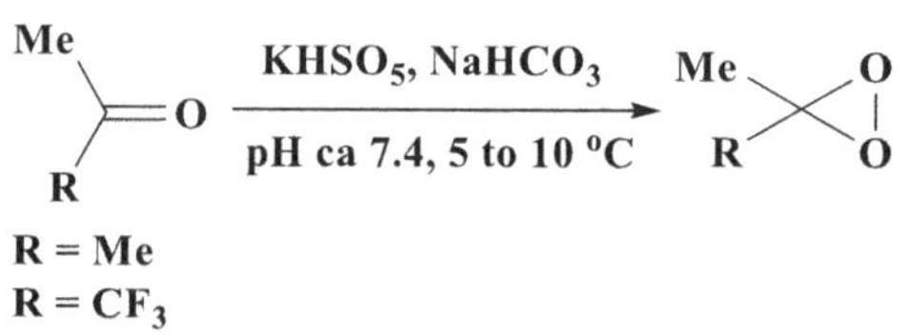

SCHEME 24

Murray and Jeyeraman [43] isolated the first dioxirane, i.e., DMD (Scheme 25). It was prepared from CH_3COCH_3 and potassium peroxymonosulfate (or known as potassium caroate, which is commercially sold as Oxone under the buffered conditions and is isolated by low temperature distillation as CH_3COCH_3 solution) [44–46].

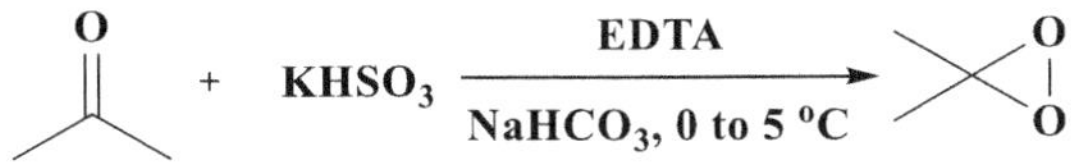

SCHEME 25

The use of dimethyldioxirane provided good yields and clean reactions; it needs cheap and commercially accessible reagents. However, the dimethyldioxirane is unstable and has to be freshly prepared, and poses severe safety problems associated with its potential explosiveness. Therefore, the development of practical substitutes for this conversion is highly required, particularly for large-scale production. Therefore, Cheshev et al. [47] reported glycal epoxidation with in situ−formed dimethyldioxirane from Oxone/acetone in a biphasic system (dichloromethane/aqueous sodium bicarbonate) (Scheme 26).

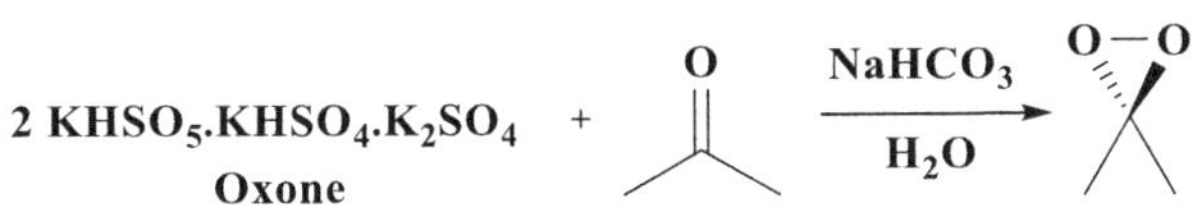

SCHEME 26

Extraordinary growth has been observed in the development of chiral ketones (through the dioxirane intermediates) based on asymmetric epoxidations (Scheme 27). Although this type of first asymmetric epoxidation was performed by Curci et al. [48,49], in the past decade, excellent enantioselectivity of these epoxidations has been obtained. Two of the most prominent workers in the field are Shi [50] (by utilizing chiral sugar-based ketone) and Yang [51] (by using chiral binapthalene derivative). Frequently, these reactions are achieved utilizing Oxone under the aqueous conditions. Many other chiral ketones have been synthesized, and these approaches have been utilized in several syntheses. Several chemists [52,53] have reported that although long reaction times were required, the amount of ketone could be reduced to 20 mol% without reducing neither enantioselectivities nor conversions. Therefore, dioxiranes could be synthesized in situ from Oxone and ketones in a catalytic reaction.

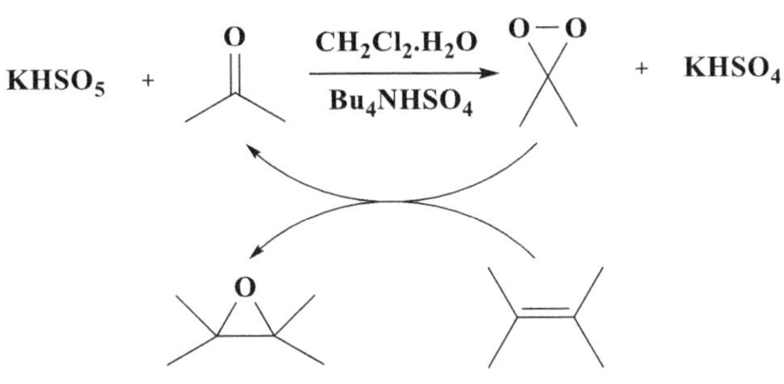

SCHEME 27

First prepared and characterized, the dioxirane ((trifluoromethyl)methyldioxirane) is usually prepared by an oxidation of 1,1,1-trifluoroacetone with potassium monoperoxosulfate (caroate). Yellow solutions of (trifluoromethyl)methyldioxirane are fairly stable at −20°C. It was also produced in homogenous organic solutions up to 1 M by an oxidation with arenesulfonic peracids synthesized in situ from the sulfonic acid, hydrogen peroxide, and sodium hydroxide (Scheme 28) [36,37].

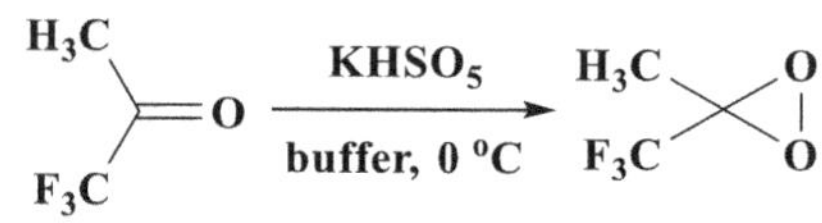

SCHEME 28

The chiral ketones, particularly the carbocyclic analogs of fructose [50], were efficient organocatalysts for an asymmetric epoxidation of *cis*- and *trans*-alkenes via peroxo ketone with Oxone (Scheme 29) [54].

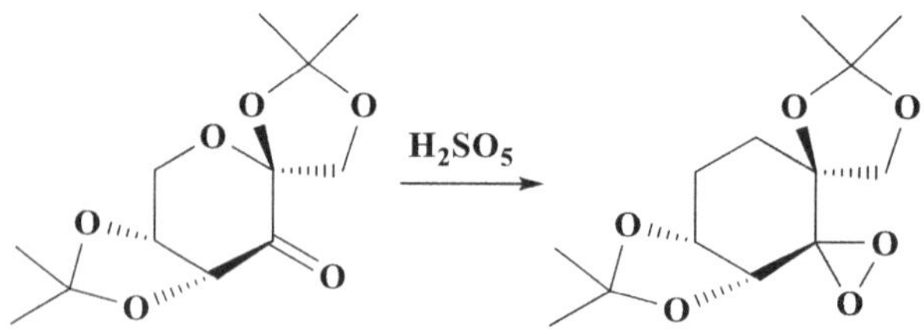

SCHEME 29

Wang et al. [55] reported that changing the buffer solution to K_2CO_3 afforded increased rate of reaction. Denmark et al. [56] reported that this increased reaction rate was due to higher concentration of peroxymonosulfate dianion at higher pH, and therefore, the concentration of conjugate base would also be higher, which would enable the formation of dioxirane (Scheme 30). Increasing the pH, by changing the base from sodium bicarbonate to potassium carbonate afforded 55% conversion of alkene after 8 h in the presence of 10 mol% catalyst.

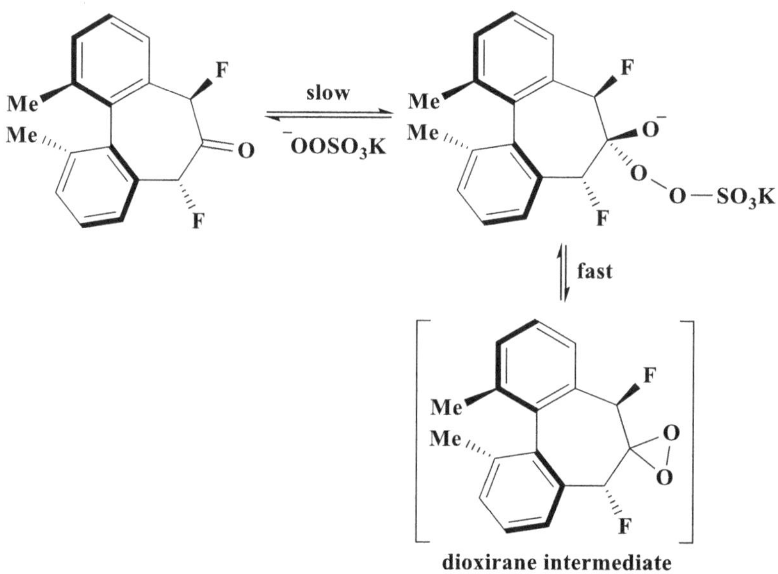

SCHEME 30

The ketones were epoxidized using Oxone reagent. The caroate ion was the actual oxidizing agent (Scheme 31) [57].

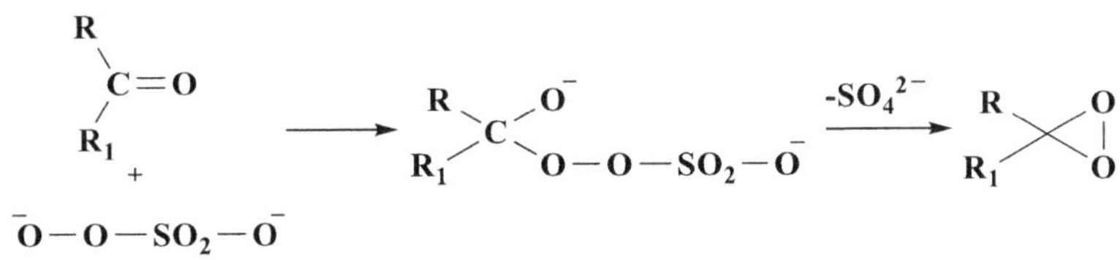

SCHEME 31

7.4 Synthesis of dioxiranes using Oxone

The chiral dioxiranes prepared from chiral ketones and Oxone ($2KHSO_5.KHSO_4.K_2SO_4$) are appropriate catalysts for the enantioselective epoxidation of cinnamate esters [58,59] and cinnamic acid derivatives [60]. These catalytic systems do not have any metal, and the oxygen transfer proceeded through the peroxocarbocyclic species [61]. With these catalysts, the *trans*-olefins are smoothly transformed than *cis*-olefins. However, they were less active than Jacobsen's catalyst. Additionally, the chiral dioxiranes were readily destroyed under the reaction conditions because of the Baeyer—Villiger oxidation reaction, which was favored at high pH ($pH > 10.5$) [62]. In contrast, atomic oxygen sources obtained from Oxone such as DMDO and tetrabutyl ammonium monoperoxosulfate (Bu_4NHSO_5) were excellent oxidants in consideration with Jacobsen's catalyst for an epoxidation of different functionalized and unfunctionalized olefins [63,64]. The dimethyldioxirane, synthesized from CH_3COCH_3 and Oxone, was a good oxidant for an asymmetric epoxidation of isoflavones and chromenes [65], as the utilization of tetrabutyl ammonium monoperoxosulfate has only been examined for unfunctionalized olefins. The mechanism was suggested for dimethyldioxirane in which the synthesis occurred in a buffered medium at pH 7—7.5 (Scheme 32) [66].

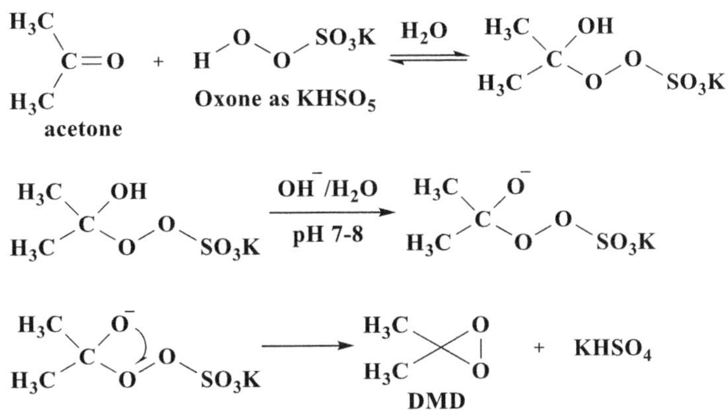

SCHEME 32

The DMDO is the reagent of choice for most of the epoxidations. The reaction occurred at rt or below and in neutral solution. The reaction was stereospecific and rapid and provided quantitative yield. In most of the cases, pure product was obtained by the removal of solvent. The method was applicable to many unsaturated systems. In nearly every case, the use of dimethyldioxirane provided a higher yield than *m*-chloroperoxybenzoic acid. The difference was dramatic in several cases. The utilization of *m*-chloroperoxybenzoic acid resulted in the opening of sensitive epoxides under the acidic reaction conditions. This problem was solved utilizing dimethyldioxirane. In many cases, dimethyldioxirane provided an epoxide when other approaches failed. A particularly pleasing instance [67] of this was the formation of 8,9-epoxide of moiety present in aflatoxin B1 and mycotoxin. Linear free energy relationship studies [68] have shown that dimethyldioxirane is an electrophilic reagent. The substrates with electron-withdrawing substituents needed longer reaction time. The earlier reported [69] faster rates for dimethyldioxirane epoxidation of *cis*-alkenes compared with their *trans*-stereoisomers were seen in the extended reaction times needed for the *trans*-isomers. This difference in rates was taken as a support for a spiro transition state in an epoxidation. Many reviews [70–73] are published on the chemistry of dimethyldioxirane including its applications as epoxidizing reagent (Scheme 33).

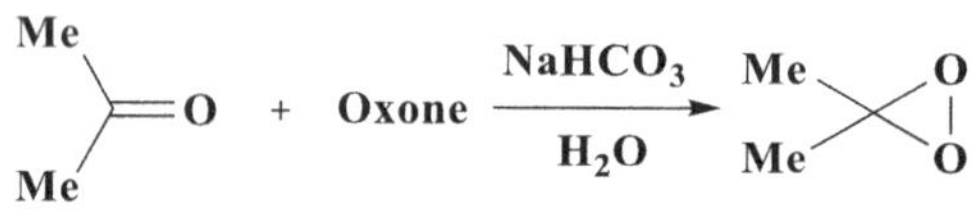

SCHEME 33

The dimethyldioxirane was prepared through an oxidation of CH_3COCH_3 with Oxone in an aqueous $NaHCO_3$ buffer solution (Scheme 34) [74,75].

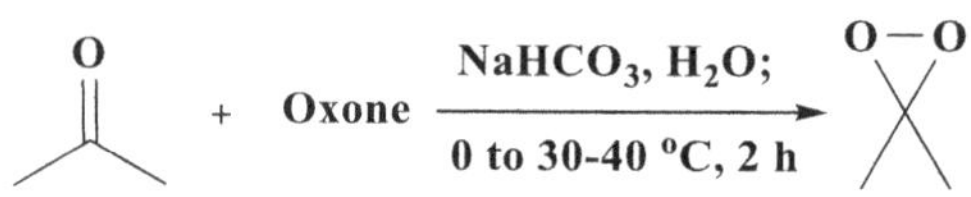

SCHEME 34

Many ketones were transformed into three-membered cyclic peroxides, i.e., dioxiranes by treatment with buffered aqueous solutions of Oxone (Scheme 35) [76].

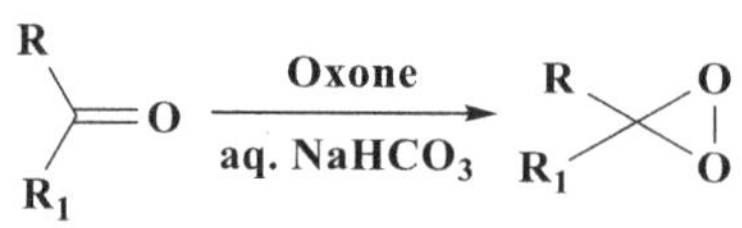

SCHEME 35

A paper by Armstrong and coworkers [77] provided the outcomes of ^{18}O-labeling study, which indicated that the dioxirane intermediate was possibly not responsible for alkene epoxidation in the ketone/Oxone system. Armstrong's group [77] presumed that an addition of Oxone to ketone was nonstereoselective and that a dioxirane intermediate would result in 50% label incorporation into the epoxide showing either of the diastereotopic oxygen atoms was geometrically capable of being transferred to the alkene. However, no such label transfer to epoxide occurred when a labeled 4-*t*-butylcyclohexanone was submitted to reaction conditions (Scheme 36). Also, there was no loss of ^{18}O from the ketone carbonyl. An explanation they provided for this was that dioxirane was perhaps not involved in an epoxidation and that the tetrahedral species resulting from an addition of HSO to the CO group was capable of alkene epoxidation. The ring closure of tetrahedral species was the rate-determining step in dioxirane synthesis; therefore, it was possible that an epoxidation by tetrahedral species was faster than ring closure to the dioxirane.

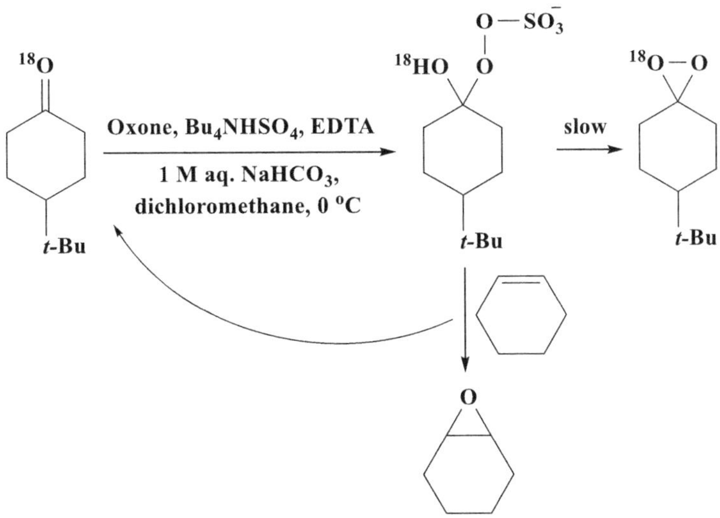

SCHEME 36

References

1. (a) N. Kaur and D. Kishore. 2014. Solid-phase synthetic approach toward the synthesis of oxygen-containing heterocycles. Synth. Commun. 44: 1019-1042. (b) N. Kaur. 2018. Green synthesis of three to five-membered *O*-heterocycles using ionic liquids. Synth. Commun. 48: 1588-1613. (c) N. Kaur. 2019. Ionic liquid assisted synthesis of six-membered oxygen heterocycles. SN Appl. Sci. 1: 1-20. (d) N. Kaur. 2019. Ionic liquid: an efficient and recyclable medium for the synthesis of fused six-membered oxygen heterocycles. Synth. Commun. 49: 1679-1707. (e) N. Kaur and D. Kishore. 2013. An insight into hexamethylenetetramine: a versatile reagent in organic synthesis. J. Iran. Chem. Soc. 10: 1193-1228. (f) N. Kaur and D. Kishore. 2012. Montmorillonite: an efficient, heterogeneous, and green catalyst for organic synthesis. J. Chem. Pharmaceut. Res. 4: 991-1015. (g) N. Kaur and D. Kishore. 2014. Peroxy acids: role in organic synthesis. Synth. Commun. 44: 721-747. (h) N. Kaur. 2018. Ruthenium catalysis in six-membered *O*-heterocycles synthesis. Synth. Commun. 48: 1551-1587.

2. (a) N. Kaur, P. Bhardwaj, M. Devi, Y. Verma and P. Grewal. 2019. Gold-catalyzed C-O bond forming reactions for the synthesis of six-membered *O*-heterocycles. SN Appl. Sci. 1: 1-37. (b) N. Kaur. 2017. Methods for metal and non-metal-catalyzed synthesis of six-membered oxygen-containing polyheterocycles. Curr. Org. Synth. 14: 531-556. (c) N. Kaur and D. Kishore. 2014. Synthetic strategies applicable in the synthesis of privileged scaffold: 1,4-benzodiazepine. Synth. Commun. 44: 1375-1413. (d) N. Kaur. 2013. An insight into medicinal and biological significance of privileged scaffold: 1,4-benzodiazepine. Int. J. Pharm. Biol. Sci. 4: 318-337. (e) N. Kaur, M. Devi, Y. Verma, P. Grewal, N.K. Jangid and J. Dwivedi. 2019. Seven and higher-membered oxygen heterocycles: metal and non-metal. Synth. Commun. 49: 1508-1542. (f) N. Kaur, M. Devi, Y. Verma, P. Grewal, P. Bhardwaj, N. Ahlawat and N.K. Jangid. 2019. Applications of metal and non-metal catalysts for the synthesis of oxygen containing five-membered polyheterocycles: a mini review. SN Appl. Sci. 1: 1-32. (g) N. Kaur, N. Ahlawat, P. Grewal, P. Bhardwaj and Y. Verma. 2019. Organo or metal complex catalyzed synthesis of five-membered oxygen heterocycles. Curr. Org. Chem. 23: 2822-2847. (h) N. Kaur, N. Ahlawat, Y. Verma, P. Grewal, P. Bhardwaj and N.K. Jangid. 2020. Metal and organo-complex promoted synthesis of fused five-membered *O*-heterocycles. Synth. Commun. 50: 457-505. (i) M. Devi, S. Jaiswal, S. Jain, N. Kaur and J. Dwivedi. 2021. Synthetic and biological attributes of pyrimidine derivatives: a recent update. Curr. Org. Synth. 18: 790-825. (j) M. Devi, S. Jaiswal, J. Dwivedi and N. Kaur. 2021. Synthetic aspects of condensed pyrimidine derivatives. Curr. Org. Chem. 25: 2625-2649.
3. (a) N. Kaur. 2017. Applications of gold catalysts for the synthesis of five-membered *O*-heterocycles. Inorg. Nano Met. Chem. 47: 163-187. (b) N. Kaur. 2014. Palladium-catalyzed approach to the synthesis of five-membered *O*-heterocycles. Inorg. Chem. Commun. 49: 86-119. (c) N. Kaur. 2019. Application of silver-promoted reactions in the synthesis of five-membered *O*-heterocycles. Synth. Commun. 49: 743-789. (d) N. Kaur and D. Kishore. 2013. Metal and non-metal based catalysts for oxidation of organic compounds. Catal. Surv. Asia 17: 20-42. (e) K. Rao, R. Tyagi, N. Kaur and D. Kishore. 2013. An expedient protocol to the synthesis of benzo(*b*)furans by palladium induced heterocyclization of corresponding 2-allylphenols containing electron rich and electron capturing substituents in the arene ring. J. Chem. 1-5. (f) N. Kaur, Y. Verma, N. Ahlawat, P. Grewal, P. Bhardwaj and N.K. Jangid. 2020. Copper-assisted synthesis of five-membered *O*-heterocycles. Inorg. Nano Met. Chem. 50: 705-740. (g) N. Kaur. 2018. Ruthenium-catalyzed synthesis of five-membered *O*-heterocycles. Inorg. Chem. Commun. 99: 82-107. (h) N. Kaur. 2014. Microwave-assisted synthesis of five-membered *O*-heterocycles. Synth. Commun. 44: 3483-3508. (i) N. Kaur, P. Grewal and K. Poonia. 2021. Dicarbonyl compounds in *O*-heterocycle synthesis. Synth. Commun. 51: 2423-2444. (j) N. Kaur, N. Ahlawat, Y. Verma, P. Grewal, P. Bhardwaj and N.K. Jangid. 2021. Crown ethers for the synthesis of heterocycles. Curr. Org. Chem. 25: 1270-1297.
4. (a) N. Kaur. 2018. Photochemical irradiation: seven- and higher-membered *O*-heterocycles. Synth. Commun. 48: 2935-2964. (b) N. Kaur and D. Kishore. 2014. Microwave-assisted synthesis of seven- and higher-membered *O*-heterocycles. Synth. Commun. 44: 2739-2755. (c) N. Kaur. 2019. Nickel catalysis: six membered heterocycle syntheses. Synth. Commun. 49: 1103-1133. (d) N. Kaur. 2019. Synthesis of six-membered *N*-heterocycles using ruthenium catalysts. Catal. Lett. 14: 1513-1539. (e) N. Kaur, Aditi and D. Kishore. 2016. A facile synthesis of face 'D' quinolino annulated benzazepinone analogues with its quinoline framework appended to oxadiazole, triazole and pyrazole heterocycles. J. Heterocycl. Chem. 53: 457-460. (f) A. Anand, N. Kaur and D. Kishore. 2014. An efficient one pot protocol to the annulation of face 'd' of benzazepinone ring with pyrazole, isoxazole and pyrimidine nucleus through the corresponding oxoketene dithioacetal derivative. Adv. Chem. 1-5. (g) N. Kaur. 2019. Seven-membered *N*-heterocycles: metal and non-metal assisted synthesis. Synth. Commun. 49: 987-1030. (h) N. Kaur. 2015. Metal catalysts: applications in higher-membered *N*-heterocycles synthesis. J. Iran. Chem. Soc. 12: 9-45. (i) N. Kaur, N. Ahlawat, Y. Verma, P. Grewal, P. Bhardwaj and N.K. Jangid. 2021. Silver-assisted syntheses of fused five-membered *N*-heterocycles. Curr. Org. Chem. 25: 2232-2256. (j) N. Kaur, P. Bhardwaj and M. Gupta. 2021. Recent

developments in the synthesis of five- and six-membered *N*-heterocycles from dicarbonyl compounds. Curr. Org. Chem. 25: 2765-2790.

5. (a) N. Kaur. 2019. Synthesis of five-membered heterocycles containing nitrogen heteroatom under ultrasonic irradiation. Mini Rev. Org. Chem. 16: 481-503. (b) N. Kaur. 2018. Ultrasound assisted synthesis of six-membered *N*-heterocycles. Mini Rev. Org. Chem. 15: 520-536. (c) N. Kaur. 2018. Synthesis of six- and seven-membered heterocycles under ultrasound irradiation. Synth. Commun. 48: 1235-1258. (d) N. Kaur, P. Bhardwaj, M. Devi, Y. Verma, N. Ahlawat and P. Grewal. 2019. Ionic liquids in the synthesis of five-membered *N,N*-, *N,N,N*- and *N,N,N,N*-heterocycles. Curr. Org. Chem. 23: 1214-1238. (e) N. Kaur, R. Tyagi and D. Kishore. 2016. Expedient protocols for the installation of 1,5-benzoazepino based privileged templates on 2-position of 1,4-benzodiazepine through a phenoxyl spacer. J. Heterocycl. Chem. 53: 643-646. (f) N. Kaur, R. Tyagi and D. Kishore. 2014. Expedient protocols for the installation of 1,5-benzoazepino-based privileged templates on the 2-position of 1,4-benzodiazepine through a phenoxyl spacer. J. Heterocycl. Chem. 51: E340-E343. (g) N. Kaur. 2018. Ionic liquid promoted eco-friendly and efficient synthesis of six-membered *N*-polyheterocycles. Curr. Org. Synth. 15: 1124-1146.
6. (a) R. Curci, L. D'Accolti and C. Fusco. 2006. A novel approach to the efficient oxygenation of hydrocarbons under mild conditions. Superior oxo transfer selectivity using dioxiranes. Acc. Chem. Res. 39: 1-9. (b) L. Cassidei, M. Fiorentino, R. Mello, O. Sciacovelli and R. Curci. 1987. Oxygen-17 and carbon-13 identification of the dimethyldioxirane intermediate arising in the reaction of potassium caroate with acetone. J. Org. Chem. 52: 699-700. (c) L. D'Accolti, C. Fusco, M.R. Rella and R. Curci. 2003. Concerning synthesis of ring-A fluorinated anthracyclines. The dioxirane shunt. Synth. Commun. 33: 3009-3016. (d) W. Adam, R. Curci, M.E. Gonzalés-Nunez and R. Mello. 1991. Thermally and photochemically initiated radical chain decomposition of ketone-free methyl(trifluoromethyl)dioxirane. J. Am. Chem. Soc. 113: 7654-7658. (e) R.W. Murray and D. Gu. 1993. Dioxirane chemistry. Part 23. The effect of solvent on the dimethyldioxirane epoxidation reaction. J. Chem. Soc. Perkin Trans. 2 11: 2203-2207. (f) R. Curci, A. Dinoi and M.F. Rubino. 1995. Dioxirane oxidations: taming the reactivity-selectivity principle. Pure Appl. Chem. 67: 811-822.
7. J. Liu, K.N. Houk, A. Dinoi, C. Fusco and R. Curci. 1998. Dioxirane epoxidations of 1,1-disubstituted ethylenes. Probing for radical pathways by computations and experiments. J. Org. Chem. 63: 8565-8569.
8. C. Annese, L. D'Accolti, A. Dinoi, C. Fusco, R. Gandolfi and R. Curci. 2008. Concerning the reactivity of dioxiranes. Observations from experiments and theory. J. Am. Chem. Soc. 130: 1197-1204.
9. W. Adam and L. Hadjiarapoglou. 1993. Dioxiranes: oxidation chemistry made easy. Top. Curr. Chem. 164: 45-62.
10. E.L. Clennan and M.-F. Chen. 1995. Photooxidations of sulfenic acid derivatives. 5. The reaction of singlet oxygen with ethyl phenyl sulfonate. J. Org. Chem. 60: 6444-6447.
11. W. Adam, A.K. Smerz and C.G. Zhao. 1997. Dioxiranes - highly reactive oxidants for stereoselective oxyfunctionalizations. J. Prakt. Chem. 339: 298-300.
12. S.E. Denmark and Z. Wu. 1999. The development of chiral, nonracemic dioxiranes for the catalytic, enantioselective epoxidation of alkenes. Synlett 1: 847-859.
13. V.P. Kazakov, A.I. Voloshin and D.V. Kazakov. 1999. Dioxiranes: from oxidative transformations to chemiluminescence. Russ. Chem. Rev. 68: 253-286.
14. W. Adam, C.R. Saha-Möller and P.A. Ganeshpure. 2001. Synthetic applications of nonmetal catalysts for homogeneous oxidations. Chem. Rev. 101: 3499-3548.
15. R.D. Bach, M.N. Glukhovtsev and C. Canepa. 1998. Oxidation of alkenes, sulfides, amines, and phosphines with peroxynitrous acid: comparison with other oxidants such as peroxyformic acid and dimethyldioxirane. J. Am. Chem. Soc. 120: 775-783.
16. D.V. Deaubel. 2001. Are peroxyformic acid and dioxirane electrophilic or nucleophilic oxidants? J. Org. Chem. 66: 3790-3796.
17. A. Greer. 2003. A view of unusual peroxides. Science 302: 235-236.

18. R.D. Bach and O. Dmitrenko. 2002. Effect of geminal substitution on the strain energy of dioxiranes. Origin of the low ring strain of dimethyldioxirane. J. Org. Chem. 67: 3884-3896.
19. R.D. Bach and O. Dmitrenko. 2002. The effect of substituents on the strain energies of small ring compounds. J. Org. Chem. 67: 2588-2599.
20. M. Freccero, R. Gandolfi, M. Sarzi-Amade and A. Rastelli. 2003. Novel pathways for oxygen insertion into unactivated C-H bonds by dioxiranes. Transition structures for stepwise routes via radical pairs and comparison with the concerted pathway. J. Org. Chem. 68: 811-823.
21. D. Yang, X.-Y. Ye, S. Gu and M. Xu. 1999. Lanthanide triflates catalyze Mn(III)-based oxidative radical cyclization reactions. Enantioselective synthesis of (-)-triptolide, (-)-triptonide, and (+)-triptophenolide. J. Am. Chem. Soc. 121: 5579-5580.
22. R.G. Harvey. 1986. Synthesis of oxidized metabolites of polycyclic aromatic hydrocarbons. Synthesis 8: 605-619.
23. T. Fukuyama, C.-L.J. Wang and Y. Kishi. 1979. Synthetic studies on polyether antibiotics. 5. Total synthesis of monensin. 2. Stereocontrolled synthesis of the right half of monensin. J. Am. Chem. Soc. 101: 260-262.
24. J.P. Marino, R. Fernández de LaPradilla and E. Laborde. 1984. Stereocontrolled synthesis of prostaglandins from cyclopentadiene monoepoxide. J. Org. Chem. 49: 5279-5280.
25. T. Nemoto, T. Ohshima and M. Shibasaki. 2000. Enantioselective total syntheses of novel PKC activator (+)-decursin and its derivatives using catalytic asymmetric epoxidation of an enone. Tetrahedron Lett. 41: 9569-9574.
26. J. Rebek, L. Marshall, J. McManis and R. Wolak. 1986. Convergent functional groups. 2. Structure and selectivity in olefin epoxidation with peracids. J. Org. Chem. 51: 1649-1653.
27. T. Koerner, H. Slebocka-Tilk and R.S. Brown. 1999. Experimental investigation of the primary and secondary deuterium kinetic isotope effects for epoxidation of alkenes and ethylene with *m*-chloroperoxybenzoic acid. J. Org. Chem. 64: 196-201.
28. W. Adam, A. Pastor, K. Peeters and E. Peeters. 2000. Opposite π-face selectivity for the DMD and *m*-CPBA epoxidations of chiral 2,2-dimethyloxazolidine derivatives of tiglic amides: control by steric interactions versus hydrogen bonding. Org. Lett. 2: 1019-1022.
29. R. Mello, M. Fiorentino, O. Sciacovelli and R. Curci. 1988. On the isolation and characterization of methyl(trifluoromethyl)dioxirane. J. Org. Chem. 53: 3890-3891.
30. R. Mello, M. Fiorentino, C. Fusco and R. Curci. 1989. Oxidations by methyl(trifluoromethyl)dioxirane. 2. Oxyfunctionalization of saturated hydrocarbons. J. Am. Chem. Soc. 111: 6749-6757.
31. M. Kurihara, S. Ito, N. Tsutsumi and N. Miyata. 1994. Stereoselective epoxidation with dioxiranes generated from ketones. Tetrahedron Lett. 35: 1577-1580.
32. S. Hajra and M. Bhowmick. 2010. Asymmetric epoxidation of cinnamic acid derivatives by in situ generated dioxiranes of chloroacetones: scope and limitations. Tetrahedron: Asymmetry 21: 2223-2229.
33. W. Adam and A.K. Smerz. 1996. Chemistry of dioxiranes - selective oxidations. Bull. Soc. Chim. Belg. 105: 581-599.
34. W. Adam, C.R. Saha-Moller and C.-G. Zhao. 2002. Dioxirane epoxidation of alkenes. Org. React. 61: 219-516.
35. A. Sjong and K. Anderson. 2013. Active polymeric filters. United States Patent US 8,524,919 B2.
36. T. Eicher and S. Hauptmann. 2003. The chemistry of heterocycles: structures, reactions, and applications. John Wiley & Sons: Chichester, UK, 5-16.
37. R.E. Banks and J.C. Tatlow. 1994. Organofluorine chemistry: nomenclature and history. R.E. Banks, B.E. Smart and J.C. Tatlow (Eds.). Organofluorine chemistry: principles and commercial applications. Plenum Press: New York, 2-4.
38. R.W. Murray, M. Singh and R. Jeyaraman. 1992. Dioxiranes. 20. Preparation and properties of some new dioxiranes. J. Am. Chem. Soc. 114: 1346-1351.

39. M. Gibert, M. Ferrer, F. Sánchez-Baeza and A. Messeguer. 1997. Availability and reactivity of concentrated dimethyldioxirane solutions in solvents other than acetone. Tetrahedron 53: 8643-8650.
40. S.E. Denmark and Z. Wu. 1997. Dioxiranes are the active agents in ketone-catalyzed epoxidations with Oxone. J. Org. Chem. 62: 8964-8965.
41. J.O. Edwards, R.H. Pater, R. Curci and F.D. Furia. 1979. On the formation and reactivity of dioxirane intermediates in the reaction of peroxoanions with organic substrates. Photochem. Photobiol. 30: 63-70.
42. R. Curci, M. Fiorentino, L. Troisi, J.O. Edwards and R.H. Pater. 1980. Epoxidation of alkenes by dioxirane intermediates generated in the reaction of potassium caroate with ketones. J. Org. Chem. 45: 4758-4760.
43. R.W. Murray and R. Jeyaraman. 1985. Dioxiranes: synthesis and reactions of methyldioxiranes. J. Org. Chem. 50: 2847-2853.
44. W. Adam, J. Bialas and L. Hadjiarapoglou. 1991. A convenient preparation of acetone solutions of dimethyldioxirane. Chem. Ber. 124: 2377-2377.
45. M. Ferrer, M. Gibert, F. Sánchez-Baeza and A. Messeguer. 1996. Easy availability of more concentrated and versatile dimethyldioxirane solutions. Tetrahedron Lett. 37: 3585-3586.
46. A. Levai. 2003. Dioxirane oxidation of sulfur-containing organic compounds. ARKIVOC (xiv): 14-30.
47. P. Cheshev, A. Marra and A. Dondoni. 2006. Direct epoxidation of D-glucal and D-galactal derivatives with in situ generated DMDO. Carbohydr. Res. 341: 2714-2716.
48. R. Curci, M. Fiorentino and M.R. Serio. 1984. Asymmetric epoxidation of unfunctionalized alkenes by dioxirane intermediates generated from potassium peroxomonosulphate and chiral ketones. Chem. Commun. 3: 155-156.
49. R. Curci, L. D'Accolti, M. Fiorentino and A. Rosa. 1995. Enantioselective epoxidation of unfunctionalized alkenes using dioxiranes generated in situ. Tetrahedron Lett. 36: 5831-5834.
50. Y. Shi. 2004. Organocatalytic asymmetric epoxidation of olefins by chiral ketones. Acc. Chem. Res. 37: 488-496.
51. D. Yang. 2004. Ketone-catalyzed asymmetric epoxidation reactions. Acc. Chem. Res. 37: 497-505.
52. K.S. Kirshenbaum and K.B. Sharpless. 1985. Improved procedure for the tungstate-catalyzed epoxidation of alpha,beta-unsaturated acids. J. Org. Chem. 50: 1979-1982.
53. S. Torii, K. Uneyama, H. Tanaka, T. Yamanaka, T. Yasuda, M. Ono and Y. Kohmoto. 1981. Efficient conversion of olefins into epoxides, bromohydrins, and dibromides with sodium bromide in water-organic solvent electrolysis systems. J. Org. Chem. 46: 3312-3315.
54. K. Sarma, A. Goswami and B.C. Goswami. 2009. Exploration of chiral induction on epoxides in lipase-catalyzed epoxidation of alkenes using (2*R*,3*S*,4*R*,5*S*)-(-)-2,3:4,6-di-*O*-isopropylidiene-2-keto-L-gulonic acid monohydrate. Tetrahedron: Asymmetry 20: 1295-1300.
55. Z.-X. Wang, S.M. Miller, O.P. Anderson and Y. Shi. 2001. Asymmetric epoxidation by chiral ketones derived from carbocyclic analogues of fructose. J. Org. Chem. 66: 521-530.
56. S.E. Denmark, D.C. Forbes, D.S. Hay, J.S. DePue and R.G. Wilde. 1995. Catalytic epoxidation of alkenes with Oxone. J. Org. Chem. 60: 1391-1407.
57. Y. Sawaki. 1993. The chemistry of hydroxyl, ether and peroxide groups. Supplement E2, S. Patai. (Ed.). Chapter 11, Wiley, Chichester, 588.
58. Z.-X. Wang, Y. Tu and Y. Shi. 1997. A new type of ketone catalyst for asymmetric epoxidation. J. Org. Chem. 62: 8622-8623.
59. A. Armstrong and B.R. Hayter. 1998. Catalytic enantioselective epoxidation of alkenes with a tropinone-derived chiral ketone. Chem. Commun. 5: 621-622.
60. O. Bortolini, M. Fogagnolo, G. Fantin, S. Maietti and A. Medici. 2001. Asymmetric epoxidation of cinnamic acid derivatives using dioxiranes generated in situ from dehydrocholic acid. Tetrahedron: Asymmetry 12: 1113-1115.

61. Z.-X. Wang, Y. Tu, M. Frohn, J.-R. Zhang and Y. Shi. 1997. An efficient catalytic asymmetric epoxidation method. J. Am. Chem. Soc. 119: 11224-11235.
62. Z.-X. Wang, S.M. Miller, O.P. Anderson and Y. Shi. 1999. A class of C2 and pseudo C2 symmetric ketone catalysts for asymmetric epoxidation. Conformational effect on catalysis. J. Org. Chem. 64: 6443-6458.
63. W. Adam, J. Jeko, A. Levai, C. Nemes, T. Patonay and P. Sebok. 1995. Enantioselective epoxidation of 2,2-dimethyl-2*H*-chromenes by dimethyldioxirane and Jacobsen's Mn(III) salen catalysts. Tetrahedron Lett. 36: 3669-3672.
64. W. Adam, R.T. Fell, A. Levai, T. Patonay, K. Peters, A. Simon and G. Tóth. 1998. Enantioselective epoxidation of isoflavones by Jacobsen's Mn(III)salen catalysts and dimethyldioxirane oxygen-atom source. Tetrahedron: Asymmetry 9: 1121-1124.
65. P. Pietikäinen. 1998. Convenient asymmetric (salen)Mn(III)-catalyzed epoxidation of unfunctionalized alkenes with hydrogen peroxide using carboxylate salt cocatalysts. Tetrahedron 54: 4319-4326.
66. K.S. Webb and S.J. Ruszkay. 1998. Oxidation of aldehydes with Oxone® in aqueous acetone. Tetrahedron 54: 401-410.
67. S.W. Baertschi, K.D. Raney, M.P. Stone and T.M. Harris. 1988. Preparation of the 8,9-epoxide of the mycotoxin aflatoxin B1: the ultimate carcinogenic species. J. Am. Chem. Soc. 110: 7929-7931.
68. A.L. Baumstark and P.C. Vasquez. 1988. Epoxidation by dimethyldioxirane. Electronic and steric effects. J. Org. Chem. 53: 3437-3439.
69. R.W. Murray and D.L. Shiang. 1990. Dioxirane chemistry. Part15. Rate studies on epoxidations by dimethyldioxirane. J. Chem. Soc. Perkin Trans. 2 2: 349-352.
70. R.W. Murray. 1989. Dioxiranes. Chem. Rev. 89: 1187-1201.
71. R. Curci and G. Modena. 1966. Oxidation of organic sulphides - XV Tetrahedron 22: 1227-1233.
72. W. Adam, R. Curci and J.O. Edwards. 1989. Dioxiranes: a new class of powerful oxidants. Acc. Chem. Res. 22: 205-211.
73. R.W. Murray and M. Singh. 2003. Synthesis of epoxides using dimethyldioxirane: *trans*-stilbene oxide [dioxirane, dimethyl- and oxirane, 2,3-diphenyl-, *trans*]. Org. Synth. 74: 91-91.
74. P. Bovicelli, E. Mincione, R. Antonioletti, R. Bernini and M. Colombari. 2001. Selective halogenation of aromatics by dimethyldioxirane and halogen ions. Synth. Commun. 31: 2955-2963.
75. C.J. Hayes, A.E. Sherlock and M.D. Selby. 2006. Enantioselective total syntheses of (-)-clasto-lactacystin β-lactone and 7-*epi*-(-)-clasto-lactacystin β-lactone. Org. Biomol. Chem. 4: 193-195.
76. W. Adam, L.H. Poglou and A. Smerz. 1991. Dioxirane epoxidation of α,β-unsaturated ketones. Chem. Ber. 124: 227-232.
77. A. Armstrong, P.A. Clarke and A. Wood. 1996. Evidence that a dioxirane is not responsible for alkene epoxidation in a ketone-Oxone® system. Chem. Commun. 7: 849-850.

CHAPTER

8 Synthesis of thiiranes from epoxides

Chapter outline

8.1 Introduction

Nowadays the organic chemists are interested in heterocyclic chemistry because of their beneficial use in the synthesis of various biologically active derivatives. The occurrence of heterocyclic compounds in nature is very broad, and they are important to life [1a–m].

The thiiranes occur in nature and are the simplest *S*-containing heterocyclic compounds. They have been utilized in the polymer, herbicide, pesticide, and pharmaceutical industries [1n]. There are several approaches developed in the literature for the formation of thiiranes [2,3]. The general process for the formation of thiiranes is the transformation of epoxides by O–S exchange reaction. The sulfur transferring agents such as thiourea [4], inorganic thiocyanates [5], silica-supported potassium thiocyanate [6] or thiourea [7], polymer-supported thiocyanates [8], and Dowex-50WX8-supported $(NH_2)_2CS$ [9] have been utilized for this purpose. The transformation of epoxides to thiiranes with $(NH_2)_2CS$ or ammonium thiocyanate in the presence of Lewis acids or other promoters such as ruthenium(III) chloride [10], bismuth chloride [11], titanium dioxide [12], $TiCl_3(OTf)$ [13], magnesium hydrogen sulfate [14], lithium tetrafluoroborate [15], lithium perchlorate [16], iodine [17], montmorillonite K10 [18], $(NH_4)_8[CeW_{10}O_{36}].20H_2O$ [19], cerium ammonium nitrate [20], Sn(TTP)$(OTf)_2$ [21], [bmim]$[PF_6]$ [22], poly(4-vinylpyridine)-supported $Ce(OTf)_4$ [23], polystyrene-supported aluminum chloride [24], etidronic acid [25], ammonium chloride [26], tetrafluoroboric acid/silicon dioxide [27], and silica chloride [28] have been also reported. Although a number of reagents or approaches are reported for the formation of thiiranes from epoxides, low yields of products, long reaction times, high temperature, use of highly acidic catalyst, use of organic solvents, problems in the separation of product from the reactant and catalyst, high catalytic loading, formation of polymeric side products, and, in various cases, use of unrecoverable and expensive catalyst are all disadvantages of these approaches. The thiiranes are *S*-containing heterocyclic compounds utilized in several

3-Membered Heterocycle Synthesis. https://doi.org/10.1016/B978-0-443-13583-5.00009-8

industries and synthetic conversions. Due to their industrial and synthetic importance, the development of practical and convenient approaches for the formation of thiiranes is needed.

The most efficient pathway is based on the transformation of epoxides into thiiranes by O–S exchange coupling. The thiourea has also been reported as an instance of one of these sulfur agents that reacts with epoxides to synthesize the thiiranes in aqueous solvents or under the wet conditions. For instance, the reaction of styrene oxide with wet $(NH_2)_2CS$ in the absence of solvent was reported to synthesize the styrene episulfide in only 45% yield after 10 h. The reaction of epoxides with thiourea in aqueous C_2H_5OH is relatively analogous to the reaction of epoxides with potassium thiocyanate, usually with lower yields and longer reaction duration, particularly in the cases of cyclic epoxides. To overcome the problems of low yields, long reaction times, high temperatures of the reactions, and the occurrence of side reactions such as desulphurization or polymerization, novel approaches including utilization of polymeric reagents or different Lewis acids have been applied. Because of the importance of decreasing environmental pollution in chemical reactions, green procedures have attracted great attention in both industrial and laboratory synthesis [29].

8.2 Synthesis of thiiranes from epoxides

This synthetic pathway was very analogous to the route utilized to synthesize the *n*-octyl thiirane. The 4-octyl styrene is commercially accessible, and efforts were made to epoxidize it to 4-octyl styrene with *m*-chloroperoxybenzoic acid, but only ring-opened hydrolysis product 4-*n*-octyl benzylethane-1,2-diol was formed. Although the 4-octyl styrene product was reported in the reaction mixture, decomposition upon workup, purification, or just standing occurred (Scheme 1) [30].

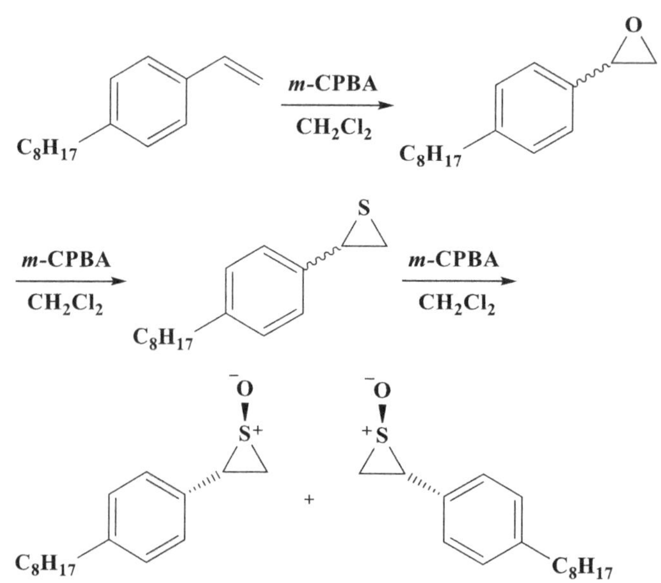

SCHEME 1

Model analysis for the formation of thiiranyl acetals through an oxirane–thiirane exchange was performed. The commercially accessible tri-9-benzyl-D-glucal was utilized as a model system instead of exocyclic fucose compound, which can be synthesized from L-fucose in five steps. The glycal was epoxidized stereospecifically and quantitatively [31] with DMDO (3,3-dimethyldioxirane) (Scheme 2). The epoxide was reacted with different reagents under the reaction conditions whereby oxirane–thiirane exchange could be achieved.

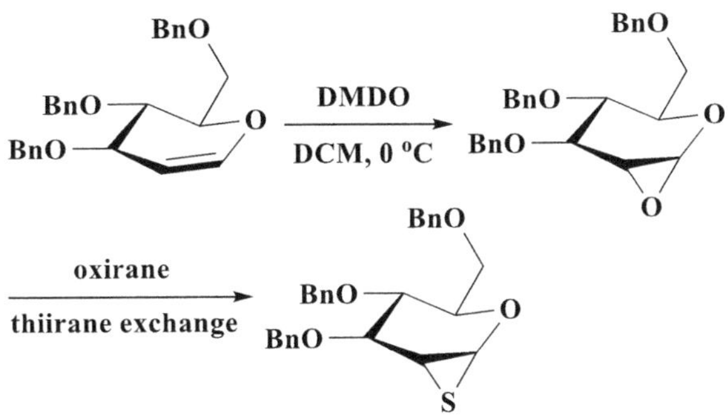

SCHEME 2

The conversion of oxiranes to thiiranes was carried out with triphenylphosphine sulfide (Scheme 3) [32].

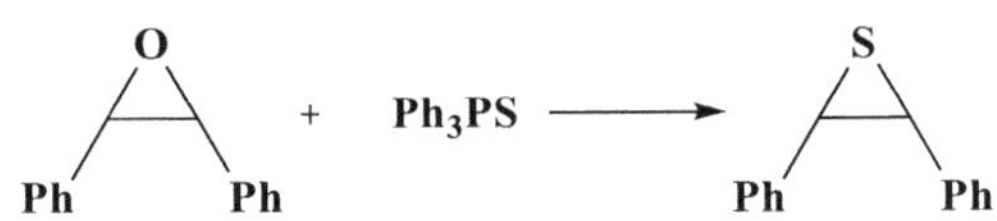

SCHEME 3

The oxazolidenes were synthesized by a thermal ring isomerization on reacting epoxides with 2-substituted imino-1,3-dithiolanes (Scheme 4) [33].

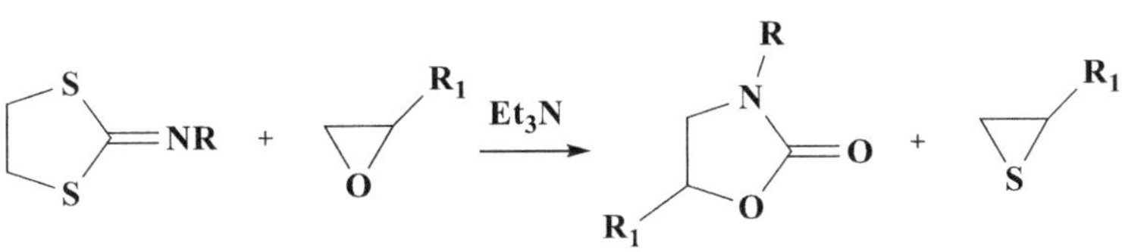

SCHEME 4

8.3 Synthesis of thiiranes from epoxides using C=S bearing moieties

The cycloaddition of epoxides with CS_2 occurred efficiently utilizing a binary catalyst system of tetradentate Schiff base aluminum complexes (assigned as salenAlX) as electrophile in connection with quaternary NH_4^+ salt as nucleophile. The steric factor of substituent groups on the aromatic rings of salenAlX and the anion of quaternary NH_4^+ salt all have considerable influence on the activity of binary catalyst system and the selectivities of formed cyclic products. The influence of time, temperature, and the molar ratio of reactants was also examined in detail with respect to propylene oxide with CS_2, and a reasonable mechanism was suggested. The cyclic thiocarbonates have received much attention due to their material science and biological activity [34–37]. One of the most efficient approaches for the formation of cyclic thiocarbonates was the synthesis of epoxide with CS_2. Depending on the reaction conditions and catalysts, the episulfide, five-membered cyclic dithiocarbonate, its regioisomers, and trithiocarbonate were obtained (Scheme 5) [38–43].

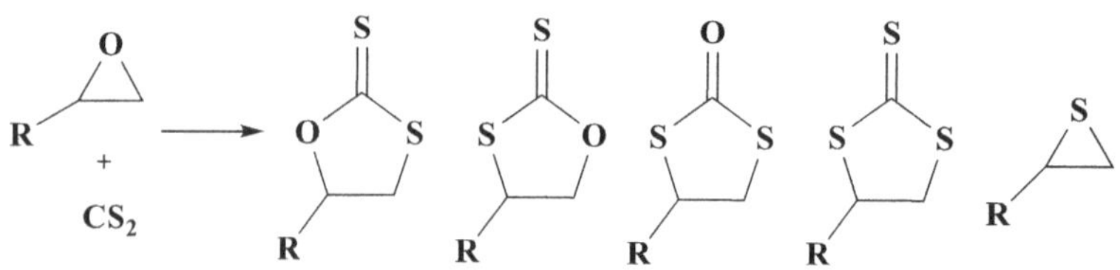

SCHEME 5

The difluorocarbene, synthesized from hexafluoropropylene oxide, was added to thiocarbonyl compounds to synthesize the thiiranes (Scheme 6). This technique provided tetrafluorothiirane in 30%–40% yield from thiocarbonyl fluoride [44,45].

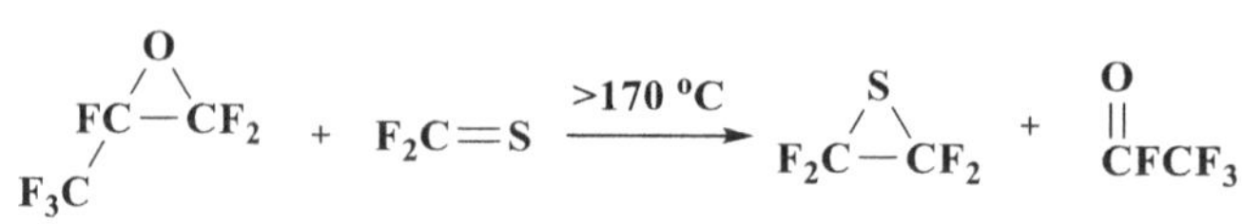

SCHEME 6

In some cases, for example, the boron trifluoride–catalyzed reaction of thiobenzophenone with 1,2-epoxycyclohexane afforded the desired 1,3-oxathiolane along with corresponding 1,3-dioxolane and 1,3-dithiolane minor products (Scheme 7) [46]. The 1,3-dioxolane and 1,3-dithiolane were secondary products, and a reaction mechanism through an intermediate synthesis of benzophenone and 1,2-epithiocyclohexane was suggested. The 1,2-epoxycyclohexane was reacted with 1,3-dimethylimidazolidine-2-thione to synthesize the 1,3-dimethylimidazolidine-2-one and 1,2-epithiocyclohexane in high yields under the milder conditions [47].

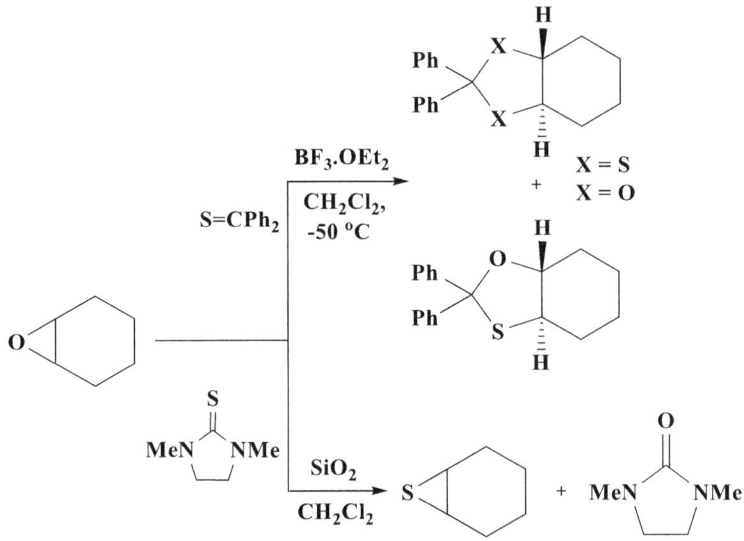

SCHEME 7

The mixture turns out to be warm after an addition of 1 eq. (*R*)-epoxide (91% *ee*) to pyrrolthione (1.25 eq.) at rt. Then, dry dichloromethane was added while stirring, and the TLC indicated the complete conversion of (*R*)-epoxide after 1 h 15 min. The (*S*)-2-phenylthiirane was obtained with 92.5% *ee* and 55% yield (high-performance liquid chromatography, Chiralcel OD-H) as a colorless oil after the evaporation of solvent and chromatographic workup (Scheme 8). In a similar research, the (*R*)-epoxide (1 eq.) was added to a solution of pyrrolthione (1.05 eq.) at 0°C in dry dichloromethane. The mixture was stirred for 1 h at 0°C and then left to warm to rt; thin-layer chromatography control exhibited only partial conversion of (*R*)-epoxide. Then, 3 mL dichloromethane and 2 g silicon dioxide were added to the mixture and stirred at rt for 30 h. After this time, no (*R*)-epoxide could be spotted by thin-layer chromatography, and usual workup by column chromatography on silica gel provided (*S*)-2-phenylthiirane with 87.6% *ee* and in 56% yield [47].

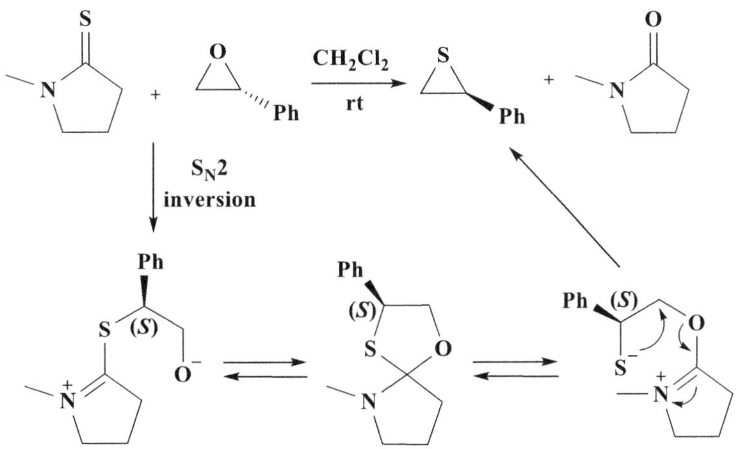

SCHEME 8

The 2-[(*t*-butyloxy)methyl]thiirane was prepared in 62.8% by the silicon dioxide–catalyzed reaction of pyrrolthione with (*RS*)-epoxide (R = methyl) (molar ratio 1.1:1) in dry dichloromethane at rt for 3.5 h after purification by column chromatography on silica gel (hexane/diethyl ether 13:1 and 10:1) (Scheme 9). The pyrrolthione was reacted with (*R*)-2-{[(triphenylmethyl)oxy]methyl}oxirane. In a similar manner, the reaction of pyrrolthione and (*R*)-epoxide (R = methyl) (molar ratio 1.2:1) in dry dichloromethane at rt using silicon dioxide afforded (*R*)-2-{[(triphenylmethyl)oxy]methyl}thiirane in 87.3% yield, stereoselectively, after 1 d, with $[\alpha]_D^{21} = +4.2$ ($c = 1$, dichloromethane). Unfortunately, the measurement of *ee* value of (*R*)-2-{[(triphenylmethyl)oxy]methyl}thiirane by high-performance liquid chromatography (Chiralcel OD-H) was unsuccessful [47].

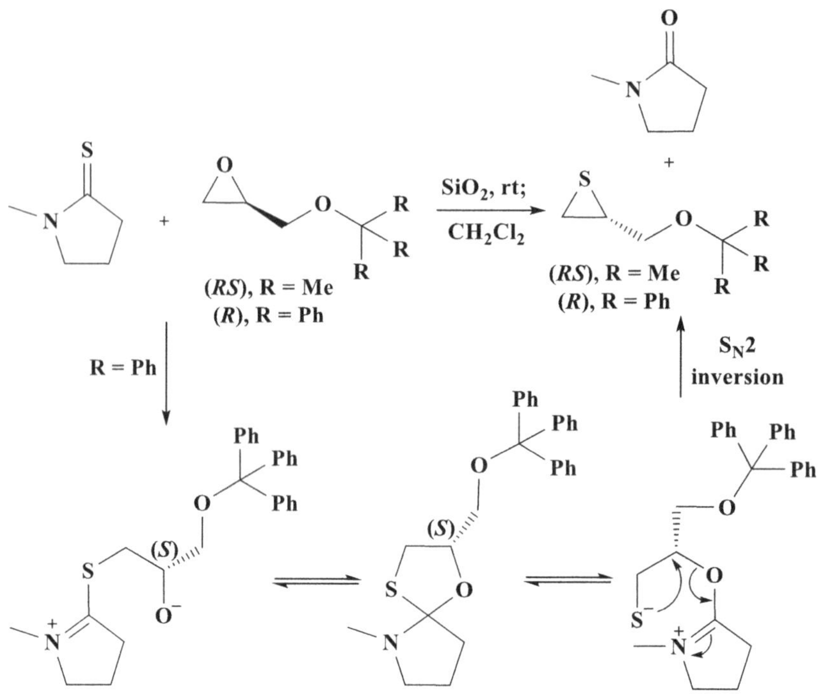

SCHEME 9

The silicon dioxide–catalyzed reaction of pyrrolthione with 1,2-epoxycyclohexane (7-oxabicyclo [4.1.0]heptane (molar ratio 1.1:1) in dry dichloromethane proceeded easily at rt for 27 h to provide the pure 7-thiabicyclo[4.1.0]heptane in 73.7% yield after purification by column chromatography on silicon dioxide (hexane/diethyl ether 23:1) (Scheme 10) [47,48].

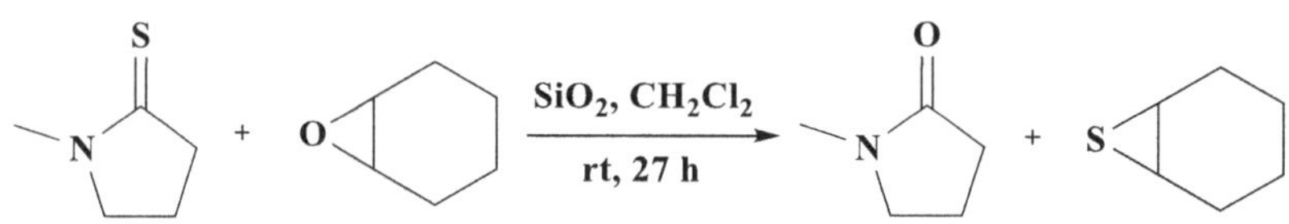

SCHEME 10

Takido et al. [49] reported a higher yielding oxirane–thiirane exchange reaction. This technique used DMTF (dimethylthioformamide) in catalytic amounts of TFA to synthesize the thiiranes in good yields (62%–95%). The reaction proceeded by the nucleophilic attack of sulfur atom in DMTF on the least substituted carbon of the epoxide. The ring closure of intermediate provided five-membered oxonium ring. The removal of DMF afforded thiirane (Scheme 11).

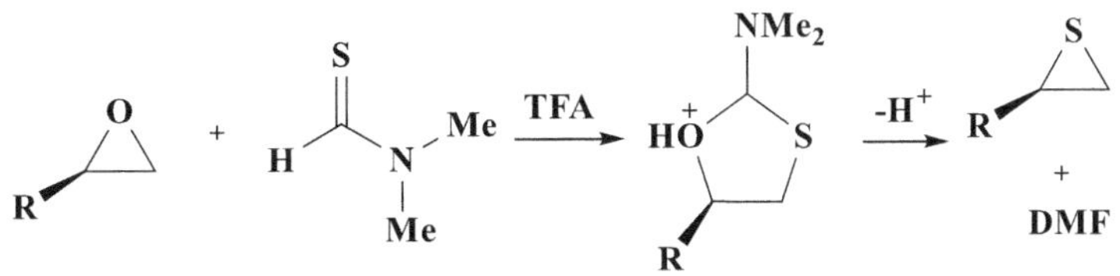

SCHEME 11

8.4 Synthesis of thiiranes from epoxides using SCN bearing moieties

The reaction of thiocyanate ion with epoxides in the absence or in the presence of catalyst is a broadly examined and appropriate technique for the formation of thiiranes [8,13,20,50–53]. The thiiranes were synthesized by the reaction of thiocyanate ion and epoxides via β-hydroxy thiocyanate intermediate (Scheme 12), but this intermediate has not been isolated because of its fast transformation to thiirane [54–56].

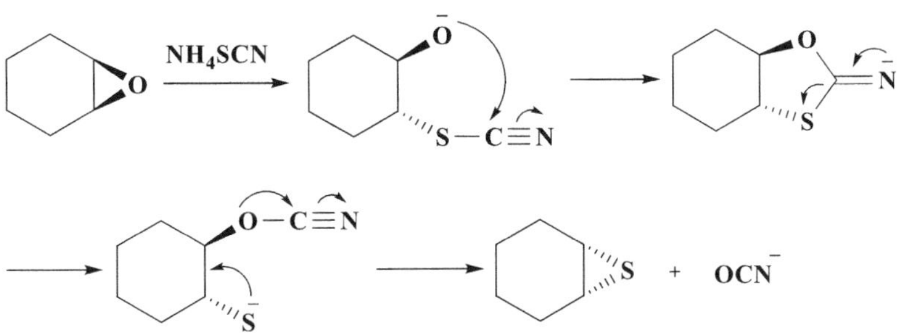

SCHEME 12

To determine the catalytic activity of solid acid, the reaction of 2,3-epoxypropyl phenyl ether with ammonium thiocyanate using catalytic amount of melamine sulfonic acid was examined under the solvent-free conditions. The reaction completed after 3 min, and the thiocyanohydrin was formed, whereas this reaction took 10 min in the absence of catalyst to provide the thiirane as the only product (Scheme 13) [56,57].

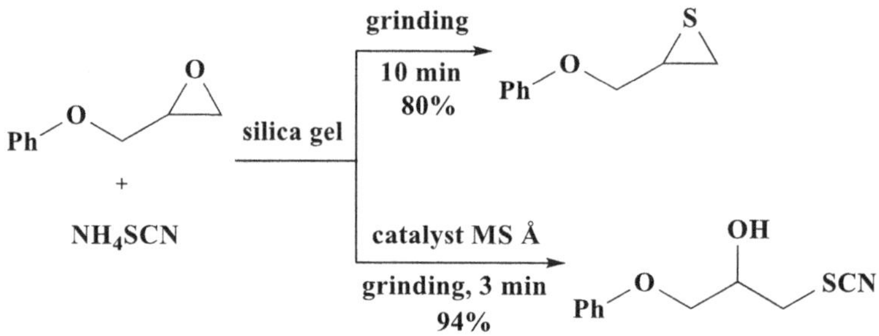

SCHEME 13

The epoxides were reacted with NH_4SCN to afford the thiiranes. The reaction occurred with an inversion of configuration and some erosion of enantiomeric purity. The (*R*)-(+)-styrene oxide provided (*S*)-(−)-styrene sulfide in 78% optical purity (Scheme 14) [10].

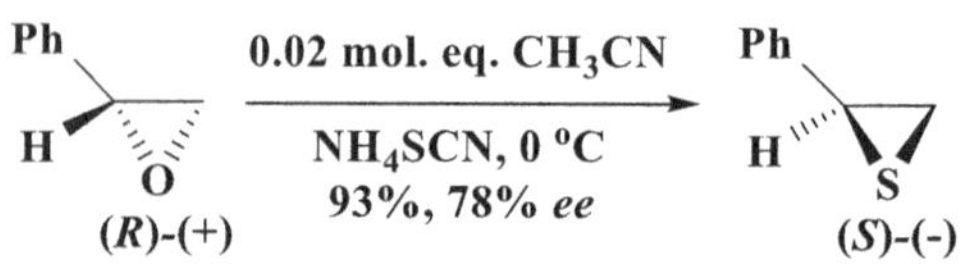

SCHEME 14

The reactions of various epoxides with ammonium thiocyanate were carried out using $Bi(OTf)_3$ (0.02 M eq.) or $Bi(TFA)_3$ (0.05 M eq.) in refluxing CH_3CN to synthesize the thiiranes in excellent yields (Scheme 15). The substrates utilized (cyclohexene and styrene oxides, epichlorohydrin, glycidyl phenyl ether, glycidyl *i*-propyl ether, allyl glycidyl ether, and 1,2-epoxyoctane) were selected as examples of aliphatic, cyclic, activated, and deactivated epoxides [58].

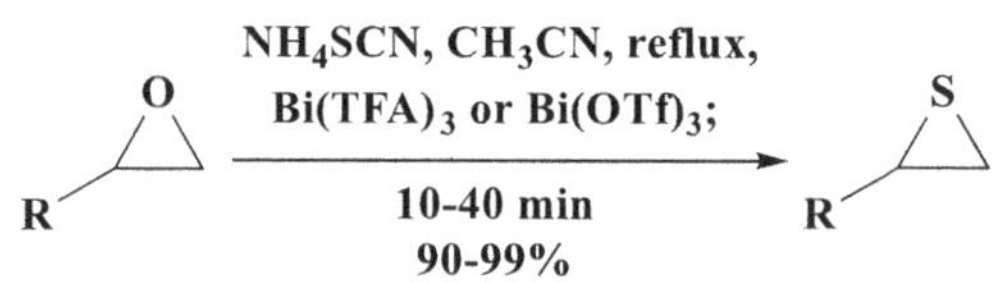

R = Ph, CH_2Cl, CH_2OPh, $CH_2OCH(CH_3)_2$, $CH_2OCH_2CH{=}CH_2$, $(CH_2)_5CH_3$

SCHEME 15

The epoxides were transformed to thiiranes using NH_4SCN in catalytic amounts of $(COOH)_2$ under the nonaqueous and mild reaction conditions with excellent isolated yields (Scheme 16). This reaction was carried out under both the MWI and conventional heating conditions. The distinct rate improvement was reported under the MWI [59].

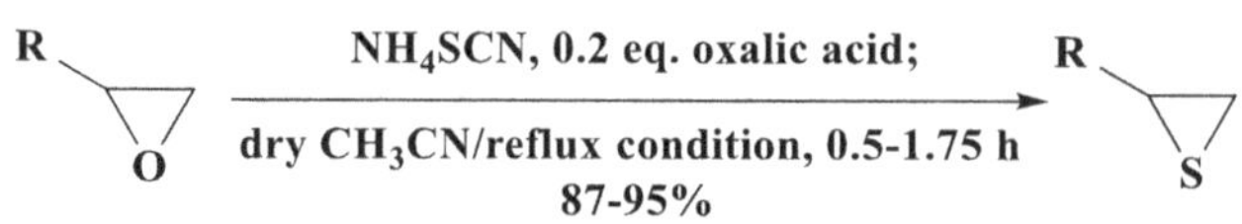

SCHEME 16

An efficient method was reported for the transformation of oxiranes to thiiranes in the presence of excess NH_4SCN under the MWI or reflux in aqueous media. The thiiranes could be prepared by the reaction of epoxide with excess amount of NH_4SCN (2 eq.) under the MWI in aqueous media. Heating under the reflux resulted in a longer reaction time and lower yields as compared to MWI. For instance, under the reflux conditions, it took 3 h to completely utilize the styrene oxide, and only trace of styrene thiirane was obtained. On the other hand, the styrene thiirane was formed in 92% yield under the MWI utilizing a 160 W source. It took 4 h to afford the 2-(naphthalene-1-oxymethyl)thiirane in 53% yield under the reflux conditions but only 45 min to provide the 2-(naphthalene-1-oxymethyl)thiirane in 90% yield under the MWI utilizing a 300 W source (Scheme 17) [60].

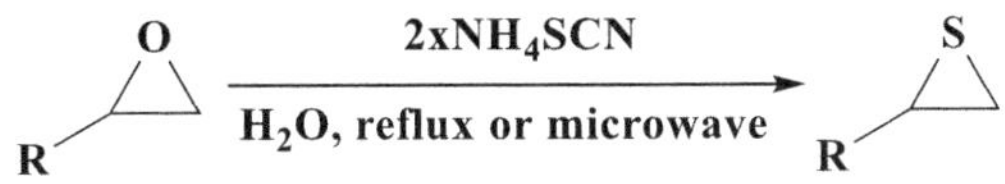

R = $C_6H_5CH_2OCH_2$
R = $C_6H_5OCH_2$
R = C_6H_5
R = 1-naphthalenoxymethyl
R = 4-methoxyphenoxymethyl
R = 3,4-dimethoxyphenoxymethyl
R = 3-isopropoxy-4-methoxyphenoxymethyl
R = 3,4-methylenedioxyphenylmethyl

SCHEME 17

The epithioundecanoates were prepared from epoxy undecanoates in ionic liquid, i.e., green reaction medium. The utilization of ionic liquid for the formation of epithio compounds helped in performing the reaction under the neutral and mild conditions to synthesize the epithioundecanoates in high yields. The performance of epithio derivative as multifunctional additives was estimated in three base oils: di-2-ethylhexyl sebacate (DOS), paraffin (S-IO5) base oil, and epoxy jatropha *n*-butyl esters (EJB). The epoxy undecanoates were utilized as starting compounds to synthesize the epithioundecanoates in the presence of NH_4SCN in stoichiometric ratio (Scheme 18). All the reactions completed within 20–24 h, providing alkyl-10-epithioundecanoatess in high yields as shown by gas chromatography (85%–96%). The products were purified by silica gel column chromatography with hexane and $CH_3COOC_2H_5$ (80/20) to provide the pure alkyl-10-epithioundecanoates in 60%–68% isolated yield [61].

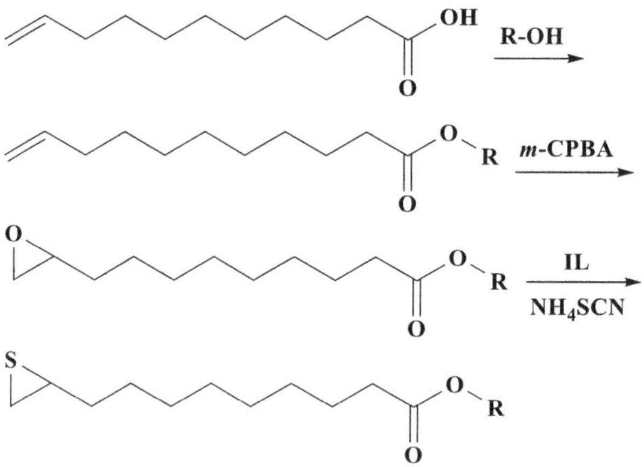

R = CH_3, $CH(CH_3)_2$, $CH_2CH(CH_3)_2$, $(CH_2)_3CH_3$, $(CH_2)_5CH_3$, $(CH_2)_5CH(CH_3)_2$

SCHEME 18

The NH_4SCN was appropriate sulfur donor for this procedure, and a very practical technique was reported utilizing cyanuric chloride as a coreagent. The reaction of styrene oxide with stoichiometric NH_4SCN and cyanuric chloride under the solvent-free conditions afforded episulfide in almost quantitative yield within 15 min (Scheme 19). The mechanism included nucleophilic ring opening of epoxide to synthesize the alkoxide intermediate. The reaction with cyanuric chloride activated the oxygen toward displacement by sulfur forming a cyanate episulfenium species, which upon hydrolysis and then loss of CO_2 provided the product [62]. The magnesium hydrogen sulfate has also been established to be an efficient activating agent [14].

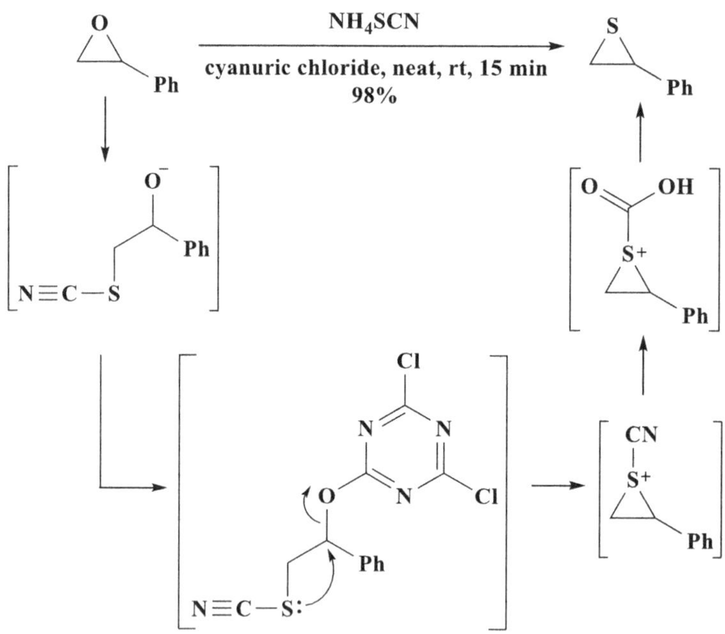

SCHEME 19

The etidronic acid is a phosphonic acid with mild acidity, is noncorrosive and nonvolatile, and is soluble in common organic solvents. It is a white crystalline solid with exceptional chemical and physical properties and is a commercially accessible inexpensive chemical. Its applications as a reaction catalyst in organic syntheses have not been much discovered. Herein, an efficient and simple reaction for the preparation of thiiranes was reported in tetrahydrofuran utilizing etidronic acid as a catalyst (Scheme 20). Originally, the reaction of styrene oxide with NH_4SCN was performed in tetrahydrofuran using different amounts of etidronic acid at reflux temperature. It was found that the best result was achieved with 10 mol% etidronic acid. It is noteworthy that the reaction did not afford any product at reflux temperature even after a long reaction time in the absence of etidronic acid. The formation of thiirane was examined in various solvents to select the best solvent for the reaction. The reaction performed in tetrahydrofuran resulted in shorter reaction time and higher yields. The tetrahydrofuran was utilized as reaction media for all the reactions. Depending on the optimized reaction conditions, a variety of thiiranes were synthesized by the reaction of 1 mmol oxiranes with 1.2 mmol NH_4SCN in tetrahydrofuran. The reaction occurred in 1 h at reflux temperature in excellent yields after an addition of 0.1 mmol etidronic acid catalyst. In these experiments, various oxiranes such as aliphatic and bicyclic oxiranes underwent a reaction easily [25].

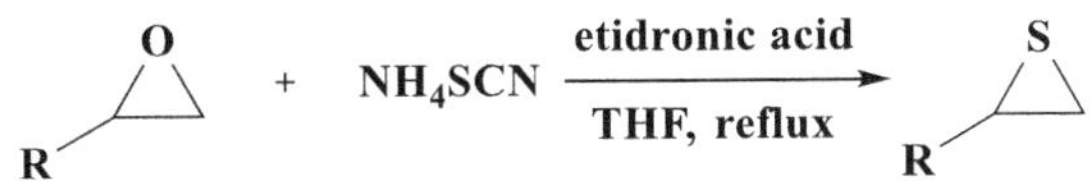

R = Ph, ClPh, MeOPh, CH_2OPh, CH_2OPhCl, CH_2OPhI, Me, CH_2Cl

SCHEME 20

The thiiranyl acetals were synthesized by oxirane–thiirane exchange. There are four reported methods for oxirane–thiirane exchange. The first formation of thiiranes from epoxides and alkali thiocyanates was reported by Price and Kirk [55]. The oxirane–thiirane exchange occurred by the *trans*-opening of epoxide with thiocyanate. Then, a second Walden inversion resulted in *trans*-closure of thiirane ring (Scheme 21).

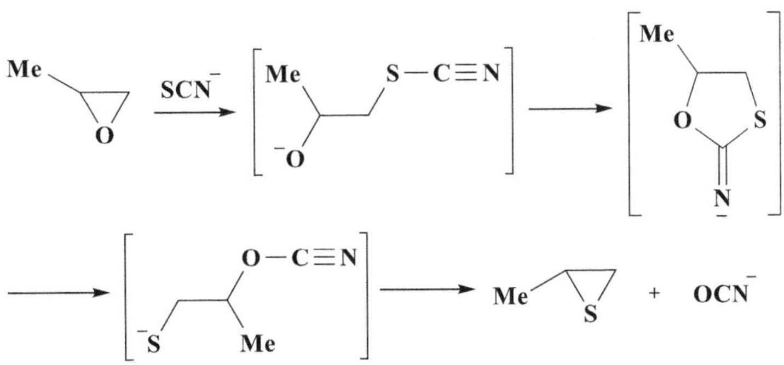

SCHEME 21

An addition intermediate was transformed into thiirane. An initial ring opening was stereoelectronically directed in a *trans*-diaxial manner. An intermediate was relaxed to diequatorial conformer before cyclizing to 1,3-oxathiolane intermediate (Scheme 22) [58].

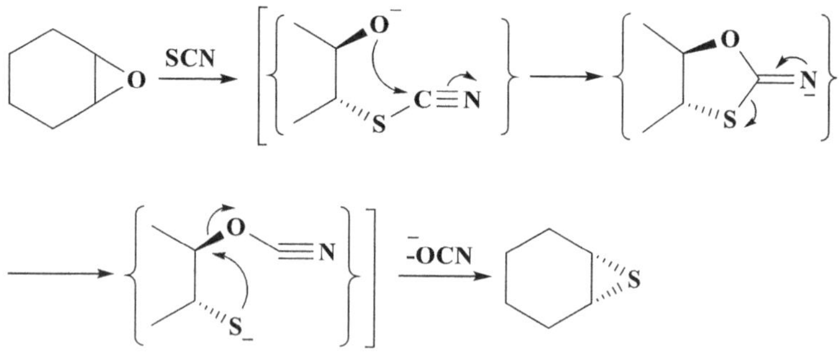

SCHEME 22

The episulfide was prepared by the substitution of the oxygen of epoxide by sulfur. The catalysts and reagents for this reaction were broadly discovered as given by the formation of cyclohexene episulfide (Scheme 23). The reaction of epoxides with triphenylphosphine sulfide using trifluoroacetic acid afforded thiiranes in moderate yields [63–65].

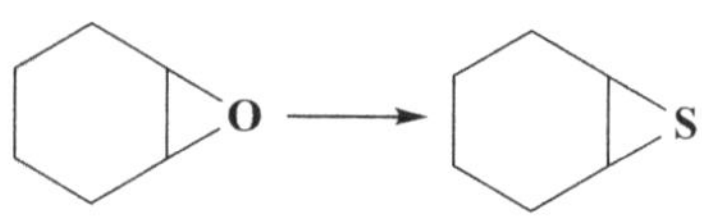

reagent	catalyst	yield (%)
KSCN	$InBr_3$	87
NH_4SCN	$LiBF_4$	94
	$Fe(OTf)_3$	93
	$SbCl_3$	98
	$BiCl_3$	98
	P_4VP-$Ce(OTf)_4$	92
	$TiO(TFA)_3$	95
	$(NH_4)_8[CeWO_{36}].20\ H_2O$	96
$H_2NC(S)NH_2$	montmorillonite K-10	98
	$Sn^{IV}(TPP)(ClO_4)_2$	93

SCHEME 23

Under the slightly basic aqueous conditions, an addition of thiocyanate was catalyzed by poly(allyl amine) (PAA). For instance, under these conditions, the phenyloxymethyl epoxide was transformed to episulfide in excellent yield (Scheme 24) [66]. Very good yields were obtained utilizing KSCN in a biphasic medium of H_2O and ionic liquid without an additional catalyst [67]. An elemental sulfur could be applied as donor in the presence of diethylphosphite, CH_3COONH_4, and Al_2O_3 under the solvent-free and MWI conditions. The mechanism included a thiophosphate species [68].

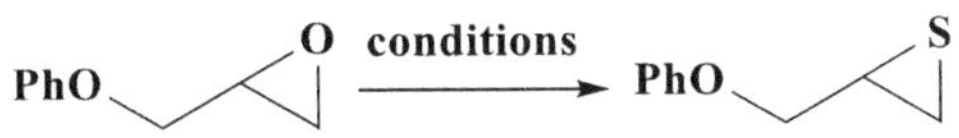

entry	S source	additives	solvent	time (°C)	time	yield (%)
1	NH_4SCN	PAA/NaOH	H_2O	45	70 min	93
2	KSCN	none	[bmim][PF_6]-H_2O	45	1.3 h	89
3	S/HP(O)$(OEt)_2$	NH_4OAc/Al_2O_3	neat	MW	2 min	68

SCHEME 24

The thiocyanohydrin was reacted under the conditions that provided the best results in the direct transformation of epoxide using modest base and ethanol—water as a solvent system. The cyclization not occurred in good yield at rt and heating the reaction mixture or increasing the amount of base resulted in the destruction of material. Additionally, a new product was obtained in this reaction, which was identified as the α-epoxide, apparently synthesized by reversion of the desired cyclization procedure with the oxyanion attacking the C-6 and the thiocyanide ion serving as a leaving group [69]. At this point, two alternatives of phase-transfer catalysis were crown ether and ammonium salt catalysis, both in a weakly polar solvent such as dichloromethane. The crown ether-catalyzed reaction provided product in 10 min at rt in 75% yield without signs of the decomposition and with only traces of epoxide. A parallel experiment without 18-crown-6 provided just recovered starting compound. This result encouraged to reevaluate a one-pot process utilizing the joint power of crown ether catalysis, and an unusual high yield of β-hydroxy thiocyanate was obtained when the reaction was carried out in a polar solvent. The direct rt oxirane—thiirane transformation of epoxides was performed using KSCN in 18-crown-6. The reaction occurred, but the desired product was formed in a little lower yield in one-pot than in the stepwise process (Scheme 25). The β-epoxide required higher pH and longer reaction times to complete the reaction because of the different stability of intermediates formed during the synthesis of thiiranes [70].

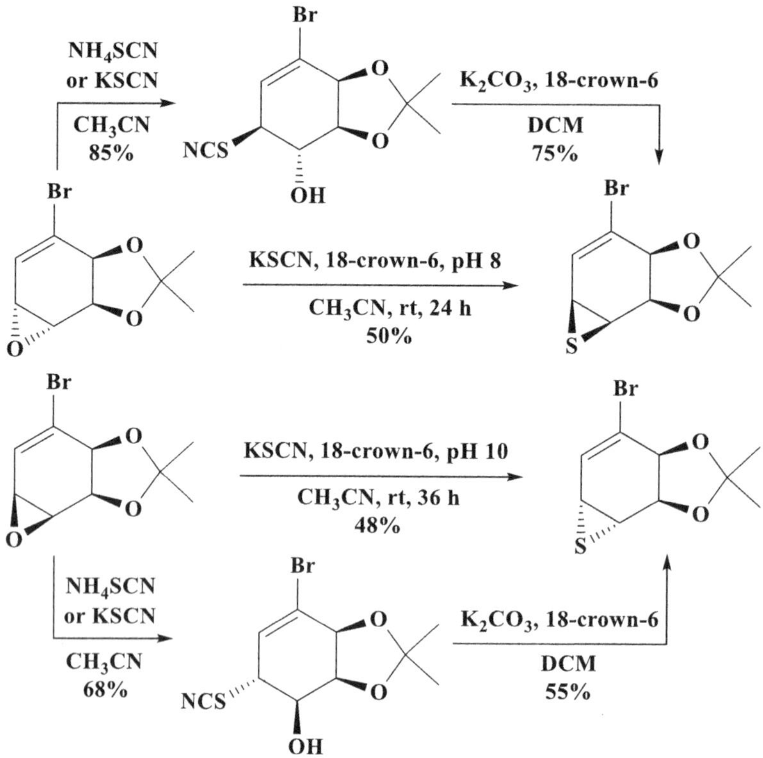

SCHEME 25

The manno-1,2-epithiosugars were synthesized by the reaction of epoxides with inorganic thiocyanates. Although this procedure was utilized onto the saccharidic framework [71], it was not used to 1,2-epoxy sugars. Unexpectedly, the use of alkali thiocyanates provided glyco-1,3-oxazolidine-2-thione exclusively (Scheme 26). Because of this unexpected reactivity, a new method was reported for the formation of carbohydrate-based 1,3-oxazolidine-2-selone derivatives and *cis*-1,2-fused 1,3-oxazolidine-2-thione.

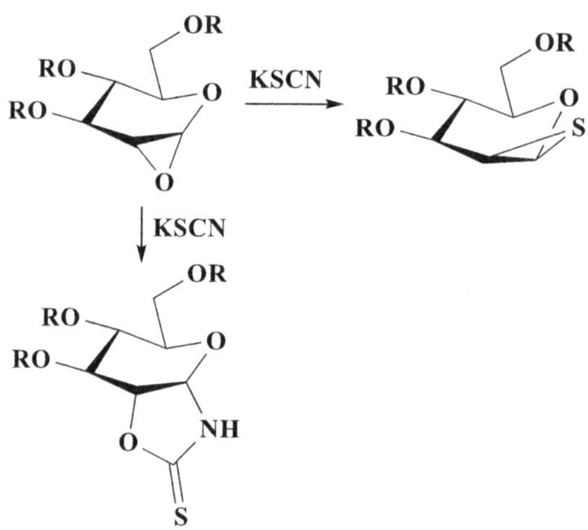

SCHEME 26

The 2,3-*cis*- and *trans*-dideuterothiirane oxide were synthesized. The *cis*- and *trans*-dideuteroethylenes, needed as starting compounds, were synthesized by procedures reported in the literature starting from dideuteroacetylene [72]. A thorough analysis of the IR of ethylene had identified diagnostic absorptions at 842 and 987 cm^{-1} for *cis*- and *trans*-dideuteroethylene [73], which could be utilized to prove the purity of ethylenes and for the quantitative analysis of ethylene formed in the decomposition of labeled thiirane oxides (Schemes 27 and 28) [74].

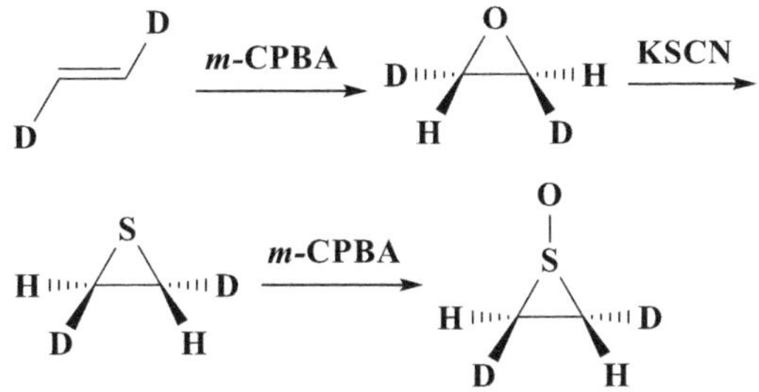

SCHEME 27

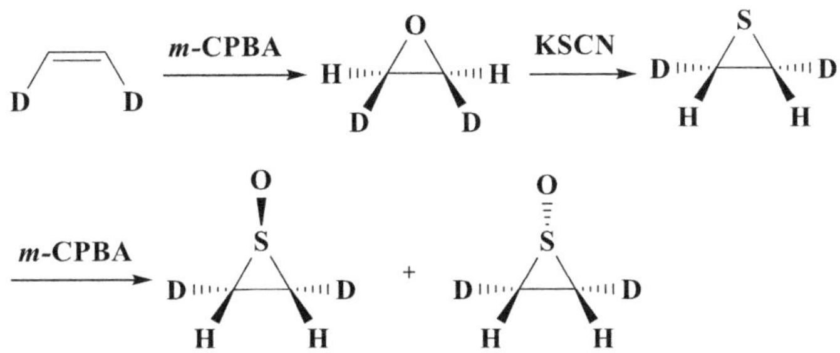

SCHEME 28

The glycals are saccharide derivatives containing double bond between the anomeric carbon and the adjacent carbon atom. These 1,2-unsaturated carbohydrates are important intermediates for the formation of many glycosyl derivatives because of their reactivity as highly electron-rich enol ethers [75]. The episulfides, sulfur analogs of epoxides, are beneficial and important building blocks in the formation of pesticides, polymers, and pharmaceuticals. Several approaches have been reported for the formation of these compounds. The most common and important technique is the conversion of oxiranes into thiiranes with various sulfur-introducing agents such as inorganic thiocyanates [10,13,76] and thiourea [15,77] (Scheme 29).

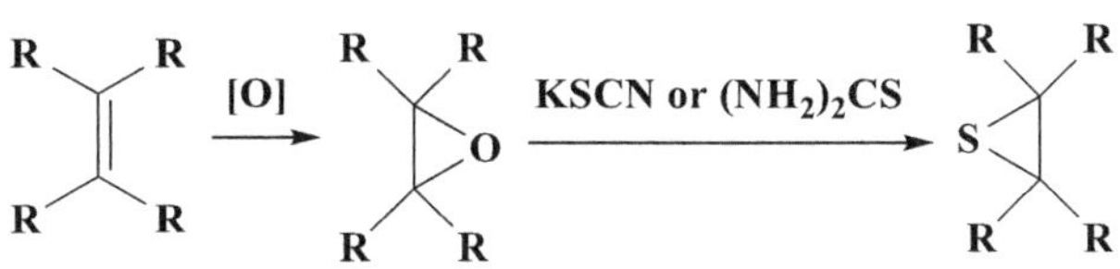

SCHEME 29

The catalyst-free, high-yielding transformation of oxiranes into thiiranes has efficiently been performed by reacting with thiourea or potassium thiocyanate at rt utilizing polyethylene glycol (PEG-400) as a reaction medium (Scheme 30). A series of thiiranes was prepared from many oxiranes. Any additional catalyst was not needed. The reaction was completed within 45 min [77].

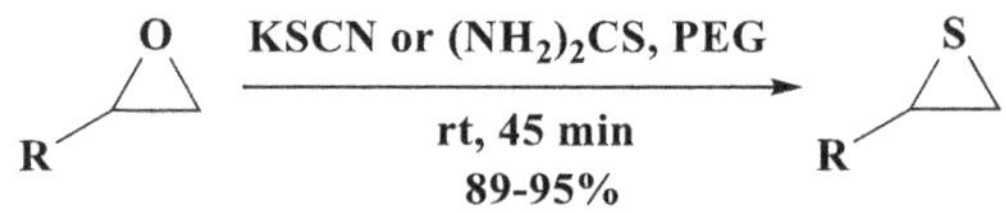

SCHEME 30

The bioactivity between different three cyclic compounds was compared on the basis of ring composition. To avoid a de novo formation of the thiiranes, the O–S exchange reaction was utilized to prepare library, thus shortening the reaction pathway by various steps. The reaction of oxirane probes with both thiourea and potassium thiocyanate was attempted under various conditions, but only starting compounds were recovered. The EWGs avoided the formation of thiirane ring as the reactions worked with donor-substituted epoxides (Scheme 31) [55,78,79].

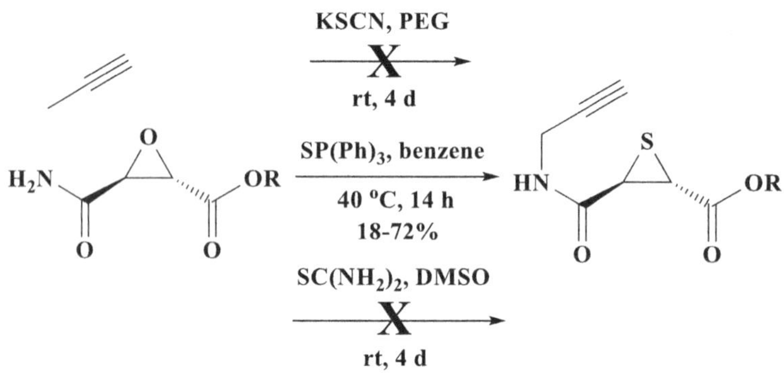

SCHEME 31

Geurts [80] reported the epoxy/amine system for latex. In that case, the materials were detached into two different phases known as "the two-in-one system." The key problem consists in incorporating amine groups in acrylic latexes. It was simple to introduce epoxide groups thanks to the GMA (glycidyl methacrylate monomer). However, the utilization of glycidyl methacrylate monomer for latex formation raises a limitation (particularly for precross-linking) because of the instability of that monomer in aqueous medium. Therefore, O'Brien and coworkers [81] utilized episulfide, an equivalent of epoxide. The episulfide is very stable toward H_2O, hence limiting precross-linking. The cross-linking of episulfides was slower in the presence of piperazine than that taking place using *O*-containing derivative, and the best conditions of cross-linking were 30 min at 65°C. However, storage stability was not much improved. This amine was utilized for the hardening of both epoxide and episulfide because it is H_2O-soluble, and it could be migrated into particles (Scheme 32) [82].

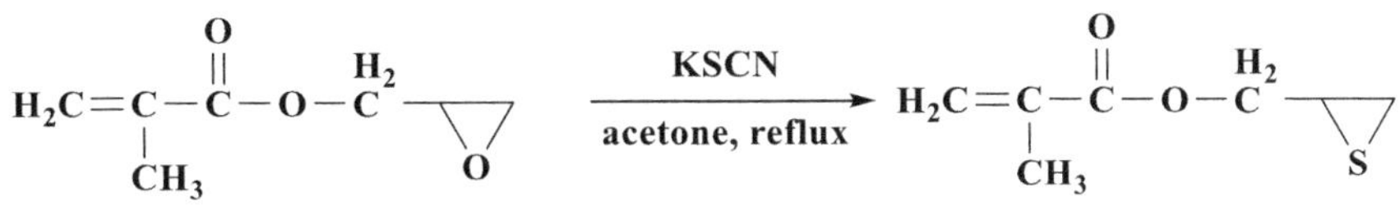

SCHEME 32

The epoxide was reacted with thiocyanate anion for the synthesis of episulfides in basic media (Scheme 33) [83].

SCHEME 33

Various epoxides responded quickly with KSCN in PF_6–H_2O (2:1) solvent system at rt under the mild and convenient conditions to synthesize the thiiranes in high to quantitative yields (Scheme 34) [22].

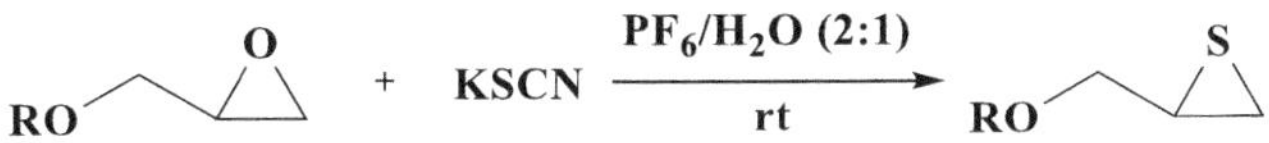

SCHEME 34

The phenyl thiirane was prepared as sulfur source. It was prepared from styrene oxide with KSCN using Lewis acid as reported in the literature (Scheme 35) [13].

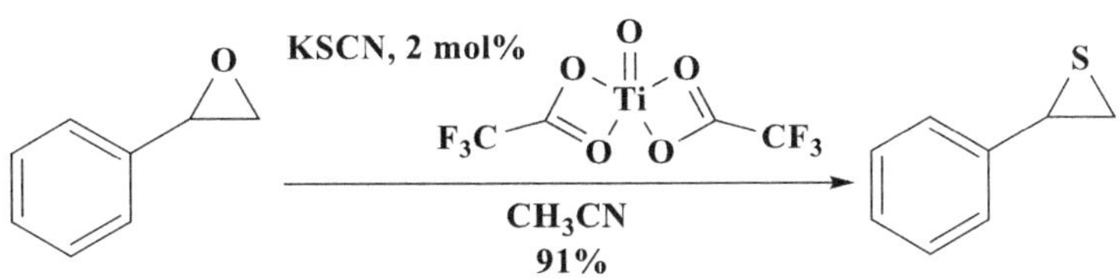

SCHEME 35

The epoxides are an ideal source and have synthetic value for diversity in organic synthesis due to their ease of synthesis, wide reactivity, and ability to undergo regioselective ring-opening reactions. The strained ring and electronegativity of oxygen atom made them good electrophile and could be smoothly opened with many nucleophiles providing functionally diverse compounds. The thiiranes, tetrahydropyrans, dihydropyrans, and *vic*-halohydrins, which can be prepared from epoxides, are beneficial precursors in the formation of other bioactive molecules and various natural products (Scheme 36) [7–10].

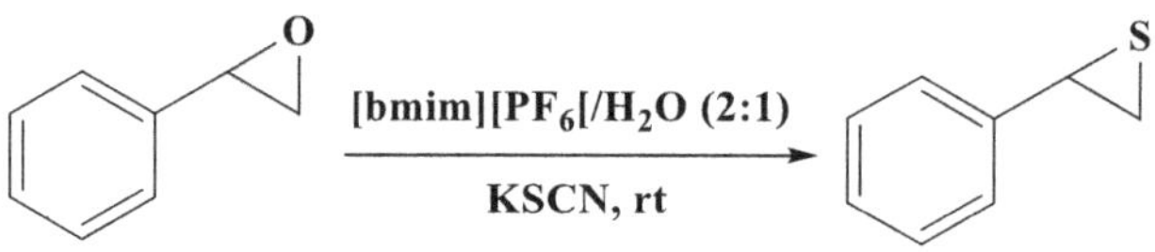

SCHEME 36

In the current years, ionic liquids are emerging as a set of green solvents with exclusive properties such as high thermal stability, tunable polarity, negligible vapor pressure, immiscibility with many organic solvents, and recyclability. Because of their green credentials and potential to improve selectivities and rates, ionic liquids are utilized for the formation of episulfides from oxiranes and KSCN under the neutral and mild conditions. Originally, the experiments were performed with 2-phenoxymethyloxirane and KSCN for 2 h in [bmim][PF_6]-H_2O (2:1) solvent system to synthesize the episulfide in 95% yield (Scheme 37). Several epoxides were reacted efficiently with KSCN under these reaction conditions to synthesize the thiiranes in excellent yields. A green approach was reported for the formation of episulfides from epoxides utilizing ionic liquids as reaction medium [7–10].

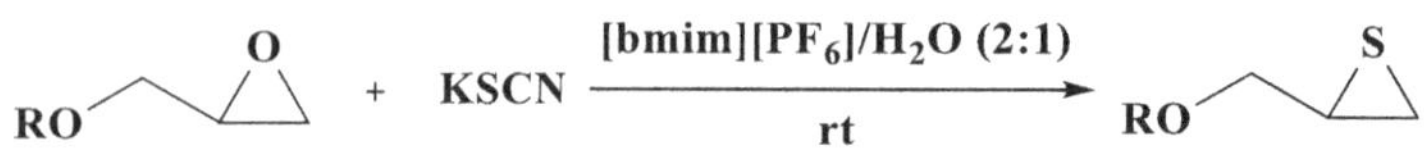

SCHEME 37

The thiocyanate ion is beneficial in transforming epoxides into sulfur analogs known as thiiranes or episulfides (Scheme 38). A complete curved-arrow mechanism was proposed for this conversion. The stereochemistry of all intermediates was clearly shown, and the stereochemistry of product was predicted [7–10].

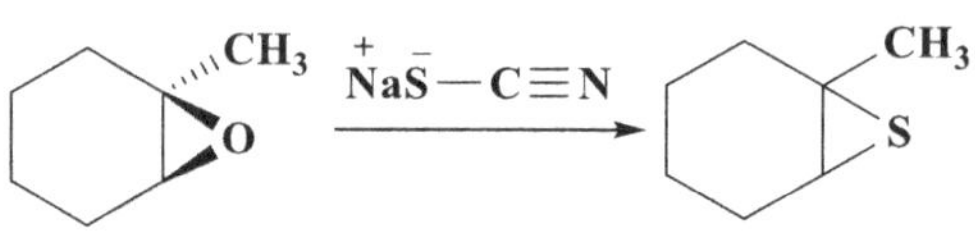

SCHEME 38

In the first step, as in any lithiation, yield was limited by exposure to H_2O. The careful drying of ether solvent provided octyl benzyl bromide in >90% isolated yields. An epoxidation using *m*-chloroperoxybenzoic acid provided epoxide in almost quantitative yield. The transformation to episulfide was also almost quantitative (~96% isolated yield). The final oxidation to *n*-octylbenzylthiirane oxide exhibited the complete transformation of episulfide; however, two products were formed (Scheme 39). The *p*-allyl octyl benzene was also formed in ~36% yield together with the expected 4-(*n*-octyl)benzyl thiirane oxide. The reaction was carried out at 0°C and protected from light. Since the decomposition was still noticed, it was suspected that this compound might be too unstable for additional analysis. Although this material could be prepared, the retro-cheletropic equilibrium lies too much toward the right and prohibited its use as a surfactant and made characterization of its CMC and decomposition analysis unwarranted [84].

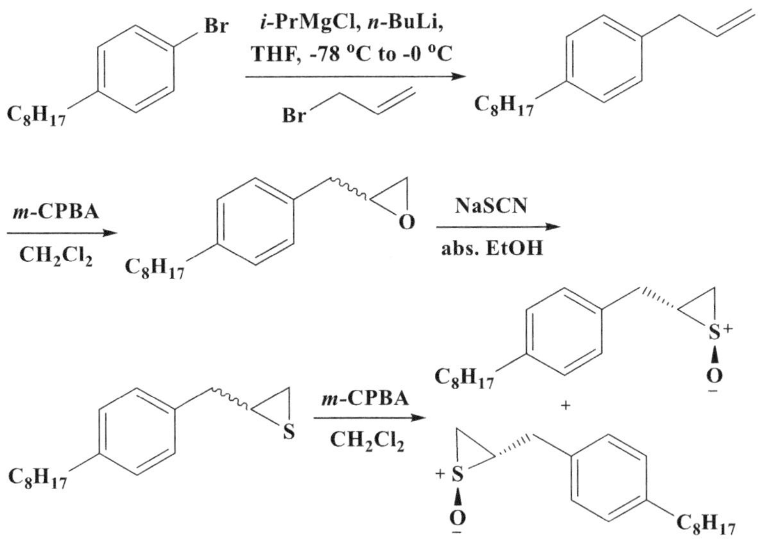

SCHEME 39

8.5 Synthesis of thiiranes from epoxides using thiourea

The conversion of oxiranes into thiiranes has been known for over 60 years [78,85–87], and many instances have been reported. The thiourea was utilized as "sulfur transfer reagent" in most of the cases. Various new synthetic modifications have been reported in the literature [9,88–90]. The stereochemical course of this conversion, which occurred with an inversion of configuration of both stereogenic centers, was analyzed broadly [91–96]. For instance, the reaction of optically active oxirane with thiourea in methanol at rt provided thiirane stereoselectively (Scheme 40). Instead of thiourea, heterocyclic thiones possessing the *N*–C(S)-X functionality (X = N, O, and S) [47,97,98] as well as thiourethanes [99] can be utilized as "sulfur transfer reagent."

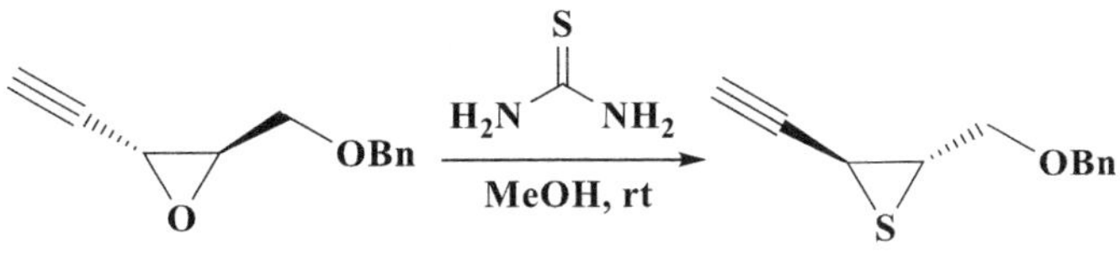

SCHEME 40

To overcome the severe disadvantages of reported procedures, an attractive process for the formation of thiiranes different from the classical method was reported under the supramolecular catalysis in H_2O using β-cyclodextrin. The reactions were performed by in situ synthesis of β-cyclodextrin complex of epoxide in H_2O followed by an addition of thiourea and stirring at rt to synthesize the thiiranes in impressive yields (Schemes 41 and 42). These reactions occurred at rt, and no side products or rearrangements were reported [100].

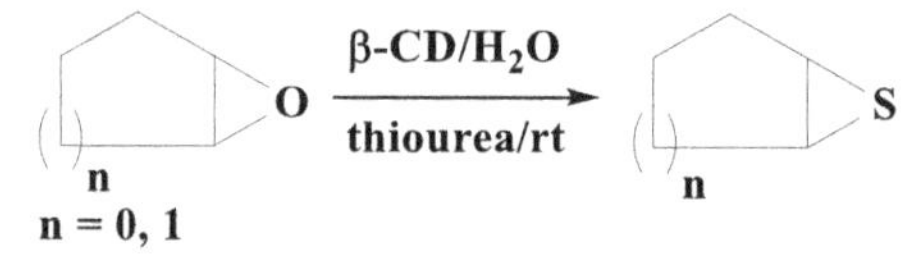

SCHEME 41

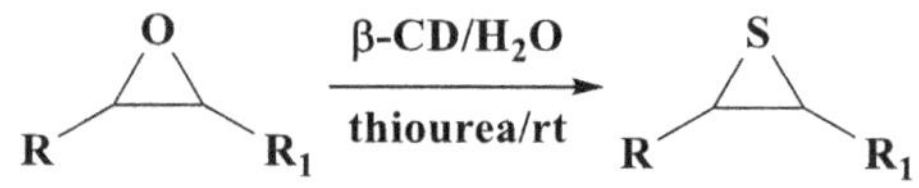

SCHEME 42

Kaboudin and Norouzi [68] reported an efficient process for the formation of thiiranes from epoxides by a one-pot reaction of epoxides with diethyl phosphite in the presence of CH_3COONH_4 or $(NH_4)HCO_3$/S/acidic alumina under the solvent-free and MWI conditions. This reaction was extended by Zeynizadeh and Yeghaneh [9] utilizing Dowex 50WX8 ion-exchange resin under the MWI in 75%–98% yield (Scheme 43) [101].

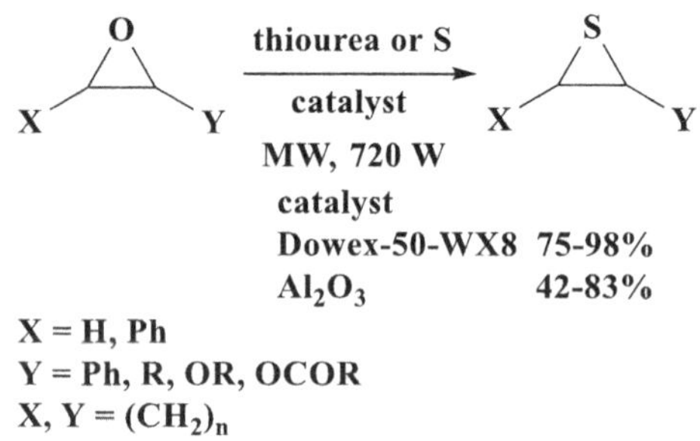

SCHEME 43

The reaction has not been tested on 1,1-disubstituted, trisubstituted, or tetrasubstituted epoxides at this time depending on the experience with mono- and 1,2-disubstituted ones. The catalytic abilities of these two catalysts have also been examined for the transformation of epoxides to thiiranes with thiourea as sulfur-introducing reagent (Scheme 44). The thiourea was less reactive and more stable sulfurating agent in comparison with NH_4SCN, and its handling was convenient. Therefore, the proposed technique based on utilizing thiourea can be considered as a practical and beneficial accomplishment in the formation of thiiranes from epoxides [58].

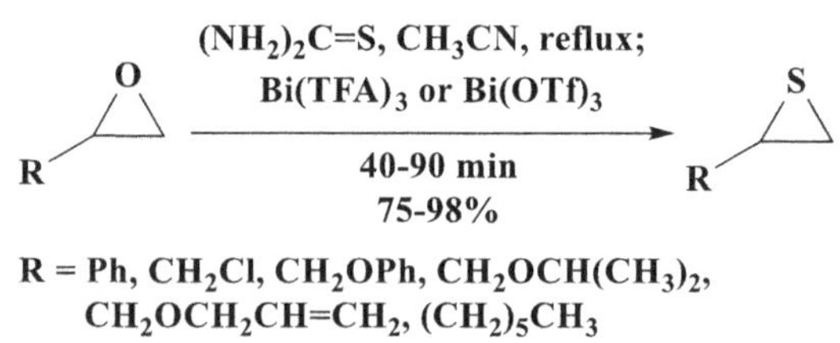

SCHEME 44

The solvent-free transformation of several epoxides to thiiranes was performed efficiently with immobilized thiourea on calcium carbonate. The thiiranes were formed in 88%–98% yield by reactions under the oil bath (60–70°C) conditions within 1–12 min. A literature review exhibited that solvent-free transformation of epoxides to thiiranes has been rarely examined, and this aim was achieved with thiourea at 120°C [4] and thiourea/silicon dioxide system [7]. Nevertheless, these approaches suffered from modest yields and high reaction temperature. In this context, Dowex-50WX8-supported thiourea has been reported as an efficient polymeric supporting system for solvent-free transformation of epoxides to thiiranes [9]. The lack of information related to the use of alkali and alkali metal sulfates and carbonates in the formation of thiiranes from epoxides inspired to examine the ability of some eco-friendly and costless reagents such as sodium carbonate, potassium carbonate, cesium carbonate, magnesium carbonate, calcium carbonate, strontium carbonate, barium carbonate, sodium bicarbonate, potassium bicarbonate, sodium bisulfate, potassium bisulfate, and sodium sulfate for the aimed conversion. The reaction of styrene oxide with immobilized thiourea on the sulfates and carbonates was performed in the presence and absence of solvent. The results exhibited that among the reagents, only calcium carbonate offered the perfect ability in the transformation of styrene oxide to styrene episulfide. Additionally, the efficiency and rate enhancement under the solvent-free conditions were higher than the solution phase. A practical and green method was reported for the formation of thiiranes by solvent-free transformation of epoxides to thiiranes with calcium carbonate–supported thiourea under the oil bath conditions (60–70°C) (Scheme 45) [102].

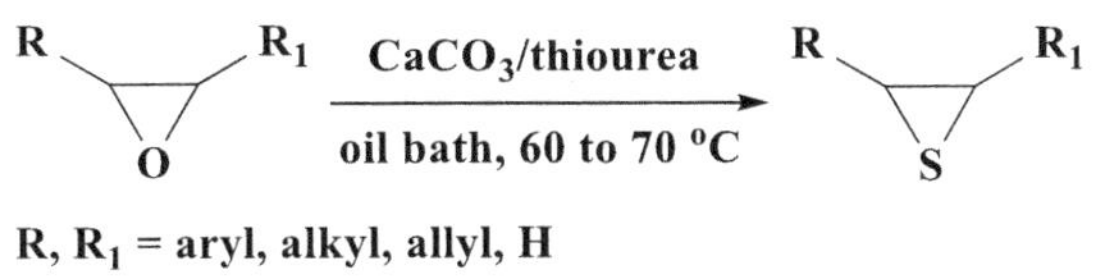

SCHEME 45

The oxiranes were transformed to thiiranes using thiourea in catalytic amount of ruthenium(III) or lithium tetrafluoroborate (Scheme 46) [15,103].

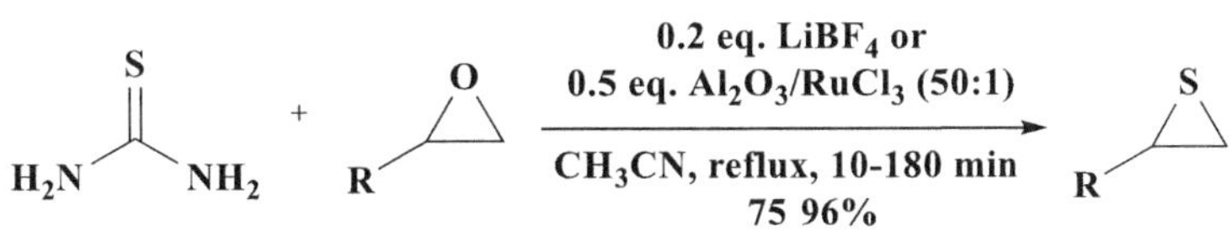

SCHEME 46

The utilization of silica gel–supported potassium thiocyanate as a source of sulfur in the reaction with epoxides to synthesize the thiiranes suffered from long reaction times (4.5–18 h) and high reaction temperature (90°C) [7]. However, when silica gel–supported thiourea was reacted with epoxides without solvent, a rapid reaction was reported (3–5 min) and quantitative conversion to thiiranes was noticed. The vacuum distillation methods utilizing a bulb-to-bulb system were applied to crude reaction mixture for an isolation of thiiranes. The possibility of utilizing this transformation was examined in solvent or doing solvent workup for those reactions in which distillation was not possible (e.g., small-scale reactions). Among the solvents utilized for this study such as dichloromethane, chloroform, diethyl ether, tetrahydrofuran, and acetone, the dichloromethane was observed to be the most appropriate. The reactions in dichloromethane occurred slower (0.5–3 h) as compared with reactions under the solvent-free conditions, but the conversion yields were analogous to those obtained in the presence of solvent. The cyclohexene oxide was the only substrate, which afforded product in higher yield in solution. Although the formation of thiiranes from starting compounds other than epoxides is broadly analyzed, reports on the formation of optically active thiiranes are rare [104]. Because of the low optical purity and low yields of thiiranes obtained by these approaches, epoxides have gained favor as an alternative starting compound for the formation of optically active thiiranes. An enantioselective preparation of (2*S*,3*S*)-(+)-*trans*-2,3-epithiosuccinate in 48% yield from epoxide has been reported in the presence of triphenylphosphine sulfide and TFA as a catalyst [63]. The transformation of optically active styreneoxide to optically active styrene episulfide utilizing ammonium thiocyanate and ruthenium(III) or cerium(IV) as catalysts was also reported [6]. The utilization of thiourea/silica gel under the solvent-free conditions resulted in the conversion of (*R*)-(+)-styrene oxide into (*S*)-(+)-styrene sulfide with high stereospecificity and yield (Scheme 47). The reaction of optically active (*R*)-(+)-styrene oxide with this reagent at 0°C without solvent afforded (*S*)-(+)-styrene episulfide in 90% optical purity and 96% yield.

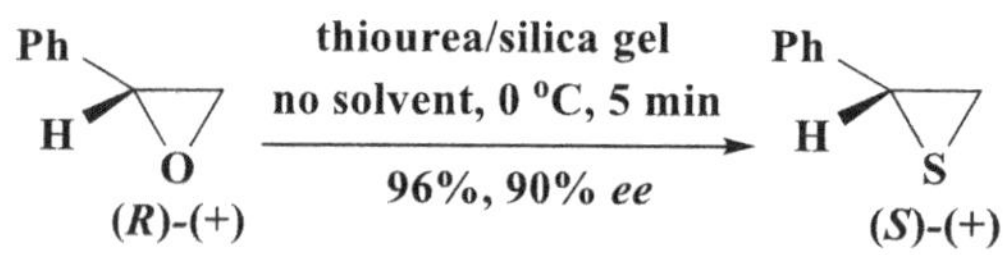

SCHEME 47

This sulphurization occurred with the inversion of configuration, and the reaction conditions affected the degree of stereospecificity. The reaction of (*R*)-styrene oxide with thiourea in the absence of solvent resulted in very little loose of optical purity during the reaction (Scheme 48). However, the *ee* dropped to 70% in dichloromethane at 0°C [7].

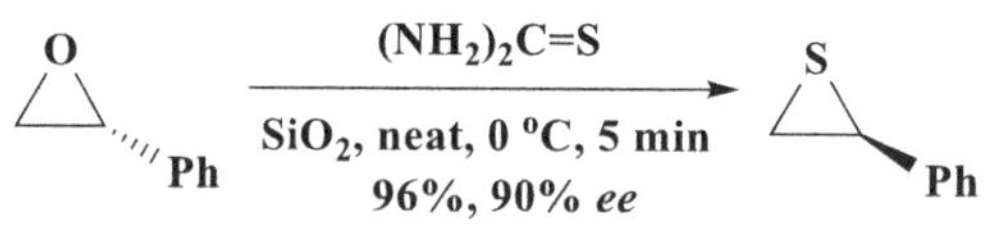

SCHEME 48

The widest and most beneficial procedure for the formation of episulfides is from epoxide involving the substitution of oxygen atom with sulfur. One common reagent utilized to influence this conversion is thiourea, which is favored for its ease of handling and relative stability, and quite a small number of reaction conditions have been proposed for this purpose. For instance, the cyclohexene oxide was cleanly transformed into episulfide in the presence of thiourea at higher temperatures under the solvent-free conditions (entry 1) (Scheme 49) [4]. The reaction also occurred in dichloromethane under the catalysis of silica gel in CH_3CN with silica-supported $AlCl_3$ [105] and in H_2O utilizing β-cyclodextrin as a catalyst [7,100].

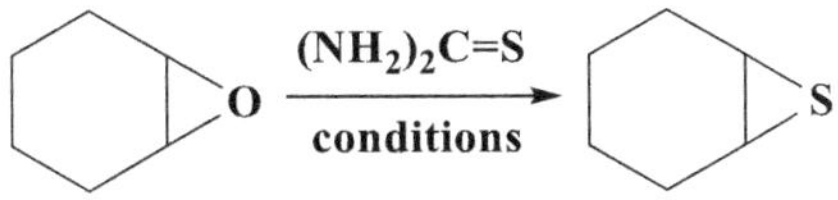

entry	activator	solvent	temp (°C)	time	yield (%)
1	none	neat	120	25 min	92
2	SiO_2	CH_2Cl_2	rt	30 min	92
3	SiO_2-$AlCl_3$	MeCN	45	1.3 h	89
4	β-cyclodextrin	H_2O	rt	6 h	82

SCHEME 49

The 2,3-epithio alcohols were prepared by reacting 2,3-epoxy alcohols with thiourea at 0°C or rt using Ti(O*i*-Pr)$_4$ in tetrahydrofuran solvent (Scheme 50) [106]. The reactions of *trans*-substituted 2,3-epoxy alcohols occurred with high stereo- and regioselectivity to provide the *trans*-2,3-epithio alcohols only with complete inversion of configuration at both stereogenic centers. However, the yields of epithio alcohols were low and thiodiols were also obtained when *cis*-2,3-epoxy alcohols were utilized as starting compounds. The epithiocinnamyl alcohols can also be prepared from epoxycinnamyl alcohols at 0°C.

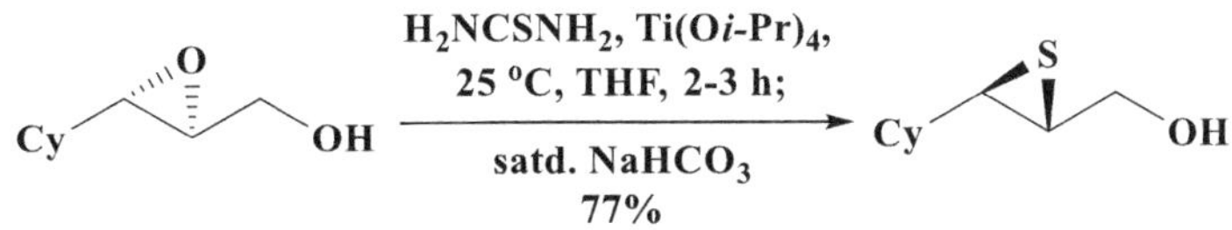

SCHEME 50

References

1. (a) N. Kaur. 2019. Synthetic routes to seven and higher membered *S*-heterocycles by use of metal and nonmetal catalyzed reactions. Phosphorus, Sulfur, Silicon Relat. Elem. 194: 186-209. (b) N. Kaur. 2019. Cobalt-catalyzed C-N, C-O, C-S bond formation: synthesis of heterocycles. J. Iran. Chem. Soc. 16: 2525-2553. (c) N. Kaur. 2018. Ultrasound-assisted green synthesis of five-membered *O*- and *S*-heterocycles. Synth. Commun. 48: 1715-1738. (d) N. Kaur. 2018. Solid-phase synthesis of sulfur-containing heterocycles. J. Sulfur Chem. 39: 544-577. (e) N. Kaur. 2019. Ionic liquid assisted synthesis of *S*-heterocycles. Phosphorus, Sulfur, Silicon Relat. Elem. 194: 165-185. (f) N. Kaur. 2014. Microwave-assisted synthesis of five-membered *S*-heterocycles. J. Iran. Chem. Soc. 11: 523-564. (g) N. Kaur. 2019. Metal and non-metal catalysts in the synthesis of five-membered *S*-heterocycles. Curr. Org. Synth. 16: 258-275. (h) N. Kaur. 2015. Palladium-catalyzed approach to the synthesis of *S*-heterocycles. Catal. Rev. 57: 478-564. (i) N. Kaur, N.K. Jangid and V. Sharma. 2018. Metal- and nonmetal-catalyzed synthesis of five-membered *S,N*-heterocycles. J. Sulfur Chem. 39: 193-236. (j) N. Kaur. 2014. Microwave-assisted synthesis of seven-membered *S*-heterocycles. Synth. Commun. 44: 3201-3228. (k) N. Kaur and D. Kishore. 2014. Microwave-assisted synthesis of six-membered *S*-heterocycles. Synth. Commun. 44: 2615-2644. (l) N. Kaur, M. Devi, P. Grewal, N. Ahlawat, P. Bhardwaj, Y. Verma and N.K. Jangid. 2022. Synthesis of five-membered nitrogen-containing heterocycles using copper. J. Iran. Chem. Soc. 19: 679-727. (m) N. Kaur,

M. Rajput and P. Bhardwaj. 2022. Synthesis of heterocycles using guanidine: an overview. Synth. Commun. 52: 1547-1580. (n) T.H. Chan and J.R. Finkenbine. 1972. Facile conversion of oxiranes to thiiranes by phosphine sulfides. Scope, stereochemistry, and mechanism. J. Am. Chem. Soc. 94: 2880-2882.

2. N. Iranpoor, B. Tamami and K. Niknam. 1997. Iodine and iodine supported on polyvinylpyrrolidone as catalysts and reagents for alcoholysis, hydrolysis, and acetolysis of epoxides and thiiranes. Can. J. Chem. 75: 1913-1919.
3. W. Adam and R.M. Bargon. 2004. Synthesis of thiiranes by direct sulfur transfer: the challenge of developing effective sulfur donors and metal catalysts. Chem. Rev. 104: 251-262.
4. A.R. Kiasat, F. Kazemi and M.F. Mehrjardi. 2004. Solvent-free conversion of oxiranes to thiiranes with thiourea. Phosphorus, Sulfur, Silicon Relat. Elem. 179: 1841-1844.
5. H. Bouda, M.E. Borredon, M. Delmas and A. Gaset. 1987. Reaction in low hydrated solid-liquid heterogeneous medium: thiiranes synthesis from epoxides. Synth. Commun. 17: 943-951.
6. M.O. Brimeyer, A. Mehrota, S. Quici, A. Nigam and S.L. Regen. 1980. Silica gel assisted synthesis of thiiranes. J. Org. Chem. 45: 4254-4255.
7. N. Iranpoor, H. Firouzabadi and A.A. Jafari. 2005. A green protocol for the easy synthesis of thiiranes from epoxides using thiourea/silica gel in the absence of solvent. Phosphorus, Sulfur, Silicon Relat. Elem. 180: 1809-1814.
8. B. Tamami and A.R. Kiasat. 1996. Synthesis of thiiranes from oxiranes under mild and nonaqueous conditions using polymer supported thiocyanate. Synth. Commun. 26: 3953-3958.
9. B. Zeynizadeh and S. Yeghaneh. 2009. A green protocol for solvent-free conversion of epoxides to thiiranes with Dowex-50WX8-supported thiourea. Phosphorus, Sulfur, Silicon Relat. Elem. 184: 362-368.
10. S. Kazemi, A.B. Martin, D. Sordi, M.C. Kroon, C.J. Peters and I.W.C.E. Arends. 2012. Vanadium-catalyzed epoxidation reaction of cinnamyl alcohol in ionic liquids. Green Process. Synth. 1: 509-516.
11. I. Mohammadpoor-Baltork and H. Aliyan. 1998. Bismuth(III) chloride; a mild and efficient catalyst for synthesis of thiiranes from oxiranes. Synth. Commun. 28: 3943-3947.
12. B. Yadollahi, S. Tangestaninejad and M.H. Habibi. 2004. Titanium dioxide as a mild and efficient catalyst for conversion of epoxides to thiiranes. Synth. Commun. 34: 2823-2827.
13. N. Iranpoor and B. Zeynizadeh. 1998. Conversion of epoxides to thiiranes catalyzed with $TiO(TFA)_2$ and $TiCl_3(Otf)$ in the presence of ammonium thiocyanate or thiourea. Synth. Commun. 28: 3913-3918.
14. P. Salehi, M.M. Khodaei, M.A. Zolfigol and A. Keyvan. 2003. Magnesium hydrogensulfate: a cheap and efficient catalyst for the conversion of epoxides into β-alkoxy alcohols, vicinal-diols, and thiiranes. Synth. Commun. 33: 3041-3048.
15. F. Kazemi, A.R. Kiasat and S. Ebrahimi. 2003. $LiBF_4$: a mild and efficient catalyst for conversion of oxiranes to thiiranes with thiourea. Synth. Commun. 33: 595-600.
16. C.S. Reddy and S. Nagavani. 2008. $LiClO_4$ Catalyzed mild and efficient method for the synthesis of thiiranes from oxiranes. Heteroat. Chem. 19: 97-99.
17. J.S. Yadav, B.V.S. Reddy, S. Sengupta, M.K. Gupta, G. Baishya, S.J. Harshavardhana and U. Dash. 2008. Iodine as a mild, efficient, and cost-effective catalyst for the synthesis of thiiranes from oxiranes. Monatsh. Chem. 139: 1363-1367.
18. I. Mohammadpoor-Baltork and H. Aliyan. 2000. Montmorillonite K-10: a mild, inexpensive and convenient catalyst for synthesis of thiiranes from oxiranes under non-aqueous conditions. J. Chem. Res. 3: 122-123.
19. V. Mirkhani, S. Tangestaninejad and L. Alipanah. 2002. Mild, efficient, and convenient conversion of oxiranes to thiiranes with ammonium thiocyanate and thiourea in the presence of cerium(IV) polyoxometallate. Synth. Commun. 32: 621-626.
20. N. Iranpoor and F. Kazemi. 1996. Cerium(IV) catalysis: efficient conversion of epoxides to thiiranes. Synthesis 7: 821-822.
21. M. Moghadam, S. Tangestaninejad, V. Mirkhani and R. Shaibani. 2004. Rapid and efficient ring opening of epoxides catalyzed by a new electron deficient tin(IV) porphyrin. Tetrahedron 60: 6105-6111.

22. J.S. Yadav, B.V.S. Reddy, C.S. Reddy and K. Rajasekhar. 2003. [bmim]PF_6: a novel and recyclable ionic liquid for conversion of oxiranes to thiiranes in aqueous media. J. Org. Chem. 68: 2525-2527.
23. N. Iranpoor, B. Tamami and M. Shekarriz. 1999. Conversion of epoxides to thiiranes with thiourea or ammonium thiocyanate catalyzed with poly(4-vinyl pyridine)-Ce$(OTf)_4$. Synth. Commun. 29: 3313-3321.
24. B. Tamami and K.P. Borujeny. 2004. Synthesis of thiiranes from oxiranes using cross-linked polystyrene supported aluminium chloride as a catalyst. Synth. Commun. 34: 65-70.
25. L. Wu, Y. Wang, F. Yan and C. Yang. 2010. Facile conversion of epoxides to thiiranes with ammonium thiocyanate catalyzed with etidronic acid. Bull. Korean Chem. Soc. 31: 1419-1420.
26. B. Zeynizadeh and S. Yeghaneh. 2008. Solvent-free conversion of epoxides to thiiranes by thiourea/NH_4Cl system. Phosphorus, Sulfur, Silicon Relat. Elem. 183: 2280-2286.
27. B.P. Bandgar, A.V. Patil, V.T. Kamble and J.V. Totre. 2007. An efficient synthesis of thiiranes from oxiranes using fluoroboric acid adsorbed on silica gel (HBF_4-SiO_2) as a catalyst under mild conditions in the absence of solvent. J. Mol. Catal. A: Chem. 273: 114-117.
28. L. Wu, L. Yang, L. Fang, C. Yang and F. Yan. 2010. Silica chloride-catalyzed synthesis of thiiranes from oxiranes under solvent-free conditions. Phosphorus, Sulfur, Silicon Relat. Elem. 185: 2159-2164.
29. R. Ketcham and V.P. Shah. 1963. *Cis*- and *trans*-stilbene sulfides. J. Org. Chem. 28: 229-230.
30. G.S. Andrade, J.E. Berkner, C.L. Liotta, C. Eckert, D.A. Schiraldi, A. Andersen and D.M. Collard. 2003. The one-pot synthesis and Diels-Alder reactivity of 2,5-dihydrothiophene-1,1-dioxide-3-carboxylic acid. Synth. Commun. 33: 3643-3650.
31. R.L. Halcomb and S.J. Danishefsky. 1989. On the direct epoxidation of glycals: application of a reiterative strategy for the synthesis of beta-linked oligosaccharides. J. Am. Chem. Soc. 111: 6661-6666.
32. V. Calo, L. Lopez, L. Marchese and G. Pesce. 1975. A simple method for converting oxirans into thiirans stereospecifically. J. Am. Chem. Soc. Commun. 15: 621-622.
33. Y. Ueno, T. Nakai and M. Okawara. 1970. 2-Substituted imino-1,3-dithiolanes. Their thermal ring-isomerizations and reactions with epoxides. Bull. Chem. Soc. Jpn. 43: 162-167.
34. S. Kubota, E. Taguchi, M. Eto and K. Maekawa. 1972. Synthese und biologische aktivität von 1,3-dithiolen, 1,2-dithiolanen und 1,3-dithian-2-thionen. J. Fac. Agric. Kyusyu Univ. 22: 1-13.
35. W. Choi, F. Sanda and T. Endo. 1998. Dependence of ring-opening reaction of five-membered dithiocarbonates on cationic catalyst: control of isomerization and polymerization. Macromolecules 31: 2454-2460.
36. W. Choi, F. Sanda and T. Endo. 1998. A novel construction of living polymerization by neighboring group participation: living cationic ring-opening polymerization of a five-membered cyclic dithiocarbonate. Macromolecules 31: 9093-9095.
37. W. Choi, F. Sanda and T. Endo. 2000. Substituent effects on the control of cationic isomerization and polymerization of five-membered cyclic dithiocarbonates. Heterocycles 52: 125-128.
38. Y. Taguchi, M. Yasumoto, I. Shibuya and Y. Suhara. 1989. The synthesis of 1,3-dithiolan-2-ones on the reaction of oxiranes with carbon disulfide under high pressure. Bull. Chem. Soc. Jpn. 62: 474-478.
39. Y. Taguchi, K. Yanagiya, I. Shibuya and Y. Suhara. 1988. The reaction of oxiranes with carbon disulfide under high pressure. Bull. Chem. Soc. Jpn. 61: 921-925.
40. M.V. Jesudason and L.N. Owen. 1974. Dithiols. Part XXIV. Proof of the epoxide chirality in 2,3-epoxypropyl β-D-glucopyranoside by correlation with (*R*)-2,3-thiocarbonyldithiopropanol. J. Chem. Soc. Perkin. Trans. 1 0: 1443-1446.
41. J.A. Durden, H.A. Stambury and W.H. Catlette. 1960. The preparation of alkene trithiocarbonates. J. Am. Chem. Soc. 82: 3082-3084.
42. N. Adachi, Y. Kida and K. Shikata. 1977. Copolymerization of propylene oxide and carbon disulfide with diethyl zinc-electron-donor catalyst. J. Polym. Sci. Polym. Chem. 15: 937-944.
43. S. Hayashi, M. Furukawa, Y. Wino, T. Nakao and K. Nagato. 1971. Reactions of epoxides. I. Synthesis of trithiocarbonate by the reaction of epoxide with alkali xanthate. Chem. Pharm. Bull. 19: 1594-1597.

44. T. Eicher and S. Hauptmann. 2003. The chemistry of heterocycles: structures, reactions, and applications. John Wiley & Sons: Chichester, UK, 5-16.
45. R.E. Banks and J.C. Tatlow. 1994. Organofluorine chemistry: nomenclature and history. R.E. Banks, B.E. Smart and J.C. Tatlow (Eds.). Organofluorine chemistry: principles and commercial applications. Plenum Press: New York, 2-4.
46. M. Blagoev, A. Linden and H. Heimgartner. 1999. Lewis acid catalyzed reactions of thioketones with 1,2-epoxycyclohexane and 1,2-epoxycyclopentane. Helv. Chim. Acta 82: 2316-2335.
47. S. Malaschichin, C. Fu, A. Linden and H. Heimgartner. 2005. Ring enlargement and sulfur-transfer processes in SiO_2-catalyzed reactions of thiocarbonyl compounds with optically active oxiranes. Helv. Chim. Acta 88: 3253-3262.
48. C. Fu, A. Linden and H. Heimgartner. 2011. Reaction of optically active oxiranes with thiofenchone and 1-methylpyrrolidine-2-thione: formation of 1,3-oxathiolanes and thiiranes. Helv. Chim. Acta 94: 773-784.
49. T. Takido, Y. Kobayashi and K. Itabashi. 1986. A convenient synthesis of thiirans by the use of dimethylthioformamide. Synthesis 9: 779-780.
50. M. Sander. 1966. Thiiranes. Chem. Rev. 66: 297-339.
51. E. Vedejs and G.-A. Krafft. 1982. Cyclic sulfides in organic synthesis. Tetrahedron 38: 2857-2881.
52. K. Kolc, E. Kubicz and J. Mlochowski. 1984. A novel approach to functionalization of azines. Oxiranyl and thiiranyl derivatives of pyridine, quinoline and isoquinoline. J. Heterocycl. Chem. 22: 2517-2522.
53. K. Jankowski and R. Harvey. 1972. Stereoisomerization of episulfides. Synthesis 11: 627-628.
54. N. Iranpoor and F. Kazemi. 1997. Stereospecific formation of *S*-(-)-styrene sulfide: efficient conversion of epoxides to thiiranes catalysed with Ru(III). Tetrahedron 53: 11377-11382.
55. C.C. Price and P.F. Kirk. 1953. Some observations on the reaction of alkali thiocyanates with epoxides. J. Am. Chem. Soc. 75: 2396-2400.
56. F. Heidarizadeh and S. Elahi. 2011. Some versatile and efficient methods for the regioselective ring opening of epoxides to corresponding β-hydroxy thiocyanates. Orient. J. Chem. 27: 1531-1553.
57. A.-R. Kiasat and M.-F. Mehrjardi. 2010. Melamine sulfonic acid: a recoverable catalyst for the ecofriendly synthesis of thiocyanohydrins under solvent-free conditions. Synth. Commun. 40: 1551-1558.
58. I. Mohammadpoor-Baltork and A.R. Khosropour. 2001. $Bi(TFA)_3$ and $Bi(OTf)_3$ catalyzed conversions of epoxides to thiiranes with ammonium thiocyanate and thiourea under non-aqueous conditions. Molecules 6: 996-1000.
59. F. Kazemi and A.R. Kiasat. 2003. Facile conversion of epoxides to thiiranes with ammonium thiocyanate catalyzed with oxalic acid. Phosphorus, Sulfur, Silicon Relat. Elem. 178: 1333-1337.
60. H.-M. Chen, P.-Y. Chen, C.-F. Wang, J.-C. Tsai and E.-C. Wang. 2012. An improved method for the conversion of oxiranes to thiiranes and the discrimination of their base peaks in EI-MS. ARKIVOC (viii) 157-168.
61. G. Gorla, K.V. Padmaja, S. Arukali and R.B.N. Prasad. 2014. Synthesis, characterization, and evaluation of 10-undecenoic acid-based epithio derivatives as multifunctional additives. J. Agric. Food Chem. 62: 11505-11511.
62. B.P. Bandgar, N.S. Joshi and V.T. Kamble. 2006. 2,4,6-Trichloro-1,3,5-triazine catalyzed synthesis of thiiranes from oxiranes under solvent-free and mild conditions. Tetrahedron Lett. 47: 4775-4777.
63. A.C. Korn and L. Moroder. 1995. Enantioselective synthesis of diethyl (+)-(2*S*,3*S*)-*trans*-2,3-epithiosuccinate. J. Chem. Res. (S) 3: 102-103.
64. J. Gay and G. Scherowsky. 1995. Chiral thietanes and thiethane-1,1-dioxides - novel building blocks for liquid crystals. Synth. Commun. 25: 2665-2672.
65. A.J. Blake, P.A. Cooke, J.D. Kendall, N.S. Simpkins and S.M. Westway. 2000. Concerning the synthesis and enantioselective rearrangements of episulfoxides. J. Chem. Soc. Perkin Trans. 1 2: 153-164.

66. B. Tamami and M. Kolahdoozan. 2004. Synthesis of thiiranes from oxiranes in water using polymeric cosolvents. Tetrahedron Lett. 45: 1535-1537.
67. P. Salehi, B. Seddighi, M. Irandoost and K.F. Behbahani. 2000. Ferric perchlorate: an efficient reagent for regio- and stereoselective alcoholysis and hydrolysis of epoxides. Synth. Commun. 30: 2967-2973.
68. B. Kaboudin and H. Norouzi. 2004. A new, efficient, and simple method for the synthesis of thiiranes from epoxides under solvent-free conditions. Tetrahedron Lett. 45: 1283-1285.
69. D.A. Lighter and C. Djerassi. 1965. Studies in organic sulphur compounds - XVI. Tetrahedron 21: 583-601.
70. A. Bellomo, D.G.A. Bellomo and D. Gonzalez. 2007. Diastereodivergent synthesis of optically pure vinyl episulfides and β-hydroxy thiocyanates from a bacterial metabolite. Tetrahedron Lett. 48: 3047-3051.
71. E. Repetto, V.E. Manzano, M.L. Uhrig and O. Varela. 2012. Synthesis of branched dithiotrisaccharides via ring-opening reaction of sugar thiiranes. J. Org. Chem. 77: 253-265.
72. P.P. Nicholas and R.T. Carroll. 1968. Chlorination of olefins with cupric chloride. Comparative study of *trans*-ethylene-d2 and *cis*- and *trans*-2-butene. J. Org. Chem. 33: 2345-2349.
73. R.L. Arnett and B.L. Crawford. 1950. The vibrational frequencies of ethylene. J. Chem. Phys. 18: 118-126.
74. P.P. Nicholas. 1975. Thermolysis of ethyl azidoformate in 2,3-dimethyl-2-butene. Example of a simple olefin giving a dominant 1,2,3-Δ^2-triazoline intermediate. J. Org. Chem. 40: 3396-3398.
75. S.J. Danishefsky and M.T. Bilodeau. 1996. Glycals in organic synthesis: the evolution of comprehensive strategies for the assembly of oligosaccharides and glycoconjugates of biological consequence. Angew. Chem. Int. Ed. 35: 1380-1419.
76. B. Das, V.S. Reddy and M. Krishnaiah. 2006. An efficient catalyst-free synthesis of thiiranes from oxiranes using polyethylene glycol as the reaction medium. Tetrahedron Lett. 47: 8471-8473.
77. H. Bouda, M.E. Borredon, E. Delmas and A. Gaset. 1989. Synthesis of thiiranes from epoxides in the presence of thiourea in a weakly hydrated organic medium. Synth. Commun. 19: 491-500.
78. F.G. Bordwell and H.M. Andersen. 1953. The reaction of epoxides with thiourea. J. Am. Chem. Soc. 75: 4959-4962.
79. F.G. Bordwell, H.M. Andersen and B.M. Pitt. 1954. The reaction of thiacyclopropanes (olefin sulfides) and thiacyclobutanes with organolithium compounds. J. Am. Chem. Soc. 76: 1082-1085.
80. J.M. Geurts. 1997. Latexes with intrinsic crosslink activity. Double liaison. Phys. Chim. Econ. Peint. Adhes. 44: 59-67.
81. R.M. O'Brien, S.A. Brown, B.G. Bufkin and J.R. Grawe. 1981. Use of thiirane-functional monomers as a means of developing cross linkable emulsions. J. Coat. Technol. 53: 49-58.
82. G. Tillet, B. Boutevin and B. Ameduri. 2011. Chemical reactions of polymer cross linking and post-cross linking at room and medium temperature. Prog. Polym. Sci. 36: 191-217.
83. E.E. van Tamelen. 1972. Valenzisomere aromatischer syst. Acc. Chem. Res. 5: 186-192.
84. K. Soai and T. Mukaiyama. 1979. Enantioface-differentiating reactions using (2*S*,2'*S*)-2-hydroxymethyl-1-[(1-alkyl-2-pyrrolidinyl)methyl]pyrrolidines as chiral ligands. Addition of lithium derivatives of methyl phenyl sulfide, acetonitrile, *N*-nitrosodimethylamine, and 2-methylthiothiazoline to aldehydes. Bull. Chem. Soc. Jpn. 52: 3371-3376.
85. C.C.J. Culvenor, W. Davies and N.S. Heath. 1949. Reactions of ethylene oxides. Part II. Reactions with thioamides, thiols, and inorganic sulphur salts. J. Chem. Soc. 0: 278-282.
86. C.G. Moore and M. Porter. 1959. The reaction of sulphur and sulphur compounds with olefinic substances - XIII: the mechanism of interaction of sulphur with (+)-limonene. Tetrahedron 6: 10-15.
87. J.A. Durden, H.A. Stansbury and W.H. Catlette. 1959. The reaction of glycidic esters with thiourea in aqueous sulfuric acid solution. J. Am. Chem. Soc. 81: 1943-1946.
88. H. Yu, S. Cao, L. Zhang, G. Liu and J. Xu. 2009. Synthesis of α-aliphatic and β-aromatic substituted taurines via regioselective ring opening of thiiranes with ammonia. Synthesis 13: 2205-2209.

89. A.T. Guseinova, K.I. Alieva, I.A. Rzaeva, A.M. Magerramov, I.A. Aliev, V.M. Farzaliev and M.A. Allakhverdiev. 2009. Synthesis and antioxidant activity of 1-amino-4-phenyl-2-butanethiols. Russ. J. Appl. Chem. 82: 1587-1591.
90. C. Forbes, Q. Shi, J.F. Fisher, M. Lee, D. Hesek, L.I. Llarrull, M. Toth, M. Gossing, R. Fridman and S. Mobashery. 2009. Active site ring-opening of a thiirane moiety and picomolar inhibition of gelatinases. Chem. Biol. Drug Design 74: 527-534.
91. H. Jiang, N. Holub, M.W. Paixao, C. Tiberi, A. Falcicchio and K.A. Joergensen. 2009. Iminium-ion activation as an efficient strategy for divergent synthesis of optically active propargylic, homopropargylic, and allenic compounds. Chem. Eur. J. 15: 9638-9641.
92. Z.-H. Sun and B. Wang. 2008. A facile and practical synthesis of peracylated 4-thio-D-ribofuranoses from D-glucose. J. Org. Chem. 73: 2462-2465.
93. M. Fukudome, T. Onizuka, S. Kawamura, D.-Q. Yuan and K. Fujita. 2007. Selective functionalization of β-cyclodextrin: efficient conversions of 2,3-alloepoxypyranosides to 2,3-mannoepithiopyranosides. Tetrahedron Lett. 48: 6665-6668.
94. S. Escher, Y. Niclass, M. van de Waal and C. Starkenmann. 2006. Combinatorial synthesis by nature: volatile organic sulfur-containing constituents of *Ruta chalepensis*. L. Chem. Biodivers. 3: 943-957.
95. E. Elhalem, M.J. Comin and J.B. Rodriguez. 2006. Synthesis of conformationally locked carbocyclic nucleosides built on a thiabicyclo[3.1.0]hexane system as a pseudosugar surrogate. Eur. J. Org. Chem. 19: 4473-4482.
96. M. Lee, M.M. Bernardo, S.O. Meroueh, S. Brown, R. Fridman and S. Mobashery. 2005. Synthesis of chiral 2-(4-phenoxyphenylsulfonylmethyl)thiiranes as selective gelatinase inhibitors. Org. Lett. 7: 4463-4465.
97. N.K. Rozhkova, K. Sabirov and K.L. Seitanidi. 1983. Reaction of benzothiazoline-2-thione with 1-chloro-2,3-epoxypropane. Chem. Heterocycl. Compd. 19: 1170-1173.
98. R.C. Cambie, P.S. Rutledge, G.A. Strange and P.D. Woodgate. 1982. *vic*-Iodothiocynates and iodoisothiocyanates. Part 6. The synthesis of thiirans. Heterocycles 19: 1501-1510.
99. P. Kalicki, M. Karchier, K. Michalak and J. Wicha. 2010. Rearrangement of 2,3-epoxy alcohol dimethylthiocarbamate derivatives. Synthesis of 2,3-epithio alcohol derivatives under mild conditions. J. Org. Chem. 75: 5388-5391.
100. K. Surendra, N.S. Krishnaveni and K.R. Rao. 2004. A new and efficient method for the synthesis of thiiranes from oxirane-β-cyclodextrin complexes and thiourea in water. Tetrahedron Lett. 45: 6523-6526.
101. A. Dastan, A. Kulkarni and B. Torok. 2012. Environmentally benign synthesis of heterocyclic compounds by combined microwave-assisted heterogeneous catalytic approaches. Green Chem. 14: 17-37.
102. B. Zeynizadeh, M.M. Baradarani and R. Eisavi. 2011. A practical and eco-friendly method for conversion of epoxides to thiiranes with immobilized thiourea on $CaCO_3$. Phosphorus, Sulfur, Silicon Relat. Elem. 186: 2208-2215.
103. F. Kazemi, A.R. Kiasat and S. Ebrahimi. 2001. Efficient conversion of oxiranes to thiiranes with thiourea catalyzed with ruthenium trichloride and alumina. Phosphorus, Sulfur, Silicon Relat. Elem. 176: 135-140.
104. C.C.J. Culvenor, W. Davies and W.E. Savige. 1952. The conversion of ethylene oxides into ethylene sulphides. J. Chem. Soc. 0: 4480-4486.
105. K.P. Borujeni. 2005. Conversion of oxiranes to thiiranes catalyzed with silica gel-supported aluminium chloride. Synth. Commun. 35: 2575-2579.
106. Y. Gao and K.B. Sharpless. 1988. Titanium isopropoxide mediated formation of 2,3-epithio alcohols from 2,3-epoxy alcohols. J. Org. Chem. 53: 4114-4116.

CHAPTER

Synthesis of thiiranes, thiirenes, dithiiranes, and thiaziridines

9

Chapter outline

9.1 Introduction

The organic sulfur compounds have become increasingly beneficial and important in organic synthesis. The thiiranes, which are the simplest sulfur heterocyclic compounds, are beneficial and important building blocks for the formation of medicines, herbicides, pesticides, and polymers. A number of approaches have been developed for the formation of thiiranes [1].

The cyclic organic sulfur compounds occur naturally in some plant products and are of great importance as components of dye-stuff and synthetic medicines [2–6]. Since organic sulfur products have become increasingly beneficial, development of practical and convenient synthetic approaches for these compounds is highly required. The thiiranes are the simplest *S*-containing heterocyclic compounds that occur in nature [7–9].

The thiiranes are valuable synthetic building blocks with industrial potential for the synthesis of polymers [10–12], liquid crystals [13,14], and adhesives [15,16], as many others [17]. Additionally, the thiirane is an important heterocyclic core of several compounds with biological activities such as selective gelatinase inhibitors [18], selective A1 adenosine receptor agonists [19], potential topoisomerase I inhibitors [20], and estrogen synthetase inhibitor [21]. As a result, researchers have made efforts to find an environmentally benign and efficient process for the formation of thiiranes. The methods for the formation of thiiranes include the (1) reaction of alkenes with several reagents such as dichlorodisulfides [22] and iodothiocyanate [23], (2) reaction of ketones with sulfuryl lithium [24], (3) reduction of 1-(benzthiazol-2-yl-sulfanyl)-2-alkanones with NaH [25], (4) reaction of epoxides with NH_4SCN [26–37] or KSN [38–41], (5) transformation of epoxides with thiourea [42–47] as well as other approaches [48–52].

3-Membered Heterocycle Synthesis. https://doi.org/10.1016/B978-0-443-13583-5.00004-9

9.2 Synthesis of thiiranes

The reaction of precursor with sulfur dichloride (1.25 M eq.) for 5 min at 25°C in chloroformal provided thiirane (R = H) (80°C) [53]. The product was thiirane (R = Cl) (quantitative yield) when the reaction was carried out in dichloromethane at 0°C for 80 min (Scheme 1). The difference in the reactions was ascribed to the fate of an addition compound, which decomposed with loss of Cl^- to provide the thiirane (R = H) or with loss of hydrogen chloride to synthesize the ring-chlorinated compounds.

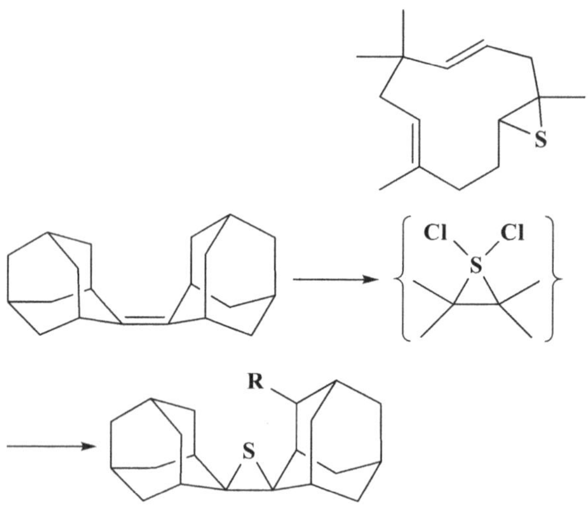

SCHEME 1

In a single instance by Ando et al. [54], the episulfidation of bis-adamantylidene allene was carried out by heating with elemental sulfur in dimethylformamide. The thietanethione side product was also obtained (Scheme 2). Although this process worked moderately well for this precursor, it was not a common technique for the formation of methylene thiiranes [55].

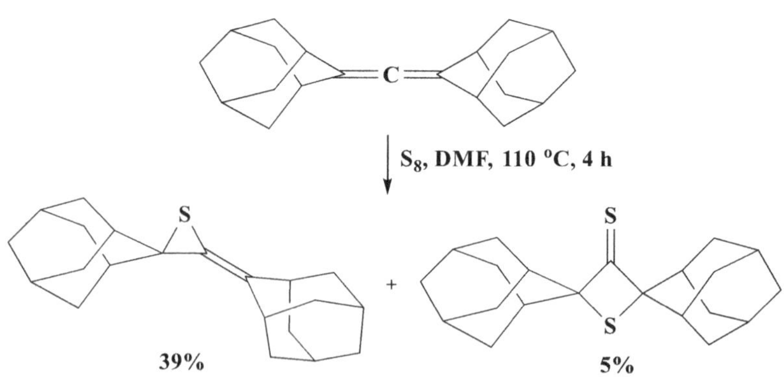

SCHEME 2

Zhou et al. [56] reported the serendipitous synthesis of methylene thiiranes from allenyl sulfones. The allene was reacted with Br_2 under the aqueous conditions, and then an aqueous sodium thiosulfate work-up was carried out (Scheme 3). The reaction was general for numerous 1,1,3-trisubstituted and 1,3-disubstituted allenes and provided an efficient chirality transfer for enantioenriched precursors. The methylene thiiranes were stable to column chromatography [55].

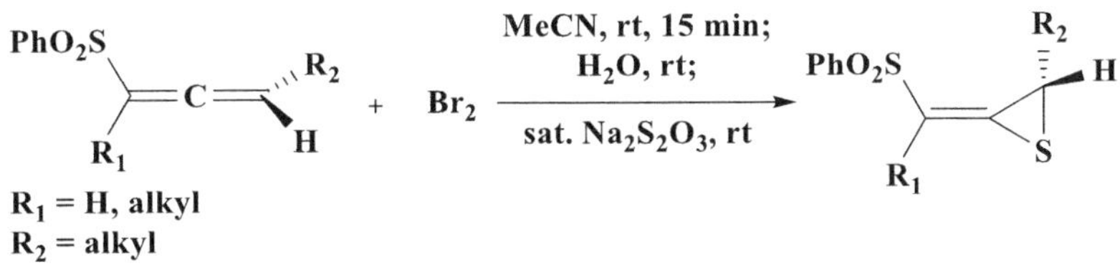

SCHEME 3

The reaction of thiones with diazo compounds provided thione *S*-methylides or thiiranes and thiadiazolidines, while lactones, olefins, and dithiolanes were formed in a few cases. The variety of products was dependent on both the electronic character of the diazo compound and the steric bulk of the thione substituents. The unhindered thiones were reacted with diazo compounds to provide the thiiranes (Scheme 4) [57–59].

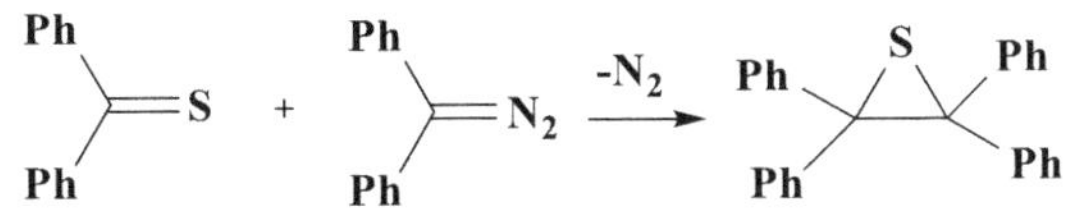

SCHEME 4

The carbene, prepared from trimethylsilyldiazomethane, was added to sulfenes to synthesize the three-membered ring sulfones (Scheme 5) [60].

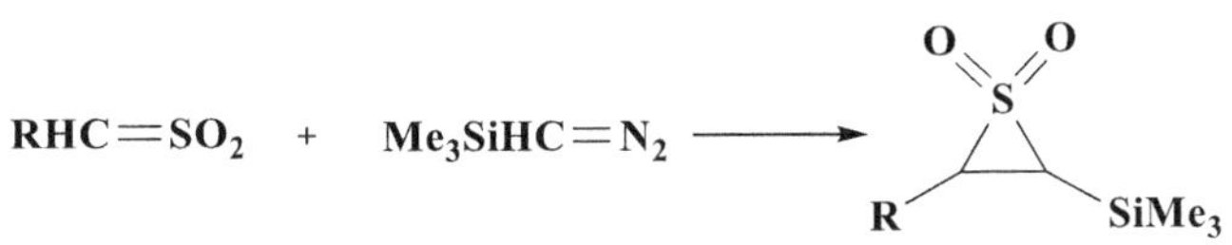

SCHEME 5

The diazo compounds were reacted with adamantanethione at rt in pentane solution. The adamantanethione was less stable and underwent dimerization or/and trimerization [61]. Therefore, the reactions of adamantanethione were studied utilizing the most reactive diazodicarbonyl compounds (Scheme 6). The experiments demonstrated that the reactions of adamantanethione with diazoacetoacetate and diazoacetylacetone carried out at room temperature occurred to synthesize the spirocyclic tetrasubstituted 1,3-oxathioles in moderate yields (39%–48%). The thiirane was isolated as an only compound from the mixture obtained after the reaction of diazomalonate with adamantanethione in low yield (13%) [62].

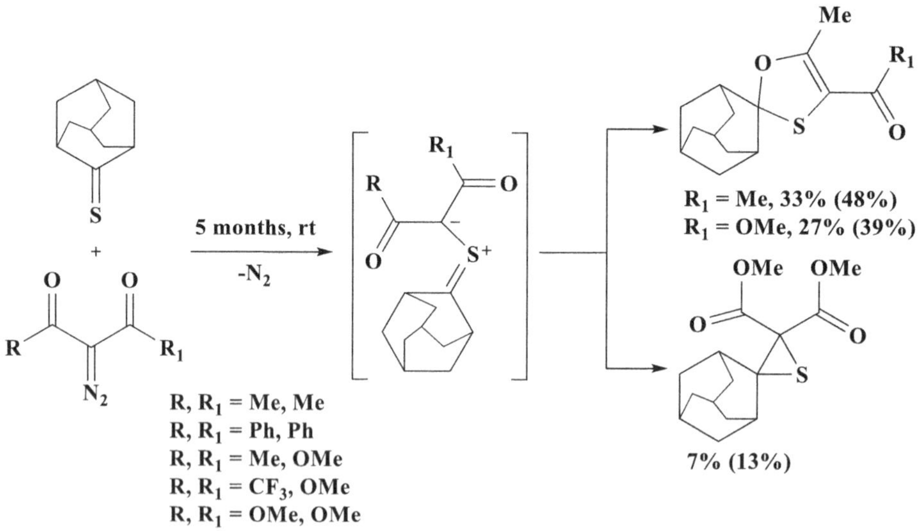

SCHEME 6

The reaction of diazo compounds with thiocarbonyl compounds provided thiiranes [63]; for instance, the diazopropionate was reacted with thiobenzophenone to synthesize the thiirane (Scheme 7) [64].

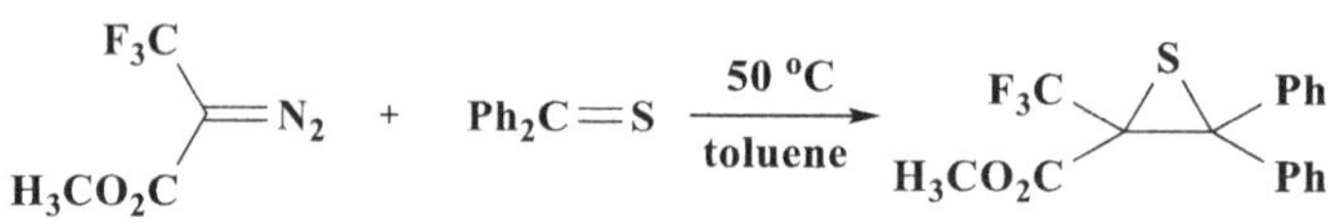

SCHEME 7

The reactions that occurred through the thiadiazoline intermediates received direct support when thiadiazole was isolated in 92% yield by the reaction of hexafluorothioacetone with bis(trifluoromethyl)diazomethane. The perfluorotetramethylthiirane was formed in 95% yield upon refluxing thiadiazole (Scheme 8) [63,64].

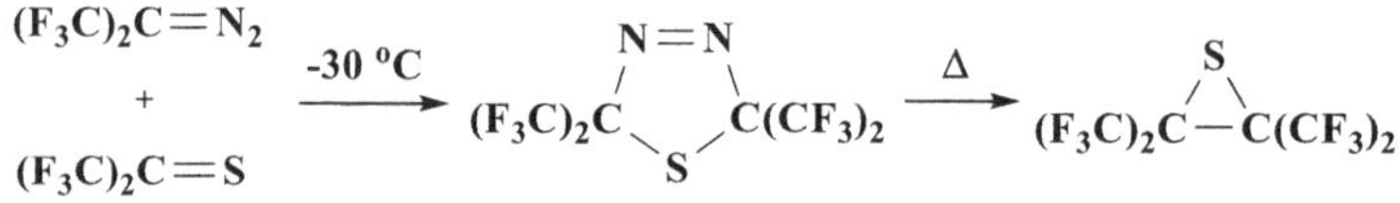

SCHEME 8

In contrast to the triplet atomic sulfur [$S(^3P)$], the singlet state [$S(^1D)$] was inappropriate for episulfidation both in solution and vapor phase. Already with ethylene, an insertion of carbon–hydrogen occurred to synthesize the vinylthiol in vapor phase (Scheme 9) [65–67].

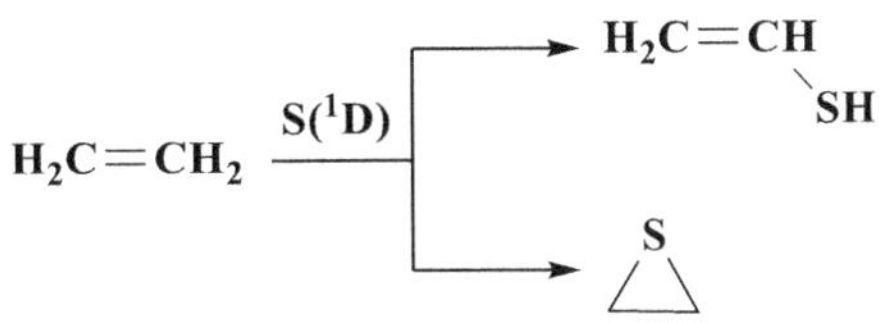

SCHEME 9

The cheapest direct source of sulfur, elemental sulfur itself without chemical activation, was utilized successfully only in a small number of instances. The episulfide was obtained stereoselectively in 83% yield on heating *syn*-bibenzonorbornenylidene with S_8 under the reflux (Scheme 10). The *anti*-episulfide was obtained in a comparable yield (72%) in the case of *anti*-isomer. The retention of configuration was explained in terms of an intermediary diradical oligosulfide, which collapsed to episulfide before carbon–carbon bond rotation occurred [67,68].

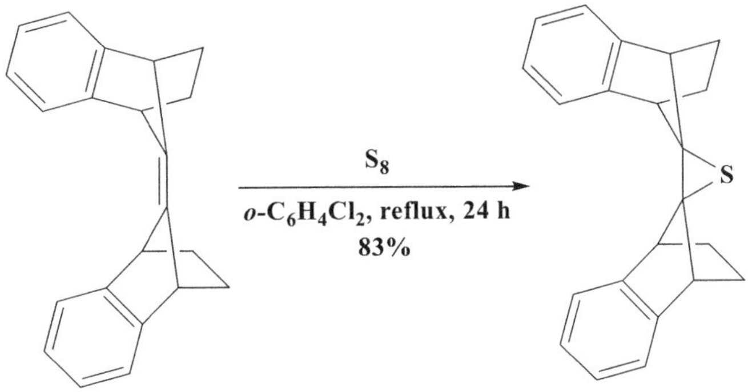

SCHEME 10

The episulfides were obtained in only exceptional cases; for instance, the thiiranes of sesquiterpenes humulene and caryophyllene [69]. Furthermore, the adamantylidene adamantine [70] and norbornadiene [71,72] were converted into episulfides in 25%–58% yield. In other cases, the alkenes were either inactive for episulfidation or oligosulfides were obtained. When norbornene was reacted with photochemically or thermally activated sulfur (Scheme 11), 6% norbornene pentathiepane and 31% norbornene trithiolane were isolated in the thermal reaction, while in the photochemical reaction, at least 8% norbornene episulfide as a minor product along with 77% norbornene trithiolane was obtained [73]. Nevertheless, the alkene (Scheme 12) and adamantylidene adamantine were episulfidated in quite good yields [67,74].

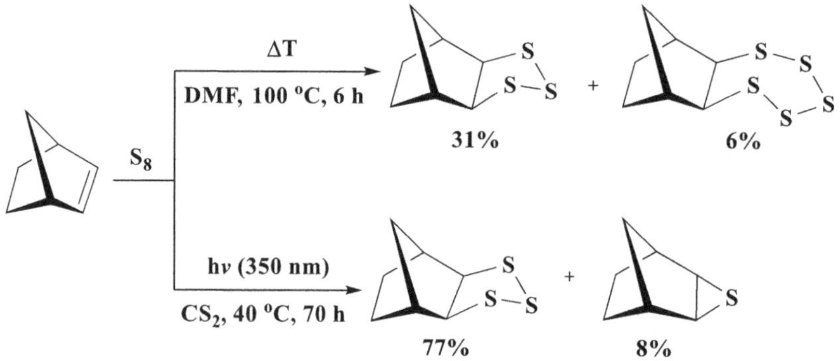

SCHEME 11

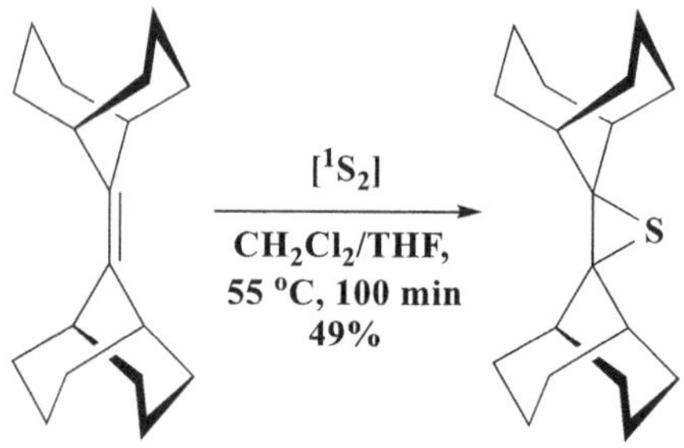

SCHEME 12

The silylsulfenyl bromides were used as episulfidizing agents. Quite generally, the sulfenyl bromides were more reactive but also more selective in addition reactions with alkenes [75]. The reactivity of sulfenyl bromides increased by the lower strength of the sulfur–bromine bond in comparison with the sulfur–chlorine bond, while the larger bromide ion attacked preferentially the sterically less crowded substituent of the exocyclic sulfonium sulfur atom. If a silyl group was utilized as an exocyclic substituent, the higher affinity of bromide ion for silicon than for carbon completely avoided an unwanted 1,2-addition. The episulfide, formed in 20% yield by the reaction of sulfenyl bromide with an excess of cyclohexene, was detected by gas chromatography–mass spectrometry but not isolated [76]. Capozzi et al. [77] reported a one-pot formation of thiiranes for trimethylsilylsulfenyl bromide as a direct source of sulfur at −78°C (Scheme 13), the lowest reaction temperature so far used in episulfidations. The thiiranes can be synthesized stereospecifically and directly from 1,2-disubstituted alkenes by adding trimethylsilylsulfenyl bromide, which was prepared by the reaction of bromine with bis(trimethylsilyl)sulfide at −78°C [67].

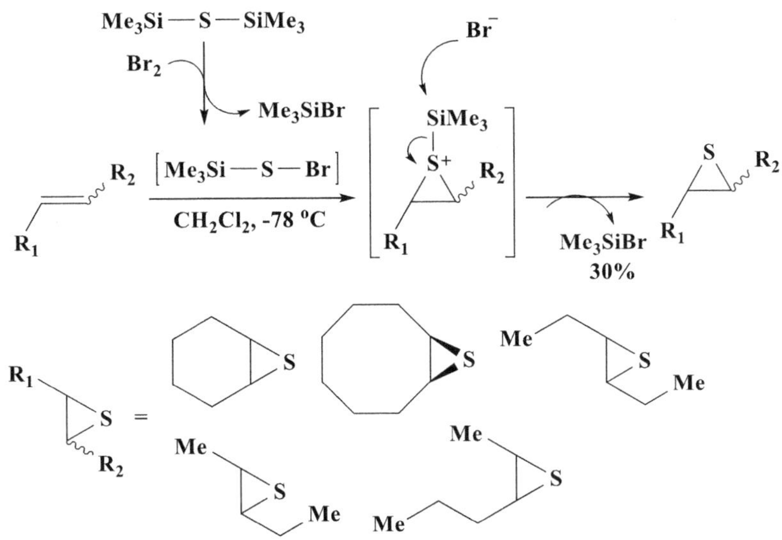

SCHEME 13

The cathodic cleavage of dihalides afforded a variety of heterocyclic compounds (Scheme 14) [78–80].

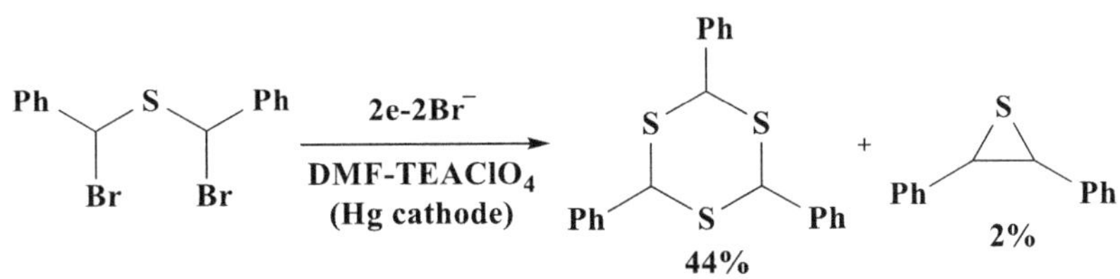

SCHEME 14

The reaction was carried out using benzaldehyde, which was substituted in lieu of the expensive *n*-octyl-styrene. No sign of formation of episulfide was reported after two attempts (Scheme 15) [81].

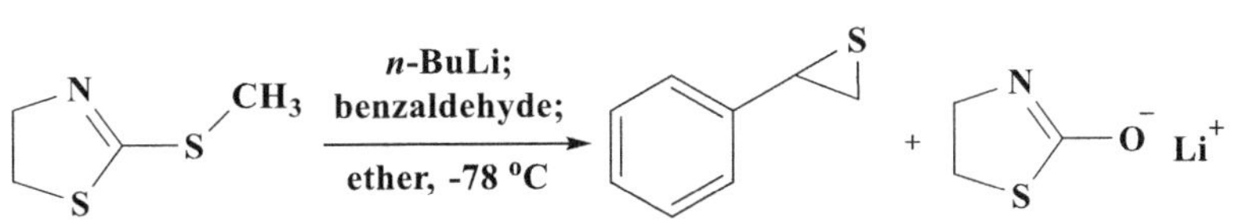

SCHEME 15

The SCl_2 was utilized instead of S_2Cl_2 in an addition to alkynes. As expected, the SCl_2 was added to acetylenes to synthesize the thiiranes almost quantitatively (Scheme 16). In this case, no elemental sulfur was formed. Because of this isolation process, it was much easier to isolate thiiranes; mere evaporation of solvent afforded thiiranes in practically pure form [82].

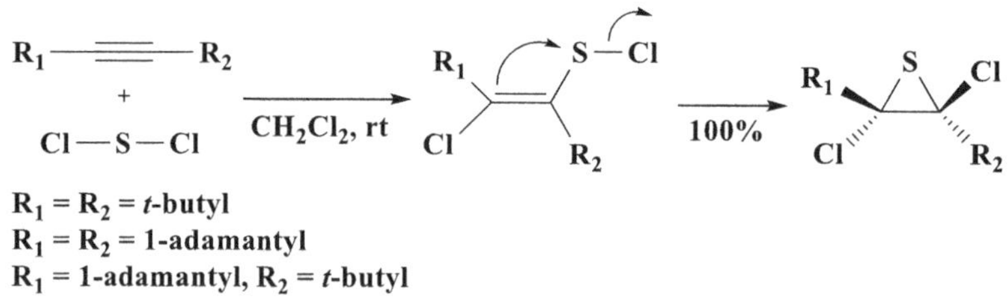

SCHEME 16

The olefins afforded episulfides by a new, two-step synthetic method. The disulfur dichloride was added to olefin in the first step for the formation of mixtures of chloroalkylmono-, di-, and trisulfides. The adducts were transformed to episulfide using aluminum amalgam or Na_2S (Scheme 17) [22,83,84].

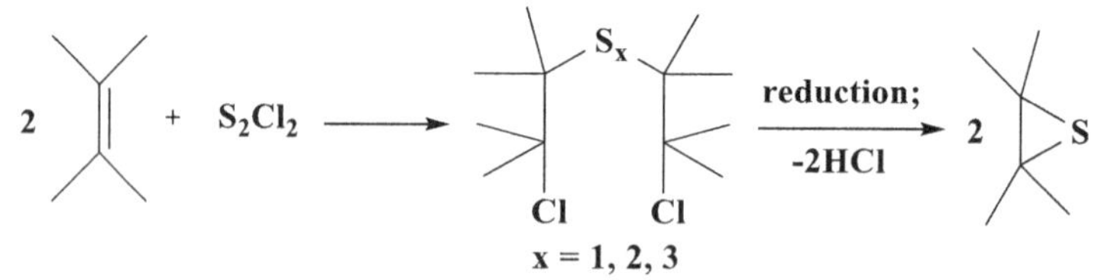

SCHEME 17

In the case of addition of S_2Cl_2 to allyl derivatives, the distribution of isomer in the adduct influenced the nature of formed cyclic sulfide. The possibilities for additions to allyl bromide, allyl chloride, and allyl acetate are shown in Scheme 18. Markovnikov addition occurred; then reduction and dehydrohalogenation took place. Likewise, Markovnikov addition of disulfur dichloride took place. The determination of isomer distributions by nuclear magnetic resonance data from addition of S_2Cl_2 at ambient temperature in excess olefin showed the formation of *anti*-Markovnikov adducts exclusively. The reaction of both the allyl bromide and allyl chloride-S_2Cl_2 adducts with aluminum amalgam occurred from allyl acetate [22,85,86].

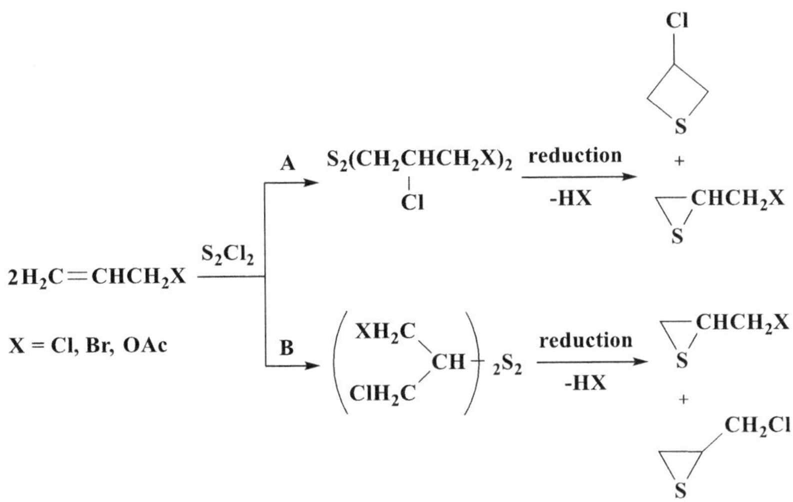

SCHEME 18

The cyclization of open-chain substrates synthesized fluorothiiranes. The thiirane was prepared in 68% yield by the reaction of sulfenyl chloride with TEA-BF_3 complex (Scheme 19). The disulfide, synthesized from alkene and S_2Cl_2, was converted into thiirane using tris(diethylamino)phosphine (Scheme 20). An indirect pathway from chlorodisulfide that has the benefit of using both halves of the molecule was exemplified by the conversion of chlorodisulfide into thiirane (Scheme 21) [63,64].

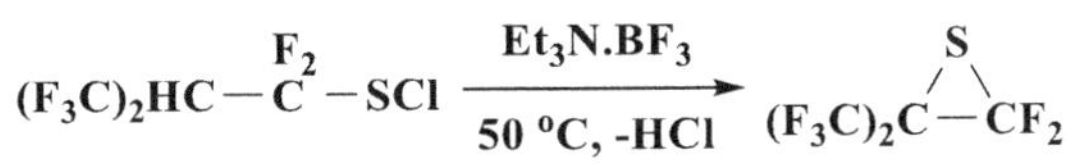

SCHEME 19

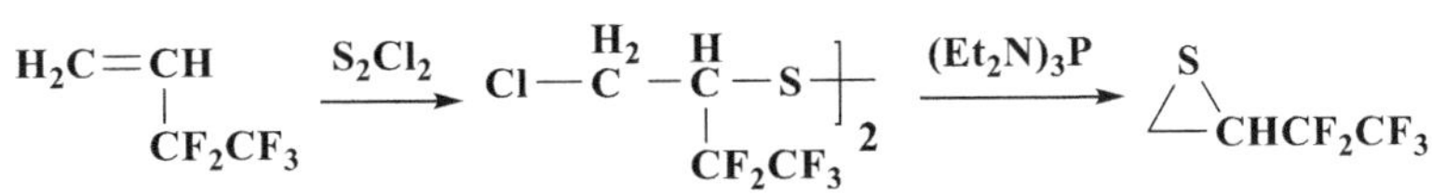

SCHEME 20

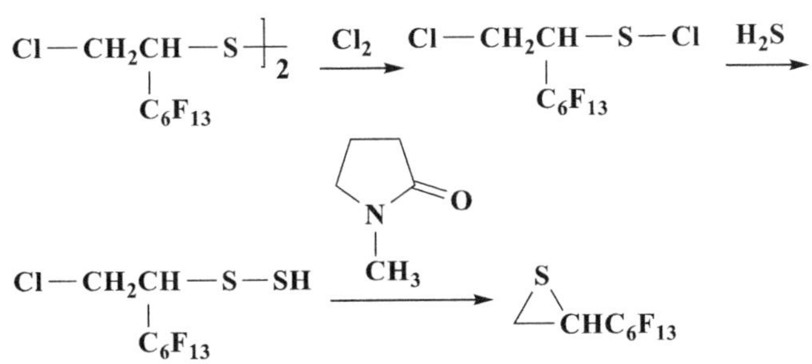

SCHEME 21

Under the conservation of the alkene configuration, this episulfidation was highly selective since only 1,2-disubstituted alkenes provided episulfides; however, the episulfide was formed only in 30% yield. As confirmed by means of a control experiment, the reversibility of episulfidation was the reason for the low yield, in which the episulfides were desulfurized by reacting with TMSBr [77]. In contrast to the sulfenyl halides, the episulfidation by sulfur chlorides, disulfur dichloride [87] and sulfur dichloride [68,88,89], was not stereoselective. The lack of stereoselectivity was rationalized in terms of carbocations synthesized by reversible ring opening of intermediary thiiranium ion. As a further evidence for the intermediacy of carbocations, the synthesis of chlorinated thiirane shown in Scheme 22 was conducted with disulfur dichloride and sulfur dichloride possibly via a nonclassical carbocation [53,67].

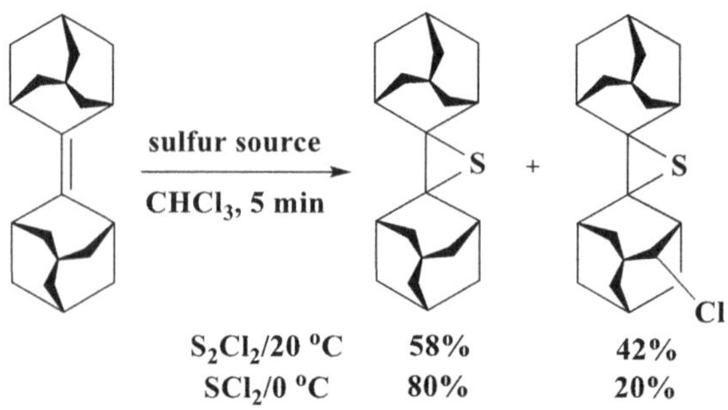

SCHEME 22

The di-*t*-butyl- and di-1-adamantylacetylenes were reacted with S_2Cl_2 at rt. Surprisingly, the reaction provided dichlorothiiranes in good yields (Scheme 23). The expected 1,2-dithietane was not obtained even in very small amounts. The thiiranes were sensitive to hydrolysis [82].

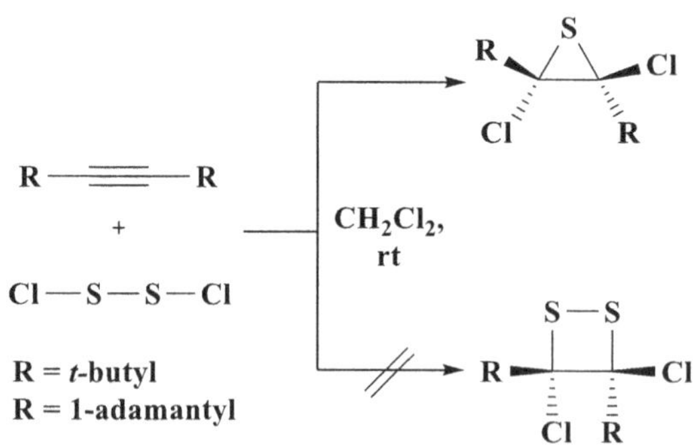

SCHEME 23

The bromomethane was removed from adamantylidene adamantane for the formation of thiirane in 82% yield (Scheme 24) [90]. In this case, the sulfenyl chlorides were used to afford the episulfides in 93% yield [91,92]. In the case of sterically less crowded alkenes, for example, norbornene, 1,2-addition via a ring opening of intermediary thiirane by another molecule of sulfenyl chloride was reported as a side reaction [67,93].

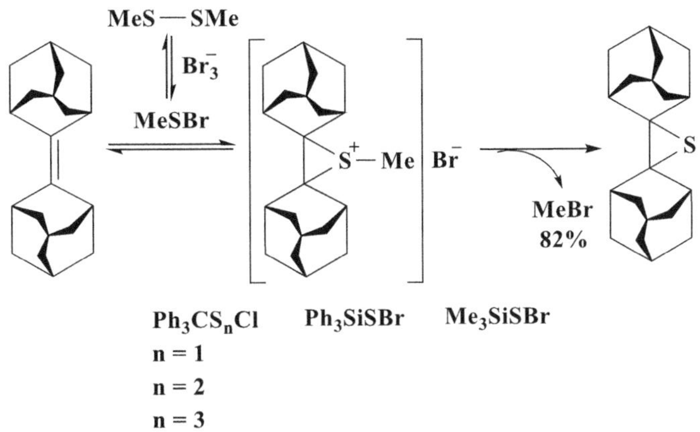

SCHEME 24

A two-step formation of thiiranes was performed by adding succinimide-*N*-sulfenyl chloride or phthalimide-*N*-sulfenyl chloride to alkenes followed by $LiAlH_4$ cleavage of the adducts (Scheme 25) [94,95].

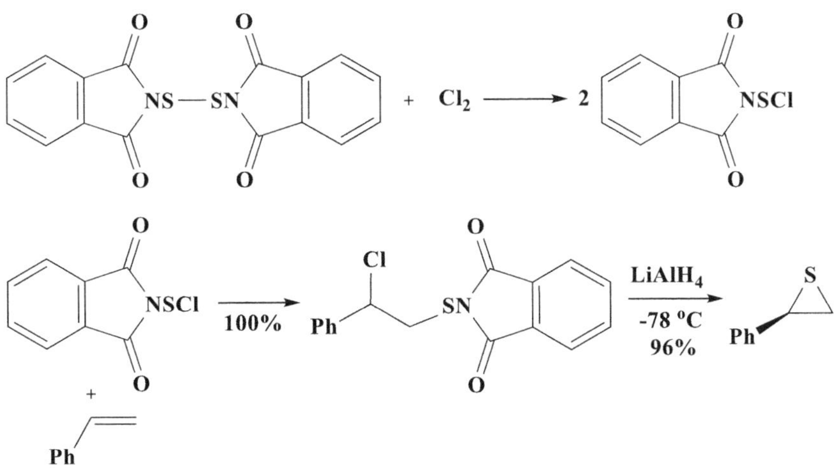

SCHEME 25

The 1,1′-dithiobis(1*H*-1,2,4-triazole) was reacted with alkenes without activating agents to synthesize the thiiranes in satisfactory yield under the mild conditions with the retention of starting alkene geometry (Scheme 26) [96,97].

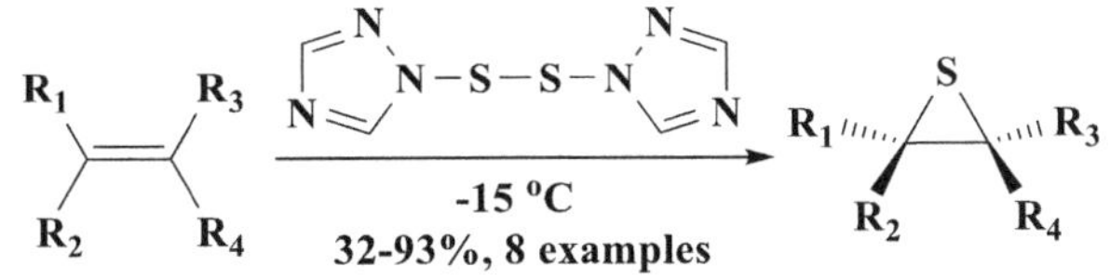

SCHEME 26

The reaction using either trimethylsilyldiazomethane or CH_2N_2 was proved to be ineffective, whereas the success of former reaction was highly dependent on the precursor. The metalation of ethylpropyl sulfone, reaction with I_2, and base treatment of the formed mixture of iodo sulfones provided methylethyl episulfone in 37% overall yield (Scheme 27). Unfortunately, use of this synthetic sequence to sulfones, which led to episulfones with a more clearly defined regiochemical bias, provided very low yields and was not pursued further [98,99].

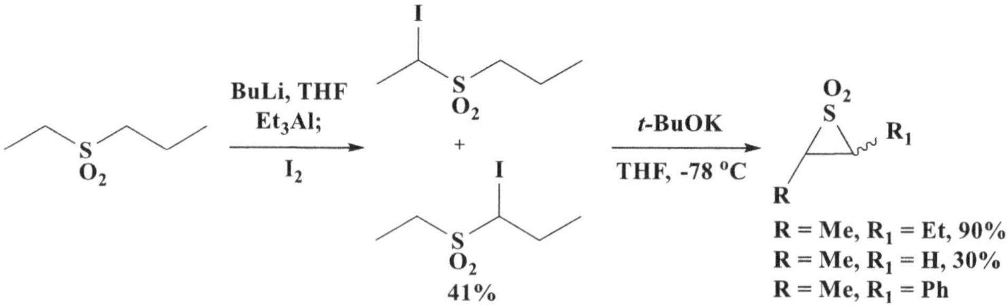

SCHEME 27

Opitz et al. [100,101] synthesized a variety of simple, symmetrically substituted episulfones by the reaction of alkanesulfonyl chlorides with triethylamine in acetonitrile. The episulfones were formed as mixtures of *trans*- and *cis*-isomers, mostly in the form of colorless oils, which could be stored under the refrigeration (Scheme 28). It was possible to separate the stereoisomeric mixtures by flash chromatography, enabling isolation of samples of pure *cis*- and *trans*-isomers, although in some cases mixtures of stereoisomeric episulfones were used. The stilbene episulfone, accessible by this technique, was proved to be much more unstable than the aliphatic episulfones and was not utilized in the subsequent work [102]. Although most of the studies used simple episulfones, the unsymmetrically substituted systems were also required, which would be of great interest in addressing aspects of regioselectivity in subsequent reactions. Well-known pathways to such compounds included the reaction of diazoalkanes with sulfenes (synthesized in situ from the suitable alkanesulfonyl halides) [99,103] or the modified Ramberg–Bäcklund procedure introduced by Taylor et al. [104–106].

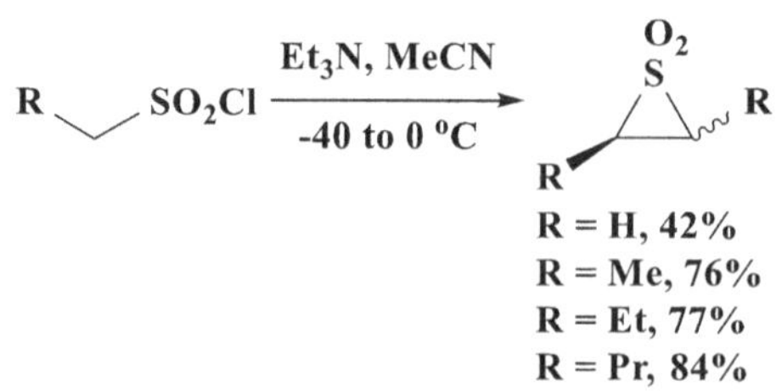

SCHEME 28

The reaction of alkyl halides and metal sulfides afforded symmetrical sulfides with a variety of substituents. This method was surveyed in comprehensive organic functional group transformations, and no additional advances were reported on noncatalyzed reactions of alkyl halides with metal sulfides. This technique was also efficient for the formation of cyclic sulfides in higher yields (Scheme 29) [107].

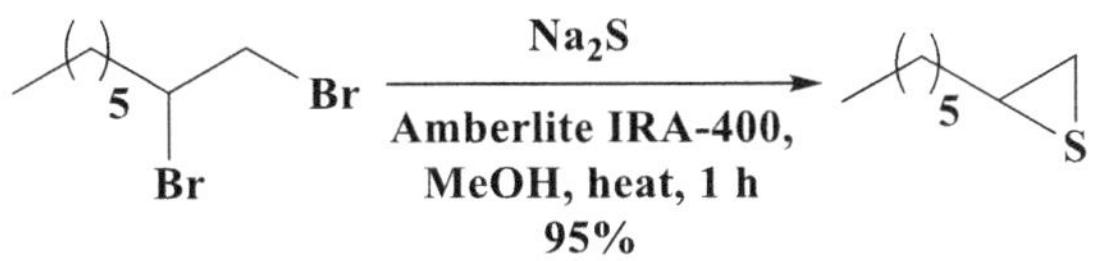

SCHEME 29

The thiirane oximes were synthesized by two approaches. At first, the thiirane oxime was synthesized by the reaction of ketone with hydroxylammonium chloride using sodium acetate/ethanol/water (Scheme 30) [108]. The 2-(α-hydroxyiminoethyl)-2-methylthiirane was prepared in 40% yield by the reaction of 3,4-dibromo-3-methyl-2-hydroxyiminobutane with $Na_2S.H_2O$ in $(CH_3)_2CO$ (Scheme 31) [109,110].

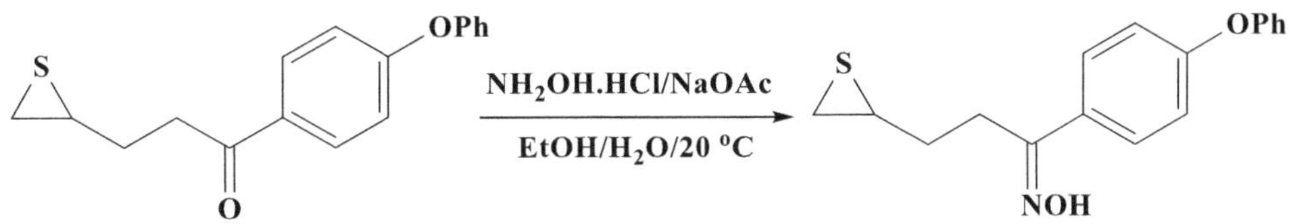

SCHEME 30

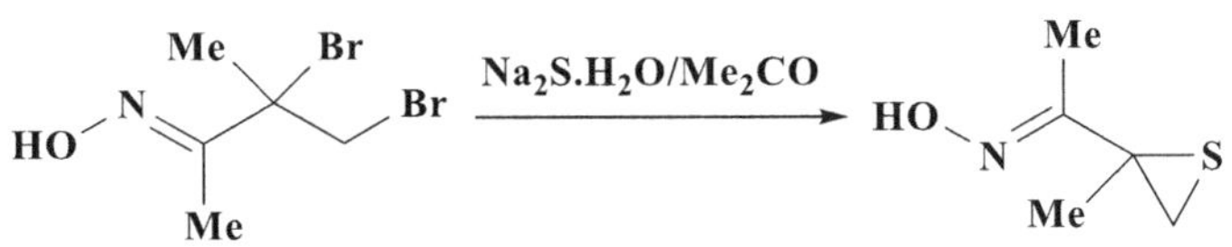

SCHEME 31

An oxidation of thioketene with *m*-chloroperoxybenzoic acid to synthesize the thioketene-*S*-oxide and then reaction with Lawesson's reagent at rt in dichloromethane for 12 h afforded dithiolactone bearing a three-membered ring (Scheme 32) [111,112].

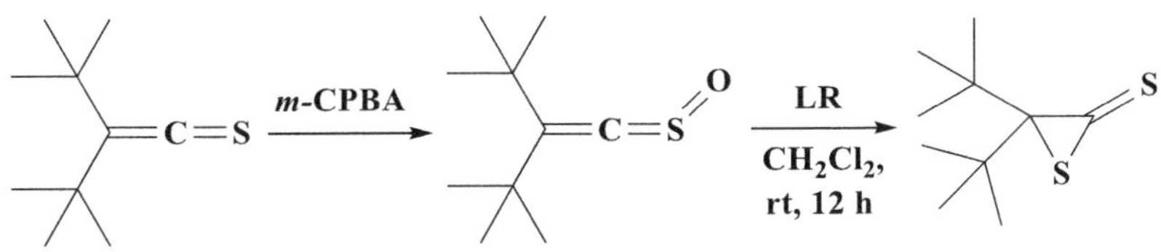

SCHEME 32

The promising outcomes obtained with in situ formation of oxathiiranes led to develop an additional and more general access to reactive oxathiiranes (Scheme 33). The episulfide was obtained in 18% yield by reacting indanetrione (ninhydrin) with an excess of potassium *p*-thiotosylate and (*E*)-cyclooctene [67,113].

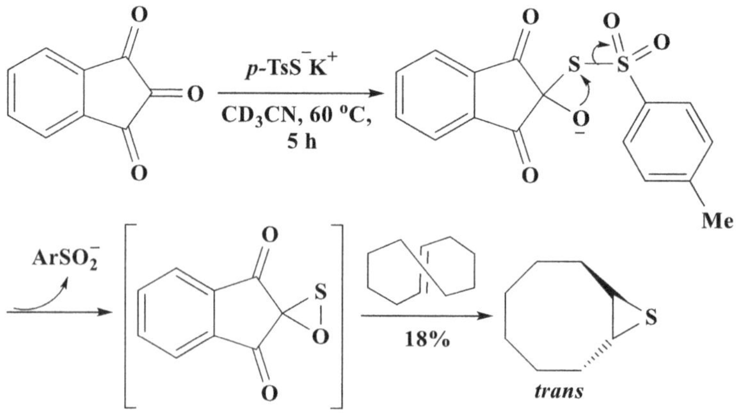

SCHEME 33

One very interesting technique that has received little attention is the transformation of haloketones into episulfides utilizing commercially accessible *O*,*O*-diethyl hydrogen phosphodithioate as a sulfur donor under the MWI. The reaction was reasonably general and provided high yield. The bromoacetophenone afforded episulfide within 3 min in excellent yield (Scheme 34). The mechanism included a sequence of nucleophilic displacement, phosphorus transfer, and cyclization. For precursors that afforded 1,2-disubstituted episulfides, diastereoselectivity was very high with *cis*/*trans* ratios usually being more than 1:10 as shown in the formation of diethylepisulfide (Schemes 35 and 36) [114].

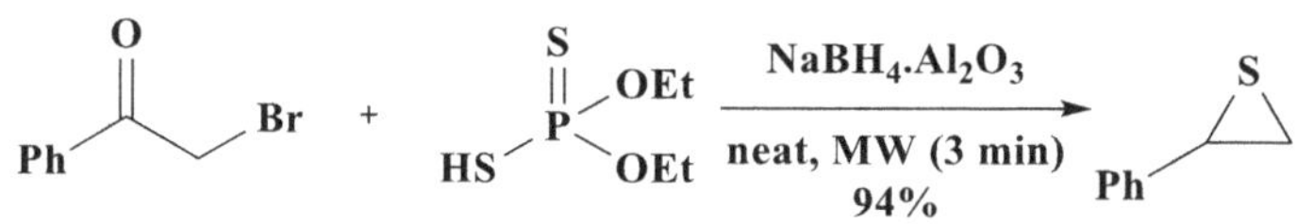

SCHEME 34

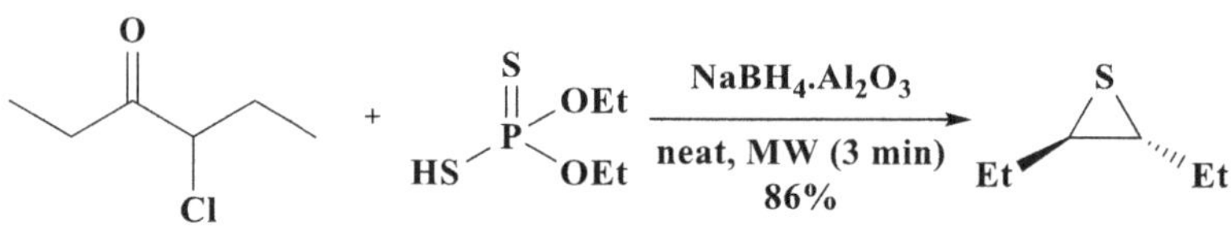

SCHEME 35

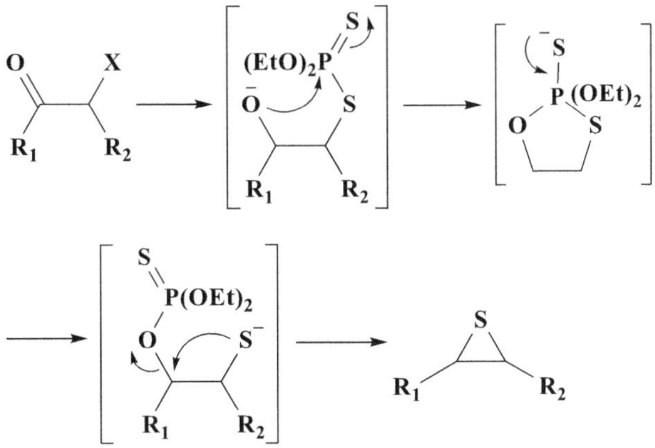

SCHEME 36

9.3 Synthesis of thiirenes

Three acetylenes were allowed to react with an equimolar amount of SCl_2 at rt, and then the reaction mixtures were evaporated under the reduced pressure. The obtained crude thiiranes were directly subjected to alkaline hydrolysis. In this way, the 2,3-di-*t*-butyl-, 2,3-di-1-adamantyl-, and 2-adamantyl 3-*t*-butylthiirene oxides were obtained in overall yields of 70%, 80%, and 90%, respectively, depending on the starting acetylenes. These were essentially one-flask reactions and afforded a most convenient formation of thiirene-1-oxides (Scheme 37) [82].

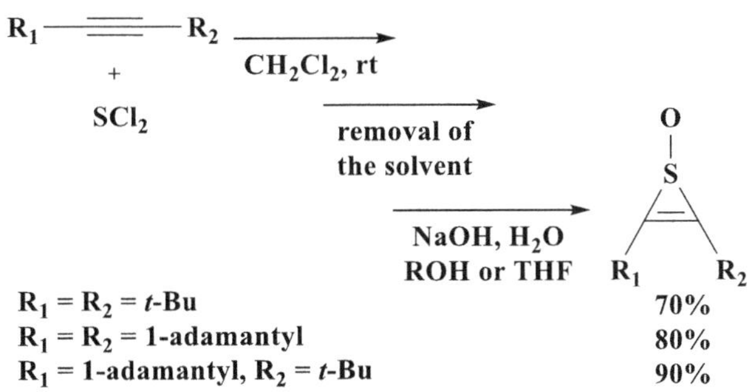

SCHEME 37

To facilitate the operation, the diphenylthiirene dioxide was made the target of initial synthetic analyses. The bromination of dibenzyl sulfide by means of bromine followed by an oxidation of intermediate dibromo sulfide provided dibromodibenzyl sulfone as a mixture of racemic and *meso*-diastereomers (Scheme 38). The reaction of mixture of racemic and *meso*-dibromosulfones with triethylamine [115,116] resulted in an isolation of diphenylthiirene dioxide [117–124].

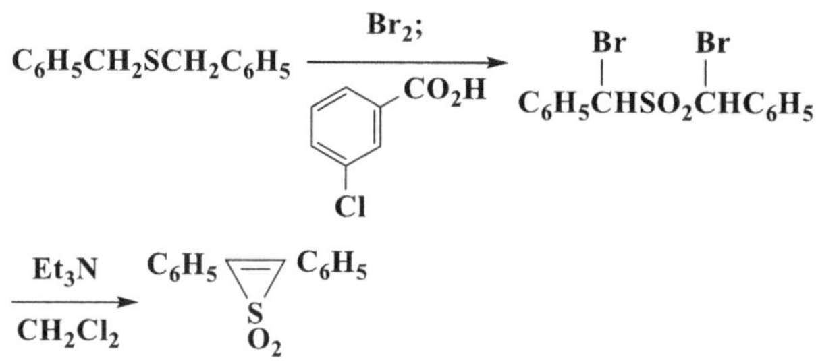

SCHEME 38

The formation of hypothetically interesting and synthetically beneficial aryl alkyl, diaryl, and dialkylthiirene 1,1-dioxides was reported. Unfortunately, the derivatives having alkyl substituents obtained so far only through the sulfene and diazoalkane intermediates and consequently are not easily accessible on a large scale either as substrates of other beneficial synthetic intermediates or for a wide range study of their properties. Due to this deficiency, a new pathway to dialkyl thiirene dioxides was reported, which made these compounds as easily available as the diaryl analogs. The central to the new method was the 1,3-elimination [122] of bromine from bis(dibromoalkyl)sulfone by means of PPh_3 or tris(dialkylamino)phosphine, R = CH_3CH_2 (89%). To be beneficial, this technique needed an easy formation of tetrabromosulfones, and this type of process has now been developed based on a novel but obscure reaction first reported in the patent literature. Adapting Szabo's technique for the formation of tetrachloro analogs, the reaction of triethyl- and trimethylphosphite with bis(tribromomethyl)sulfone provided sulfone (R = Et or Me). Along with its importance for the formation of vinylene sulfones, this unique process of C–C bond formation combined with transformation to dialkyl thiirene dioxides followed by facile thermal cycloelimination of SO_2 showed a beneficial pathway to internal acetylenes (Scheme 39) [125,126].

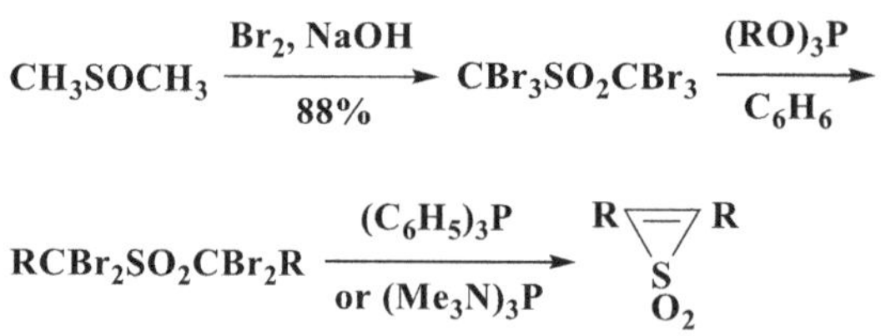

SCHEME 39

In extending novel thiirene dioxide formation to dimethyl analog, it was observed that the alcohol-free solutions of diazoethane must be utilized to avoid the destructions of an intermediate sulfene and a stronger base such as 1,5-diazabicyclo[4.3.0]non-5-ene was needed for an elimination (Scheme 40) [124,127–129].

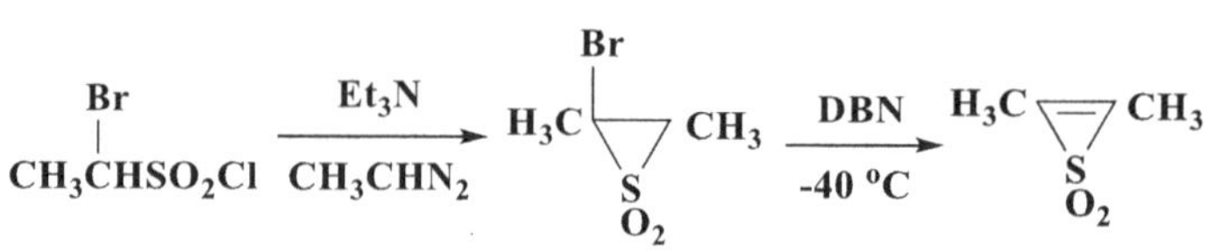

SCHEME 40

The modification of Ramberg–Backlund reaction afforded thiirene-1-oxide from dibromo compound in 50% yield (Scheme 41) [82,130,131].

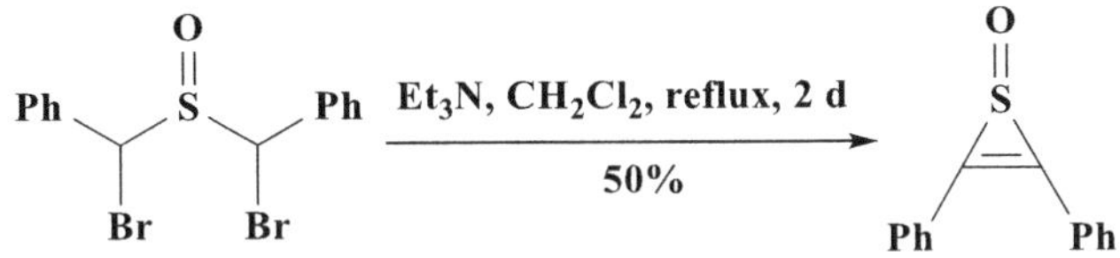

SCHEME 41

Nakayama et al. [82] reacted alkyne with alkynes and alkenes at rt in poor to modest yields (Scheme 42). This rt extrusion was proposed to be concerted and therefore, the first instance of singlet SO.

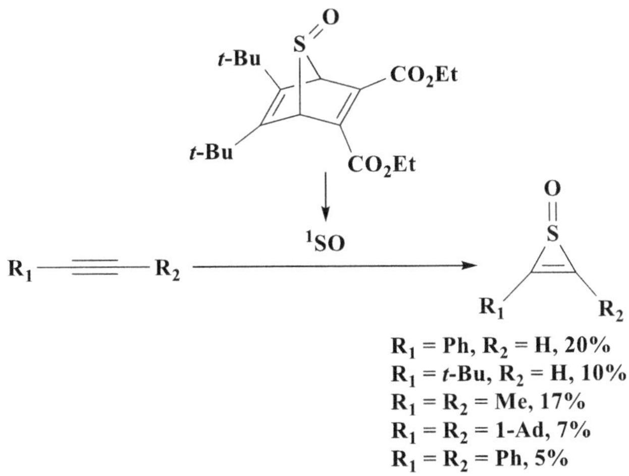

SCHEME 42

9.4 Synthesis of dithiiranes and thiaziridines

An oxidation of tetrathiolanes with dimethyldioxirane provided mixtures of thioketones and dithiirane-1-oxides (Scheme 43). The presence of tetrathiolane-1-oxides intermediate was verified by nuclear magnetic resonance analysis of the cooled and evaporated reaction mixture [132].

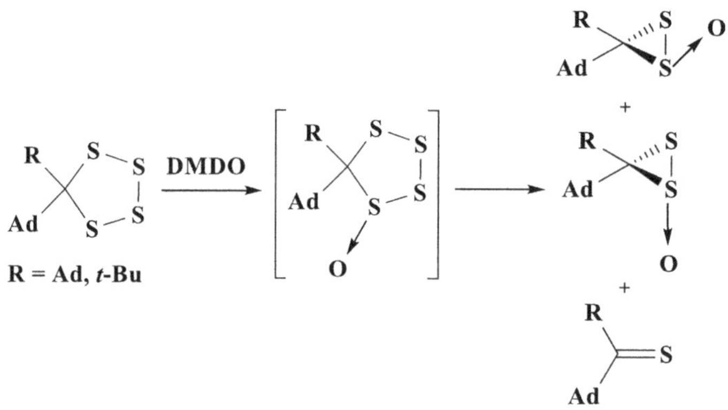

SCHEME 43

The *vic*-disulfoxide exhibited a high thermal stability as compared with that of acyclic *vic*-disulfoxides, which were previously only noticed at −40°C, and bridged *vic*-disulfoxides [82]. However, in solution, the α-disulfoxide decomposed cleanly to diastereomeric dithiirane-1-oxides, elemental sulfur, and tetrathiolane (Scheme 44).

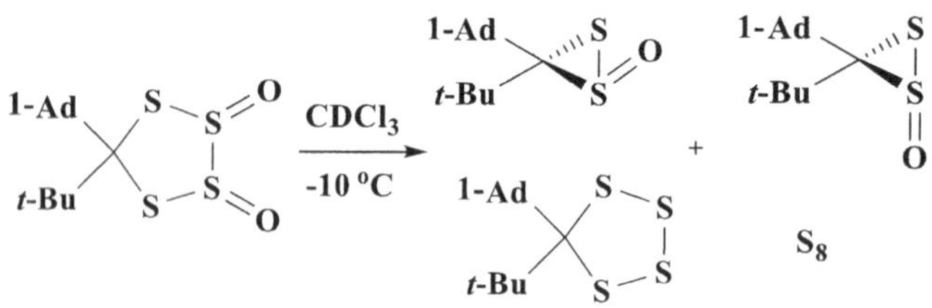

SCHEME 44

The thermal synthesis of nitrene has been invoked to account for the result in the reaction between thioketene and sterically hindered benzyl azide (Scheme 45) [133]. Although initially a 1,3-dipolar cycloaddition between the azide and the C=S bond was expected, the temperature of reaction (138°C) resulted in the decomposition of benzyl azide into benzyl nitrene followed by [2 + 1]-cycloaddition.

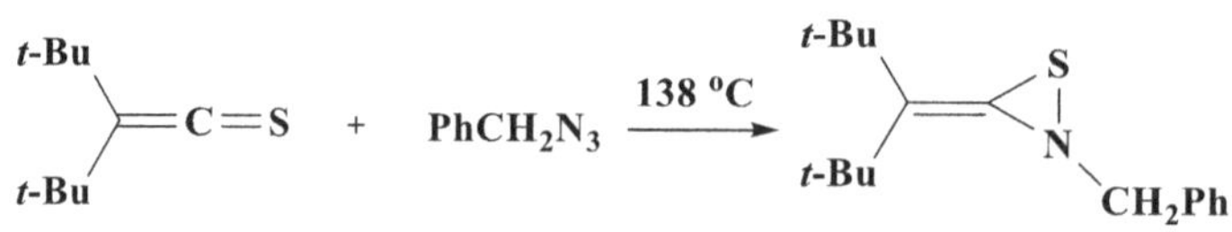

SCHEME 45

Kidwai and Batra [134] reported the reaction of *N*-alkylsulfonylimines with a variety of simple alkyl-substituted diazoalkanes (Scheme 46). The obtained thiaziridine-1,2-dioxides were all found to be unstable to prolonged storage.

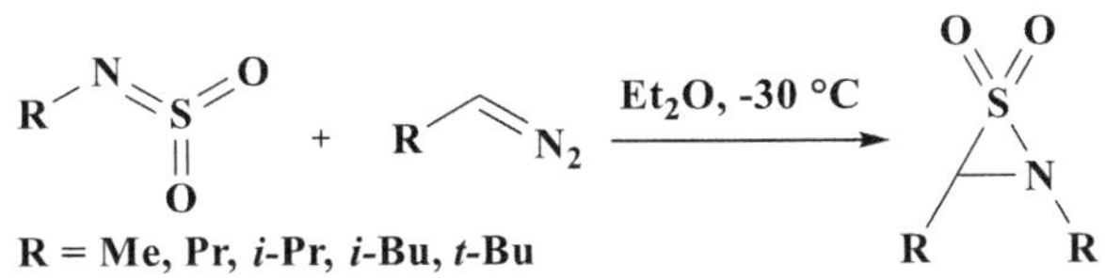

SCHEME 46

References

1. (a) N. Iranpoor, B. Tamami and K. Niknam. 1997. Iodine and iodine supported on polyvinylpyrrolidone as catalysts and reagents for alcoholysis, hydrolysis, and acetolysis of epoxides and thiiranes. Can. J. Chem. 75: 1913-1919. (b) N. Kaur. 2019. Synthesis of three-membered and four-membered heterocycles with the assistance of photochemical reactions. J. Heterocycl. Chem. 56: 1141-1167.
2. G.W. Gribble. 2000. Recent developments in indole ring synthesis - methodology and applications. J. Chem. Soc. Perkin Trans. 1 7: 1045-1075.
3. T.A. Gilchrist. 1998. Synthesis of aromatic heterocycles. J. Chem. Soc. Perkin Trans. 1 3: 615-628.
4. A. Nobuyoshi, O. Akihiko, M. Chikara, T. Tatsuya, O. Masami and S. Hiromitsu. 1999. Synthesis and antitumor activity of duocarmycin derivatives: modification of segment-A of A-ring pyrrole compounds. J. Med. Chem. 42: 2946-2960.
5. M.S. Al-Said, M.S. Bashandy, S.I. Al-Qasoumi and M.M. Ghorab. 2001. Anti-breast cancer activity of some novel 1,2-dihydropyridine, thiophene and thiazole derivatives. Eur. J. Med. Chem. 46: 137-141.
6. R. Romagnoli, P.G. Baraldi, O. Cruz-Lopez, M. Tolomeo, A. DiCristina, R.M. Pipitone, S. Grimaudo, J. Balzarini, A. Brancale and E. Hamel. 2011. Synthesis of novel antimitotic agents based on 2-amino-3-aroyl-5-(hetero)arylethynyl thiophene derivatives. Bioorg. Med. Chem. Lett. 21: 2746-2751.
7. B. Kaboudin and H. Norouzi. 2004. A new, efficient, and simple method for the synthesis of thiiranes from epoxides under solvent-free conditions. Tetrahedron Lett. 45: 1283-1285.
8. H. Staudinger and J. Siegwart. 1920. Einwirkungen von aliphatischen diazoverbindungen auf thioketone. Helv. Chim. Acta 3: 833-840.
9. W. Chew and D.N. Harpp. 1993. Recent aspects of thiirane chemistry. J. Sulfur Chem. 15: 1-39.
10. H. Kudo, S. Makino, A. Kameyama and T. Nishikubo. 2005. Synthesis of cyclic polymers: ring-expansion reaction of cyclic *S*-dithioester with thiiranes. Macromolecules 38: 5964-5969.
11. A. Kameyama, Y. Murakami and T. Nishikubo. 1999. Novel sequence-ordered polymers by transformation of polymer backbone: quantitative and regioselective insertion of thiiranes into poly(*S*-aryl thioester). Macromolecules 35: 1407-1412.
12. T. Imai, K. Hayakawa, T. Satoh, H. Kaga and T. Kakuchi. 2002. Enantiomer-selective polymerization of (*RS*)-(phenoxymethyl)thiirane with diethylzinc/L-amino acid. J. Polym. Sci. Pol. Chem. 40: 3443-3448.
13. G. Gottarelli, P. Mariani, G.P. Spada, B. Samori, A. Forni, G. Solladie and M. Hibert. 1983. Induction of cholesteric mesophases in nematic liquid crystals, and correlation of absolute configurations of some chiral oxiranes and thiiranes. Tetrahedron 39: 1337-1344.
14. G. Scherowsky and J. Gay. 1989. Ferroelectric liquid crystals and dopants containing the chiral thiirane unit. A comparison with analogous oxiranes. Liq. Cryst. 5: 1253-1258.
15. K. Chino, K. Suga, M. Ikawa and H. Satoh. 2001. Novel rapid-cure adhesives for low temperature using thiirane compound. J. Appl. Polym. Sci. 82: 2953-2957.
16. Y. Kadoma. 2002. Surface treatment agent for dental metals using a thiirane monomer and a phosphoric acid monomer. Dent. Mater. J. 21: 156-169.
17. E. Goethals. 1968. Sulfur-containing polymers. J. Polym. Sci. Pol. Rev. 2: 73-144.

18. S.A. Testero, M. Lee, R.T. Staran, M. Espahbodi, L.I. Llarrull, M. Toth, S. Mobashery and M. Chang. 2011. Sulfonate-containing thiiranes as selective gelatinase inhibitors. ACS Med. Chem. Lett. 2: 177-181.
19. A.H. Sally, P.B. Stephen, L. Joel and J.S. Peter. 2004. New potent and selective A1 adenosine receptor agonists. Bioorg. Med. Chem. 12: 4877-4884.
20. H.J. Cho, M.J. Jung, Y. Kwon and Y. Na. 2009. Oxiranylmethyloxy or thiiranylmethyloxy-azaxanthones and -acridone analogues as potential topoisomerase I inhibitors. Bioorg. Med. Chem. Lett. 19: 6766-6769.
21. J.T. Kellis, W.E. Childers, C.H. Robinson and L.E. Vickery. 1987. Inhibition of aromatase cytochrome P-450 by 10-oxirane and 10-thiirane substituted androgens. Implications for the structure of the active site. J. Biol. Chem. 262: 4421-4426.
22. F. Lautenschlaeger and N.V. Schwartz. 1969. Synthesis of episulfides from olefins and sulfur monochloride. J. Org. Chem. 34: 3991-3998.
23. P.D. Woodgate, H.H. Lee, P.S. Rutledge and R.C. Cambie. 1976. Reaction of alkenes with iodine(I) thiocyanate. Tetrahedron Lett. 18: 1531-1534.
24. A.I. Meyers and M.E. Ford. 1975. Synthesis via oxazolines - one carbon homologation of aldehydes and ketones to thiiranes and olefins. Tetrahedron Lett. 16: 2861-2864.
25. L. DiNunno, C. Franchini, A. Nacci, A. Scilimati and M.S. Sinicropi. 1999. Baker's yeast-induced asymmetric reduction of α-ketosulfides: synthesis of optically active 1-(benzothiazol-2-ylsulfanyl)-2-alkanols, 2-alkanols, and thiiranes. Tetrahedron: Asymmetry 10: 1913-1926.
26. B. Tamami and M. Kolahdoozan. 2004. Synthesis of thiiranes from oxiranes in water using polymeric co-solvents. Tetrahedron Lett. 45: 1535-1537.
27. N. Iranpoor and F. Kazemi. 1997. Stereospecific formation of *S*-(-)-styrene sulfide: efficient conversion of epoxides to thiiranes catalysed with Ru(III). Tetrahedron Lett. 53: 11377-11382.
28. N. Iranpoor and F. Kazemi. 1996. Cerium(IV) catalysis: efficient conversion of epoxides to thiiranes. Synthesis 7: 821-822.
29. F. Kazemi and A.R. Kiasat. 2003. Facile conversion of epoxides to thiiranes with ammonium thiocyanate catalyzed with oxalic acid. Phosphorus, Sulfur, Silicon Relat. Elem. 178: 1333-1337.
30. F. Kazemi, A.R. Kiasat and S. Ebrahimi. 2002. Efficient conversion of oxiranes to thiiranes with ammonium thiocyanate catalysed with $LiBF_4$. J. Chem. Res. Synop. 4: 176-177.
31. I. Mohammadpoor-Baltork and A.R. Khosropour. 2001. $Bi(TFA)_3$ and $Bi(OTf)_3$ catalyzed conversions of epoxides to thiiranes with ammonium thiocyanate and thiourea under non-aqueous conditions. Molecules 6: 996-1000.
32. V. Mirkhani, S. Tangestaninejad and L. Alipanah. 2002. Mild, efficient, and convenient conversion of oxiranes to thiiranes with ammonium thiocyanate and thiourea in the presence of cerium(IV) poly-oxometallate. Synth. Commun. 32: 621-626.
33. N. Iranpoor and H. Adibi. 2000. Iron(III) trifluoroacetate as an efficient catalyst for solvolytic and non-solvolytic nucleophilic ring opening of epoxides. Bull. Chem. Soc. Jpn. 73: 675-680.
34. I. Mohammadpoor-Baltork and A.R. Khosropoor. 1999. Antimony trichloride; a mild, efficient and convenient catalyst for conversion of oxiranes to thiiranes under non-aqueous conditions. Indian J. Chem. Sect. B 38: 605-606.
35. N. Iranpoor, B. Tamami and M. Shekarriz. 1999. Conversion of epoxides to thiiranes with thiourea or ammonium thiocyanate catalyzed with poly(4-vinylpyridine)-$Ce(OTf)_4$. Synth. Commun. 29: 3313-3321.
36. I. Mohammadpoor-Baltork and H. Aliyan. 1998. Bismuth(III) chloride; a mild and efficient catalyst for synthesis of thiiranes from oxiranes. Synth. Commun. 28: 3943-3947.
37. N. Iranpoor and B. Zeynizadeh. 1988. Conversion of epoxides to thiiranes catalyzed with $TiO(TFA)_2$ and $TiCl_3(OTf)$ in the presence of ammonium thiocyanate or thiourea. Synth. Commun. 28: 3913-3918.
38. J.S. Yadav, B.V.S. Reddy, S. Reddy and K. Rajasekhar. 2003. [bmim]PF_6: a novel and recyclable ionic liquid for conversion of oxiranes to thiiranes in aqueous media. J. Org. Chem. 68: 2525-2527.

39. J.S. Yadav, B.V.S. Reddy and G. Baishya. 2003. Indium tribromide: a novel and highly efficient reagent for the conversion of oxiranes to thiiranes. Synth. Lett. 3: 396-398.
40. M.O. Brimeyer, A. Mehrota, S. Quici, A. Nigam and S.L. Regen. 1980. Silica gel assisted synthesis of thiiranes. J. Org. Chem. 45: 4254-4255.
41. C.S. Reddy and S. Nagavani. 2008. $LiClO_4$ Catalyzed mild and efficient method for the synthesis of thiiranes from oxiranes. Heteroat. Chem. 19: 97-99.
42. F. Kazemi, A.R. Kiasat and S. Ebrahimi. 2003. $LiBF_4$: a mild and efficient catalyst for conversion of oxiranes to thiiranes with thiourea. Synth. Commun. 33: 595-600.
43. I. Mohammadpoor-Baltork and H. Aliyan. 2000. Montmorillonite K-10: a mild, inexpensive and convenient catalyst for synthesis of thiiranes from oxiranes under non-aqueous conditions. J. Chem. Res. Synop. 3: 122-123.
44. S. Tangestaninejad and V. Mirkhani. 1999. Efficient conversion of epoxides to thiiranes with thiourea catalyzed by tin(IV) porphyrin. Synth. Commun. 29: 2079-2083.
45. F. Kazami, A.R. Kiasat and S. Ebrahim. 2001. Efficient conversion of oxiranes to thiiranes with thiourea catalyzed with ruthenium trichloride and alumina. Phosphorus, Sulfur, Silicon Relat. Elem. 176: 135-140.
46. B. Zeynizadeh, M.M. Baradarani and R. Eisavi. 2011. A practical and eco-friendly method for conversion of epoxides to thiiranes with immobilized thiourea on $CaCO_3$. Phosphorus, Sulfur, Silicon Relat. Elem. 186: 2208-2215.
47. R. Eisavi, B. Zeynizadeh and M.M. Baradarani. 2011. Zeolite molecular sieve 4Å: a reusable catalyst for fast and efficient conversion of epoxides to thiiranes with thiourea. Phosphorus, Sulfur, Silicon Relat. Elem. 186: 1902-1909.
48. N. Iranpoor, H. Firouzabadi and M. Shekarize. 2003. Micellar media for the efficient ring opening of epoxides with CN-, N_3-, NO_3-, NO_2-, SCN-, Cl- and Br-catalyzed with $Ce(OTf)_4$. Org. Biomol. Chem. 1: 724-727.
49. T. Takido, Y. Kobayashi and K. Itabashi. 1986. A convenient synthesis of thiirans by the use of dimethylthioformamide. Synthesis 9: 779-780.
50. S. Tangestaninejad and V. Mirkhani. 1999. Iron(III) 5,10,15,20-tetrakis (pentafluorophenyl)porphyrin as an efficient catalyst for conversion of epoxides to thiiranes. J. Chem. Res. Synop. 6: 370-371.
51. L. Wu, L. Yang, L. Fang, C. Yang and F. Yan. 2010. Silica chloride-catalyzed synthesis of thiiranes from oxiranes under solvent-free conditions. Phosphorus, Sulfur, Silicon Relat. Elem. 185: 2159-2164.
52. L. Wu, Y. Wang, F. Yan and C. Yang. 2010. Facile conversion of epoxides to thiiranes with ammonium thiocyanate catalyzed with etidronic acid. Bull. Korean Chem. Soc. 31: 1419-1420.
53. G.A. Tolstikov, B.M. Lerman, L.I. Umanskaya, Y.T. Struchkov, A.A. Espenbetov and A.L. Yanovskii. 1980. New reactions of sulphur dichloride with olefines. Tetrahedron Lett. 21: 4189-4192.
54. W. Ando, N. Tokitoh and Y. Kabe. 1991. Synthesis and reaction of novel sulfur containing heterocyclic systems. Phosphorus, Sulfur, Silicon Relat. Elem. 58: 179-205.
55. C.S. Adams, C.D. Weatherly, E.G. Burke and J.M. Schomaker. 2014. The conversion of allenes to strained three-membered heterocycles. Chem. Soc. Rev. 43: 3136-3163.
56. C. Zhou, C. Fu and S. Ma. 2007. Highly selective thiiranation of 1,2-allenyl sulfones with Br_2 and $Na_2S_2O_3$: mechanism and asymmetric synthesis of alkylidenethiiranes. Angew. Chem. Int. Ed. 46: 4379-4381.
57. S. Holm and A. Senning. 1973. Reactions of electron-depleted thiocarbonyl compounds with diazoalkanes. Tetrahedron Lett. 14: 2389-2392.
58. D. Paqwer. 1972. Int. J. Sulfur Chem. B 7: 269.
59. N. Latif and I. Fathy. 1962. Carbonyl and thiocarbonyl compounds. VII.1a A new method for the direct synthesis of ethylene sulfides. J. Org. Chem. 27: 1633-1636.

60. T. Aoyama, S. Toyama, N. Tamaki and T. Shioiri. 1983. New methods and reagents in organic synthesis. 37. Trimethylsilyldiazomethane. A convenient reagent for the preparation of vinylsilanes from alkanesulfonyl chlorides. Chem. Pharm. Bull. (Tokyo) 31: 2957-2959.
61. H. Heimgartner, G. Mlostoń and J. Romański. 2005. Adamantanethione. e-EROS, Encyclopedia of reagents for organic synthesis. Wiley online library, RN00504.
62. V.A. Nikolaev, A.V. Ivanov, L.L. Rodina and G. Mloston. 2013. Less reactive dipoles of diazodicarbonyl compounds in reaction with cycloaliphatic thioketones - first evidence for the 1,3-oxathiole-thiocarbonyl ylide interconversion. Beilstein J. Org. Chem. 9: 2751-2761.
63. T. Eicher and S. Hauptmann. 2003. The chemistry of heterocycles: structures, reactions, and applications. John Wiley & Sons: Chichester, UK, 5-16.
64. R.E. Banks and J.C. Tatlow. 1994. Organofluorine chemistry: nomenclature and history. R.E. Banks, B.E. Smart and J.C. Tatlow (Eds.). Organofluorine chemistry: principles and commercial applications. Plenum Press: New York, 2-4.
65. M. Hawkins, M.J. Almond and A.J. Downs. 1985. Photochemistry of low-temperature matrixes containing carbonyl sulfide: reactions of sulfur atoms with the phosphorus trihalides phosphorus trifluoride (PF_3) and phosphorus trichloride (PCl_3) and the hydrocarbons methane, ethene, and ethyne. J. Phys. Chem. 89: 3326-3334.
66. K.S. Sidhu, E.M. Lown, O.P. Strausz and H.E. Gunning. 1966. The reactions of sulfur atoms. VI. The addition to C4 olefins. A stereospecific triplet-state reaction. J. Am. Chem. Soc. 88: 254-263.
67. W. Adam and R.M. Bargon. 2004. Synthesis of thiiranes by direct sulfur transfer: the challenge of developing effective sulfur donors and metal catalysts. Chem. Rev. 104: 251-261.
68. Y. Sugihara, K. Noda and J. Nakayama. 2000. Sulfuration of 9,9'-bibenzonorbornenylidenes by elemental sulfur: exclusive episulfide formation with retention of the configuration of alkenes. Tetrahedron Lett. 41: 8913-8916.
69. T. Ashitani and S. Nagahama. 1999. Direct episulfidation of caryophyllene and humulene. Nat. Prod. Lett. 13: 163-167.
70. J. Nakayama, Y. Ito and A. Mizumura. 1992. Sulfuration of dibenzobarrelene and adamantylideneadamantane. Sulfur Lett. 14: 247-250.
71. P.D. Bartlett and T. Ghosh. 1987. Sulfuration of the norbornene double bond. J. Org. Chem. 52: 4937-4943.
72. J. Emsley, D.W. Griffiths and G.J.J. Jayne. 1979. The synthesis and spectra of polycyclic 1,2,3-trithiolans. J. Chem. Soc. Perkin Trans. 1 0: 228-232.
73. S. Inoue, T. Tezuka and S. Oae. 1978. A new type of photochemical addition reaction of elemental sulfur to olefins. Phosphorus, Sulfur, Silicon Relat. Elem. 4: 219-221.
74. W. Ando, H. Sonobe and T. Akasaka. 1990. Selective sulfurization of olefins by 9,10-epidithio-9,10-dihydroanthracene: intermediacy of diatomic sulfur. Tetrahedron 46: 5093-5100.
75. V. Luccini, G. Modena and L. Pasquato. 1997. Thiiranium and thiirenium ions. From reaction intermediates to building blocks in organic synthesis. Gazz. Chim. Ital. 127: 177-188.
76. J. Kowalski, J. Chojnowski and J. Michalski. 1983. Reactions of triorganosilysulfenyl halides with some nucleophiles. J. Organomet. Chem. 258: 1-5.
77. F. Capozzi, G. Capozzi and S. Menichetti. 1988. Thiiranes: one-pot synthesis from alkenes, and catalytic desulphurization. Tetrahedron Lett. 29: 4177-4180.
78. I. Carelli, A. Inesi, V. Carelli, M.A. Casadei, F. Liberatore and F.M. Moracci. 1986. Electrochemical studies of β-lactams. Part 21. Electrosynthesis of β-lactams via bond formation between C-3 and C-4. Synthesis 7: 591-593.
79. A.J. Fry and L.L. Chung. 1976. Oxyphosphirane intermediates in the electrochemical reduction of α,α'-dihalophosphinates. Tetrahedron Lett. 17: 645-648.

80. A.J. Fry, K. Ankner and V. Handa. 1981. Electrochemical reduction of α,α'-dibromo-dibenzylsulfone, -sulfoxide, and -sulfide. Tetrahedron Lett. 22: 1791-1794.
81. K. Soai and T. Mukaiyama. 1979. Enantioface-differentiating reactions using (2*S*,2'*S*)-2-2-(methylthio) thiazoline to aldehydes. Bull. Chem. Soc. Jpn. 52: 3371-3376.
82. J. Nakayama, K. Takahashi, Y. Ono, M. Morita, Y. Sugihara and A. Ishii. 2002. Practical synthesis of thiirene 1-oxides that possess two bulky alkyl substituents. Heteroat. Chem. 13: 424-430.
83. P. Fowles. M. de Sorgo, A.J. Yarwood, O.P. Strausz and H.E. Gunning. 1967. The reactions of sulfur atoms. IX. The flash photolysis of carbonyl sulfide and the reactions of S(^{1}D) atoms with hydrogen and methane. J. Am. Chem. Soc. 89: 1352-1362.
84. H.E. Gunningand and O.P. Strausz. 1966. The reactions of sulfur atoms. Adv. Photochem. 4: 143-194.
85. I.L. Knunyantrr and E.G. Bykhoveksya. 1955. Izv. Akad. Nauk SSSR Ser. Khim. 852.
86. L.I. Zakharkin and V.V. Korneva. 1964. Synthesis of 2-alkylidene- and 2-alkylcyclododecanones. Bull. Acad. Sci. USSR Div. Chem. Sci. 13: 2102-2104.
87. G.A. Tolstikov, B.M. Lerman and T.A. Belogaeva. 1986. Study of reaction of sulfur halides with unsaturated compounds. 24. Reaction of sulfur monochloride with adamantylidene-adamantane. Bull. Acad. Sci. USSR Div. Chem. Sci. (Engl. Transl.) 35: 563-566.
88. G.A. Tolstikov, B.M. Lerman, L.I. Umanskaya and Y.T. Struchkov. 1982. Reaction of sulfur halides with unsaturated compounds. Bull. Acad. Sci. USSR Div. Chem. Sci. (Engl. Transl.) 31: 588-595.
89. G.A. Tolstikov, B.M. Lerman and T.A. Belogaeva. 1988. Reactions of sulfur halides with unsaturated compounds. 25. Synthesis and reactions of novel sterically hindered olefins, 2,3-disubstituted 7-adamantylidenenorbornanes, with sulfur dichloride. Bull. Acad. Sci. USSR Div. Chem. Sci. (Engl. Transl.) 37: 1629-1635.
90. J. Boster and R.M. Kellogg. 1978. Preparation of some *S*-methyl thiiranium salts of adamantylideneadamantane. Demethylation and desulphurization with nucleophiles. J. Chem. Soc. Chem. Commun. 14: 630-631.
91. I.A. Abu-Yousef and D.N. Harpp. 1997. Effective precursors for sulfur monoxide formation. J. Org. Chem. 62: 8366-8371.
92. I.A. Abu-Yousef and D.N. Harpp. 1995. A useful precursor for sulfur monoxide transfer. Tetrahedron Lett. 36: 201-204.
93. I.A. Abu-Yousef, R.C. Hynes and D.N. Harpp. 1993. Sulfenyl chloride chemistry. Sulfur transfer to double bonds. Tetrahedron Lett. 34: 4289-4292.
94. M.U. Bombala and S.V. Ley. 1979. Preparation of episulphides from alkenes via succinimide-*N*-sulphenyl chloride or phthalimide-*N*-sulphenyl chloride adducts. J. Chem. Soc. Perkin Trans. 1 0: 3013-3016.
95. J. Ohishi, K. Tsuneoka, S. Ikegami and S. Akaboshi. 1978. Roles of heteroatoms in solvolytic reactions. 4. Solvolysis of the *exo* and *endo* esters of 2-thiabicyclo[2.2.1]heptan-6-ols. J. Org. Chem. 43: 4013-4020.
96. Y. Sugihara, K. Onda, M. Sato and T. Suzuki. 2010. A novel method for thiiranation of alkenes with 1,1'-dithiobis(1*H*-1,2,4-triazole). Tetrahedron Lett. 51: 4110-4112.
97. K. Hemming. 2011. Heterocyclic chemistry. Annu. Rep. Prog. Chem. Sect. B 107: 118-137.
98. T. Imamoto and H. Koto. 1985. An efficient synthesis of α-iodo derivatives of sulfones, sulfoximines, and phosphine oxides. Synthesis 10: 982-983.
99. A.P. Dishington, R.E. Douthwaite, A. Mortlock, A.B. Muccioli and N.S. Simpkins. 1997. Episulfone substitution and ring-opening reactions via α-sulfonyl carbanion intermediates. J. Chem. Soc. Perkin Trans. 1 3: 323-337.
100. G. Opitz, K. Rieth and T. Ehlis. 1990. Secondary reactions of sulfenes from sulfonyl chlorides and tertiary amines. On the competing formation of sulfene-trialkylamine *S,N*- and *C,N*-adducts. Chem. Ber. 123: 1563-1569.

101. G. Opitz, T. Ehlis and K. Rieth. 1989. Symmetrische alkene über episulfone aus primären sulfonylchloriden. Tetrahedron Lett. 30: 3131-3132.
102. J. Nakayama, M. Tanuma, Y. Honda and M. Hoshino. 1984. Reaction of arylmethanesulfonyl and styrylmethanesulfonyl chlorides with triethylamine. Tetrahedron Lett. 32: 4553-4556.
103. N.H. Fischer. 1970. The synthesis of episulfones and olefins via sulfenes. Synthesis 8: 393-404.
104. A.G. Sutherland and R.J.K. Taylor. 1989. The first isolation of an episulphone intermediate from a Ramberg-Bäcklund reaction. Tetrahedron Lett. 30: 3267-3270.
105. R.A. Ewin, W.A. Loughlin, S.M. Pyke, J.C. Morales and R.J.K. Taylor. 1993. The isolation of episulphones from the Ramberg-Bäcklund rearrangement. Part 31. Synlett 9: 660-662.
106. S.M. Jeffery, A.G. Sutherland, S.M. Pyke, A.K. Powell and R.J.K. Taylor. 1993. Isolation of episulfones from the Ramberg-Bäcklund rearrangement. Part 2. X-Ray molecular structure of 2,3-epithio-8,8-dimethyl-6,10-dioxaspiro[4.5]decane *S*,*S*-dioxide and of *R*-6-benzyl-*t*-7,*t*-8-epithio-1,4-dioxaspiro[4.4]nonane *S*,*S*-dioxide. J. Chem. Soc. Perkin Trans. 1 19: 2317-2327.
107. P.C.B. Page, R.D. Wilkes and D. Reynolds. 1995. Alkyl chalcogenides: sulfur-based functional groups. A.R. Katritzky, O. Meth-Cohn and C.W. Rees (Eds.). Comprehensive organic functional group transformations. Elsevier, Oxford, vol. 2: 113-276.
108. M. Lee, M. Ikejiri, D. Klimpel, M. Toth, M. Espahbodi, D. Hesek, C. Forbes, M. Kumarasiri, B.C. Noli, M. Chang and S. Mobashery. 2012. Structure-activity relationship for thiirane-based gelatinase inhibitors. ACS Med. Chem. Lett. 3: 490-495.
109. V.P. Tashchi, T.I. Rukasov, Y.G. Orlova, Y.A. Putsykin and Y.A. Baskakov. 1982. Synthesis of thiirane oximes from 3,4-dibromo-3-methyl-2-butanone oxime. Chem. Heterocycl. Compd. 18: 566-568.
110. E. Abele. 2013. Synthesis, reactions and biological activity of derivatives of oximes of three-membered heterocycles. Heteroletters 3: 229-245.
111. K. Okuma, T. Shigetomi, Y. Nibu, K. Shioji, M. Yoshida and Y. Yokomori. 2004. Synthesis of isolable thiirane-2-thione (α-dithiolactone) from thioketene *S*-oxide. J. Am. Chem. Soc. 126: 9508-9509.
112. T. Ozturk, E. Ertas and O. Mert. 2007. Use of Lawesson's reagent in organic syntheses. Chem. Rev. 107: 5210-5278.
113. W. Adam and B. Frohling. 2000. An α,α'-dioxothione and its [4+2] cycloaddition with *trans*-cyclooctene in the reaction of ninhydrin with potassium thiotosylate. Org. Lett. 16: 2519-2522.
114. L.D.S. Yadav and R. Kapoor. 2002. An expeditious one-pot synthesis of thiiranes from α-halo ketones in solvent-free conditions using microwaves. Synthesis 16: 2344-2346.
115. L.A. Carpino, P.H. Terry and S.D. Thatte. 1964. Interception of an intermediate in the alkaline degradation of 4-halo-2-pyrazolin-5-ones. Tetrahedron Lett. 5: 3329-3332.
116. R. Breslow, J. Posner and A. Krebs. 1963. Synthesis of cyclopropenones by a modified Favorskii reaction. J. Am. Chem. Soc. 85: 234-234.
117. D. Barnard, J.M. Fabian and H.P. Koch. 1949. Valence vibration frequencies and hydrogen bond formation of sulphoxide and sulphone groups. (Absorption spectra and structure of organic sulphur compounds. Part V). J. Chem. Soc. 0: 2442-2454.
118. E.A. Robinson. 1961. Characteristic vibrations of the sulphuryl group. Can. J. Chem. 39: 247-255.
119. P.M.G. Bavin, G.W. Gray and A. Stephenson. 1960. Infrared absorptions of sulphones. Spectrochim. Acta 16: 1312-1317.
120. R. Breslow and C. Yuan. 1958. The *sym*-triphenylcyclopropenyl cation, a novel aromatic system. J. Am. Chem. Soc. 80: 5991-5994.
121. R. Breslow, T. Eicher, A. Krebs, R.A. Peterson and J. Posner. 1965. Diphenylcyclopropenone. J. Am. Chem. Soc. 87: 1320-1325.
122. F.G. Bordwell, B.E. Jarvis and P.W.R. Corfleld. 1968. A 1,3-elimination proceeding by double inversion. J. Am. Chem. Soc. 90: 5298-5299.

123. M. Nishio and T. Ito. 1965. Nuclear magnetic resonance studies of sulfur compounds. I. The preparation of α-(alkylsulfinyl)cinnamic acid derivatives and their nuclear magnetic resonance spectra. Chem. Pharm. Bull. 13: 1392-1398.
124. L.A. Carpino, L.V. McAdams, R.H. Rynbrandt and J.W. Spiewak. 1971. Thiirene dioxides. Synthesis, characterization, and reactivity. J. Am. Chem. Soc. 93: 476-484.
125. B.B. Jarvis, S.D. Dutky and H.L. Ammon. 1972. Stereochemistry of 1,3-eliminations from dibromo sulfoxides. J. Org. Chem. 94: 2136-2138.
126. Y. Kamano, G.R. Pettit and M. Tozawa. 1974. Steroids and related natural products. 88. Synthesis of periplogenin. J. Org. Chem. 39: 2319-2320.
127. C.D. Kubik and V.I. Sternberg. 1966. Chem. Ind. (London) 248.
128. G. Opitz. 1967. Sulfines and sulfenes - the *S*-oxides and *S,S*-dioxides of thioaldehydes and thioketones. Angew. Chem. Int. Ed. 6: 107-123.
129. H. Oediger, H.J. Kabbe, F. Moller and K. Eiter. 1966. 1.5-Diaza-bicyclo[4.3.0]nonen-(5). Ein neues reagenz zur einführung von doppelbindungen. Chem. Ber. 99: 2012-2016.
130. L.A. Carpino and H.-W. Chen. 1971. 2,3-Diphenylthiirene 1-oxide. J. Am. Chem. Soc. 93: 785-786.
131. L.A. Carpino and H.-W. Chen. 1979. Diphenylthiirene 1-oxide. Synthesis, characterization, and reactivity. J. Am. Chem. Soc. 101: 390-394.
132. Y.-N. Jin, A. Ishii, Y. Sugihara and J. Nakayama. 1998. A new route to dithiirane 1-oxides: oxidation of tetrathiolanes with dimethyldioxirane. Tetrahedron Lett. 39: 3525-3528.
133. E. Schaumann, O. Bolte and H. Behr. 1990. *S*-Vinyl thio-oximes from thioketenes and benzylnitrene. J. Chem. Soc. Perkin Trans. 1 1: 182-183.
134. M. Kidwai and R. Batra. 1991. A new route to thiaziridine 1,1-dioxides. Indian J. Chem. Sect. B 30: 784-785.

CHAPTER

Synthesis of thiiranes from sulfur bearing moieties

10

Chapter outline

10.1 Introduction

Heterocyclic compounds are the biggest and most diverse family of organic compounds. Among them, aromatic heterocyclic compounds represent structural motifs found in a great number of biologically active synthetic and natural compounds, agrochemicals, and medicines. Moreover, aromatic heterocyclic compounds are extensively utilized for the formation of dyes and polymeric materials of great importance. In organic synthesis, there are several reports on the use of aromatic heterocyclic compounds as intermediates. Although a variety of highly efficient approaches have been described in the past for the formation of aromatic heterocyclic compounds and their derivatives, the development of new procedures is in continuous demand. Particularly, development of novel synthetic methodologies toward heterocyclic compounds, aiming at attaining better levels of molecular complexity and improved functional group compatibilities in a convergent and atom economical fashion from easily available starting compounds under mild reaction conditions, is one of the major research activities in synthetic organic chemistry [1].

Sulfur has been known for more than 100 years, and it is broadly spread in nature. Sulfur is considered as an important biogenic element because of the presence of sulfur in all living organisms and vegetables in the form of inorganic or organic compounds. A variety of food stuffs such as truffle, mushroom, coffee, leek, garlic, chive, broccoli, mustard, onion, radish, asparagus, cabbage, and pineapple possess many organosulfur compounds [2].

Staudinger and Siegwart [3] discovered the thiiranes. The limited stability of thiiranes related to their oxirane counterparts has resulted in a limited number of synthetic pathways and even less information regarding their pharmacological activity [4]. The chemistry of thiiranes is very similar to their oxirane counterparts in various ways, which in contrast to the episulfides have been studied in detail. The main differences between both compound classes can be derived from the physical and structural properties of the tricycle.

3-Membered Heterocycle Synthesis. https://doi.org/10.1016/B978-0-443-13583-5.00002-5

10.2 Synthesis of thiiranes from C=S bearing moieties

Both the sulfur ylides, DIMSY (dimethylsulfonium methylide) and DIMSOY (dimethylsulfoxonium methylide), are known to act as nucleophile and transfer methylene group to certain electrophilic unsaturated linkages such as C=N, C=O, C=S, and activated C=C bonds. In these reactions, the ylide acted as a nucleophile in the first step, and then subsequent loss of dimethylsulfoxide afforded cyclic compounds (Scheme 1) [5].

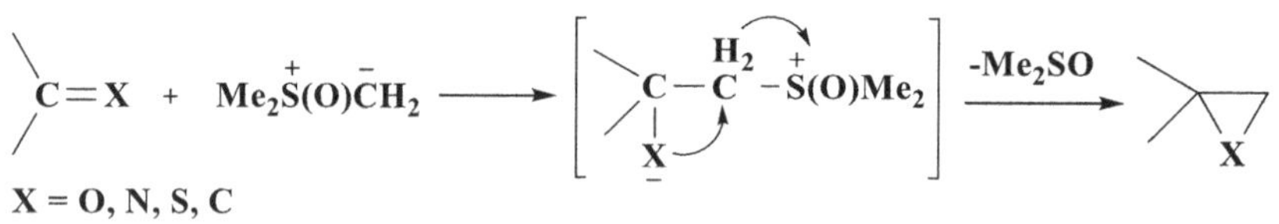

SCHEME 1

The transition between thiaphosphetane- and *gauche*-betaine-type structures of intermediates in the thio-Wittig reaction of ylides $R_3P{=}CR_1R_2$ with $S{=}C(R_3)_2$ has been discovered by ^{31}P nuclear magnetic resonance spectroscopy and predicted by computational analysis [6]. Thus, ab initio calculations for the reaction of precursors revealed the synthesis of a conventional oxaphosphetane intermediate that featured a planar four-membered ring, while thio-Wittig reaction (X = S) intermediate was characterized by a large P−S separation and departure from planarity. The betaine character of an intermediate reduced on decreasing the phosphonium stabilization electronically by replacing the trimethylphosphine functionality consecutively with phosphine and $(CF_3)_3P$. Both products decomposed at slightly elevated temperatures to synthesize the triphenylphosphine sulfide and $Ph_2C{=}CH_2$ through the 2,2-diphenylthiirane intermediate and triphenylphosphine (Scheme 2) [7].

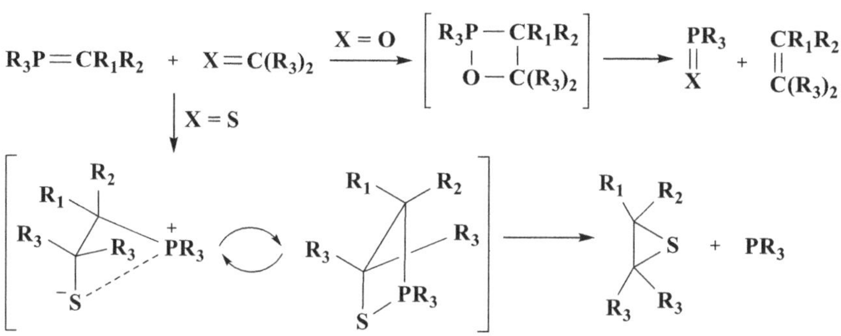

$R_1 = R_2 = H, R = R_3 = Ph$
R = Me
R = H
$R = CF_3$ where $R_1 = R_2 = H, R_3 = Me$
$R_3 = p\text{-}Me_2NC_6H_4$
$R_3 = p\text{-}OMeC_6H_4$ where $R_1 = H, R_2 = Me, R = Et$

SCHEME 2

Among the few methods for the formation of *S*-containing dipolar intermediates, the reaction of diazomethane with C=S dipolarophiles followed by removal of nitrogen was used mostly [8]. The reactive 1,3-dipoles (thiocarbonyl ylides) prepared in situ could be trapped by different electron-poor dipolarophiles, but aromatic thioketones were verified to be the most reactive. In contrast to thioketones, the dithioesters have been utilized less often as substrates of sulfomethanides (thiocarbonyl ylides). However, the reports [9] on the synthesis of thiiranes have been described starting with methyl propanedithioate. The intermediate underwent a thermal cleavage for the formation of thiocarbonyl *S*-methanides, which were subjected to an intramolecular rearrangement to provide the thiirane. In addition, the sterically more hindered 1,3-dithiolane isomers were formed when intermediate thiocarbonyl *S*-methanides were trapped with methyl dithioesters in a regioselective manner. The thiocarbonyl *S*-methanide was proposed as a reactive intermediate in these reactions (Scheme 3).

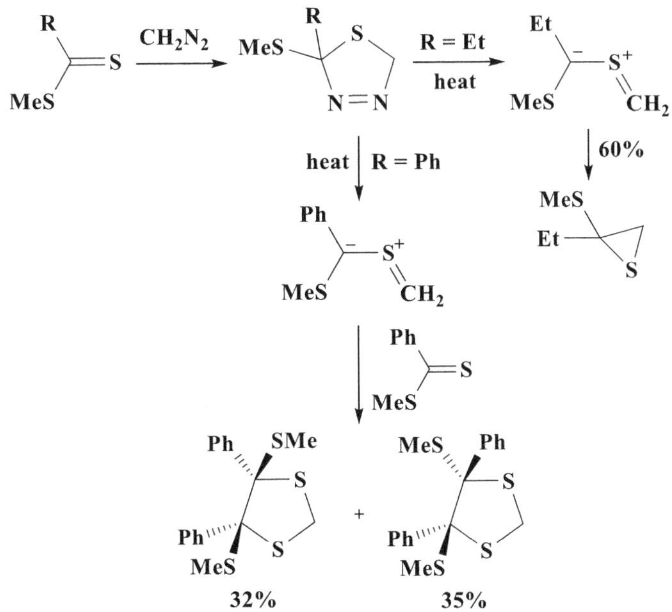

SCHEME 3

The dichlorocarbene was utilized to transform the imines and alkenes into dichloroaziridines and dichlorocyclopropane. The 2,2-dichlorothiiranes were prepared from nonenolizable, sterically congested dichlorocarbene and thioketones (Scheme 4) [10].

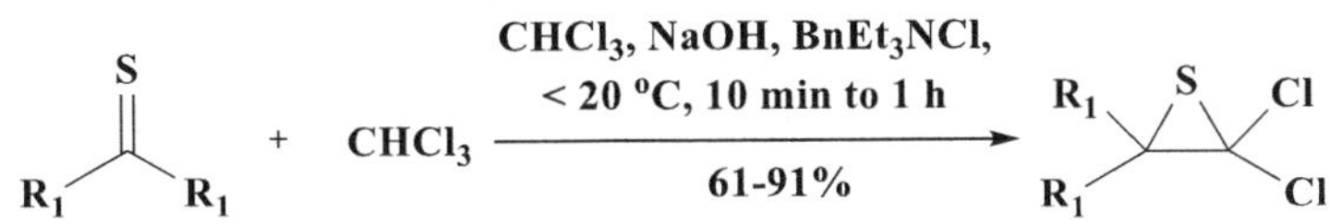

SCHEME 4

The [2 + 1]-cycloaddition involving an addition of carbenes to CS bond in diarylthiones afforded thiiranes. The carbene species may be prepared from diazo compounds, from phenyliodonium bis-(alkyl- or aryl-sulfonyl) methylides, or from organomercury compounds. Scheme 5 represents the use of thiobenzophenones, although in low yields [11], and many by-products such as benzothiophenes were obtained [12].

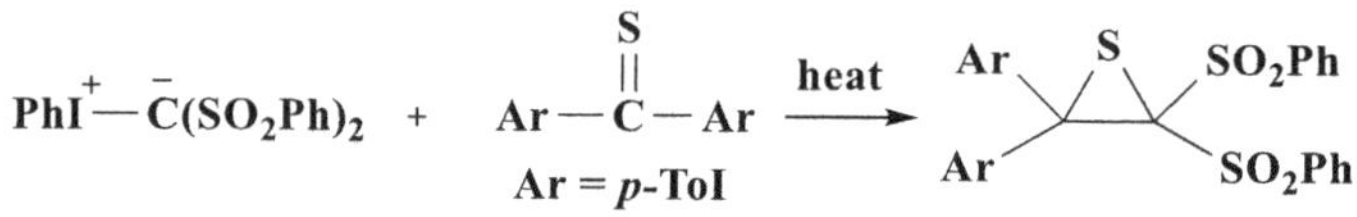

SCHEME 5

Along with Diels–Alder reactions, the [2 + 3]-cycloadditions are the most familiar approaches utilized both in the formation of a broad range of heterocyclic systems and in the trapping of unstable thiocarbonyl species [13]. The first instances of dipolar cycloadditions between thioketones and diazoalkanes date back to 1920 [3], but it was not until Huisgen and Langhals [14] conducted a systematic study that thiones achieved broad appreciation as versatile synthons. Huisgen and Langhals [14] proposed the terminology "superdipolarophiles" to highlight the high reactivity of the C=S bond. The 1,3-cycloadditions of diphenyldiazomethane to thioketones were much faster than those to unsaturated compounds and nitriles previously regarded as the fastest dipolarophiles. The primary cycloadducts, 1,3,4-thiadiazolines, extruded nitrogen to synthesize the thiiranes through the formation of thiocarbonyl ylides (Scheme 6), which were not isolable.

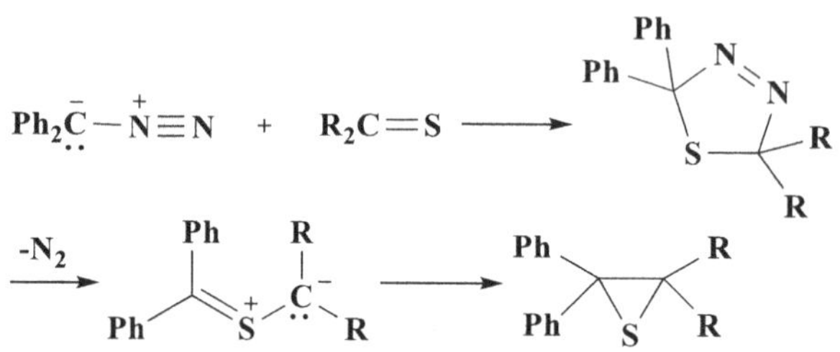

SCHEME 6

The Corey-Chaykovsky reaction involved the reaction of sulfur ylide, either dimethylsulfonium methylide or dimethylsulfoxonium methylide (Corey's ylide, sometimes known as DMSY), with electrophile such as olefin, carbonyl, thiocarbonyl, or imine to afford the cyclopropane, epoxide, thiirane, or aziridine [15,16]. For an unsaturated carbonyl compound, the dimethylsulfoxonium methylide was added preferentially to olefin to synthesize the cyclopropane derivative, while the more reactive dimethylsulfonium methylide underwent a methylene transfer to carbonyl to give the epoxide. Also because of the difference of reactivities, reactions utilizing dimethylsulfoxonium methylide needed elevated temperature to some extent usually around 50–60°C, while reactions utilizing the more reactive dimethylsulfonium methylide could be performed at colder temperature varying from −15°C to rt. Moreover, while it was favorable to freshly prepare both ylides in situ, dimethylsulfonium methylide was not as stable as dimethylsulfoxonium methylide, which could be stored at rt for several days (Scheme 7).

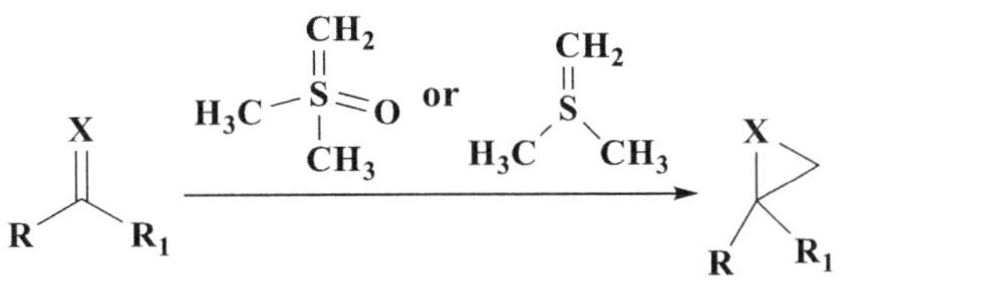

X = O, CH_2, NR_2, S, $CHCOR_3$, $CHCO_2R_3$, $CHCONR_2$, CHCN

SCHEME 7

The direct formation of thiiranes from ketohydrazones occurred via generation of diazoalkane intermediate. The diazoalkane afforded thioketone via its nucleophilic attack on sulfur and the simultaneous removal of nitrogen. The thioketone was reacted with another molecule of diazo compound to synthesize the cyclic sulfide (Scheme 8). It was highlighted [17a] that the sulfur–sulfur bonds in sulfur and organic polysulfides were easily broken by nucleophilic reagents. The formation of diazo compound as an intermediate was proved by the fact that its characteristic color was developed during the reaction, and in some experiments, it has been isolated from the reaction mixture.

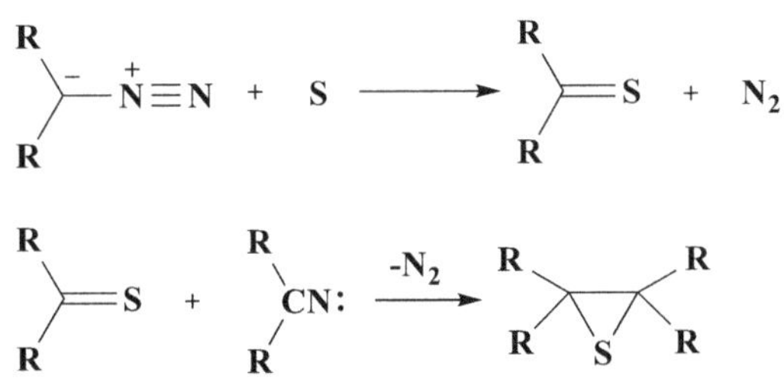

SCHEME 8

The dimerization of thiocarbonyl compounds provided dithietanes, which were oxidized to monosulfones to afford the thiiranes on heating. Both stereoisomers of dithetane, formed by the dimerization of trifluorothioacetyl chloride, afforded sulfone, which was oxidized with fuming HNO_3 and CrO_3. The vacuum pyrolysis of sulfone provided thiirane as a *cis/trans* mixture (Scheme 9) [17b,c].

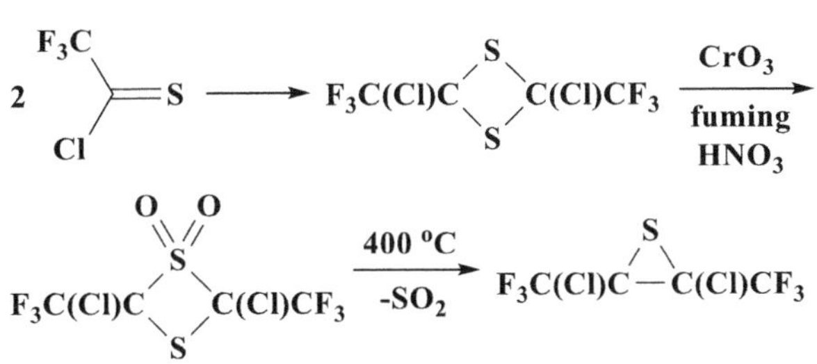

SCHEME 9

The reaction of di-*t*-butyl thioketone with trimethylsilyldiazomethane provided episulfide with the evolution of N_2, while its reaction with TMSC(Li)N_2 resulted in the removal of one *t*-butyl group to synthesize the 1,2,3-thiadiazole (Scheme 10) [18].

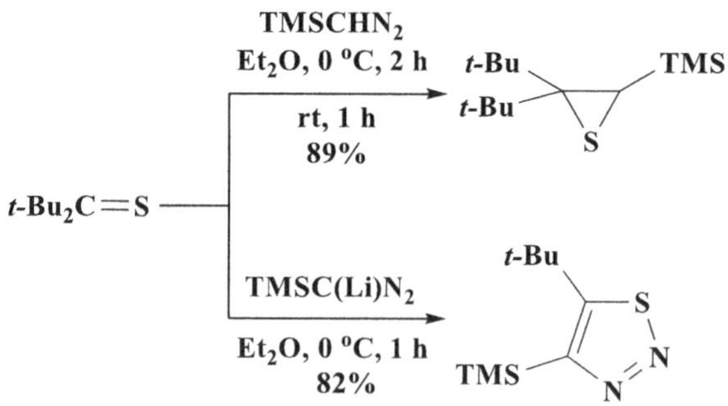

SCHEME 10

Of the strained heterocyclic compounds, MTs (methylene thiiranes) are less frequently prepared from allenes [19]. Generally, the methylene thiiranes are synthesized by the decomposition of hydrazones [20] or diazolines [21] (Schemes 11–13). An addition of diazo compounds to thioketenes was also reported [22]. Ando et al. [23–32] examined the chemistry of thiiranoradialenes and methylene thiiranes in significant detail, in similarity to their work on methylene silacyclopropanes. However, none of their methylene thiirane precursors were derived from allenes [17d].

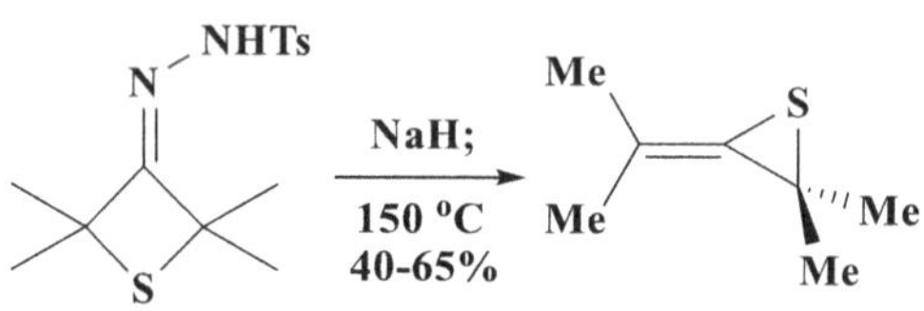

SCHEME 11

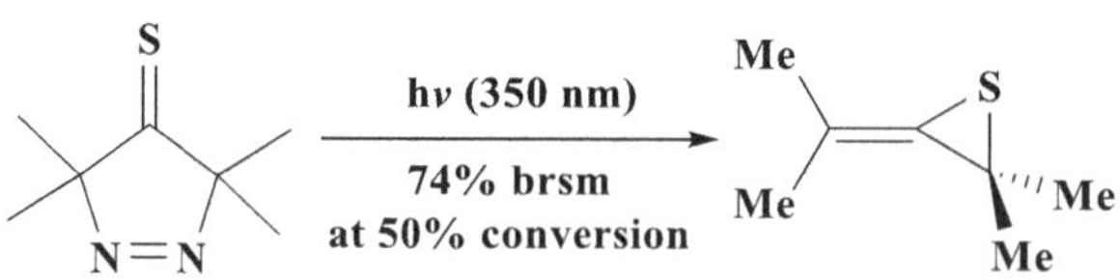

SCHEME 12

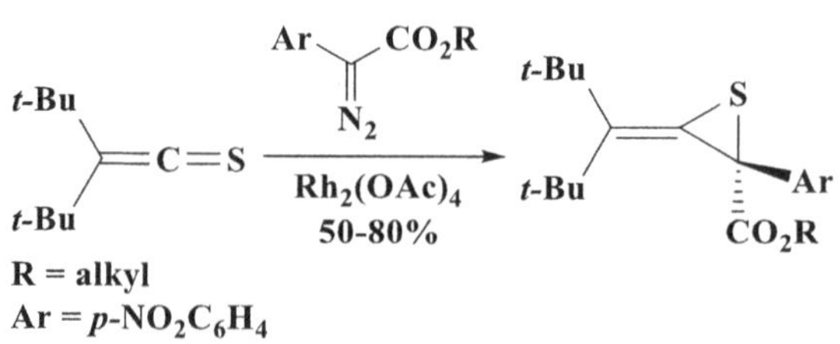

SCHEME 13

It was found that the transformation of ketone to TBS (*t*-butyldimethylsilyl)hydrazone using 1,2-bis(*t*-butyldimethylsilyl)hydrazine resulted in much lower degree of racemization. This work was based on a broad analysis by Furrow and Myers [33,34] who reported the transformation of a series of aldehydes and ketones to *t*-butyldimethylsilylhydrazones under the relatively mild conditions. The deprotonation with NH_2NH_2 at α-position afforded planar enolate followed by nonstereoselective reprotonation being the cause of racemization under these conditions. The 1,2-bis(*t*-butyldimethylsilyl)hydrazine was much more sterically hindered as a base owing to bulky *t*-butyldimethylsilyl groups, which was the origin of decreased amount of racemization. The bulky *t*-butyldimethylsilyl groups also hampered its nucleophilicity; however, both the fact that a silicon-oxygen bond was formed during the reaction and the presence of a Lewis acid catalyst resulted in its high reactivity toward the synthesis of *t*-butyldimethylsilyl hydrazone. In a preliminary attempt, the enantiopure (*S*)-ketone was transformed into *t*-butyldimethylsilyl hydrazone, which was immediately reacted with [bis(acetoxy)iodo]benzene resulting in an oxidation to diazo compound and then the thioxanthone was added (Scheme 14). A large amount of crystals was obtained during workup when the reaction mixture was poured in H_2O, which were examined and demonstrated to be consisted of enantiomerically pure (*S*)-episulfide. The small amount of episulfide that had not crystallized during workup was isolated and revealed to be completely racemic. Apparently, the racemization occurred only to a minor extent.

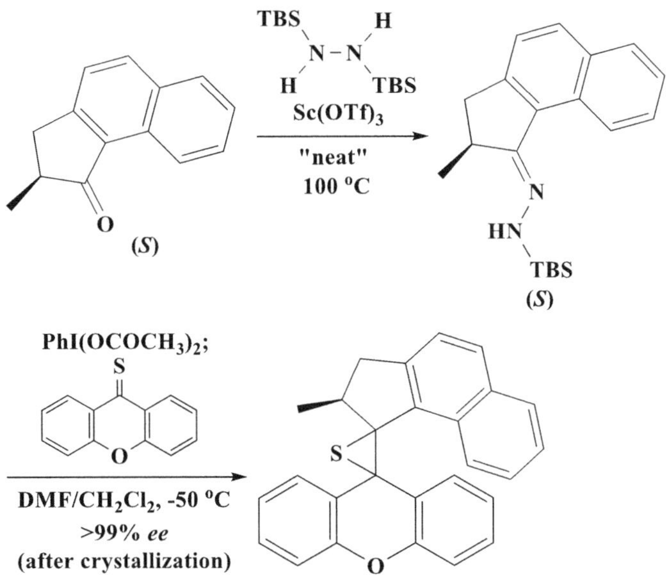

SCHEME 14

In light of the promising outcomes obtained with sultene as episulfidation agent, it was found that such direct sulfur-transfer procedure should be extendable to other *S*-heterocyclic compounds. The 1,2,4-oxadithiolane containing reactive sulfenate moiety may be formed in situ from a reactive thioketone and sulfine (Scheme 15). Indeed, the *trans*-episulfide was isolated in 58% yield along with thermolysis compounds (the dione and thiobenzophenone) when the reaction was performed in the presence of (*E*)-cyclooctene [35].

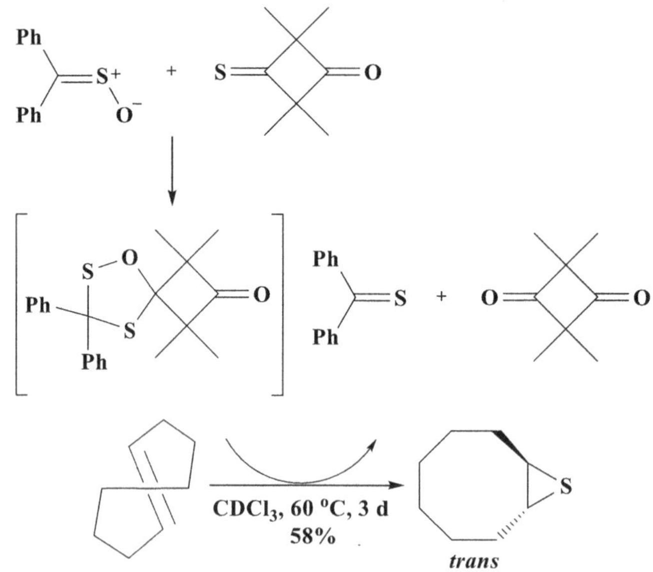

SCHEME 15

An isolation of thioaldehyde-*S*-oxide, prepared from methyl acetate, was unsuccessful. Although the sulfine was obtained, the reaction led to a dark colored unidentifiable mixture of products. Baltas and coworkers [36–38] failed in an isolation of *t*-butoxycarbonyl thioaldehyde-*S*-oxide, while the methyl thioketone-*S*-oxide prepared from methyl propionate was very stable. There are strong indications that the thioaldehyde-*S*-oxides occur in a dimeric form. Himbert and coworkers [39–41] reported that these dimeric thioaldehyde-*S*-oxides decompose to synthesize the fumaric or/and maleic acid derivatives and in some cases also thiiranes (Scheme 16). In the present case, the mixture of compounds was not examined further.

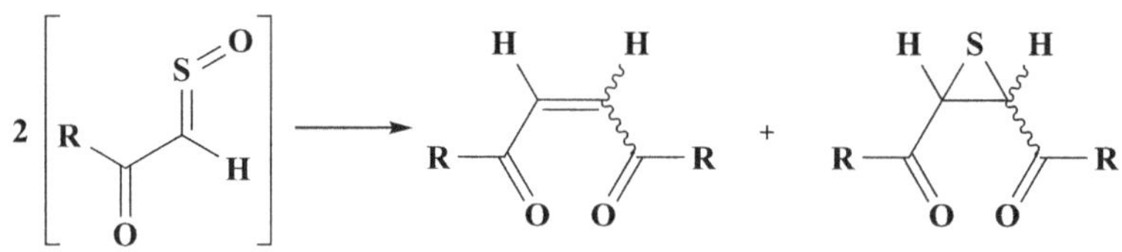

SCHEME 16

The [2 + 1] cycloadduct was synthesized in 18% yield by reacting diphenylsulfine with dichlorocarbene (Scheme 17) [42].

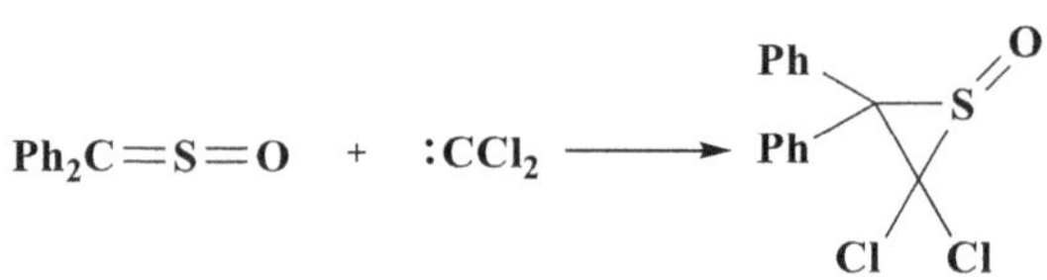

SCHEME 17

The quaternary phosphonium salt was formed when sulfene was reacted with PPh_3. The reaction appeared to occur through an initial [2 + 1]-cycloaddition to synthesize the thiirane, which was rearranged to synthesize the thiaphosphoridene dioxide intermediate (Scheme 18) [43].

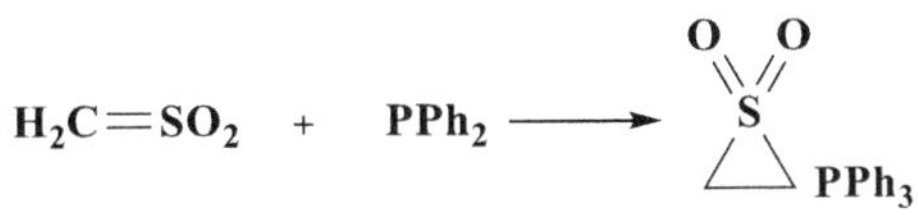

SCHEME 18

The carbon cumulenes underwent [2 + 1]-, [2 + 2]-, [3 + 2]-, and [4 + 2]-cycloaddition reactions with appropriate precursors to synthesize the cyclic products. Generally, the [2 + 1]-cycloaddition reactions provided three-membered ring products with an attached double bond, and sometimes the initially formed cycloadducts rearrange to give an isomeric three-membered ring cycloadduct. An instance was an addition of diphenylcarbene to dialkylthioketenes where the initially prepared cycloadduct on photolysis provided an isomer with bulky substituents on the three-membered ring (Scheme 19) [44].

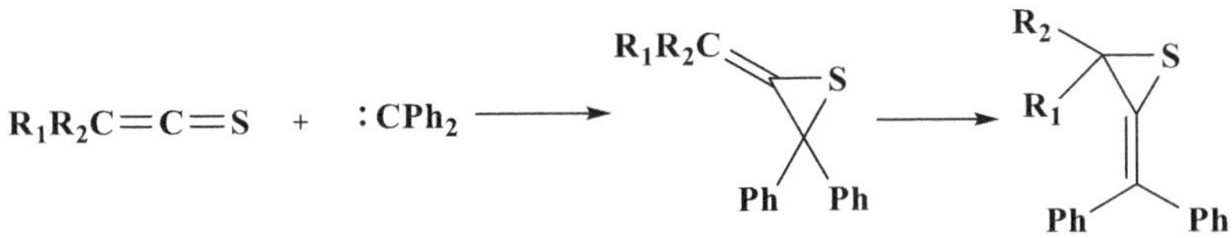

SCHEME 19

As an instance, the dithiiranone and oxathiiranone were proposed as possible sulfur donors in the reaction of *N,N*-dimethylaniline-*N*-oxide with CS_2 (Scheme 20) [45]. This oxidative desulfuration of CS_2 in the presence of electron-rich tetramethylethylene provided as much as 38% thiirane. The synthetic application of this unusual reaction was limited due to the 100-fold excess of the CS_2 and alkene [35b].

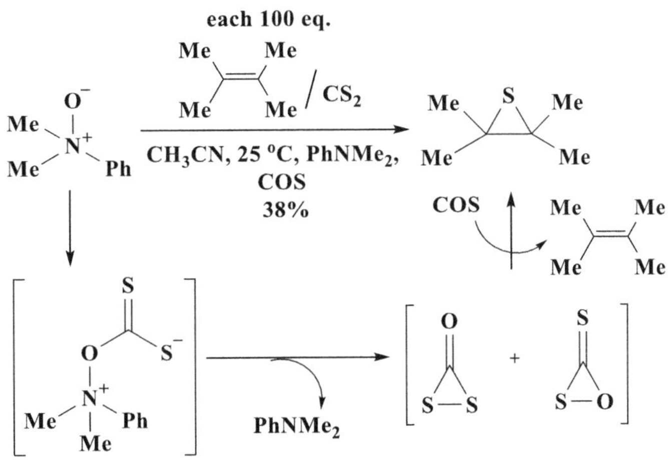

SCHEME 20

10.3 Synthesis of thiiranes from SCN bearing moieties

The known epoxides [46–48] are accessible in two or three steps from the homochiral metabolite formed by whole-cell fermentation of bromobenzene with *Pseudomonas putida* F39/D [49–51]. The *cis*-dienediol was protected as its acetonide with 2,2-dimethoxypropane, and the more reactive double bond was either oxidized directly with *m*-CPBA to give the α-oxirane or subjected to bromination and ring-closure sequence to synthesize the β-epoxide. Both epoxides were reacted under various ring-opening conditions to optimize the formation of deoxycyclitols substrates [52,53]. The first two steps in the sequence were fast, whereas the last two steps occurred at a very slow rate. Literature reports [54] in the regiospecific ring opening of vinyl epoxides showed that the regiochemistry of attack was controlled by stereoelectronic effects directed to the allylic position as experimentally detected. Overall, the complete reaction was accompanied by Walden inversion at both of the two carbon atoms of the ring (Scheme 21) [55–57].

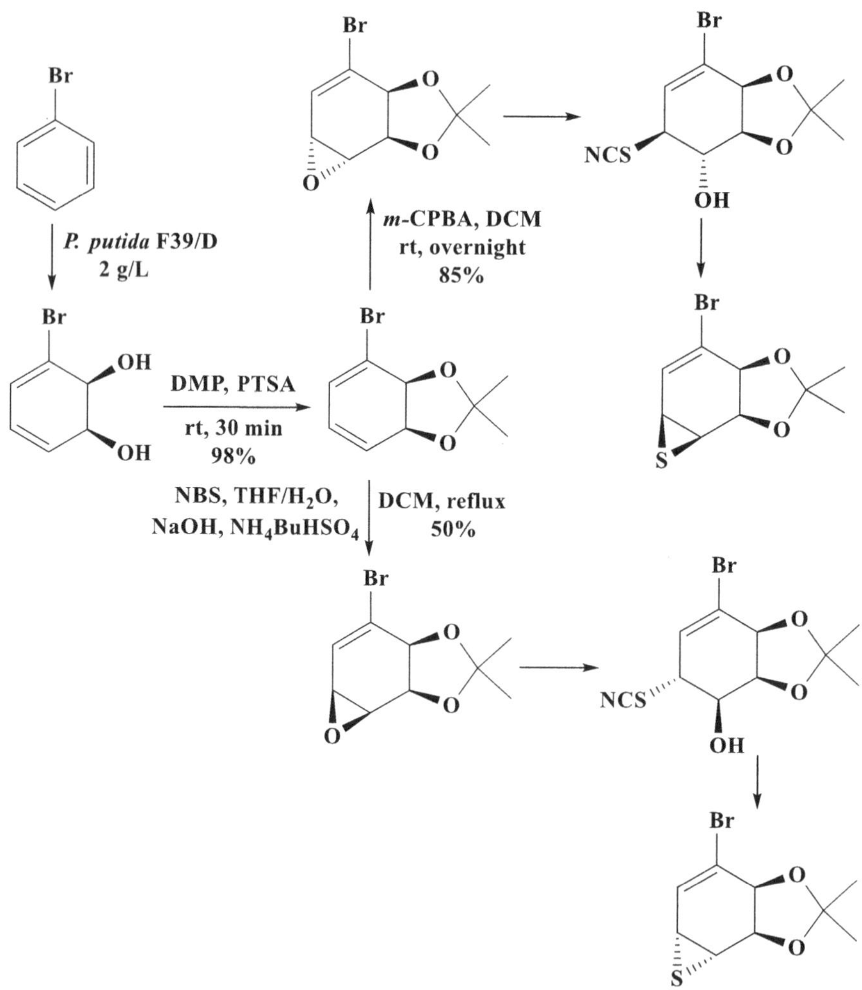

SCHEME 21

An acid-promoted addition of thiophenol to cyclohexadiene monoxide synthesized 1,2-addition products *syn*- and *anti*-cyclohexadiene hydroxide, where the utilization of ytterbium(III) trifluoromethanesulfonate as a catalyst can attain a diastereomeric ratio of *syn*/*anti* at 1:7.5 (Scheme 22) [55]. The *anti*-cyclohexadiene hydroxide was synthesized by bimolecular nucleophilic substitution ring opening of cyclohexadiene monoxide, whereas *syn*-cyclohexadiene hydroxide perhaps arose from unimolecular nucleophilic substitution reaction induced by acid-promoted epoxide cleavage. In the absence of acid, nucleophilic ring opening with thiocyanate furnished only *anti*-cyclohexadiene hydroxide thiocyanate by bimolecular nucleophilic substitution ring opening. The alkoxide from oxirane cleavage was activated in the presence of 18-crown-6 to undergo further reactions with thiocyanate to afford the episulfide. Same transformation was induced by reacting *anti*-cyclohexadiene hydroxide thiocyanate with base [57,58].

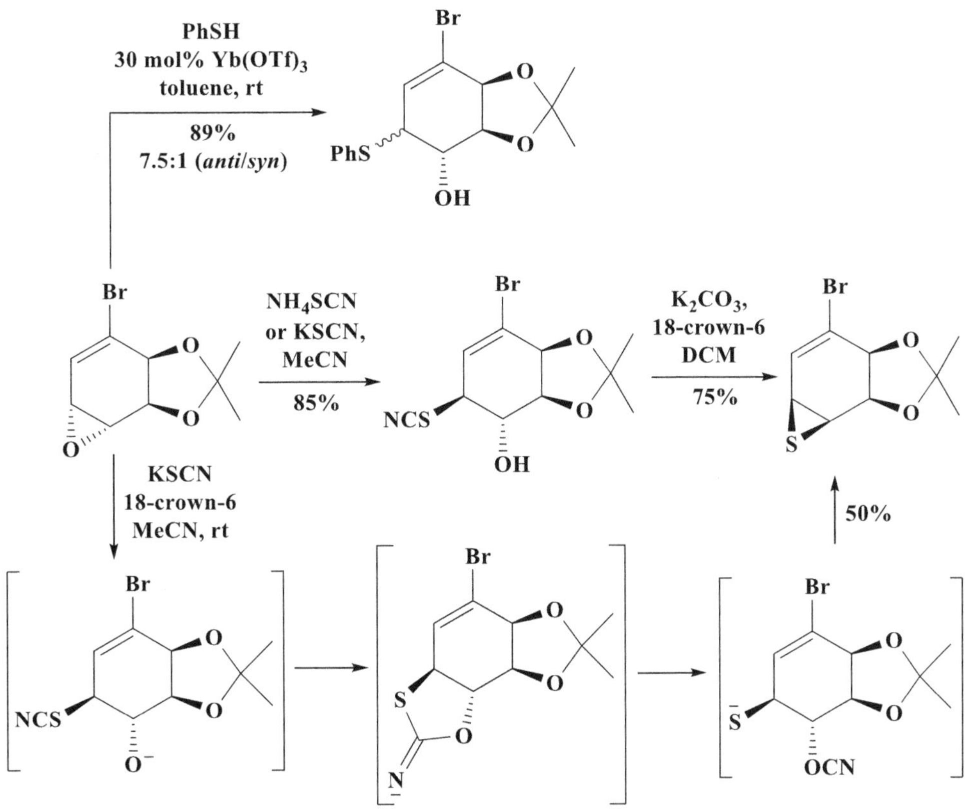

SCHEME 22

The thiocyanogen addition to silylated alkene was utilized in the formation of *trans*-2,3-bis(trimethylsilyl)thiirane and several silylated thiols (Schemes 23 and 24) [59,60].

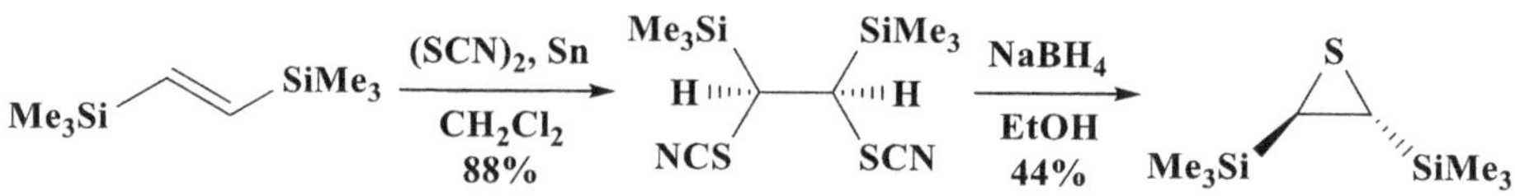

SCHEME 23

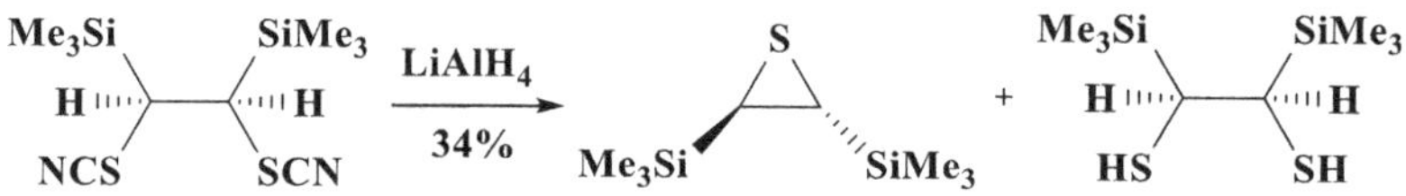

SCHEME 24

Guillerm and Guillerm [61] used allylglycine derivatives as substrates for the formation of conformationally restricted methionine analogs possessing an episulfide functionality. These analogs are considered as useful inhibitors of *S*-adenosyl transferase enzyme. Appropriately protected (*Z*)- and (*E*)-crotylglycine underwent an iodolactonization to synthesize the lactones, which in turn were reacted with sodium thiocyanate (Scheme 25) [62].

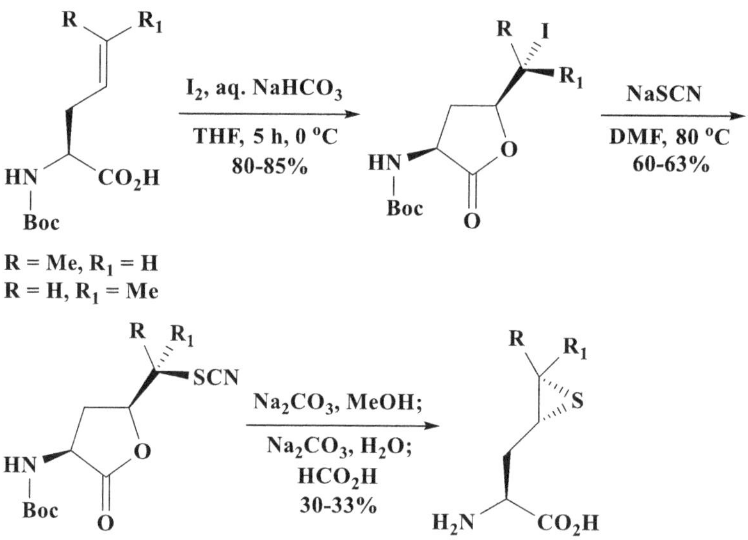

SCHEME 25

The *vic*-diols were chosen for this reaction. The acyclic polyols derived from D-mannitol and chiral glycerine were selected for their structural simplicity. The 1,2-*O*-*i*-propylidenehexofuranoses with diverse substituents at C-3 were selected for the formation of 5-thio sugars. The dimethyl acetals prepared from hexoses and pentoses were chosen for the formation of 4-thiofuranoses and 5-thiopyranoses. The cyclic sulfates of these diols were smoothly obtained in high yields through the cyclic sulfites following the general process reported by Gao and Sharpless [63]. Generally, the diastereomeric mixtures of cyclic sulfites were not separated and were directly utilized in an oxidation reaction with $NaIO_4$–$RuCl_3.3H_2O$. All of these sulfates were stable compounds for 2 months at low temperature (−5°C). The commercial cyclic sulfate prepared from 1,2:5,6-di-*O*-*i*-propylidene-D-mannitol was also utilized along with these cyclic sulfates. The episulfides were synthesized by the nucleophilic ring opening of cyclic sulfates with potassium thioacetate and KSCN. As expected, the nucleophilic attack occurred in terminal cyclic sulfates regiospecifically to the less hindered primary position affording β-thiocyanate or β-acetylthio sulfates. The reactions of cyclic sulfates were carried out with potassium thioacetate or KSCN in dry CH_3COCH_3 at rt, and the corresponding salts were easily isolated in high yields (72%–100%) as stable solids that can be stored under the vacuum for several months. These salts were purified to perform the complete characterization of these compounds, but they were directly utilized without purification for preparative purposes. The reaction of these potassium salts with sodium methoxide–methanol afforded sodium thiolates that resulted in an intramolecular displacement of β-sulfate groups to provide the episulfides in good to high yields (63%–95%). The nucleophilic displacement took place through an inversion of configuration at β-chiral atom (Schemes 26–28) [64].

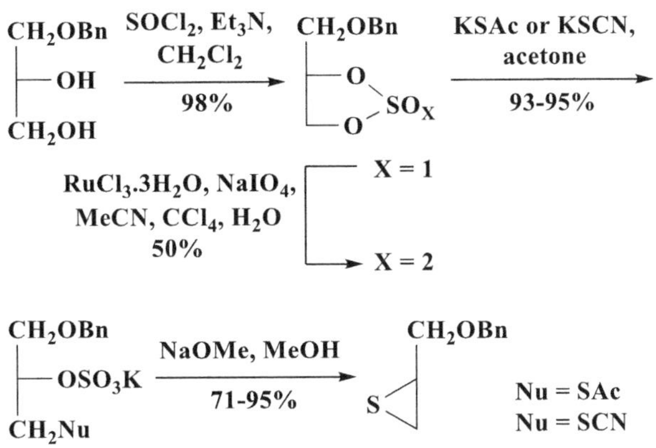

SCHEME 26

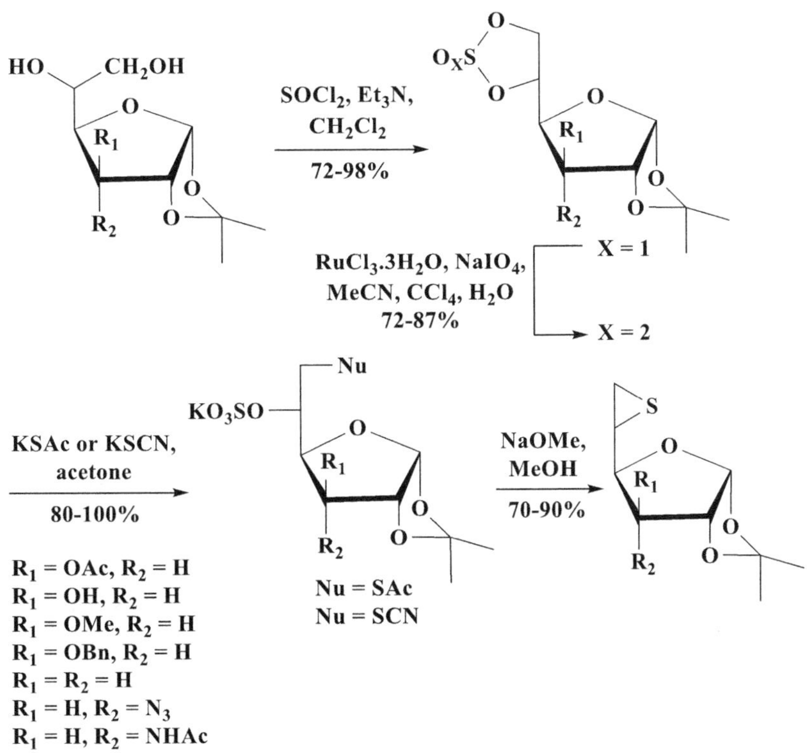

SCHEME 27

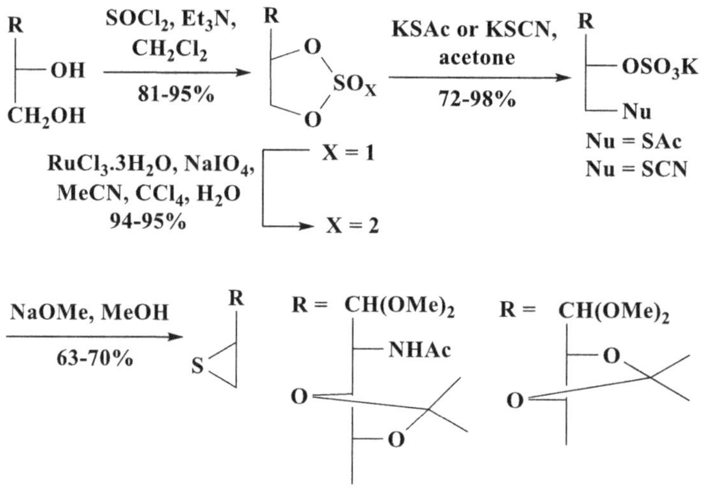

SCHEME 28

An efficient method was developed for the synthesis of thiiranes from oxiranes utilizing I_2 in catalytic amount. The noteworthy characteristics of this process were high conversions, mild reaction conditions, economic feasibility of the reagents, short reaction durations, simple experimental product isolation processes, and compatibility with a number of moieties, which make it a beneficial and attractive procedure for the formation of various thiiranes. In continuation of interest in the catalytic use of elemental iodine [65–67], here a mild, efficient, and applied procedure was reported for the formation of thiiranes from oxiranes and NH_4SCN utilizing molecular iodine as a catalyst. Initially, the reaction of 2-(phenoxymethyl)oxirane was attempted with NH_4SCN using 10 mol% elemental iodine in CH_3CN at rt. The reaction was completed in 2.5 h forming thiirane in 94% yield (Scheme 29) [68].

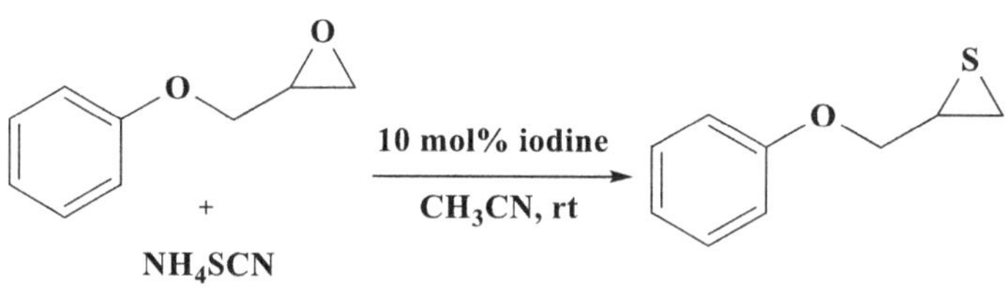

SCHEME 29

The catalytic action of I_2 delivered the incentive for reactions with other epoxides. Various epoxides were reacted efficiently with NH_4SCN to synthesize the thiiranes in good yields. Under these reaction conditions, the functionalized oxiranes also worked well without influencing the protecting groups (Scheme 30). This technique was compatible with precursors having acid-sensitive protecting groups such as tetrahydropyran, benzyl, *t*-butyldimethylsilyl, and methyl ethers. This process was also beneficial for the selective transformation of epoxide into thiirane without influencing an acid labile primary acetonide [68].

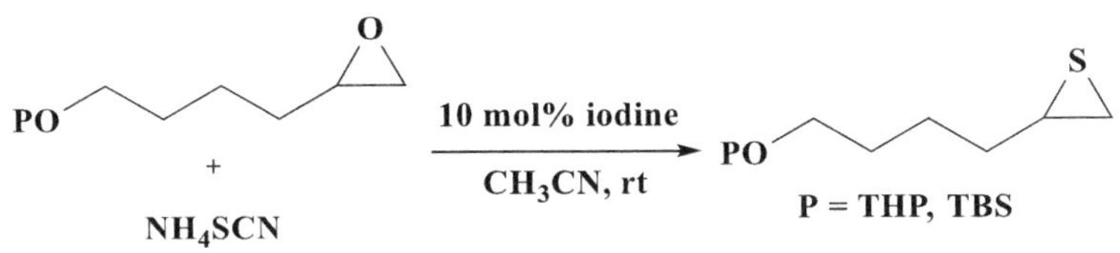

SCHEME 30

An easy and efficient technique for the transformation of oxiranes to thiiranes utilizing NH_4SCN under the catalyst- and solvent-free conditions was reported. These conditions allowed fast and clean transformation of oxiranes to thiirane. The optimum reaction parameters were selected after analyzing various factors of reaction (Scheme 31). In a set of experiments, in the absence of any catalyst, the reaction of styrene oxide as a model precursor with NH_4SCN was examined in various solvents. The 2-phenylthiirane was not formed when the reaction was carried out in DCM, $CHCl_3$, CH_3NO_2, and 1,4-dioxane at rt. The 2-phenylthiirane was formed in 20%–50% yield when the reaction was carried out in $(CH_3)_2CO$, CH_3CN, and THF under the same reaction conditions. The styrene oxide was completely transformed to 2-phenylthiirane after 6.5 h under the solvent-free conditions at rt. The influence of temperature was examined in all the solvents to improve the styrene oxide transformation. The product was not obtained after 24 h even at reflux in DCM, $CHCl_3$, and 1,4-dioxane. However, the transformation of styrene oxide to 2-phenylthiirane was improved considerably in refluxing CH_3COCH_3, CH_3CN, and THF. The refluxing CH_3NO_2 unexpectedly afforded an efficient medium for the transformation of styrene oxide to 2-phenylthiirane in a short reaction time. Therefore, the influence of temperature and molar ratio of styrene oxide/ammonium thiocyanate was examined in CH_3NO_2. Lowering the temperature increased the reaction time when the molar ratios of styrene oxide/ammonium thiocyanate were increased, the reaction completed in a shorter reaction time. The best result was achieved at 90°C with a 1:2 M ratio of styrene oxide/ ammonium thiocyanate in CH_3NO_2. Finally, the optimized conditions (90°C and 1:2 M ratio of styrene oxide/ammonium thiocyanate) were utilized under the solvent-free reaction conditions. The styrene oxide was completely transformed to 2-phenylthiirane with a 1:2 M ratio of styrene oxide/ ammonium thiocyanate at 90°C in the presence of ammonium thiocyanate under the catalyst- and solvent-free conditions. The reaction time was not influenced upon reducing the molar ratio of ammonium thiocyanate/styrene oxide at the same temperature. Depending upon this data, the best molar ratio of styrene oxide/ammonium thiocyanate was 1:1 at 90°C, under the catalyst- and solvent-free conditions, for the synthesis of 2-phenylthiirane. The transformation of styrene oxide to 2-phenylthiiranes under the catalyst- and solvent-free conditions was very sensitive to temperature. The reaction of various oxiranes with NH_4SCN was also examined under the catalyst- and solvent-free conditions at 90°C. The reaction temperatures were lowered to 60–65°C because of the low boiling points of aliphatic oxiranes. A broad range of diverse alkyl and aryl oxiranes were reacted with NH_4SCN to afford an efficient transformation to thiiranes with 100% conversion under these optimized catalyst- and solvent-free conditions. This technique was efficient in the transformation of mono-, di-, and trisubstituted oxiranes to thiiranes [69].

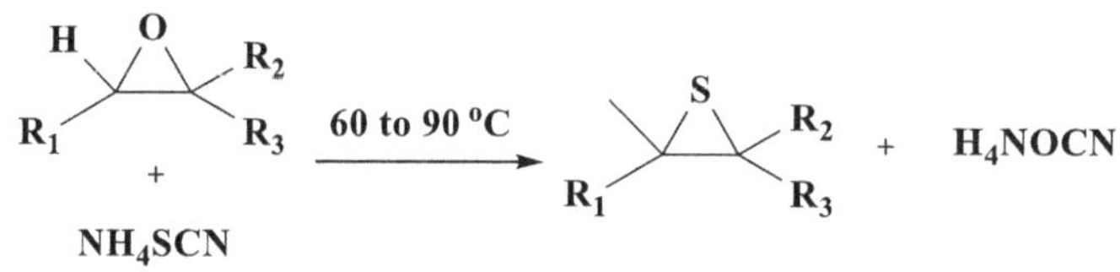

R_1 = Ph, $PhOCH_2$, 4-$CNC_6H_4OCH_2$, 4-$NO_2C_6H_4OCH_2$, 4-$FC_6H_4OCH_2$, 4-$ClC_6H_4OCH_2$, 4-$BrC_6H_4OCH_2$, 4-*t*-$BuC_6H_4OCH_2$, 2-*t*-Bu-4-$MeC_6H_4OCH_2$, 2-Me-4-*t*-$BuC_6H_4OCH_2$, 1-naphthylOCH_2, 2-naphthylOCH_2, $CH_3(CH_2)_3$, $ClCH_2$, $CH_2CHCH_2OCH_2$, $(CH_3)_2CHOCH_2$, $CH_2CH(CH_3)CO(CH_2)_2$

$R_2 = R_3$ = H

SCHEME 31

The silica chloride (SiO_2—Cl), which is easily prepared from $SOCl_2$ and silica gel, is a good solid acid in terms of low cost, convenience, insolubility, and easy synthesis in all organic solvents. Due to its insolubility in organic solvents, excellent in situ proton formation, and its acceptance of various nucleophiles, it could be utilized for various purposes in organic chemistry such as silica chloride-catalyzed protection of carbonyl compounds [70]; acetylation of alcohols [71]; formation of 1,8-dioxo-octahydroxanthenes [72], 14-substituted 14*H*-dibenzo[*a*,*j*]xanthenes, and 2-aminothiazoles [73]; and silica chloride as starting compounds for the formation of other silica-bonded reagents [74]. An easy, efficient, and eco-friendly formation of thiiranes under the solvent-free conditions in the presence of silica chloride is shown in Scheme 32. Initially, under several conditions, a systematic analysis was performed for the estimation of silica chloride as a catalyst for the reaction of NH_4SCN and styrene oxide. The reaction was slow without catalyst, and low yields were reported using only silicon dioxide as a catalyst. The yield increased when silica chloride was utilized as a catalyst. Then, the amount of catalyst (silica chloride) was optimized under the solvent-free conditions at rt. However, inferior results were observed in the presence of solvents. Various thiiranes were synthesized by the reaction of oxiranes with NH_4SCN without solvent depending upon the optimized reaction conditions.

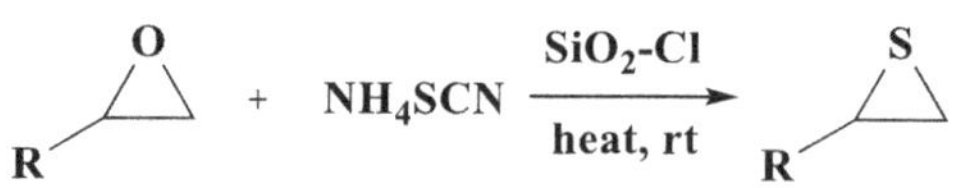

R = Ph, PhCl, PhOMe, CH_2Cl, CH_2OPh, CH_3, CH_2OPhI, $CH_2OPhOMe$

SCHEME 32

10.4 Synthesis of thiiranes from other heterocycles

The thiiranes were synthesized by the pyrolysis of lithium salts of hydrazones derived from thietan-3-ones (Scheme 33) [25,75].

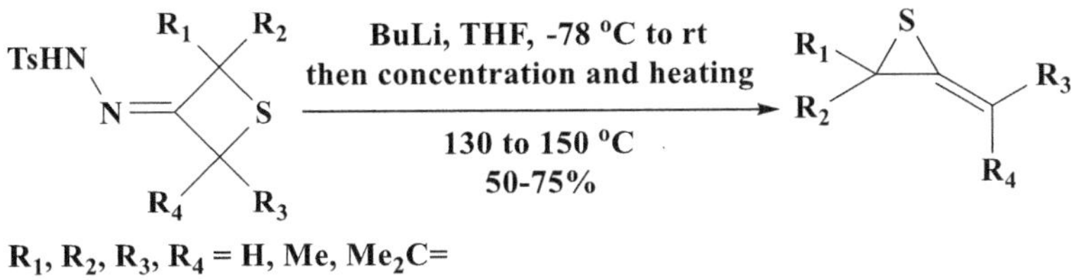

SCHEME 33

The fundamental synthetic methods to episulfides have been reviewed. The ketohydrazones were reacted with sulfur to synthesize the *trans*-3,4-diphenyl-1,2-epithiocyclobutane in 34% yield by short irradiation of *cis*- and *trans*-phenylvinyl sulfide as the first synthesis of cyclobutene episulfide, the pyrolysis of alkyl- or aryl-2-hydroxyethyl thiolcarbonates, the reduction of 2-hydroxyethyl disulfide with phosphines and secondary phosphine oxides, and the reduction of sulfide to an unsaturated episulfide (Scheme 34) [17,52,76].

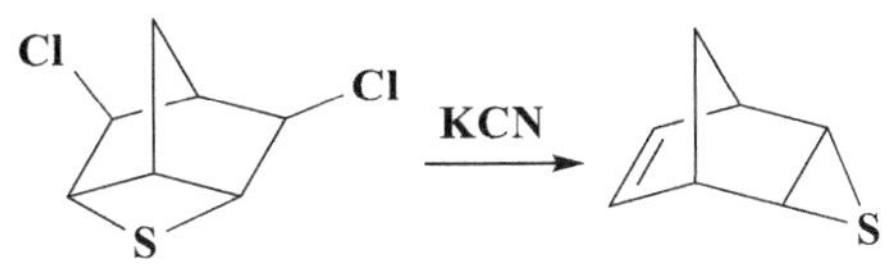

SCHEME 34

The episulfide was prepared in good yield by the thermolysis of thiophene *endo*-peroxide when the strained acetylene cyclooctyne was used as a sulfur acceptor [77]. The labile thiirene was released affording episulfide upon in situ [4 + 2]-cycloaddition with ene dione formed in the thermolysis of *endo*-peroxide (Scheme 35) [35b].

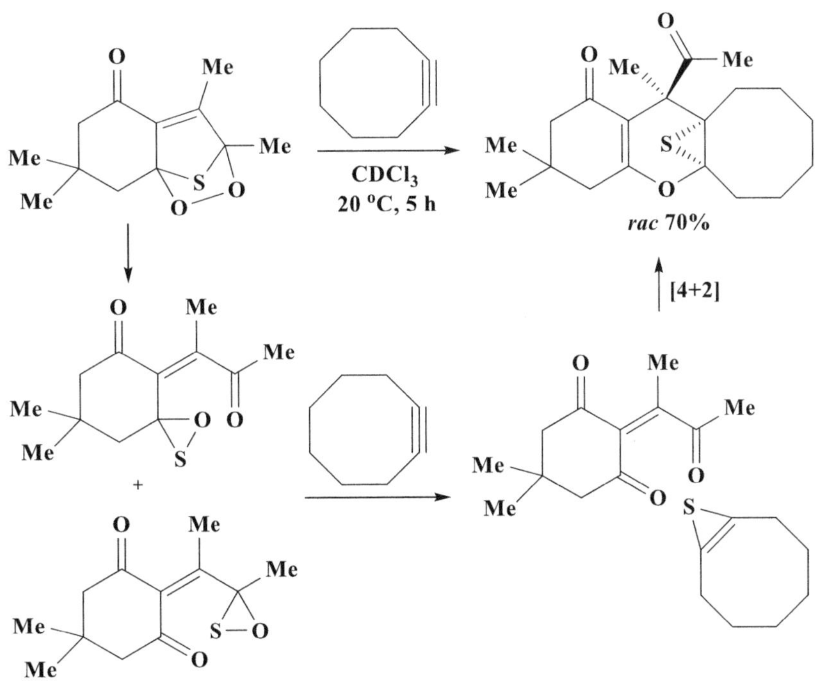

SCHEME 35

In analogous to the formation of oxirane, the halothiols were reacted with bases to synthesize the thiiranes. The β-acetoxythiols also provided thiiranes under the similar conditions. The 1,3-oxathiolan-2-one, prepared by reacting 2-sulfanylethanol with phosgene in the presence of pyridine, was heated to 200°C for decarboxylation providing thiirane (Scheme 36) [78,79].

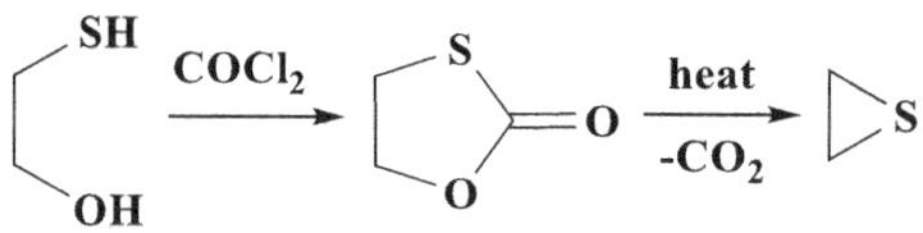

SCHEME 36

The opening of nonterminal cyclic sulfate was not efficient with potassium thioacetate even on changing the experimental conditions (solvent and temperature). However, the episulfide was directly formed in 42% yield when Na_2S was utilized as nucleophile in boiling MeOH (Scheme 37) [64,80,81].

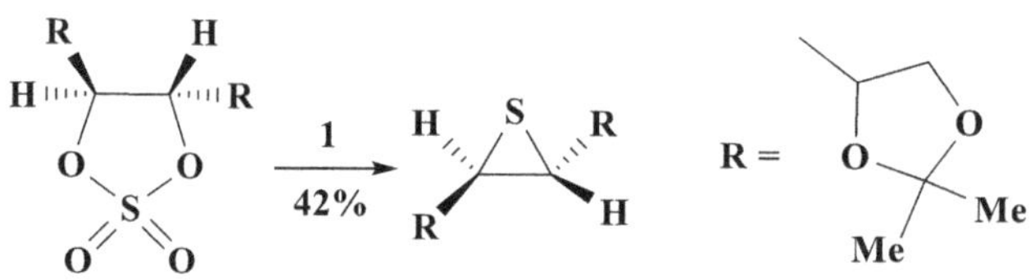

Reagents and conditions: (1) $Na_2S.9H_2O$, methanol, reflux

SCHEME 37

Scheme 38 depicts a facile formation of thiiranes. The desilylation with CsF induced the synthesis of sulfur-stabilized anion equivalent, which was added to aldehyde followed by subsequent deoxygenation to provide the thiirane [82].

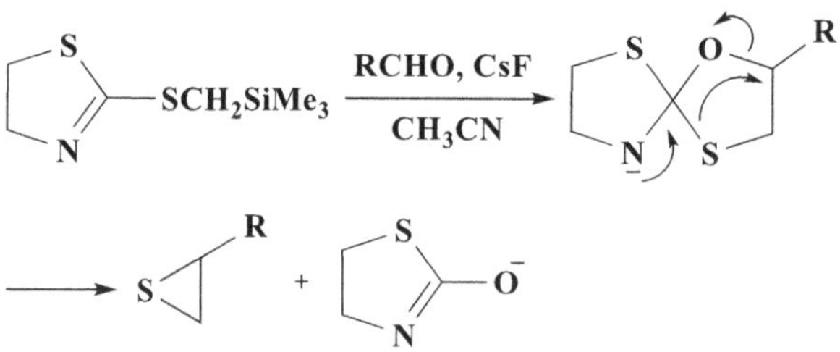

SCHEME 38

The 1,3-dipolar cycloreversion with the removal of nitrogen afforded 1,3-dithiolanes, thiirane, and dithiane (Scheme 39) [83,84].

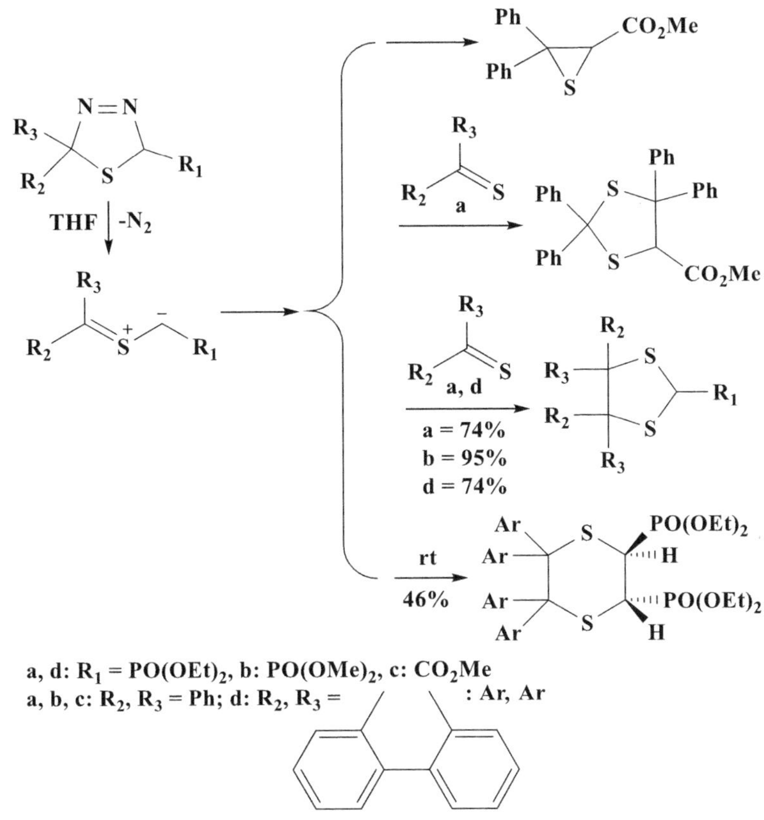

SCHEME 39

The 1,3-dipolar cycloreversion reaction of 2′,5′-dihydrospiroadamantane-2,2’-(1,3,4)-thiadiazole resulted in the removal of nitrogen to afford the adamantanethione *S*-methylide (Scheme 40) [84,85].

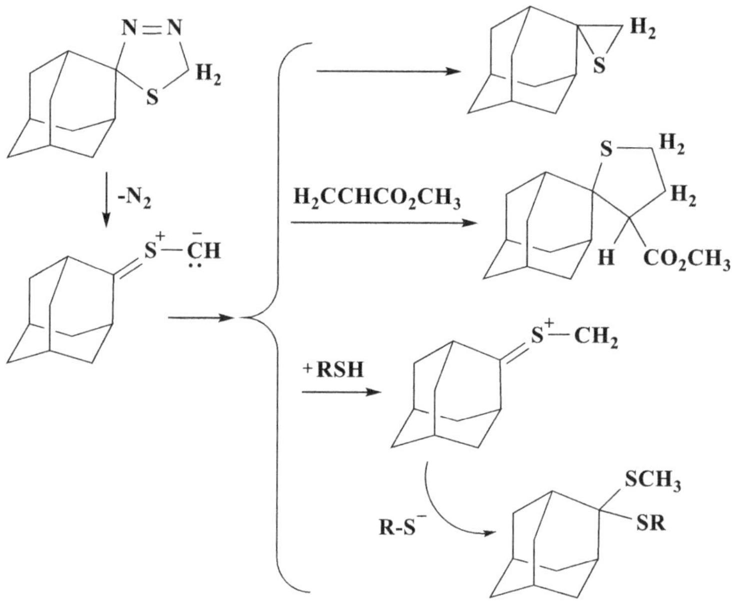

SCHEME 40

Kellogg et al. [86] reported a detailed analysis of substituted unperturbed thione *S*-methylides synthesized by the loss of N_2 from dihydrothiadiazoles (Scheme 41).

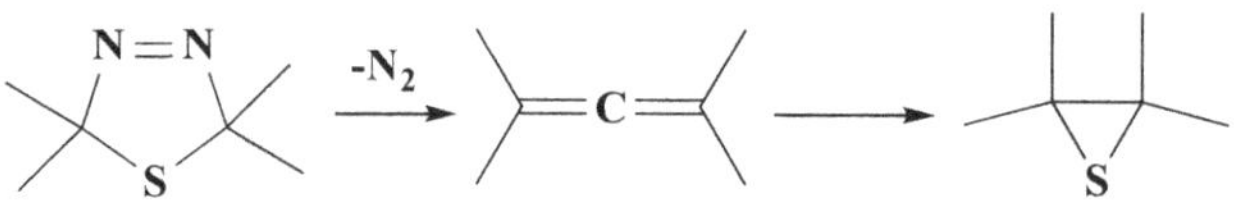

SCHEME 41

If episulfides were strictly similar to their cousin epoxides, their formation would spring mainly from the direct sulphurization of alkenes. This was, however, not the case [35b]. Nonetheless, there are some procedures that worth special attention. For instance, the sultene, an isolable cycloadduct from *trans*-cyclooctene and fluorinethione-*S*-oxide, has been shown to exhibit promising *S*-transfer abilities. The *exo*-episulfide was formed in 83% yield upon the reaction of norbornene with sultene using TFA in catalytic amounts (Scheme 42) [87]. The transfer of sulfur from one episulfide to another alkene was catalyzed with dithiophosphate molybdenum complex. This has allowed the utilization of a more easily accessible precursor as a sulfur donor as represented in the episulfidation of transcyclononene (Scheme 43) [88]. The thiatriazole also worked as a stoichiometric sulfur transfer reagent for a variety of alkene precursors [89]. The dinitrogen sulfide has been implicated as the active sulfur transfer species (Scheme 44) [90].

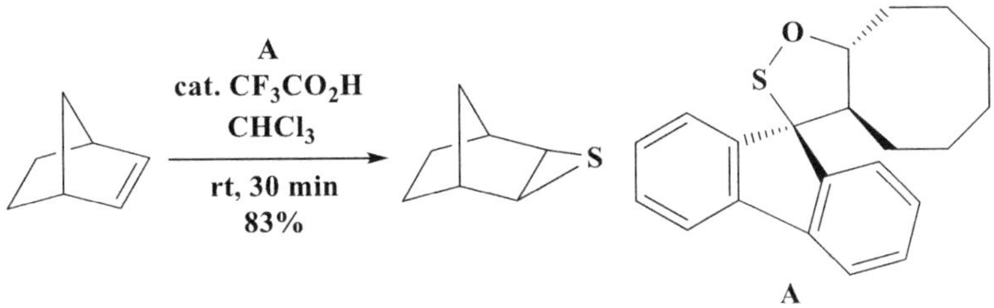

SCHEME 42

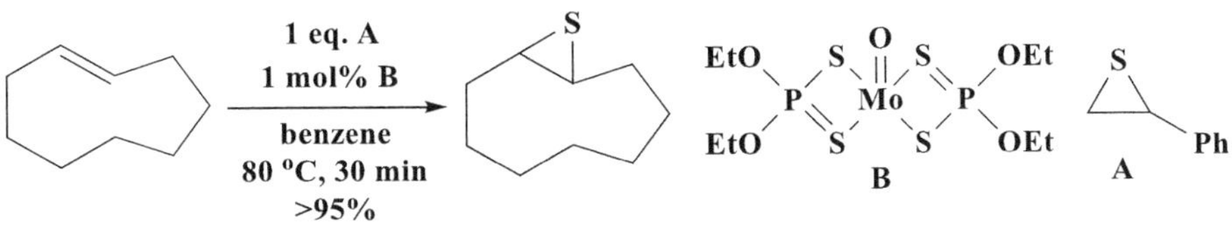

SCHEME 43

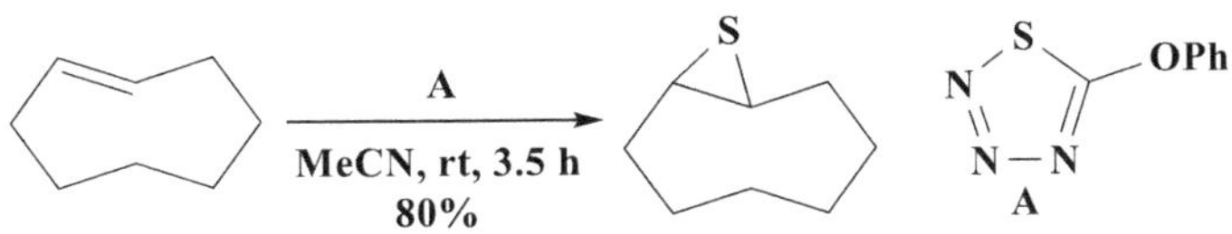

SCHEME 44

The photolysis of various *S*-containing compounds synthesized atomic sulfur, which was trapped by alkenes in the form of episulfides. The 1,2,3,4-thiatriazoles also served this purpose, when photolyzed in liquid cyclohexene [91]; however, dinitrogen sulfide was formed as a sulfur donor in the thermolysis of 1,2,3,4-thiatriazoles [92,93]. Indeed, the episulfides were obtained when (*E*)-cyclooctene and norbornene were utilized. When 5-phenyloxy-1,2,3,4-thiatriazole was thermolyzed at rt, the parent episulfides were isolated in 45% and 80% yield (Scheme 45) [89]. In a hypothetical analysis, it was reported that the released nitrogen sulfide transfers its sulfur atom not in an addition–elimination mechanism via thiiranium ion (like sulfur halides) [94], sultene [87], and thiosulfonates [95] but via a concerted bimolecular nucleophilic substitution reaction route (like DMD [96] and oxathiiranes [97]) [35b,90].

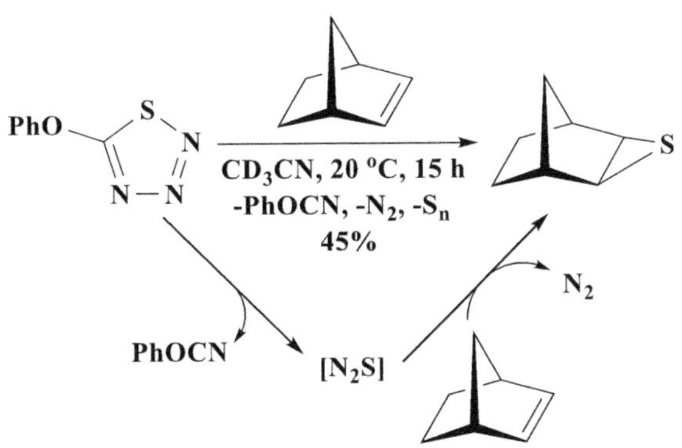

SCHEME 45

References

1. (a) N. Kaur. 2017. Ionic liquids: promising but challenging solvents for the synthesis of *N*-heterocycles. Mini Rev. Org. Chem. 14: 3-23. (b) N. Kaur. 2019. Synthesis of three-membered and four-membered heterocycles with the assistance of photochemical reactions. J. Heterocycl. Chem. 56: 1141-1167. (c) N. Kaur. 2019. Multiple nitrogen-containing heterocycles: metal and non-metal assisted synthesis. Synth. Commun. 49: 1633-1658. (d) N. Kaur. 2018. Synthesis of six- and seven-membered and larger heterocycles using Au and Ag catalysts. Inorg. Nano Met. Chem. 48: 541-568. (e) N. Kaur. 2018. Mercury-catalyzed synthesis of heterocycles. Synth. Commun. 48: 2715-2749. (f) N. Kaur, Y. Verma, P. Grewal, P. Bhardwaj and M. Devi. 2019. Application of titanium catalysts for the syntheses of heterocycles. Synth. Commun. 49: 1847-1894. (g) N. Kaur. 2019. Cobalt-catalyzed C-N, C-O, C-S bond formation: synthesis of heterocycles. J. Iran. Chem. Soc. 16: 2525-2553. (h) N. Kaur, P. Bhardwaj and M. Gupta. 2021. Recent developments in the synthesis of five- and six-membered *N*-heterocycles from dicarbonyl compounds. Curr. Org. Chem. 25: 2765-2790. (i) N. Kaur, P. Grewal and K. Poonia. 2021. Dicarbonyl compounds in *O*-heterocycle synthesis. Synth. Commun. 51: 2423-2444. (j) N. Kaur, M. Rajput and P. Bhardwaj. 2022. Synthesis of heterocycles using guanidine: an overview. Synth. Commun. 52: 1547-1580. (k) N. Kaur. 2019. Copper catalyzed synthesis of seven and higher-membered heterocycles. Synth. Commun. 49: 879-916. (l) N. Kaur. 2019. Synthesis of seven and higher-membered heterocycles using ruthenium catalysts. Synth. Commun. 49: 617-661. (m) N. Kaur. 2018. Ruthenium catalysis in six-membered *O*-heterocycles synthesis. Synth. Commun. 48: 1551-1587. (n) N. Kaur, P. Bhardwaj, M. Devi, Y. Verma and P. Grewal. 2019. Gold-catalyzed C-O bond forming reactions for the synthesis of six-membered *O*-heterocycles. SN Appl. Sci. 1: 1-37. (o) N. Kaur. 2017. Methods for metal and non-metal-catalyzed synthesis of six-membered oxygen-containing polyheterocycles. Curr. Org. Synth. 14: 531-556.
2. S.C. Sahu. 2002. Dual role of organosulfur compounds in foods: a review. J. Environ. Sci. Health C 20: 61-76.
3. H. Staudinger and J. Siegwart. 1920. Einwirkungen von aliphatischen diazoverbindungen auf thioketone. Helv. Chim. Acta 3: 833-840.
4. W. Chew and D.N. Harpp. 1993. Recent aspects of thiirane chemistry. J. Sulfur Chem. 15: 1-39.
5. P.K. Kundu, R. Singh and S.K. Ghosh. 2009. Silicon assisted diversified reaction of a β-silylmethylene malonate with dimethylsulfoxonium methylide. J. Organomet. Chem. 694: 382-388.
6. C. Puke, G. Erker, N.C. Aust, E.-U. Wurthwein and R. Frohlich. 1998. Evidence for a continuous transition between thiaphosphetane and betaine-type structures in the thio-Wittig reaction. J. Am. Chem. Soc. 120: 4863-4864.
7. B. Rickborn. 1989. Isobenzofurans. Adv. Theor. Interesting Mol. 1: 1-134.
8. J.M. Beiner, D. Lecadet, D. Paquer, A. Thuillier and J. Vialle. 1973. Organoschwefelverbindungen 42. Mitt. rk. von diazomethan mit thiocarbonylverbindungen, synthese von thiiranen 1. mitt. Bull. Soc. Chim. Fr. 6: 1979-1983.
9. G. Mloston, K. Urbaniak, M. Gulea, S. Masson, A. Linden and H. Heimgartner. 2005. [2+3]-Cycloadditions of phosphonodithioformate *S*-methanides with C=S, N=N, and C=C dipolarophiles. Helv. Chim. Acta 88: 2582-2592.
10. M.F. Ward. 2000. Heterocyclic chemistry. Annu. Rep. Prog. Chem. Sect. B 96: 157-186.
11. L. Hatjiarapoglou and A. Varvoglis. 1989. Reactivity of phenyliodonium bis(aryl- or alkyl-sulphonyl) methylides towards thiobenzophenones. J. Chem. Soc. Perkin Trans. 1 3: 379-382.
12. N. Tokitoh, T. Suzuki and W. Ando. 1989. A facile synthesis and novel reactions of 1,2,3,4-pentatetraene episulfides. Tetrahedron Lett. 30: 4271-4274.
13. E. Bergmann, M. Magat and D. Wagenberg. 1930. Beiträge zur kenntnis der doppelten bindung, III.: Über aromatische thio-ketone, insbesondere ihre reaktion mit diazomethan. Chem. Ber. 63: 2576-2584.

14. R. Huisgen and E. Langhals. 1989. Thiones as superdipolarophiles. Tetrahedron Lett. 30: 5369-5372.
15. Y.G. Gololobov and A.N. Nesmeyanov. 1987. Twenty-five years of dimethylsulfoxonium ethylide (Corey's reagent). Tetrahedron 43: 2609-2651.
16. A.-H. Li, L.-X. Dai and V.K. Aggarwal. 1997. Asymmetric ylide reactions: epoxidation, cyclopropanation, aziridination, olefination, and rearrangement. Chem. Rev. 97: 2341-2372.
17. (a) N. Latif, I. Fathy, N. Mishriky and B. Haggag. 1966. Utilization of sulfur: synthesis of symmetrical thiiranes from ketohydrazones. Can. J. Chem. 44: 629-632. (b) T. Eicher and S. Hauptmann. 2003. The chemistry of heterocycles: structures, reactions, and applications. John Wiley & Sons: Chichester, UK, 5-16. (c) R.E. Banks and J.C. Tatlow. 1994. Organofluorine chemistry: nomenclature and history. R.E. Banks, B.E. Smart and J.C. Tatlow (Eds.). Organofluorine chemistry: principles and commercial applications. Plenum Press: New York, 2-4. (d) C.S. Adams, C.D. Weatherly, E.G. Burke and J.M. Schomaker. 2014. The conversion of allenes to strained three-membered heterocycles. Chem. Soc. Rev. 43: 3136-3163. (e) F. Lautenschlaeger and N.V. Schwartz. 1969. Synthesis of episulfides from olefins and sulfur monochloride. J. Org. Chem. 34: 3991-3998.
18. T. Shioiri, Y. Iwamoto and T. Aoyama. 1987. New methods and reagents in organic synthesis. 68. Reaction of thioketones with lithium trimethylsilyldiazomethane. Heterocycles 26: 1467-1470.
19. S.K. Richardson and A.R. Howell. 2007. Synthesis and reactions of 2-alkylidene thiiranes and thietanes. Synthesis 18: 2755-2778.
20. A.G. Hortmann and A. Bhattacharjya. 1976. Heterocyclobutanylidenes. A unique synthesis of tetramethylallene episulfide. J. Am. Chem. Soc. 98: 7081-7082.
21. H. Quast and A. Fuss. 1981. Tetramethylcyclopropanone and isopropylidene(dimethyl)thiirane via photolysis of reluctant 3,3,5,5-tetramethyl-1-pyrazolines; influence of solvent and temperature on the competition between [3+2]- and [4+1]-photocycloelimination. Angew. Chem. Int. Ed. 20: 291-292.
22. H. Nakano and T. Ibata. 1995. The rhodium(II) acetate-catalyzed reaction of alkenyl and alkynyl α-diazoacetates with thioketene. Bull. Chem. Soc. Jpn. 68: 1393-1400.
23. W. Ando, Y. Hanyu, T. Furuhata and T. Takata. 1983. Thioxyallyl ion from allene episulfide with acid. J. Am. Chem. Soc. 105: 6151-6152.
24. W. Ando, T. Furuhata, Y. Hanyu and T. Takata. 1984. A new strategy to sulfur heterocycles: reactions of an allene episulfide with electrophiles. Tetrahedron Lett. 25: 4011-4014.
25. T. Furuhata and W. Ando. 1986. Nature of thioxyallyl intermediate in the decomposition of allene episulfide and pyrazoline-4-thione. Tetrahedron 42: 5301-5308.
26. T. Furuhata and W. Ando. 1986. MCSCF studies of thioxyallyl intermediate. Tetrahedron Lett. 27: 4035-4038.
27. N. Tokitoh, N. Choi and W. Ando. 1987. Effect of silyl substituent on thermal isomerization of allene episulfide. A new aspect in thioxyallyl intermediate. Chem. Lett 16: 2177-2180.
28. W. Ando, A. Itami, T. Furuhata and N. Tokitoh. 1987. Isomerization and desulfurization of allene episulfides. Tetrahedron Lett. 28: 1787-1790.
29. N. Tokitoh, A. Itami and W. Ando. 1988. Oxidation of allene episulfide. Selective sulfur oxidation and novel isomerization of resulting *S*-oxide. Chem. Lett. 17: 1501-1504.
30. N. Tokitoh, N. Choi, M. Goto and W. Ando. 1989. Formation of cyclopentenethiones via cyclization of beta-thioallyl cations. J. Org. Chem. 54: 4660-4663.
31. W. Ando, N. Choi and Y. Kabe. 1990. (Thioallyl)iron tricarbonyl complexes by the reaction of allene episulfides with diiron nonacarbonyl. J. Am. Chem. Soc. 112: 4574-4576.
32. N. Choi, Y. Kabe and W. Ando. 1991. Palladium(0)-catalyzed intra- and intermolecular cyclization reactions of allene episulfides. Tetrahedron Lett. 32: 4573-4576.

33. M.E. Furrow and A.G. Myers. 2004. Practical procedures for the preparation of *N-tert*-butyldimethylsilylhydrazones and their use in modified Wolff-Kishner reductions and in the synthesis of vinyl halides and *gem*-dihalides. J. Am. Chem. Soc. 126: 5436-5445.
34. M.E. Furrow and A.G. Myers. 2004. A general procedure for the esterification of carboxylic acids with diazoalkanes generated in situ by the oxidation of *N-tert*-butyldimethylsilylhydrazones with (difluoroiodo) benzene. J. Am. Chem. Soc. 126: 12222-12223.
35. (a) W. Adam, R.M. Bargon and G. Mloston. 2003. Episulfidation of *trans*-cyclooctene with an 1,2,4-oxadithiolane. Eur. J. Org. Chem. 20: 4012-4015. (b) W. Adam and R.M. Bargon. 2004. Synthesis of thiiranes by direct sulfur transfer: the challenge of developing effective sulfur donors and metal catalysts. Chem. Rev. 104: 251-261.
36. M. Baltas, K. Raouf-Benchekroun, A. de Blic, L. Cazaux, P. Tisnès, L. Gorrichon, K. Hussein and J.-C. Barthelat. 1996. Aminolysis of sulfinamoyl-esters, -sulfonamides and -sulfones. Thiooxamate and thiourea formation via a sulfine intermediate. Thiophilic or carbophilic reaction? Tetrahedron 52: 14865-14876.
37. M. Baltas, L. Cazaux, L. Gorrichon, P. Maroni and P. Tisnès. 1988. Sulphinamoylacetates as sulphine precursors. Mechanism of basic hydrolysis and scheme of irreversible inactivation of cinnamoyl alcohol dehydrogenase, an enzyme of the lignification process. J. Chem. Soc. Perkin Trans. 2 8: 1473-1478.
38. M. Baltas, L. Cazaux, A. de Blic, L. Gorrichon and P. Tisnès. 1990. Synthesis of thiooxamates from *t*-butyl sulfinamoyl acetates. Occurrence of a new rearrangement involving a thione *S*-imide intermediate. Tetrahedron Lett. 31: 4453-4456.
39. S. Kosack and G. Himbert. 1988. α-Sulfinylamidine aus der reaktion von inaminen mit *N*-(sulfinyl)-arensulfonamiden. Chem. Ber. 121: 833-841.
40. K.-P. Pfeifer and G. Himbert. 1990. The retro-Diels-Alder-reactivity of 1,4,3-oxathiazin-4-oxides. Tetrahedron Lett. 31: 5725-5728.
41. G. Himbert, K.-P. Pfeifer and C.E. Finkele. 1991. α-Acyl-sulfines by reaction of acetylenes with *N*-sulfinylamides. Phosphorus, Sulfur, Silicon Relat. Elem. 59: 125-128.
42. B.F. Bonini and G. MacCagnani. 1973. Episulphoxide formation from sulphines. Tetrahedron Lett. 14: 3585-3588.
43. J.F. King, E.G. Lewars and L.J. Danks. 1972. Organic sulfur mechanisms. 12. Reaction of sulfenes with triphenylphosphine. Can. J. Chem. 50: 866-870.
44. E. Schaumann. 1988. The chemistry of thioketens. Tetrahedron 44: 1827-1871.
45. M.F. Zipplies, M.J. de Vos and T.C. Bruice. 1985. The formation of thiiranes from olefins in the course of the deoxygenation of tertiary amine *N*-oxides by carbon disulfide. J. Org. Chem. 50: 3228-3230.
46. T. Hudlicky, J.D. Price, F. Rulin and T. Tsunoda. 1990. Efficient and enantiodivergent synthesis of (+)- and (-)-pinitol. J. Am. Chem. Soc. 112: 9439-9440.
47. T. Hudlicky, X. Tian, M.R. Königsberger, J. Rouden and B. Fan. 1996. Toluene dioxygenase-mediated *cis*-dihydroxylation of aromatics in enantioselective synthesis. Asymmetric total syntheses of pancratistatin and 7-deoxypancratistatin, promising antitumor agents. J. Am. Chem. Soc. 118: 10752-10765.
48. T. Hudlicky, D. Gonzalez and D.T. Gibson. 1999. Enzymatic dihydroxylation of aromatics in enantioselective synthesis: expanding asymmetric methodology. Aldrichimica Acta 32: 35-62.
49. K. Ramesh, M.S. Wolfe, Y. Lee, D.V. Velde and R.T. Borchardt. 1992. Synthesis of hydroxylated cyclohexenyl- and cyclohexanyladenines as potential inhibitors of *S*-adenosylhomocysteine hydrolase. J. Org. Chem. 57: 5861-5868.
50. T.C. Nugent and T. Hudlicky. 1998. Chemoenzymatic synthesis of all four stereoisomers of sphingosine from chlorobenzene: glycosphingolipid precursors. J. Org. Chem. 63: 510-520.
51. T. Hudlicky, H. Luna, H.F. Olivo, C. Andersen, T.C. Nugent and J.D. Price. 1991. Biocatalysis as the strategy of choice in the exhaustive enantiomerically controlled synthesis of conduritols. J. Chem. Soc. Perkin Trans. 1 12: 2907-2917.

52. M. Sander. 1966. Thiiranes. Chem. Rev. 66: 297-339.
53. E.E. van Tamelen. 1951. The formation and ring-opening of alkene sulfides. J. Am. Chem. Soc. 73: 3444-3448.
54. K.C. Nicolaou, C.V.C. Prasad, P.K. Somers and C.K. Hwang. 1989. Activation of 6-*endo* over 5-*exo* hydroxy epoxide openings. Stereoselective and ring selective synthesis of tetrahydrofuran and tetrahydropyran systems. J. Am. Chem. Soc. 111: 5330-5334.
55. A. Bellomo and D. Gonzalez. 2006. Catalytic thiolysis of chemoenzymatically derived vinylepoxides. Efficient synthesis of homochiral phenylthioconduritol F. Tetrahedron: Asymmetry 17: 474-478.
56. A. Bellomo, C. Giacomini, B. Brena, G. Seoane and D. Gonzalez. 2007. Chemoenzymatic synthesis and biological evaluation of (-)-conduramine C-4. Synth. Commun. 37: 3519-3528.
57. A. Bellomo, D.G.A. Bellomo and D. Gonzalez. 2007. Diasterodivergent synthesis of optically pure vinyl episulfides and β-hydroxy thiocyanates from a bacterial metabolite. Tetrahedron Lett. 48: 3047-3051.
58. J. He, J. Ling and P. Chiu. 2014. Vinyl epoxides in organic synthesis. Chem. Rev. 114: 8037-8128.
59. E.J. Corey, F.A. Carey and R.A.E. Winter. 1965. Stereospecific syntheses of olefins from 1,2-thionocarbonates and 1,2-trithiocarbonates. *trans*-Cycloheptene. J. Am. Chem. Soc. 87: 934-935.
60. E.J. Corry and J.I. Shulman. 1968. The direct synthesis of optically active-cyclooctene. Optically active *trans*-bicyclo[6.1.0]nonane. Tetrahedron Lett. 9: 3655-3658.
61. D. Guillerm and G. Guillerm. 1992. Enantioselective synthesis of episulfide analogues of L-methionine. Tetrahedron Lett. 33: 5047-5050.
62. H.K. Chenault, J. Dahmer and G.M. Whitesides. 1989. Kinetic resolution of unnatural and rarely occurring amino acids: enantioselective hydrolysis of *N*-acyl amino acids catalyzed by acylase I. J. Am. Chem. Soc. 111: 6354-6364.
63. Y. Gao and K.B. Sharpless. 1988. Vicinal diol cyclic sulfates. Like epoxides only more reactive. J. Am. Chem. Soc. 110: 7538-7539.
64. F.G. Calvo-Flores, P. Garcia-Mendoza, F. Hernandez-Mateo, J. Isac-Garcia and F. Santoyo-Gonzalez. 1997. Applications of cyclic sulfates of *vic*-diols: synthesis of episulfides, olefins, and thio sugars. J. Org. Chem. 62: 3944-3961.
65. J.S. Yadav, B.V.S. Reddy, U.V.S. Reddy and A.D. Krishna. 2007. Iodine/MeOH as a novel and versatile reagent system for the synthesis of α-ketothiocyanates. Tetrahedron Lett. 48: 5243-5246.
66. J.S. Yadav, B.V.S. Reddy, K. Premalatha and T. Swamy. 2005. First example of the activation of polymethylhydrosiloxane with molecular iodine: a facile synthesis of 3,6-dihydropyran derivatives. Tetrahedron Lett. 46: 2687-2690.
67. J.S. Yadav, B.V.S. Reddy, P.M.K. Reddy and M.K. Gupta. 2005. Mild and efficient method for the cleavage of cyclic and acyclic ethers by iodine under solvent-free conditions. Tetrahedron Lett. 46: 8493-8495.
68. J.S. Yadav, B.V.S. Reddy, S. Sengupta, M.K. Gupta, G. Baishya, S.J. Harshavardhana and U. Dash. 2008. Iodine as a mild, efficient, and cost-effective catalyst for the synthesis of thiiranes from oxiranes. Monatsh. Chem. 139: 1363-1367.
69. B. Akhlaghinia, M. Rahimizadeh, H. Eshghi, S. Zhaleh and S. Rezazadeh. 2012. Green synthesis of thiiranes from oxiranes under solvent- and catalyst-free conditions. J. Sulfur Chem. 33: 351-361.
70. Y. Kamitori, M. Hojo, R. Masuda, T. Kimura and T. Yoshida. 1986. Selective protection of carbonyl compounds. Silica gel treated with thionyl chloride as an effective catalyst for thioacetalization. J. Org. Chem. 51: 1427-1431.
71. H. Firouzabadi, N. Iranpoor and H. Hazarkhani. 2001. Silica chloride (SiO_2-Cl) promotes highly efficient transformation of acylals to dithianes, dithiolanesm, and oxathiolanes. Phosphorus, Sulfur, Silicon Relat. Elem. 176: 165-171.

72. L.Q. Wu, C.G. Yang, L.M. Yang and L.J. Yang. 2010. Synthesis of 14-substituted-14*H*-dibenzo[*a,j*]xanthenes and 1,8-dioxo-octahydroxanthenes using silica chloride (SiO_2-Cl) under solvent-free conditions. Phosphorus, Sulfur, Silicon Relat. Elem. 185: 903-909.
73. H. Karade, M. Sathe and M P. Kaushik. 2007. An efficient method for the synthesis of 2-aminothiazoles using silica chloride as a heterogeneous catalyst. Catal. Commun. 8: 741-746.
74. P. Salehi, M.A. Zolfigol, F. Shirini and M. Baghbanzadeh. 2006. Silica sulfuric acid and silica chloride as efficient reagents for organic reactions. Curr. Org. Chem. 10: 2171-2189.
75. W. Ando, Y. Hanyu, Y. Kumamoto and T. Takata. 1986. Synthesis of diisopropylidenethiirane (thiiranoradialene) and its reactions. Tetrahedron 42: 1989-1994.
76. M. Graysonand and C.F. Farley. 1967. Reduction of 2-hydroxyethyl disulphide to ethylene sulphide by phosphines and secondary phosphine oxides. Chem. Commun. 16: 831-832.
77. W. Adam, S.G. Bosio, B. Frohling, D. Leusser and D. Stalke. 2002. Unusual sulfur chemistry in the thermal reaction of sultene and thiophene endoperoxide sulfur donors with cyclic alkynes: reversible formation of a persistent thiirenium ion and trapping of a thiirene by [4+2] cycloaddition. J. Am. Chem. Soc. 124: 8316-8320.
78. A.V. Fokin, M.A. Allakhverdiev and A.F. Kolomiets. 1990. News in the chemistry of the thiiranes. Usp. Chim. 59: 705-737.
79. I. Mohammadpoor-Baltork, A.R. Khosropour and H. Aliyan. 2001. Efficient conversion of epoxides to 1,3-dioxolanes catalyzed by bismuth(III) salts. Synth. Commun. 31: 3411-3416.
80. A. Hercouet, B. Bessieres and M. Lecorre. 1996. First asymmetric synthesis of (-)-(2*S*,3*R*)-methanoproline. Tetrahedron: Asymmetry 7: 1267-1268.
81. A. Hercouet, B. Bessieres and M. Lecorre. 1996. Expedient synthesis of (-)-(1*S*,2*R*)-allonorcoronamic acid. Tetrahedron: Asymmetry 7: 283-284.
82. Y. Tominaga, H. Ueda, K. Ogata, S. Kohra, M. Hojo, M. Ohkuma, K. Tomita and A. Hosomi. 1992. New synthesis of thiiranes by fluoride ion-promoted reaction of *S*-methyl-*S*'-trimethylsilylmethyl *N*-(*p*-toluenesulfonyl)dithioiminocarbonate and 2-(trimethylsilylmethylthio)thiazoline with aldehydes. Tetrahedron Lett. 33: 85-88.
83. A.C. Gaumont, M. Gulea and J. Levillain. 2009. Overview of the chemistry of 2-thiazolines. Chem. Rev. 109: 1371-1401.
84. Y. Hu, C.-Y. Li, X.-M. Wang, Y.-H. Yang and H.-L. Zhu. 2014. 1,3,4-Thiadiazole: synthesis, reactions, and applications in medicinal, agricultural, and materials chemistry. Chem. Rev. 114: 5572-5610.
85. G. Mloston and R. Huisgen. 2001. Acid-base reactions of adamantanethione *S*-methylide and its spiro-1,3,4-thiadiazoline precursor. Tetrahedron 57: 145-151.
86. R.W. Kellogg, S. Wassenaar and J. Butler. 1972. Thiocarbonyl ylides. Generation, properties, and reactions. J. Org. Chem. 37: 4045-4060.
87. W. Adam and S. Weinkotz. 1998. 1,3-Dipolar cycloaddition of fluorenethiones-oxide and *trans*-cyclooctene: isolation of a sultene and its novel acid-catalyzed diastereoselective sulfur-atom transfer to cyclic olefins. J. Am. Chem. Soc. 120: 4861-4862.
88. W. Adam, R.M. Bargon and W.A. Schenk. 2003. Direct episulfidation of alkenes and allenes with elemental sulfur and thiiranes as sulfur sources, catalyzed by molybdenum oxo complexes. J. Am. Chem. Soc. 125: 3871-3876.
89. W. Adam and R.M. Bargon. 2001. Episulfidation of strained cycloalkenes in the thermolysis of 5-aryloxy-1,2,3,4-thiatriazoles. Eur. J. Org. Chem. 10: 1959-1962.
90. Z.-X. Yu and Y.-D. Wu. 2003. An S_N2-like transition state for alkene episulfidation by dinitrogen sulfide. J. Org. Chem. 68: 6049-6052.

91. R. Okazaki, K. Okawa, S. Weijiki and N. Inamoto. 1971. Photolytic and thermal decompositions of 5-substituted 1,2,3,4-thiatriazoles in the presence of olefins. Reactions of sulfur atoms. Bull. Chem. Soc. Jpn. 44: 3167-3170.
92. C. Wentrup and P. Kambouris. 1991. *N*-Sulfides. Dinitrogen sulfide, thiofulminic acid, and nitrile sulfides. Chem. Rev. 91: 363-373.
93. C. Wentrup and R. Flamang. 1998. Studies of reactive intermediates using matrix and gas-phase techniques. J. Phys. Org. Chem. 11: 350-355.
94. M. Fachini, V. Lucchini, G. Modena, M. Pasi and L. Pasquato. 1999. Nucleophilic reactions at the sulfur of thiiranium and thiirenium ions. New insight in the electrophilic additions to alkenes and alkynes. Evidence for an episulfurane intermediate. J. Am. Chem. Soc. 121: 3944-3950.
95. J.-B. Baudin, M.-G. Commenil, S.A. Julia and Y. Wang. 1996. A convergent synthesis of γ-unsaturated thioaldehyde- and thioketone-*S*-oxides. Bull. Soc. Chim. Fr. 133: 515-529.
96. W. Adam, R. Curci, L. D'Accolti, C. Dinoi, C. Fusco, F. Gasparrini, R. Kluge, R. Paredes, M. Schulz, A.K. Smerz, L.A. Veloza, S. Weinkotz and R. Winde. 1997. Epoxidation and oxygen insertion into alkane C-H bonds by dioxirane do not involve detectable radical pathways. Chem. Eur. J. 3: 105-109.
97. W. Adam, O. Deeg and S. Weinkotz. 1997. Direct thioepoxidation of strained cyclic alkenes by the photolytic sulfur-atom transfer from thiocarbonyl *S*-oxides (sulfines). J. Org. Chem. 62: 7084-7085.

CHAPTER

11 Photochemical synthesis of diaziridines and azirines

Chapter outline

11.1 Introduction

Three-membered heterocyclic rings offer an unusual combination of synthetic flexibility, reactivity, and atom economy. The properties of three-membered heterocyclic compounds are mostly a result of the great bond angle strain (Baeyer strain). This ring strain imparts to the high chemical reactivity of product. The ring opening leading to acyclic compounds is typical. The parent system of three-membered heterocyclic compounds with one oxygen atom is oxirene [1]; the saturated heterocyclic compound is oxirane.

The three-membered heterocyclic compounds are invested with a special allure that is derived from their apparent simplicity and spartan architecture. Yet these systems are multifaceted, and the literature continues to afford a proof of their diversity in terms of both synthetic pathways and consequent conversions. These smallest heterocyclic compounds also show a synthetically very beneficial balance between reactivity and stability. Therefore, they are frequently used as selective and versatile intermediates. With the ability to add two adjacent chiral centers with high atom economy, this procedure deserves a prominent place in synthetic organic chemistry [2a].

While many members of the aziridine ring systems are fully characterized and known, derivatives of the azirine ring system are majorly known as important intermediates, and only limited instances of naturally occurring azirine derivatives have been reported. The big focus in azirines exists because of their high reactivity and potential employment in synthesis, their role as substrate for more complex heterocyclic compounds, and their employment in the formation of different amines and aziridines. The azirines are divided into 1*H*-azirines with a carbon—carbon double bond and 2*H*-azirines with a carbon—nitrogen double bond. The ring strain energy of unsubstituted 2*H*-azirines has been estimated to be 48 kcal/mol, while 1*H*-azirines were observed to be 33—37 kcal/mol higher in energy. Comparable values for the aziridine are 26—27 kcal/mol, which leads to a decrease of around 20 kcal/mol on reduction of the carbon—nitrogen double bond in the 2*H*-azirine into the aziridine. Due to certain inherent features within their structure, the 2*H*-azirines are versatile and reactive substrates. These

3-Membered Heterocycle Synthesis. https://doi.org/10.1016/B978-0-443-13583-5.00016-5

include a lone pair of electrons on nitrogen, an electron-rich bond, ring strain, and the ability to undergo ring cleavage on photochemical or thermal excitation to provide such reactive fugitive species such as nitrile ylides, iminocarbenes, and vinyl nitrenes. With an enhanced reactivity compared with acyclic imines, the 2*H*-azirines can be looked upon as strained imines. Many natural products contain 2*H*-azirine ring. A broad spectrum of antibiotic activity is exhibited by azirinomycin [2b], isolated from *Streptomyces aureus* and its methyl ester, in vitro, against both gram-negative and gram-positive bacteria [2c]. The azirine-bearing natural products (*R*)-(−)- [2d] and (*S*)-(+)-dysidazirine and (*S*)-(+)-antazirine [2e] are isolated from *Dysidea fragilis.*

The ability to efficiently access complex molecular frameworks has become a fundamental research area in various fields of chemistry. In subjects ranging from nanotechnology, biomolecule and pharmaceuticals design, polymer synthesis, materials science, or others, as target structures become more complex and with further refined characteristics, the traditional approaches accessible for their formation can rapidly become very involved. This can make these syntheses not only difficult to carry out (e.g., via increasing number of steps) but also complicate the product discovery, since this sequence of steps must often be repeated to modify structures. Additionally, these multistep syntheses tend to depend on reagents not fully incorporated into the product, reduced yields, and repeated solvent use, each of which results in increasing environmental issues. The aforementioned issues have created a significant challenge, which is to bring greater levels of molecular complexity with realistic access to researchers for both discovery of product and their ultimate synthesis. As such, a major emphasis has been placed on developing procedures to synthesize the products in minimum number of steps, limited waste, along with structural flexibility and atom economy [3,4], ideal syntheses [5], click reactions [6], green chemistry [7], tandem reactions [8−11], and others [12−16] to the forefront in the design of new synthetic approaches.

In today's scenario, a special attention is given to the green chemistry, the environmental aspect, formation of toxic waste, undesirable side products, and cost of production no matter how advanced chemistry is applied in the procedure. Discovery of innovative approaches to decrease waste and hazard has been the main motive of various organic chemists. It has been concentrated on decreasing, recycling, or eliminating the utilization of toxic solvents and chemicals to minimize the environmental and human influence without stifling scientific development.

The conversions in the supramolecular structures or crystals were facilitated utilizing light as a reagent. These types of delicate structures are not decomposed by heating or using destructive reagents. The stereoselectivity of photochemical reactions was controlled via conformation-demanding environment by performing these conversions. These types of ideas have been efficiently utilized in organic synthesis. Special attention is given to novel synthetic methods and new insights into existing procedures for the selective formation.

11.2 Photochemical synthesis of diaziridines

The photolysis of amino azide (Scheme 1), prepared from hexafluoro-*i*-propylidene imine by adding HN_3, afforded diaziridine (43%) through the formation of α-amino nitrene [17]. The hydrazone as a key product (60%) and diaziridine in only 11% yield were synthesized by the pyrolysis of amino azide.

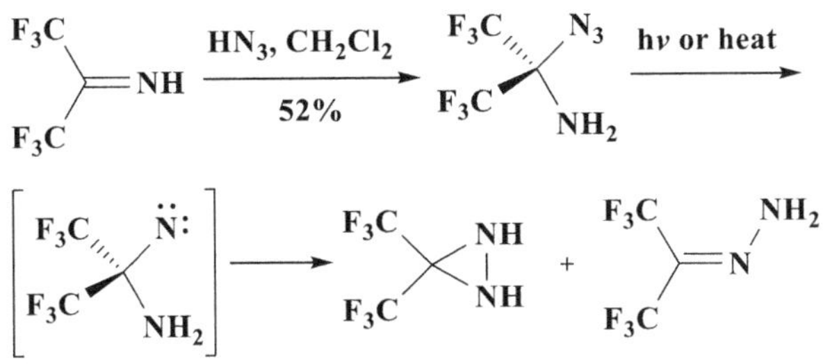

SCHEME 1

Three-membered heterocyclic rings were not synthesized using Fischer carbene complexes successfully. For example, the efforts to synthesize the aziridines through a [2 + 1]-cyclization of Fischer carbine complexes and imines failed [18]. Maxey and McElwee-White [19] reported an instance of this type of heterocyclization including azo compounds (Scheme 2). The diaziridines are accessible in low to modest yields by photochemical or thermal [2 + 1]-cyclization of tungsten carbine and azo compounds, i.e., 4-methyl-1,2,4-triazoline-3,5-dione and diethylazodicarboxylate [20].

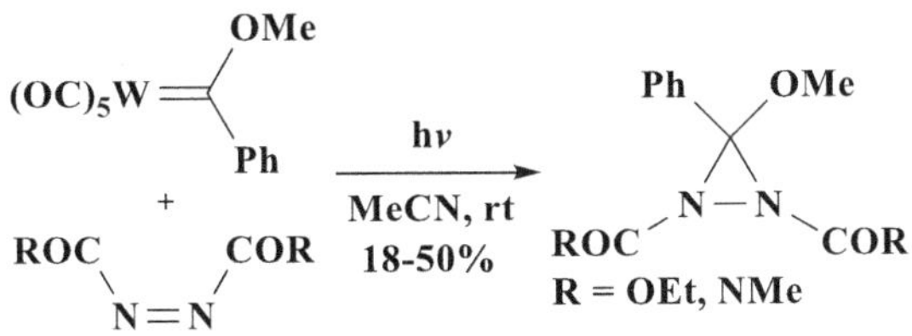

SCHEME 2

The *N*-phenyldiaziridines were synthesized in good yields (60%–70%) by the photolysis of tetrazolines (Scheme 3) [21], which are difficult to synthesize by other approaches [22]. The tetrazolines were prepared from 1,5-diphenyl- or 1-methyl-5-phenyl-5-tetrazole by methylation at N-4 and reduction with sodium borohydride.

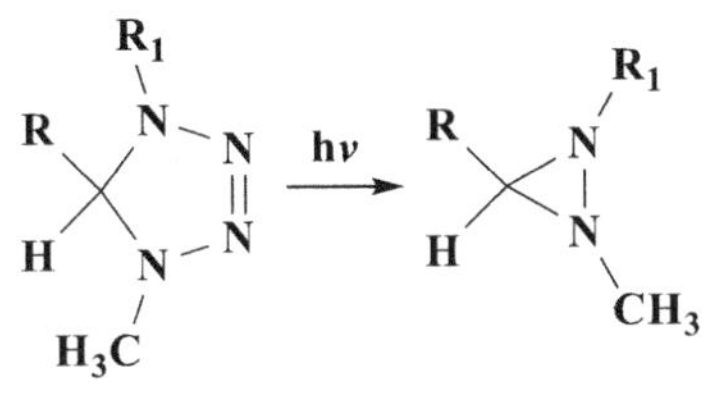

SCHEME 3

The photochemistry of 5-imino-4,5-dihydro-1*H*-tetrazoles was analyzed [23]. The diluted irradiated solutions of 5-imino-4,5-dihydro-1*H*-tetrazoles were irradiated in [D$_8$]-THF ($\lambda \geq 254$ nm; high pressure Hg lamp; $-60°C$). An irradiation was performed in sealed nuclear magnetic resonance tubes, and the reaction was monitored by nuclear magnetic resonance examination. The results have shown the interference of a key photofragmentation channel resulting in the removal of N_2 and formation of iminoaziridines in more than 80% yield, and a minor photofragmentation channel resulted in the formation of alkyl azide and *N,N*-dialkylcarbodiimide. Although this process led to a mixture of isomers and two fragmentation routes were involved, this work was very relevant in mechanistic terms, as it provided further evidence for the participation of singlet biradicals in photodegradation routes of tetrazoles. The formation of triazatrimethylenemethane diradicals was proposed as the intermediate species was generated from an initial photoinduced N_2 extrusion. The biradical intermediates have been proposed for the photofragmentation of 1,4-dihydro-1-allyl-4-phenyltetrazol-5-ones and 5-allyloxy-1-phenyltetrazoles in solution (Scheme 4) [24–27].

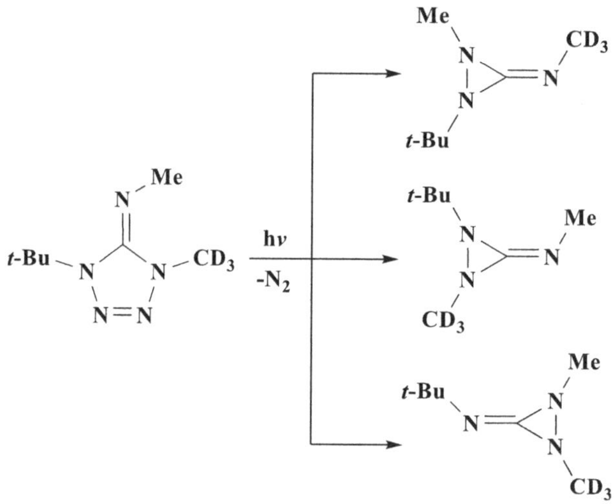

SCHEME 4

The photochemical formation of diaziridinone by the photolysis of 1,4-dimethyl-1,4-dihydro-5*H*-tetrazol-5-one was reported by Quast and Bieber [28] (Scheme 5). The photolysis of highly thermally stable tetrazolone in propan-2-ol or ether gave only secondary products from radical reactions of primary photoproducts with solvent. However, a single photoproduct, prepared by an irradiation of 1,4-dimethyl-1,4-dihydro-5*H*-tetrazol-5-one (degassed samples at 10^{-5} Torr, 150 or 450 W mercury lamp) in trideuteroacetonitrile, was characterized by proton nuclear magnetic resonance and infrared spectroscopy as 1,2-dimethyldiaziridinone [27].

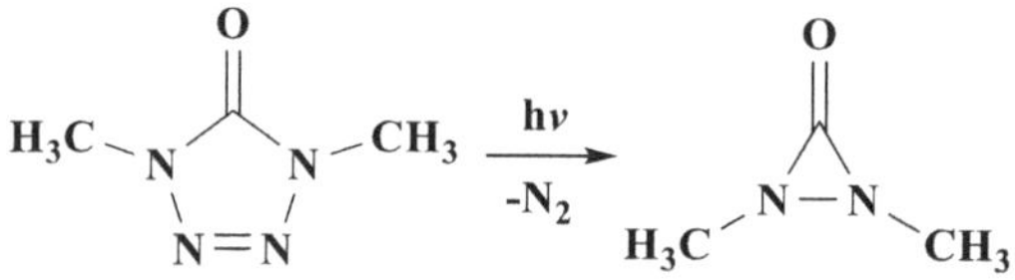

SCHEME 5

Quast and Nahr [29] reported that the photoextrusion of nitrogen from 1-alkenyl-4-alkyl-1,4-dihydro-5*H*-tetrazol-5-ones provided 1-alkenyl-2-alkyl-diaziridinones. The photolysis of tetrazolones (20°C, 254 nm, 150 W mercury lamp, degassed samples at 10^{-5} Torr.) was conducted in [D_{14}] methylcyclohexane, [D_3]acetonitrile, and [D_{12}]cyclohexane for the formation of diaziridinones in 80%–90% yield (Scheme 6) along with small amounts (<10%) of side-products. When photolysis was carried out in [D_3]acetonitrile, the side product was recognized as *N*-alkenyl-*N*-alkylurea possibly formed by the hydrolysis of primary photoproduct or from its biradical substrate [27].

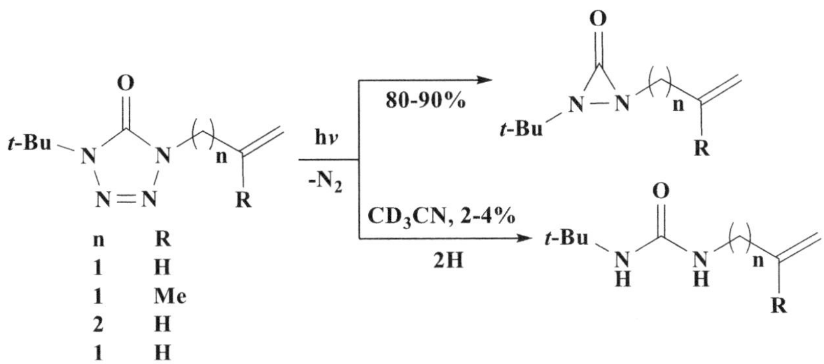

SCHEME 6

The diaziridine (70%) was synthesized by the electrocyclization of azomethine imine on an irradiation with sunlight at −80°C (Scheme 7) [30]. The first step in the photochemical ring expansion of iminopyridinium ylides to 1,2-diazepines is an electrocyclization of iminopyridinium ylides to 1,7-diazabicyclo[4.1.0]heptanes [31].

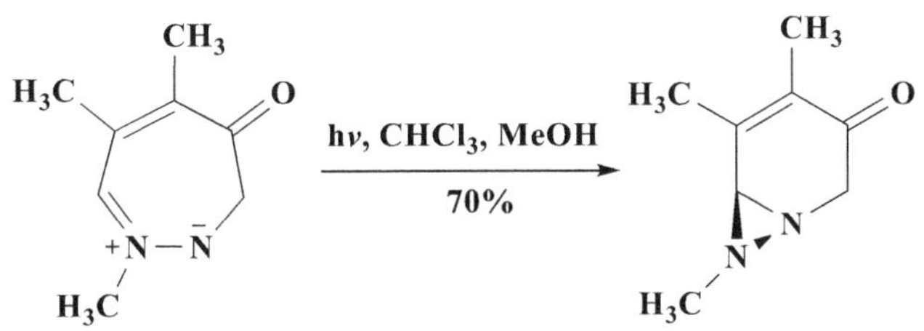

SCHEME 7

11.3 Photochemical synthesis of azirines

In these experiments, neither propionaldehyde nor benzaldehyde afforded Paternò-Büchi products. Instead, the reaction could be reported, which also occurred using a tenfold excess of aldehyde or without any aldehyde. This reaction to provide the azirines was identified as an intramolecular ring contraction of isoxazoles (Scheme 8) [32]. Surprisingly, the conversions of isoxazoles were significantly lower in the absence of aldehydes as the potential reaction partners, suggesting the chances of an energy transfer from the excited triplet or singlet aldehyde to the isoxazole [33].

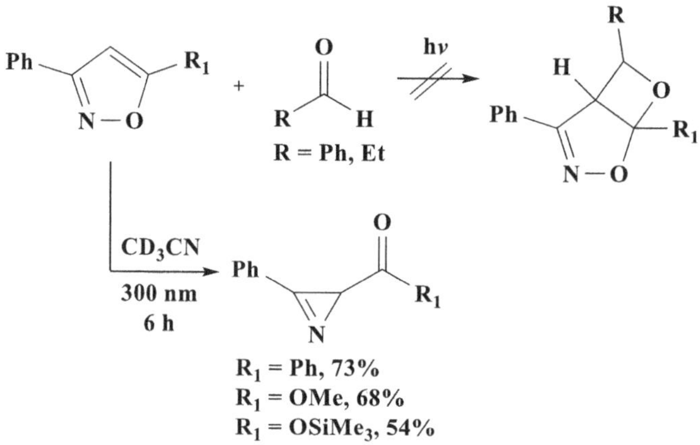

SCHEME 8

Qi et al. [34] reported that the photolysis of α-diazo oxime ethers resulted in free carbene that underwent a nitrogen–oxygen insertion to afford the 2-aryl- and 2-alkyl-oxy-2*H*-azirines (Scheme 9). In the pyrrole formation, a very beneficial intermediate was three-membered heterocyclic compound [35].

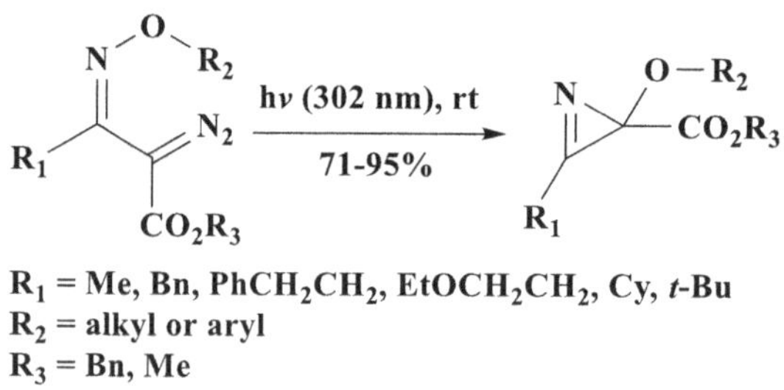

SCHEME 9

The photolysis of α-unsubstituted ($R_1 = H$) vinyl azides only at low temperature afforded isolable 1-azirines, which upon further heating underwent a decomposition (Scheme 10) [36,37].

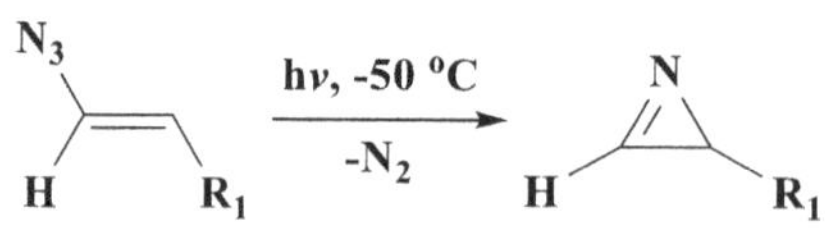

SCHEME 10

Time-resolved spectroscopy did not detect singlet vinylnitrenes, and either time-resolved or matrix spectroscopy did not detect triplet vinylnitrenes. At cryogenic temperature (8 K), an azirine was obtained on photolysis of α-azidostyrene in an argon matrix, but nitrene species was not detected even under these conditions [38]. Thus, the thermolysis and photolysis of vinyl azides resulting in concerted synthesis of azirines (without intervention of a singlet nitrene) was considered to be more reasonable (Scheme 11) [39].

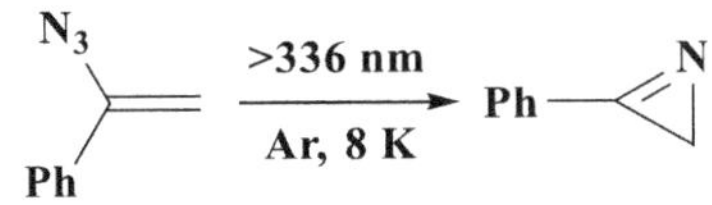

SCHEME 11

The 3-substituted 2*H*-azirines were prepared from enazides at elevated temperatures employing low boiling solvents in closed vessels, for example, DCM at 150°C [40,41]. Same reaction was also carried out under the MWI and solvent-free conditions by heating starting compounds [42]. Unfortunately, no safety hints were given for these experiments. In special cases, the azirines were prepared by the thermal reaction of vinyl azides at rt [43,44]. Both irradiation or thermolysis of enazides solutions afforded 2*H*-azirines in high yields [45–48]. Thus, heating of vinyl azides provided 2*H*-azirines in low yields due to thermal sequential reactions or degradation of the strained heterocyclic compounds. The lone-pair orbital of 2*H*-azirine nitrogen atom interacted with the π_1-orbital of cyclopentadiene ring as shown by the photoelectron spectroscopy of spirocyclic compounds. This kind of homoconjugation termed aza-spiro conjugation was observed to be compatible with quantum chemical calculations [49,50]. The fragmentation occurred to synthesize the nitriles on heating to rt or prolonged photolysis, which provided heterocyclic compounds from diazides [51], which could only be detected by infrared data in argon matrix or by nuclear magnetic resonance spectroscopy at low temperature [52]. Thus, it was reported that the 2-azido-2*H*-azirines were intermediates in the well-known thermal or photochemical conversion of *vic*-vinyl diazides into cyano compounds (Schemes 12–17) [53–55].

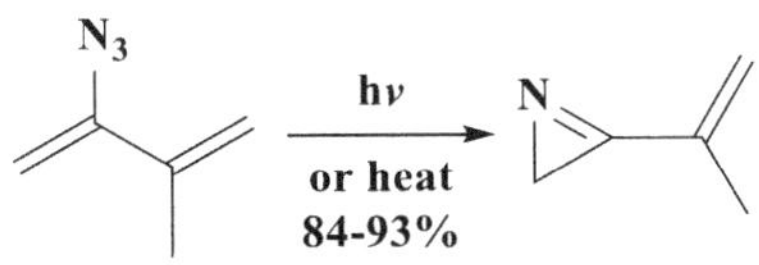

SCHEME 12

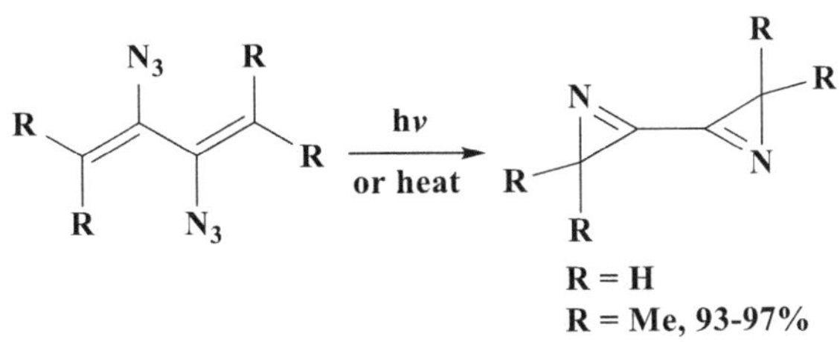

SCHEME 13

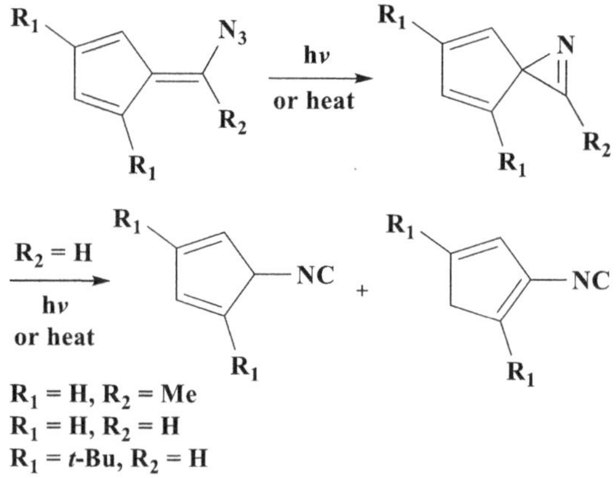

SCHEME 14

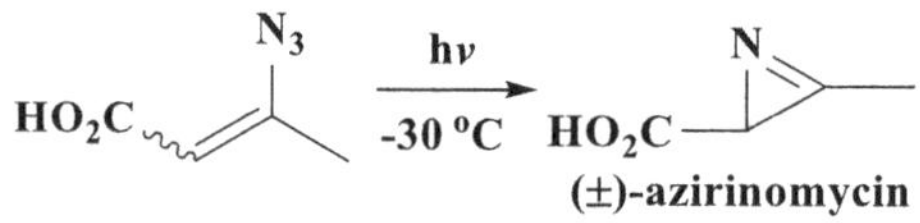

SCHEME 15

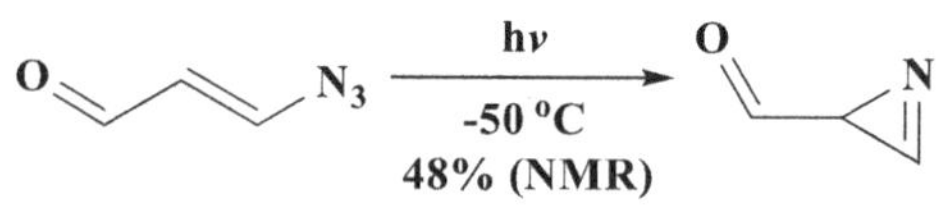

SCHEME 16

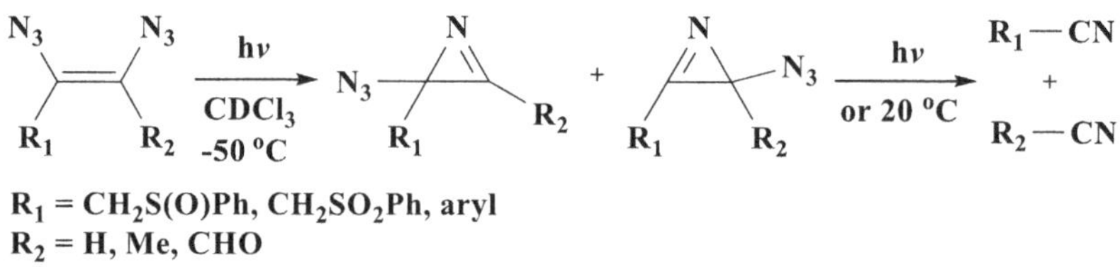

SCHEME 17

This conversion could be mediated by photolysis. Thus, an irradiation of β-phenylethyl vinyl azide induced the loss of nitrogen and synthesized azirine in excellent yield (Scheme 18) [56]. The obvious benefit of this method was that the decomposition could be performed at low temperature, thus providing more strained products including fused bicyclic azirines and bis-azirines (Scheme 19). In the latter case, the stepwise removal of nitrogen took place, and the vinyl azirine intermediate could be isolated. For both the photolytic and thermolytic approaches, one bottleneck was access to the vinyl azide precursors [57]. One promising contribution to overcoming this challenge was the Cu(I)/L-proline-assisted reaction of NaN_3 with (*Z*)-1-iodo-1-alkenes [58], which, in turn, could be accessed stereoselectively through a Wittig reaction of aldehydes with iodomethylenetriphenylphosphorane [59].

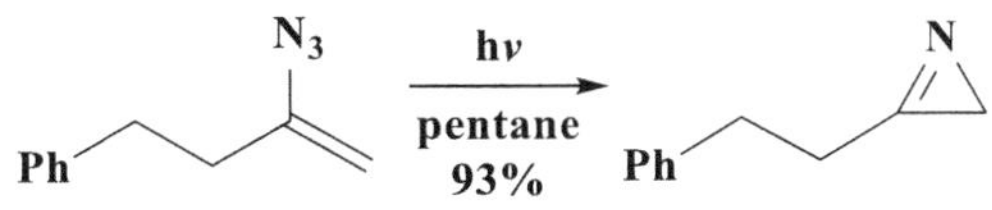

SCHEME 18

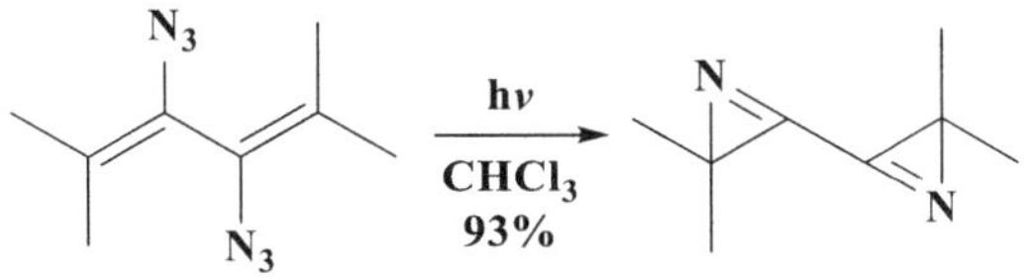

SCHEME 19

There are high-yielding approaches for the synthesis of azirines containing EWGs. Thus, the allenic phosphonates, prepared from propargyl alcohol derivatives, are themselves efficient substrates for phosphonylmethyl vinyl azides (Scheme 20). An irradiation of phosphonylmethyl vinyl azides resulted in very high-yielding photolysis to azirine [60–62].

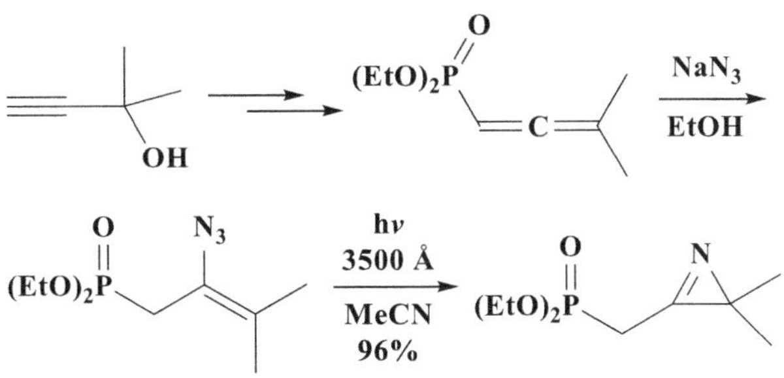

SCHEME 20

The classical syntheses of nitrogen heterocycles using organic azides included 1,3-dipolar cycloadditions to afford the 1,2,3-triazoles and intramolecular insertion reactions via short-lived nitrenes to provide the cyclic products, e.g., indoles. However, other reactions of azides, particularly of vinyl azides, also offered efficient routes to heterocycles. Thus, the allenyl azides, which were easily accessible from propargyl azides via [3,3]-sigmatropic shift of the azido group or through a prototropic (base-catalyzed) rearrangement, electrocyclize rapidly to produce the triazafulvenes. These intermediates could be trapped by nucleophiles NuH to provide the functionalized triazoles [63]. On photolysis, the allenyl azides led to methylene-2*H*-azirines, which were proved to be the first heterocyclic triazafulvenes [64,65]. Other highly strained azirines [66] and spiro compounds were also available from vinyl azides. The spiro compound could be cleaved to give unknown isocyanocyclopentadienes. The 1,4-diazido-1,3-butadienes were transformed into diazirines, which were converted to aromatic pyrimidines and pyridazines [67]. The chemistry of azidocyclopentadienes was also investigated (Scheme 21).

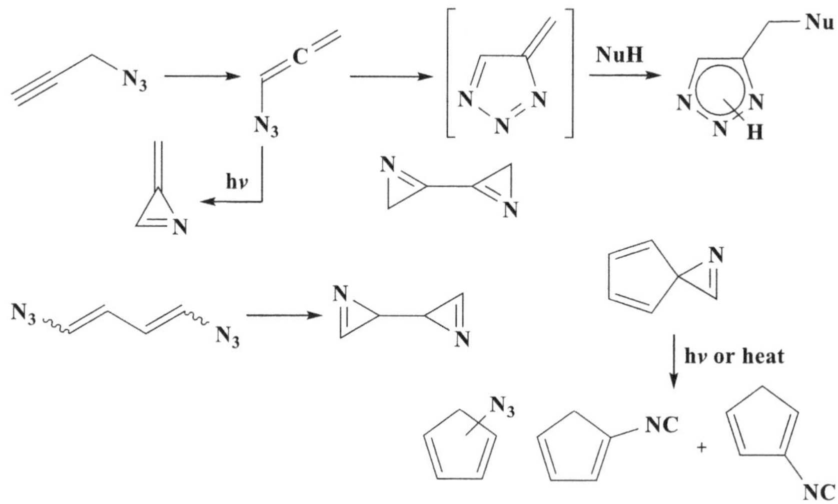

SCHEME 21

There are some instances of indirect formation of 2*H*-azirines from unsaturated azides. A pathway to methylene-2*H*-azirines was afforded by the photolysis of azidoallenes [64,66,68]. An interception of methyleneazirine ($R_1 = CH_2OMe$) by hydrogen cyanide provided relatively stable 2*H*-azirine (Scheme 22).

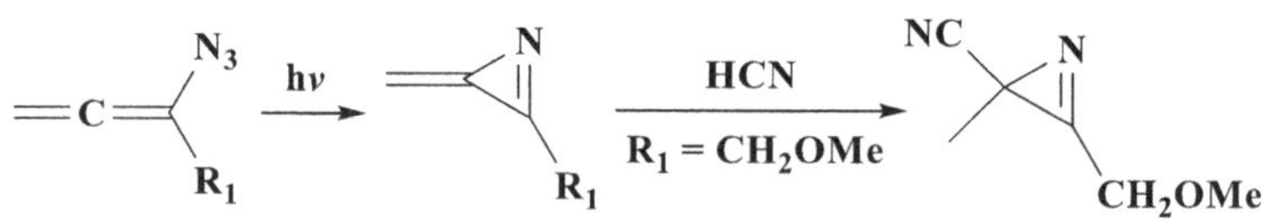

SCHEME 22

An irradiation of diazide reported the priority of vinyl azides over alkyl azides in light absorption and photochemical conversion, which afforded strained heterocyclic compound without degradation of the second azido group [69]. The bulky substituents in the 2-position stabilize the 2,3-bridged 2*H*-azirines (Scheme 23).

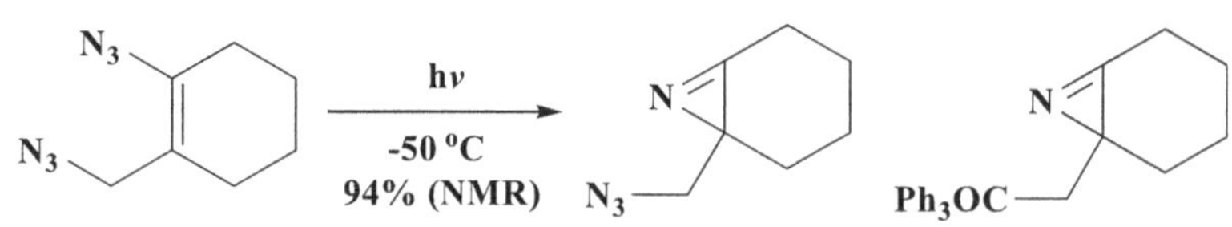

SCHEME 23

The tricyclic azirine derivative could be handled only at low temperature in solution at rt due to its conversion into ring expansion product quantitatively (Scheme 24) [69].

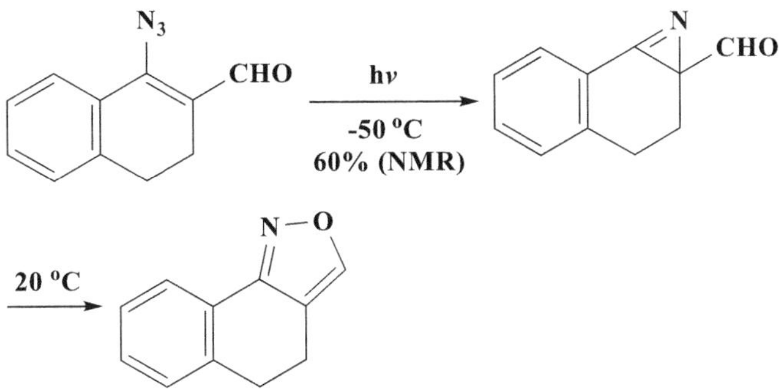

SCHEME 24

After a few microseconds of laser pulse, the transient absorption spectrum of triplet nitrene ($\lambda_{max} = 370$ nm) was obtained. At ambient temperature, with a time constant of 2.8 μs in benzene, the intensity of the transient absorption of triplet nitrene enhanced exponentially. The triplet nitrene decay monitored within 100 μs of the laser pulse was accompanied by the concurrent synthesis of 1,1′-azonaphthalene ($\lambda_{max} = 420$ nm). The tricyclic azirine was synthesized from triplet nitrene substrate (Scheme 25), a species that acted as a reservoir for singlet 1-naphthylnitrene [70]. The TRIR spectroscopy detected the anticipated azirine with lifetimes of 3.2 ± 0.6 μs [71]. The lifetime of azirine was also calculated in solvent mixtures of 1:3 and 1:1 DEA/MeCN. The slow reaction rate of azirine with DEA was consistent with the observed low yield of DEA photoadduct obtained on photolysis of azide.

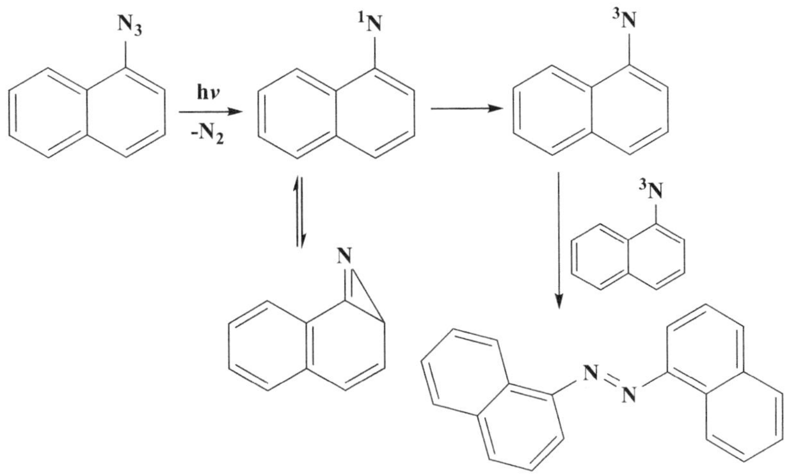

SCHEME 25

The reaction of 3-azido-2,3-dideoxypyranose and 3-azido-2,3-dideoxy-2-halohexopyranose compounds with $PhI(OAc)_2$ and I_2 afforded 2-azido-1,2-dideoxy-1-iodoalditols and 2-azido-1,2-dideoxy-1-halo-1-iodoalditols, respectively. The chemoselective dehydroiodination converted iodo azides into 2-azido-1,2-dideoxy-4-*O*-formyl-pent-1-enitols and (*Z,E*)-2-azido-1,2-dideoxy-1-halo-4-*O*-formyl-pent-1-enitols in good yields. The photochemical and thermolysis analysis of these vinyl azides and 1-halovinyl azides synthesized polyhydroxylated 3-alkyl-2*H*-azirines and the previously unknown 2-halo-3-alkyl-2*H*-azirines. It should be possible to synthesize the chiral iodo azides and hence vinyl azides and 2*H*-azirines from carbohydrates employing this approach and considering the accessibility of 3-azido-2,3-dideoxyhexopyranose. The acid-mediated reaction of 2-deoxyhex-1-enitol derivatives (glycals) with sodium azide provided 3-azido-2,3-dideoxyhexopyranose in high yield. An addition of azide anion over an α,β-unsaturated aldehyde intermediate took place via a Michael addition [72]. Starting from halohydrins and after chemoselective dehydroiodination of the 1,1-dihalo alditol intermediate provided polyhydroxylated 2-azido-1-halo-1-alkenes by extending this approach. The photolysis or thermolysis of these halovinyl azide provided an unrivaled opportunity to study the stability and properties of hitherto unknown 3-alkyl-2-halo-2*H*-azirines (Scheme 26).

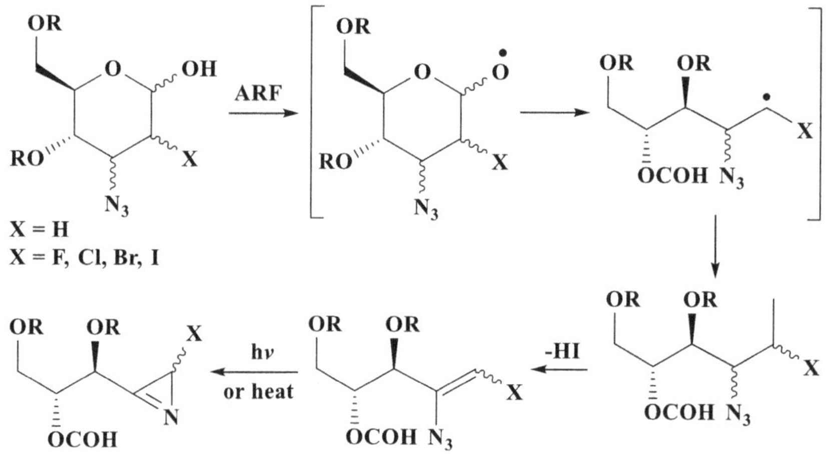

SCHEME 26

The photocyclization reactions afforded some antibacterial phosphonoaziridines, 2*H*-azirines, and their corresponding phosphonic acids (Scheme 27) [73]. The vinylphosphonate was reacted with ethyl azidoacetate and BrN_3 or NaN_3, and ring closure occurred under the UV light irradiation (Scheme 28) [74].

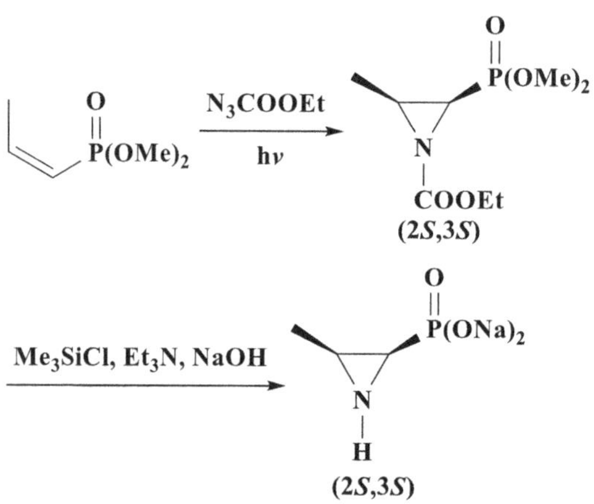

SCHEME 27

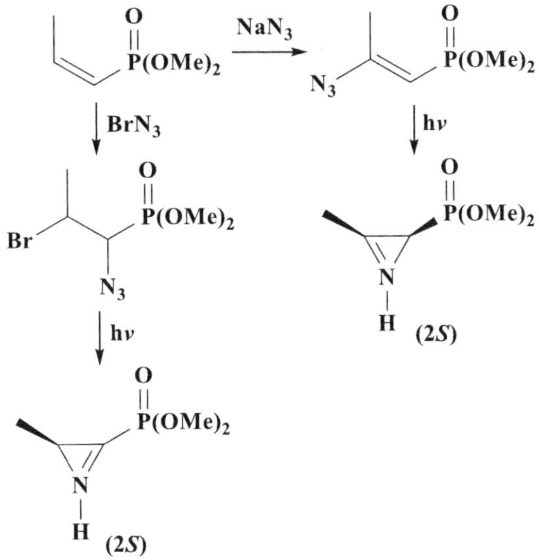

SCHEME 28

Three-membered heterocyclic compounds accessed via azide chemistry are triaziridines. An addition of nitrenes or nitrene equivalents to azo compounds provided azimines, the open-chain tautomers of triaziridines [75]. Marterer et al. [76] circumvent the synthesis of azimines by preparing a series of compounds that have an azide function in close proximity to the π-face of azoalkane. These compounds were further irradiated in dilute solution to afford the triaziridines among other products [77]. For instance, the azide delivered triaziridine (25%), a crystalline compound that was purified by sublimation (Scheme 29). In a similar manner, the cage compounds (X = CMe, CH, P=O, and P) were synthesized.

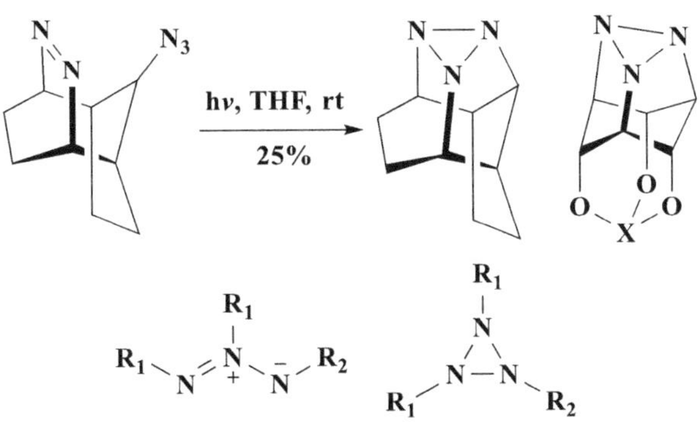

SCHEME 29

References

1. (a) K.-P. Zeller, M. Kowallik and P. Haiss. 2005. The dimethyldioxirane-mediated oxidation of phenylethyne. Org. Biomol. Chem. 3: 2310-2318. (b) N. Kaur. 2019. Synthesis of three-membered and four-membered heterocycles with the assistance of photochemical reactions. J. Heterocycl. Chem. 56: 1141-1167. (c) N. Kaur. 2017. Ionic liquids: promising but challenging solvents for the synthesis of *N*-heterocycles. Mini Rev. Org. Chem. 14: 3-23. (d) N. Kaur. 2018. Green synthesis of three to five-membered *O*-heterocycles using ionic liquids. Synth. Commun. 48: 1588-1613.
2. (a) D.J. Ramón and M. Yus. 2004. Enantioselective synthesis of oxygen-, nitrogen- and halogen-substituted quaternary carbon centers. Curr. Org. Chem. 8: 149-183. (b) T.W. Miller, E.W. Tristram and F.J. Wolf. 1971. Azirinomycin. II. J. Antibiot. 24: 48-50. (c) E.O. Stapley, D. Hendlin, M. Jackson, A.K. Miller, S. Hernandez and J.M. Mata. 1971. Azirinomycin. I. J. Antibiot. 24: 42-47. (d) T.F. Molinski and C.M. Ireland. 1988. Dysidazirine, a cytotoxic azacyclopropene from the marine sponge *Dysidea fragilis*. J. Org. Chem. 53: 2103-2105. (e) C.E. Salomon, D.H. Williams and D.J. Faulkner. 1995. New azacyclopropene derivatives from *Dysidea fragilis* collected in pohnpei. J. Nat. Prod. 58: 1463-1466.
3. B.M. Trost. 1991. The atom economy - a search for synthetic efficiency. Science 254: 1471-1477.
4. B.M. Trost. 2002. On inventing reactions for atom economy. Acc. Chem. Res. 35: 695-705.
5. P.A. Wender, S.T. Handy and D.L. Wright. 1997. Toward the ideal synthesis. Chem. Ind. 19: 765-769.
6. H.C. Kolb, M.G. Finn and K.B. Sharpless. 2001. Click chemistry: diverse chemical function from a few good reactions. Angew. Chem. Int. Ed. 40: 2004-2021.
7. N. Winterton. 2001. News and views. Green Chem. 3: 73-81.
8. J.C. Wasilke, S.J. Obrey, R.T. Baker and G.C. Bazan. 2005. Concurrent tandem catalysis. Chem. Rev. 105: 1001-1020.
9. J.M. Lee, Y. Na, H. Han and S. Chang. 2004. Cooperative multi-catalyst systems for one-pot organic transformations. Chem. Soc. Rev. 33: 302-312.
10. S.E. Denmark and A. Thorarensen. 1996. Tandem [4+2]/[3+2] cycloadditions of nitroalkenes. Chem. Rev. 96: 137-166.
11. I. Ryu, N. Sonoda and D.P. Curran. 1996. Tandem radical reactions of carbon monoxide, isonitriles, and other reagent equivalents of the geminal radical acceptor/radical precursor synthon. Chem. Rev. 96: 177-194.
12. A. de Meijere and S. Braise. 1999. Palladium in action: domino coupling and allylic substitution reactions for the efficient construction of complex organic molecules. J. Organomet. Chem. 576: 88-110.
13. M. Malarcia. 1996. Selective preparation of complex polycyclic molecules from acyclic precursors via radical mediated- or transition metal-catalyzed cascade reactions. Chem. Rev. 96: 289-306.
14. A. Padwa and M.D. Weingarten. 1996. Cascade processes of metallo carbenoids. Chem. Rev. 96: 223-270.
15. P.A. Wender, V.A. Verma, T.J. Paxton and T.H. Pillow. 2008. Function-oriented synthesis, step economy, and drug design. Acc. Chem. Res. 41: 40-49.
16. S.I. Ikeda. 2000. Nickel-catalyzed intermolecular domino reactions. Acc. Chem. Res. 33: 511-519.
17. W.J. Middleton and C.G. Krespan. 1965. Fluorimines. J. Org. Chem. 30: 1398-1402.
18. C.K. Murray, B.P. Warner, V. Dragisich, W.D. Wulff and R.D. Rogers. 1990. Thermal reactions of acyloxy and alkoxy carbene complexes with imines: metathesis, acetate rearrangements, and a new route to imino carbene complexes via Peterson type eliminations. Organometallics 9: 3142-3151.
19. C.T. Maxey and L. McElwee-White. 1991. Formation of diaziridines by reaction of $(CO)_5W:C(OMe)Ph$ with electron-deficient azo compounds. Organometallics 10: 1913-1916.
20. J. Barluenga, J. Santamaria and M. Tomas. 2004. Synthesis of heterocycles via group VI Fischer carbene complexes. Chem. Rev. 104: 2259-2284.

21. T. Akiyama, T. Kitamura, T. Isida and M. Kawanisi. 1974. Photolysis of δ^2-tetrazolines: formation of diaziridines. Chem. Lett. 3: 185-186.
22. E. Schmitz. 1979. Three-membered rings with two heteroatoms. Adv. Heterocycl. Chem. 24: 63-107.
23. H. Quast and L. Bieber. 2008. Iminodiaziridines by regio- and stereoselective cyclization of diastereomeric singlet triazatrimethylenemethane diradicals generated through photolysis of 5-imino-4,5-dihydro-1*H*-tetrazoles. J. Org. Chem. 73: 3738-3744.
24. L.M.T. Frija, I.V. Khmelinskii and M.L.S. Cristiano. 2005. Novel efficient synthesis of 3,4-dihydro-6-substituted-3-phenylpyrimidin-2(1*H*)-ones. Tetrahedron Lett. 46: 6757-6760.
25. L.M.T. Frija, I.V. Khmelinskii and M.L.S. Cristiano. 2006. Mechanistic investigations into the photochemistry of 4-allyl-tetrazolones in solution: a new approach to the synthesis of 3,4-dihydro-pyrimidinones. J. Org. Chem. 71: 3583-3591.
26. L.M.T. Frija, I.V. Khmelinskii, C. Serpa, I.D. Reva, R. Fausto and M.L.S. Cristiano. 2008. Photochemistry of 5-allyloxy-tetrazoles: steady-state and laser flash photolysis study. Org. Biomol. Chem. 6: 1046-1055.
27. L.M.T. Frija, A. Ismael and M.L.S. Cristiano. 2010. Photochemical transformations of tetrazole derivatives: applications in organic synthesis. Molecules 15: 3757-3774.
28. H. Quast and L. Bieber. 1975. Aziridinimines, diaziridinimines, diaziridinones, and carbodiimides by photolysis of 2-tetrazolines. Angew. Chem. Int. Ed. 14: 428-429.
29. H. Quast and U. Nahr. 1983. Photochemische stickstoff-eliminierung aus 1-alkenyl-4-alkyl-1,4-dihydro-5*H*-tetrazol-5-onen und -thionen. diaziridinone und carbodiimide mit alkenyl substituenten. Chem. Ber. Recl. 116: 3427-3437.
30. M.G. Pleiss and J.A. Moore. 1968. Heterocyclic studies. XXVIII. Sigmatropic and electrocyclic reactions in the 1,2-diazepine system. Formation of a 1,7-diazabicyclo[4.1.0]heptenone. J. Am. Chem. Soc. 90: 4738-4739.
31. T. Kiguchi, J.-L. Schuppiser, J.-C. Schwaller and J. Streith. 1980. Photochemical syntheses of 1,2-diazepines. 11. Regiospecific synthesis of 1,2-dihydro-1,2-diazepin-3-ones. J. Org. Chem. 45: 5095-5100.
32. B. Singh and E.F. Ullman. 1967. Photochemical transposition of ring atoms in 3,5-diarylisoxazoles. Unusual example of wavelength control in a photochemical reaction of azirines. J. Am. Chem. Soc. 89: 6911-6916.
33. A.G. Griesbeck, M. Franke, J. Neudorfl and H. Kotaka. 2011. Photocycloaddition of aromatic and aliphatic aldehydes to isoxazoles: cycloaddition reactivity and stability studies. Beilstein J. Org. Chem. 7: 127-134.
34. X. Qi, X. Xu and C.-M. Park. 2012. Facile synthesis of 2-alkyl/aryloxy-2*H*-azirines and their application in the synthesis of pyrroles. Chem. Commun. 48: 3996-3998.
35. F. Aldabbagh. 2013. Heterocyclic chemistry. Annu. Rep. Prog. Chem. Sect. B: Org. Chem. 109: 126-145.
36. K. Isomura, M. Okada and H. Taniguchi. 1969. Azirine formation by decomposition of terminal vinyl azides. Tetrahedron Lett. 10: 4073-4076.
37. W. Bauer and K. Hafner. 1969. Synthesis and reactions of a 2*H*-azirine unsubstituted on C-3. Angew. Chem. Int. Ed. 8: 772-773.
38. O.L. Chapman and J.-P. LeRoux. 1978. 1-Aza-1,2,4,6-cycloheptatetraene. J. Am. Chem. Soc. 100: 282-285.
39. F.W. Fowler, A. Hassner and L.A. Levy. 1967. Stereospecific introduction of azide functions into organic molecules. J. Am. Chem. Soc. 89: 2077-2082.
40. W. Zhao and E.M. Carreira. 2006. Conformationally restricted aza-BODIPY: highly fluorescent, stable near-infrared absorbing dyes. Chem. Eur. J. 12: 7254-7263.
41. A.S. Timén, E. Risberg and P. Somfai. 2003. Improved procedure for cyclization of vinyl azides into 3-substituted-2*H*-azirines. Tetrahedron Lett. 44: 5339-5341.
42. P.N.D. Singh, C.L. Carter and A.D. Gudmundsdóttir. 2003. A simple green procedure for the synthesis of 2*H*-azirines. Tetrahedron Lett. 44: 6763-6765.
43. M. Rens and L. Ghosez. 1970. Synthesis and reactions of 2-amino-1-azirines. Tetrahedron Lett. 11: 3765-3768.

44. H. Heimgartner. 1991. 3-Amino-2*H*-azirines. Synthons for α,α-disubstituted α-amino acids in heterocycle and peptide synthesis [new analytical methods (43)]. Angew. Chem. Int. Ed. 30: 238-264.
45. K. Banert, M. Hagedorn, C. Liedtke, A. Melzer and C. Schöffler. 2000. Azido-1,2,3-triphenylpropenes of varying stabilities: a corrigendum of structure assignment. Eur. J. Org. Chem. 2: 257-267.
46. K. Banert. 1985. Synthesis and reactions of 4-azido-3-methyl-1,2-butadiene. Angew. Chem. Int. Ed. 24: 216-217.
47. K. Banert. 1985. Synthesis of new bi-2*H*-azirin-3-yl compounds from diazides. Tetrahedron Lett. 26: 5261-5264.
48. K. Banert. 1987. Reaktionen ungesättigter azide, 31. Synthese, strukturzuordnung, photolyse und thermolyse von 2,3-diazido-1,3-butadienen. Chem. Ber. 120: 1891-1896.
49. K. Banert, F. Köhler, K. Kowski, B. Meier, B. Müller and P. Rademacher. 2002. Synthesis of 1-azaspiro[2.4] hepta-1,4,6-trienes and azaspiroconjugation studied by photoelectron spectroscopy. Chem. Eur. J. 8: 5089-5093.
50. K. Banert, F. Köhler and B. Meier. 2003. Synthesis and reactions of the first cyclopentadienyl isonitriles. Tetrahedron Lett. 44: 3781-3783.
51. J.R. Fotsing, M. Hagedorn and K. Banert. 2005. First successful synthesis, isolation and characterization of open-chain 1,2-diazidoethenes. Tetrahedron 61: 8904-8909.
52. K. Banert, J.R. Fotsing, M. Hagedorn, H.P. Reisenauer and G. Maier. 2008. Photolysis of open-chain 1,2-diazidoalkenes: generation of 2-azido-2*H*-azirines, formyl cyanide, and formyl isocyanide. Tetrahedron 64: 5645-5648.
53. M. Alajarin, R.-A. Orenes, A. Vidal and A. Pastor. 2003. A convenient and expedient synthesis of 3-aryl-2*H*-azirine-2-carboxaldehydes. Synthesis 1: 49-52.
54. T. Sakai, I. Kawabata, T. Kishimoto, T. Ema and M. Utaka. 1997. Enhancement of the enantioselectivity in lipase-catalyzed kinetic resolutions of 3-phenyl-2*H*-azirine-2-methanol by lowering the temperature to -40 °C. J. Org. Chem. 62: 4906-4907.
55. W. Bauer and K. Hafner. 1969. Synthese und reaktionen eines an C-3 unsubstituierten 2*H*-azirins. Angew. Chem. 81: 787-788.
56. A. Hassner and F.W. Fowler. 1968. Stereochemistry. XXXII. Synthesis and reactions of 1-azirines. J. Am. Chem. Soc. 90: 2869-2875.
57. A. Hassner. 1971. Regiospecific and stereospecific introduction of azide functions into organic molecules. Acc. Chem. Res. 4: 9-16.
58. W. Zhu and D. Ma. 2004. Synthesis of aryl azides and vinyl azides via proline-promoted CuI-catalyzed coupling reactions. Chem. Commun. 7: 888-889.
59. G. Stork and K. Zhao. 1989. A stereoselective synthesis of (*Z*)-1-iodo-1-alkenes. Tetrahedron Lett. 30: 2173-2174.
60. R.A. Abramovitch, M. Konieczny, W. Pennington, S. Kanamathareddy and M. Vedachalam. 1990. 2-Azidoallylphosphonates: synthesis and application to formation of 4-amino-2-ethoxy-1,2-oxaphosphol-3-ene 2-oxides. J. Chem. Soc. Chem. Commun. 3: 270-296.
61. T.M.V.D. Pinho e Melo, C.S.J. Lopes, A.L. Cardoso and A.M.R. Gonsalves. 2001. Synthesis of 2-halo-2*H*-azirines. Tetrahedron 57: 6203-6208.
62. T.M.V.D. Pinho e Melo, C.S.J. Lopes, A.L. Cardoso and A.M.R. Gonsalves. 2000. Synthesis and reactivity of 2-halo-2*H*-azirines towards nucleophiles. Tetrahedron Lett. 41: 7217-7220.
63. K. Banert. 1997. Rearrangement reactions, 6. New functionalized allenes: synthesis using sigmatropic rearrangements and unusual reactivity. Liebigs Ann. Chem. 10: 2005-2018.
64. K. Banert, M. Hagedorn, E. Knözinger, A. Becker and E.-U. Würthwein. 1994. Reactions of unsaturated azides. 11. Direct observation of 2-methylene-2*H*-azirine. J. Am. Chem. Soc. 116: 60-62.

65. K. Banert and M. Hagedorn. 1990. 2-Methylene-2*H*-azirine by photolysis of 1-azidoallenes. Angew. Chem. 102: 90-92.
66. K. Banert and M. Hagedorn. 1990. 2-Methylene-2*H*-azirines by photolysis of 1-azidoallenes. Angew. Chem. Int. Ed. 29: 103-105.
67. K. Banert and F. Köhler. 2001. Synthesis of 1,4-diazidobuta-1,3-dienes by electrocyclic ring opening: precursors for bi-2*H*-azirin-2-yls and their valence isomerization to diazabenzenes. Angew. Chem. Int. Ed. 40: 174-177.
68. J.R. Fotsing and K. Banert. 2006. New way to methylene-2*H*-azirines and their use as powerful intermediates for the stereo- and regioselective synthesis of compounds with vinylamine substructure. Eur. J. Org. Chem. 16: 3617-3625.
69. K. Banert, A. Ihle, A. Kuhtz, E. Penk, B. Saha and E.-U. Würthwein. 2013. Generation of highly strained 2,3-bridged 2*H*-azirines via cycloaddition reactions of 2-azidobuta-1,3-dienes and photolysis of the resulting cyclic vinyl azides. Tetrahedron 69: 2501-2508.
70. A.K. Schrock and G.B. Schuster. 1984. Photochemistry of naphthyl and pyrenyl azides: chemical properties of the transient intermediates probed by laser spectroscopy. J. Am. Chem. Soc. 106: 5234-5240.
71. S.E. Carroll, B. Nay, E.F.V. Scriven and H. Suschitzky. 1977. Decomposition of arylazides in piperidine: the effect of tetramethylethylenediamine on the nature of the products. Tetrahedron Lett. 18: 943-946.
72. C.R. Alonso-Cruz, A.R. Kennedy, M.S. Rodriguez and E. Suarez. 2008. Synthesis of polyhydroxylated 2*H*-azirines and 2-halo-2*H*-azirines from 3-azido-2,3-dideoxyhexopyranoses by alkoxyl radical fragmentation. J. Org. Chem. 73: 4116-4122.
73. B.G. Christensen and T.R. Beattie. 1970. Apparatus and method for optically examining security documents. Ger. Offen. 2011092A1.
74. K. Moonen, I. Laureyn and C.V. Stevens. 2004. Synthetic methods for azaheterocyclic phosphonates and their biological activity. Chem. Rev. 104: 6177-6216.
75. C. Leuenberger, L. Hoesch and A.S. Dreiding. 1982. Azimine. VI. 1-Alkoxycarbonyl-2,3-dialkyl- und -2,3-diaryl-azimine. Helv. Chim. Acta 65: 217-228.
76. W. Marterer, O. Klingler, R. Thiergardt, L. Knothe and H. Prinzbach. 1991. The azo/nitrene route to *cis,cis*-trialkyltriaziridines, 1. Synthesis of *syn*-azo azides of defined proximity. Chem. Ber. 124: 609-620.
77. W. Marterer, O. Klingler, R. Thiergardt, E. Beckmann, H. Fritz and H. Prinzbach. 1991. Photochemical transformations, 75. The azo/nitrene route to *cis,cis*-trialkyltriaziridines, photolysis of *syn*-azo azides of defined proximity - attempts for $N_3 \rightarrow N_3X$ ring enlargement. Chem. Ber. 124: 621-633.

CHAPTER

Photochemical synthesis of epoxides, dioxiranes, oxaziridines, and thiiranes 12

Chapter outline

12.1 Introduction

Several natural products (hormones, vitamins, alkaloids, antibiotics, etc.), biologically active compounds (pharmaceuticals), herbicides, insecticides, pesticides, weed killers, and rodenticides contain heterocyclic moieties. Heterocyclic rings have also been utilized as chiral catalysts, chiral auxiliaries, and ligands for an asymmetric catalysis. The formation of these compounds has been a center of focus since a very long time. Therefore, newer and efficient methods have been discovered for their synthesis [1−3].

The epoxides are very important synthetic intermediates for a variety of biologically active and synthetic molecules such as β-amino alcohols [4] utilized as β-blockers, antibiotics [5], neuroprotective agents [6], and antidepressants [7], along with other clinical and natural compounds. The chiral epoxides whether prepared from alkenes or other sources have the same benefits as electrophilic intermediates for stereochemical formation including reactions with nucleophiles.

The world is encountering with various problems related to the environment. Therefore, a good chemist should use neat and environmentally friendly procedures for the formation of molecules. Because of the increase in environmental awareness in industry and chemical research, the challenge for a sustainable environment demands for clean processes that avoid the use of hazardous organic solvents. Today's chemistry has accepted the concept of "Green Chemistry" and adopted it broadly to meet the challenges of protecting the environment and human health, a requirement that has arisen because of the indiscriminate utilization of hazardous solvents and chemicals. It has been realized that operation of any synthesis today has to take into account the generation of toxic waste, the ecological aspect, cost of manufacture, and formation of undesirable side products no matter how innovative chemistry is applied in the process. To meet this demand, chemists have resorted utilizing photochemical methods. One of the most stimulating developments in chemistry is the development of

3-Membered Heterocycle Synthesis. https://doi.org/10.1016/B978-0-443-13583-5.00013-X

reactions using visible light photocatalysts. For years, photochemical reactions have been important for the formation of various macromolecules. For several decades, photochemistry has been facing challenges, i.e., scale, reproducibility, and efficiency. Many of these problems may be overcome using flow chemistry platforms, which have become a powerful and widely accepted tool for chemists. Although photochemistry can be considered as one of the simplest manifolds of chemical reactivity, it is used to allow reactions that are, sometimes, inconceivable (or even impossible) by other pathways. Some reactions can invoke the imagination of chemists as does photochemistry [8–11].

12.2 Photochemical synthesis of epoxides

Prakash et al. [12] synthesized carbonyl ylide by irradiating dideuteriodiazomethane with monomeric formaldehyde in an ether solution. The photolysis of 9-diazofluorene provided a carbonyl ylide (Scheme 1) [13].

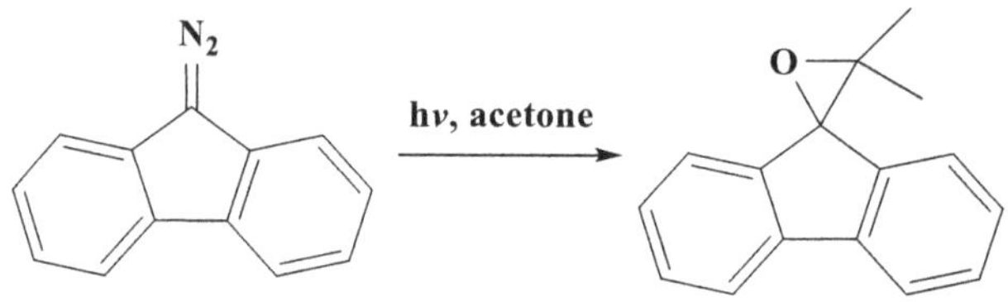

SCHEME 1

The photolysis of starting material at 7 K in a carbon dioxide–doped argon matrix that was annealed at 30 K provided difluoroketene (Scheme 2) [14]. The formation of difluoroketene from starting material involved an electrophilic removal of oxygen atom without the formation of lactone. An irradiation of difluoroketene with light more than 550 nm led to decarbonylation, but the resulting CF_2 was not observed to react with CO.

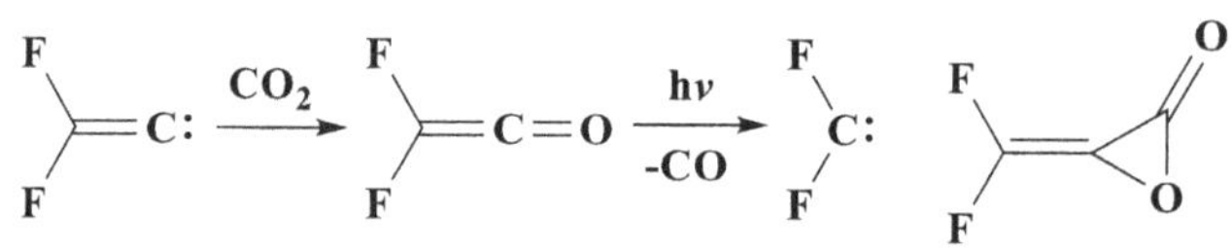

SCHEME 2

The mechanistic characteristics of photoinduced fragmentation of carbonate esters have been examined. As evidenced by an interception with suitable trapping reagents, Griffin and coworkers [15] reported that an irradiation of cyclic carbonate esters provided carbene intermediates, while the synthesis persistent to primary photoproducts has not been detected. On the other hand, the cyclic carbonates afforded products derived from 1,3-diradical intermediates under the milder photochemical conditions (3 versus 15 h) (Scheme 3) [16]. The photoinduced fragmentation reactions of *N*- and *O*-containing moieties have been studied intensively, while reports on photoextrusion reactions of *S*-containing moieties are rather rare. For example, the photoextrusion of SO_2 was performed to facilitate the formation of cyclophane derivatives [17]. The photochemistry of cyclic sulfite esters is even less well known [18–20]. Few decades ago, a report on the photochemistry of *meso*-hydrobenzoin sulfite has declared that these compounds are also carbene substrates [21]. To provide further understanding of mechanism of photoinduced cleavage of cyclic sulfites, the hydrobenzoin sulfite and styrene glycol sulfite were examined, and diradical intermediates were synthesized during photoextrusion procedures.

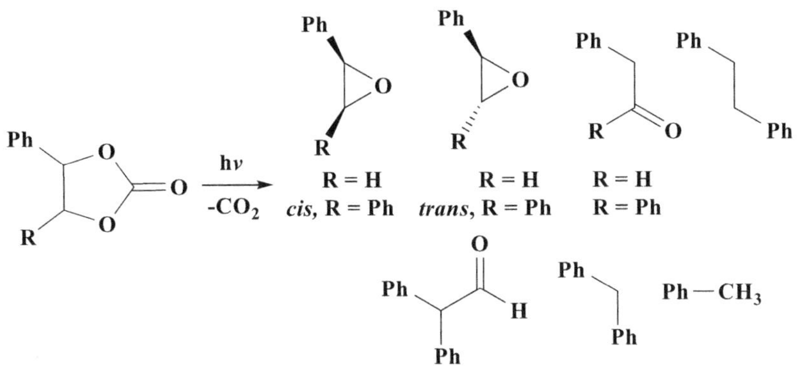

SCHEME 3

The photolysis of malonyl peroxide provided stable α-lactone bis(trifluoromethyl)acetolactone (Scheme 4). The α-lactone bis(trifluoromethyl)acetolactone, having a half-life of 8 h at 24°C in the gas phase, was decarbonylated to provide the hexafluoroacetone. Its relatively high stability was another instance of perfluoroalkyl effect [22,23].

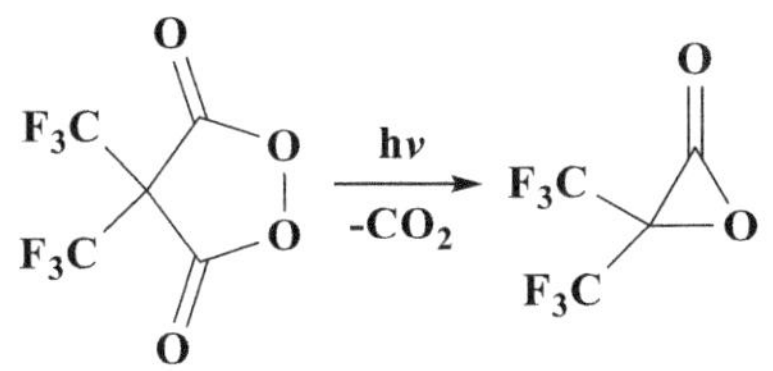

SCHEME 4

The oxirane was formed through the free radical mechanisms as in the reaction of certain fluoroalkenes with oxygen. Under the pressure at higher temperatures, oxygen alone could suffice, but activation was frequently provided in the form of radical initiators (e.g., CBr_3F) and UV light. The thermolysis of dioxole, comonomer from which DuPont's Teflon-AF was made, presented a rare pathway to an oxirane. The rearrangement of heterocyclic compounds occurred through a biradical intermediate (Schemes 5 and 6) [22,23].

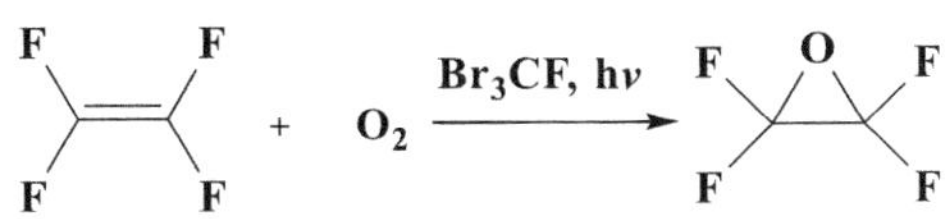

SCHEME 5

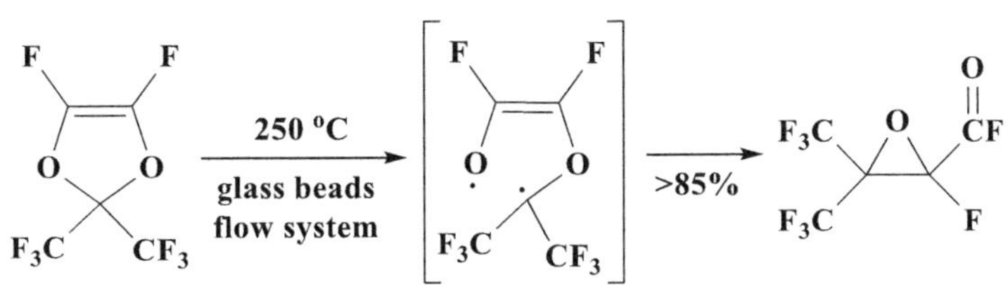

SCHEME 6

The diphosphines such as tetraethyl- and tetraphenyl-phosphines were reacted with benzophenone under the ultraviolet irradiation. The secondary products after further reaction with sulfur were thiobenzophenone, tetraphenylethylene, diphenylmethyldiphenylphosphine oxide, tetraaphenyloxirane, phosphinic acid, and tetraphenylthiirane (Scheme 7) [24].

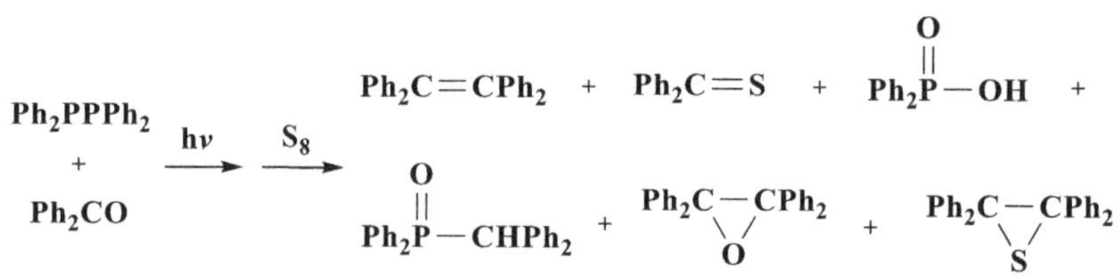

SCHEME 7

Tanaka et al. [25] reported a photopromoted ruthenium-catalyzed asymmetric aerobic sulfide oxidation and epoxidation. An oxidation of several sulfides utilizing Ru-salen complexes as catalysts occurred with enantioselectivities up to 98% *ee*. Modest to high enantioselectivities (76%–92% *ee*) were also showed by an epoxidation of conjugated olefins (Scheme 8). The role of H_2O as a proton-transfer mediator was studied [26].

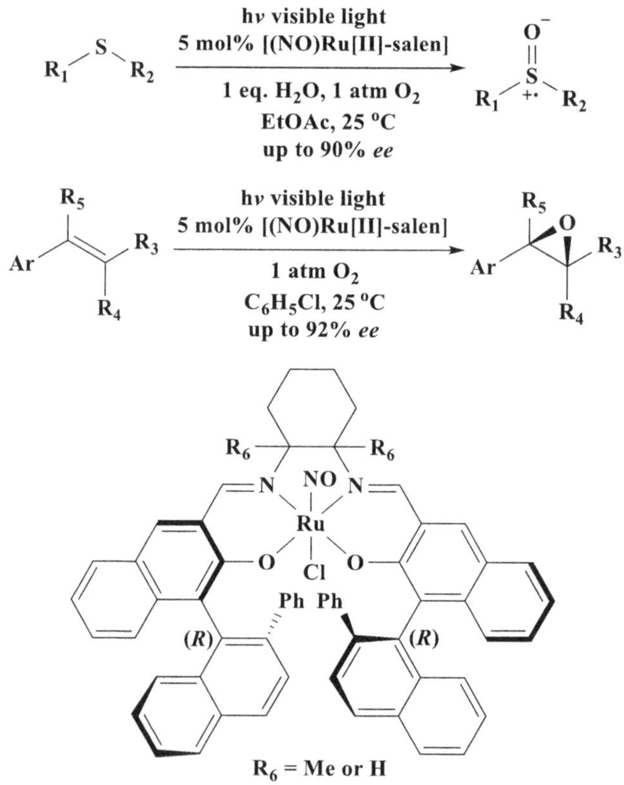

SCHEME 8

During an examination of photochemistry of cyclopentadienone-*O*-oxide, it became apparent that one of the photoproducts was an unknown cyclopentadienone epoxide. The aryl and alkyl epoxide derivatives are well-known intermediates in the photochemical transformation of 4-pyrones to 2-pyrones [27]. The reversible opening of epoxide ring of aryl-substituted cyclopentadienone epoxides occurred to afford the pyrylium salt, and irreversible isomerization of both alkyl- and aryl-substituted cyclopentadienone epoxides provided 2-furaldehydes and 2-pyrone. All efforts to synthesize the parent cyclopentadienone epoxide have been frustrated by the sensitivity of this system to acid-catalyzed rearrangement and nucleophilic attack (Scheme 9) [28].

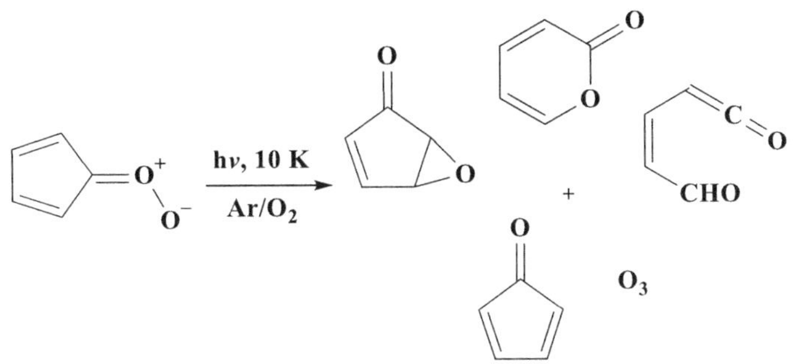

SCHEME 9

The α-iodo epoxides were formed in 72%–83% yield by alkoxyl radical rearrangement when *t*-allylic alcohols were reacted with DIB/iodine in cyclohexane under the irradiation (Scheme 10) [29,30].

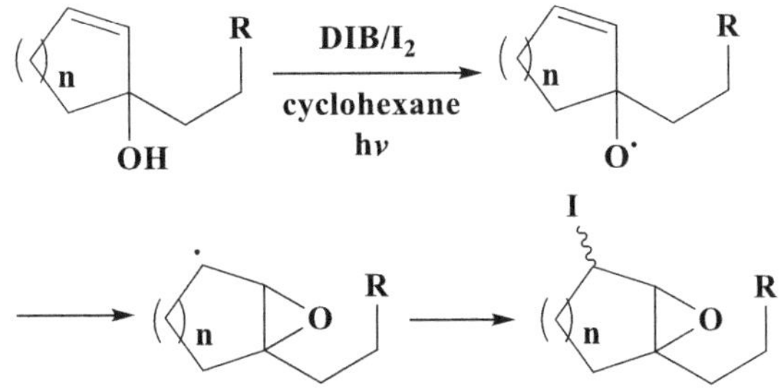

n	R	η	ratio
1	Et	72%	3.5:1
2	*i*-Pr	81%	4.4:1
3	*i*-Pr	83%	>19:1

SCHEME 10

The photochemical [6π + 2π + 2π]-cycloaddition of Cr-complexed cycloheptatriene and Si-tethered diyne gave diol cycloadduct after deprotection in 67% yield. The epoxide was synthesized in 67% yield by a Sharpless epoxidation utilizing vanadyl acetylacetonate and *t*-butyl hydroperoxide and molecular sieves in dry dichloromethane. The diol was protected as di-*t*-butyldimethylsilyl ether with *t*-butyldimethylsilyl chloride, 4-dimethylaminopyridine, and triethylamine in dichloromethane. The dihydroxylation of double bond using osmium tetraoxide provided diol, which was then oxidized to diketone in excellent yield utilizing dimethylsulfoxide and trifluoroacetic anhydride (Scheme 11) [31,32].

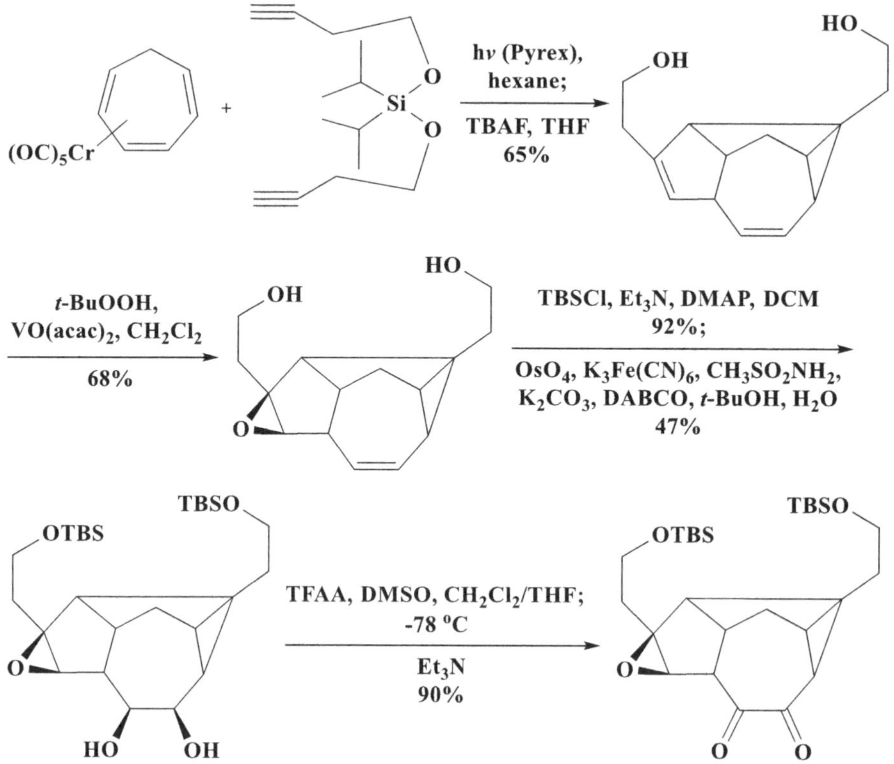

SCHEME 11

After screening various reaction conditions, the pathway in Scheme 12 provided the best results, where triquinane *C*-skeleton was obtained in fairly good yield and comparatively in a small number of steps. The expected Cr complex was synthesized in excellent yield by the photochemical [6π + 2π]-cycloaddition of 1-trimethylsilyl-1-propyne and (cycloheptatriene)-tricarbonyl-chromium (0). A crude mixture of desired cycloadduct as a single regioisomer, obtained by an irradiation of Cr complex when *t*-butyldimethylsilyloxy-3-butyne was present, was further treated with IM solution of tetra-*n*-butyl ammonium fluoride for the deprotection of silyl group. The alcohol cycloadduct was obtained in 72% yield over two steps. Under the Sharpless epoxidation conditions, an epoxidation of homo-allylic alcohol resulted in the formation of epoxide in excellent yield [33].

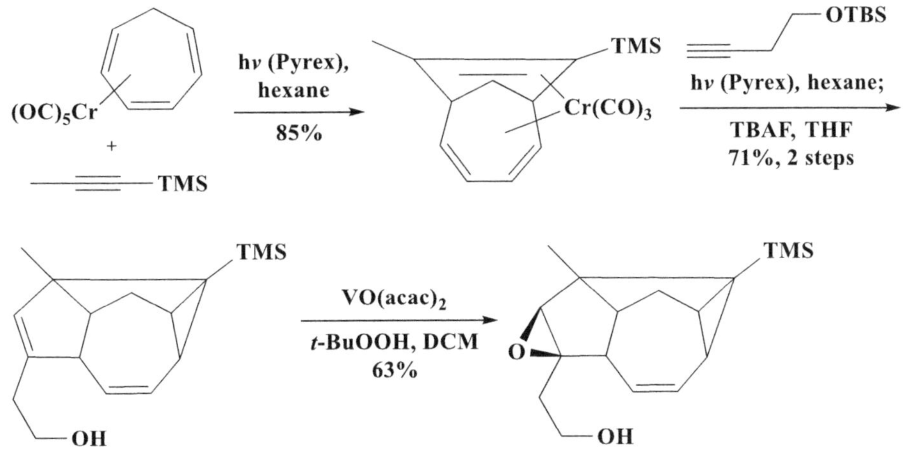

SCHEME 12

The trehazolamine is the cyclopentenol component of trehalase inhibitor trehazolamine [34]. The separable bridged aziridines were synthesized by an irradiation of *m*-hydroxymethylpyridinium bromide. The aziridine was transformed into oxazolidinone with complete retention of configuration of allylic carbon, which was present α to nitrogen; hence, inversion of this stereocenter was followed by *t*-butyloxycarbonyl-protection of the formed amine and cyclization. The allylic alcohol was synthesized by the basic cleavage of oxazolidinone, which resulted in an epoxidation to synthesize the *syn*-diastereomeric epoxide (Scheme 13) [35].

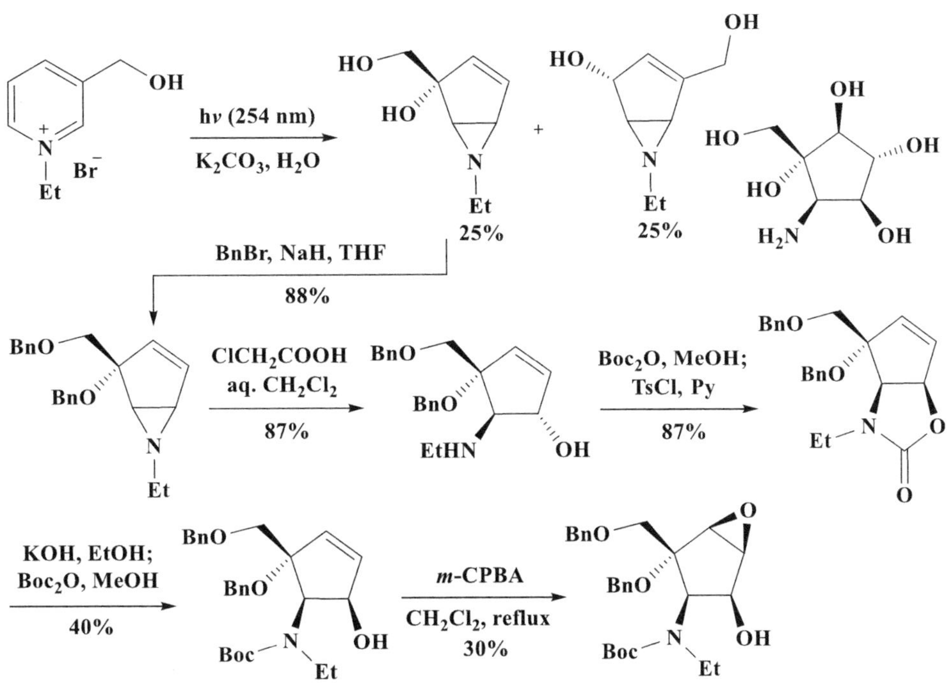

SCHEME 13

For enzymatic oxygen atom transfer reactions, the photolysis of heteroaromatic *N*-oxides has been verified to be beneficial as a mechanistic model. The naphthol and 1,2-naphthalene oxide were prepared by an irradiation of solutions containing pyridine-*N*-oxide and naphthalene under the ultraviolet light (Scheme 14) through the intermediacy of arene oxides during oxygenation of aromatic compounds [36]. This system also affected aliphatic sulfoxidation and hydroxylation reactions typical of monooxygenases. The aromatic hydroxylations were also brought about when pyrazine-*N*-oxide or pyridazine-*N*-oxide acted as oxygen donors in the photolytic procedure.

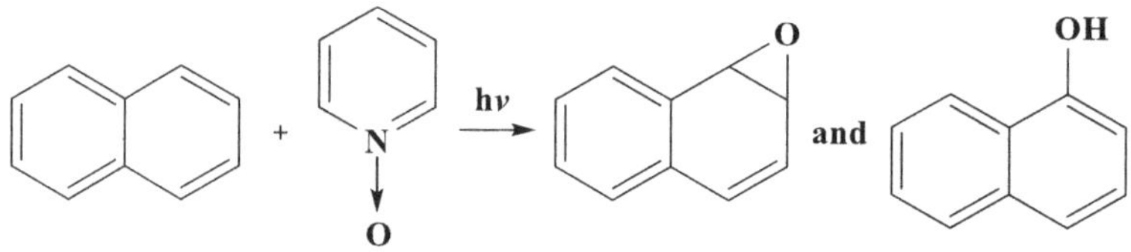

SCHEME 14

An epoxidation of 1,2-dioxenes, synthesized by an addition of singlet oxygen to dienes, was carried out by reaction with *m*-CPBA (Scheme 15). It was shown that the epoxidized dioxanes and dioxenes have inhibiting activity against the causative agents of candidiasis infections *Candida krusei*, *Candida tropicalis*, and *Candida albicans* that are in some cases comparable with the activity of currently utilized nystatin, amphotericin B, and ketonazole [37–44]. Additionally, these compounds show pronounced antimalarial activity, although lower than that of artemisinin [45–47].

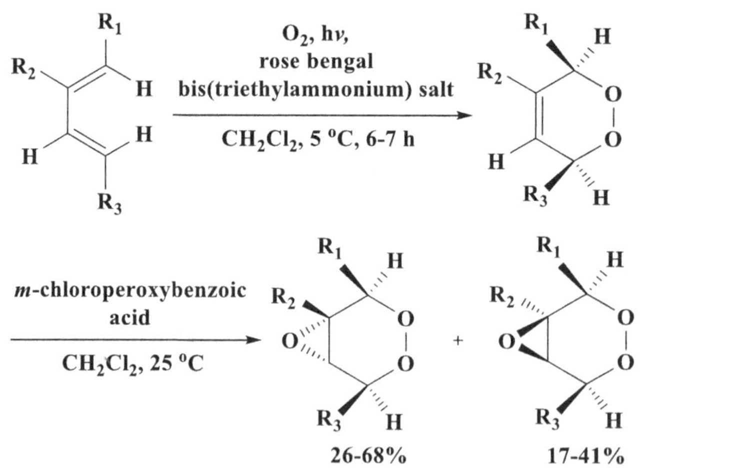

R_1, R_2, R_3 = H, Me, adamantyl (Ad), CH_2Ad, Ph, alkyl, cycloalkyl, CH_2OH, CH_2OCOAd, CH_2OCOCH_2Ad, $(CH_2)_3OH$, CH_2OCH_2Ph, F-Ph, Cl-Ph, Br-Ph, I-Ph, MeOPh

SCHEME 15

Matturro et al. [48] reported an alternative formation of thiiranes directly from alkenes of synthetic potential. The episulfide was formed in a low (ca. 7%) yield upon heating thiophene *endo*-peroxide, which was obtained by the photooxygenation of thiophene using norbornene (Scheme 16). This mechanistic analysis encouraged [49] to use this direct episulfidation for synthetic objectives by optimizing the structure of thiophene *endo*-peroxide and the reaction conditions to obtain the episulfide in better yields. The thermolysis of thiophene *endo*-peroxide was performed using more reactive olefinic substrates as sulfur acceptors. The yields of episulfide were fairly good (53%–64%) for (*Z*)-cyclooctene, norbornene, cyclooctenols, cycloheptene, and cyclopentene and excellent (>90%) for (*E*)-cyclooctene. An initial configuration of alkene was conserved in the obtained episulfide in all cases [50].

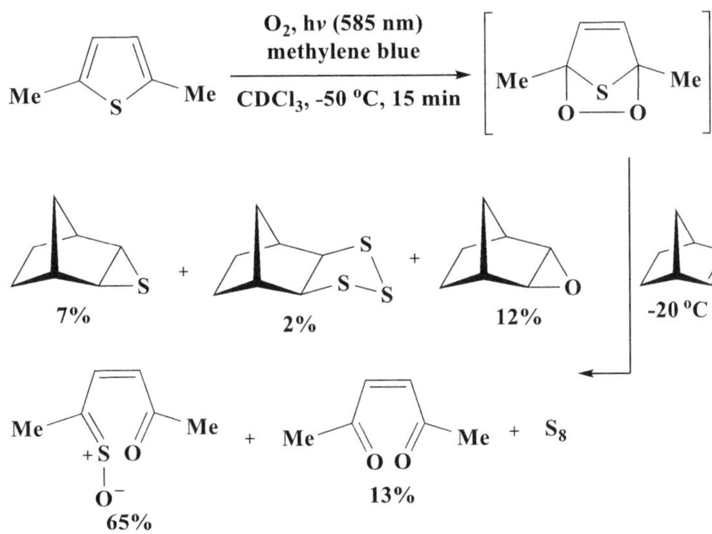

SCHEME 16

Paquette and coworkers [51] explored the direction of addition of singlet oxygen to 1,4-dimethoxynaphthalenes in which bridged bicyclic systems were fused across C-2 and C-3 in an ongoing analysis of electronic control of stereoselectivity. Utilizing rose bengal as a sensitizer, photochemical oxidation of starting compound (n = 1) provided primarily *endo*-epoxide (77%) along with 7% *exo*-epoxide, whereas for starting material (n = 2), the major product was *exo*-epoxide (77%) with only 13% *endo*-epoxide. These epoxidations have the reverse of the stereochemistry that was achieved by utilizing alkaline hydroperoxide as an oxidant (Schemes 17–20).

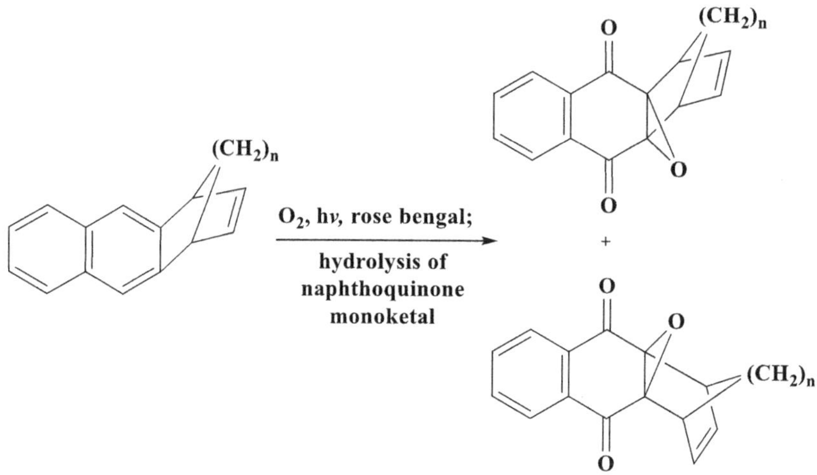

SCHEME 17

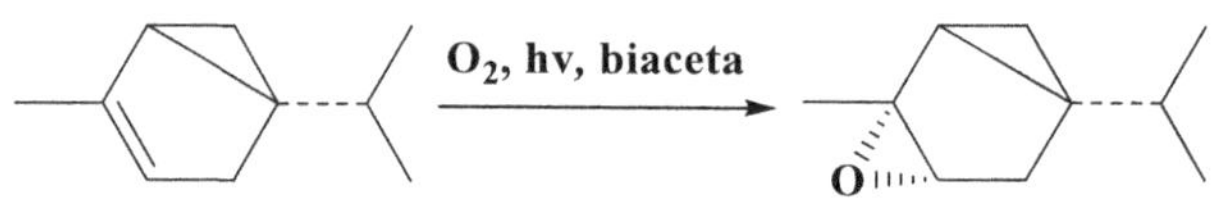

SCHEME 18

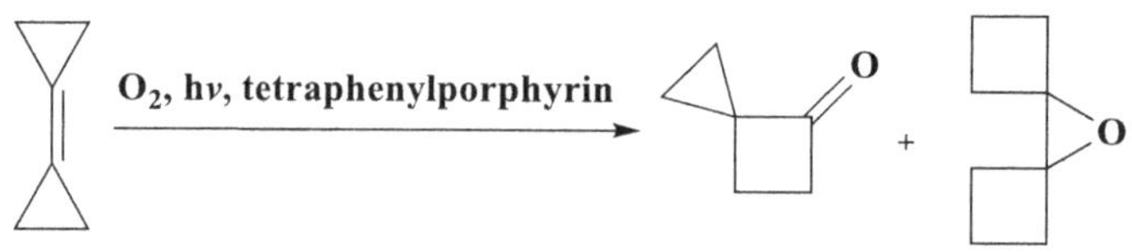

SCHEME 19

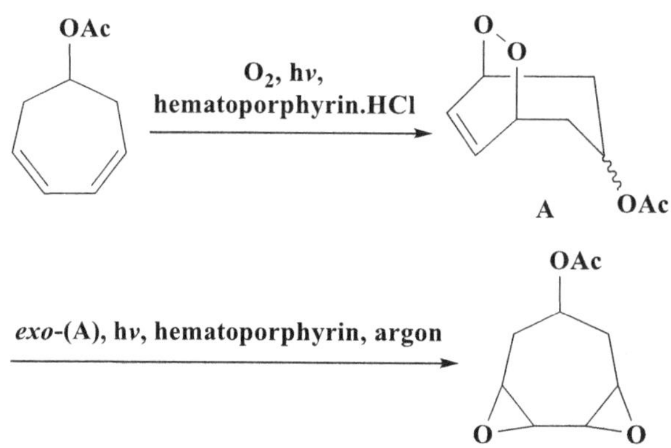

SCHEME 20

Fryer and Walser [52] synthesized 1,2-diazepine epoxides by the rearrangement of *endo*-peroxide with KOH in MeOH at 0°C (Scheme 21). The *endo*-peroxides were prepared in 3% and 60% yield by the dye-sensitized photooxygenation of ethoxycarbonyl-1,2-diazepines.

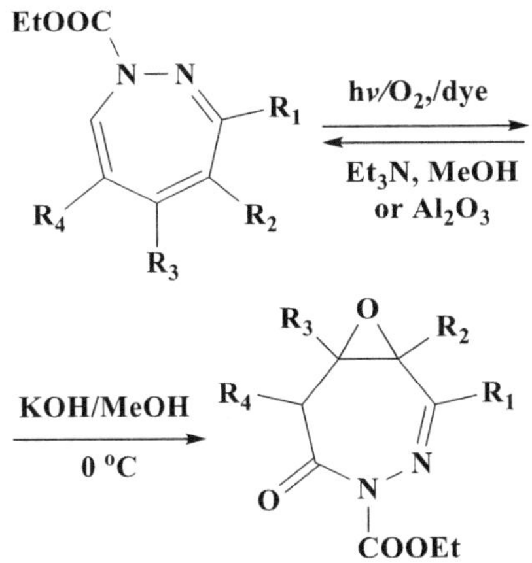

SCHEME 21

The reactions of 2,3-dihydrodiazepine afforded 2,3,4,5,6,6a-hexahydro-6a-methyl-1a*H*-oxireno[*d*][1,2]diazepine. In this case, the reaction of *endo*-peroxide with Al_2O_3 in CH_2Cl_2 provided intermediate in 80% isolated yield (Scheme 22). The nature of substituents affected the course of their rearrangement and the reactivity of these *endo*-peroxides [52].

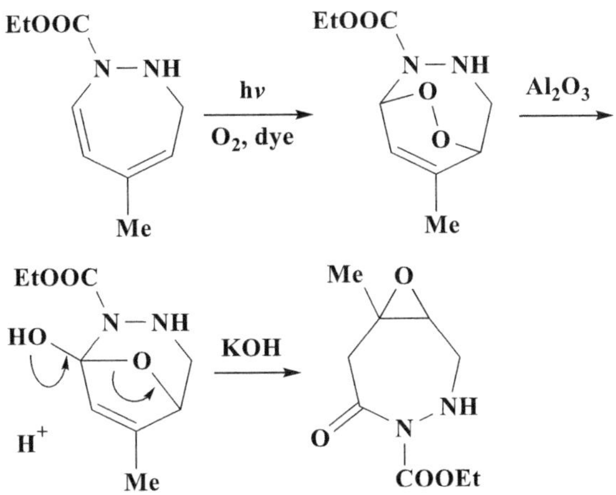

SCHEME 22

12.3 Photochemical synthesis of dioxiranes

The trifluoromethylphenylcarbene was obtained upon an irradiation of trifluoromethylphenyldiazomethane at 10 K in an Ar matrix. If the matrix was doped with oxygen, heating to 35 K led to trapping of carbene as trifluoroacetophenone oxide. The trifluoromethylphenyldioxirane and acetophenone were prepared by a subsequent photolysis of trifluoroacetophenone oxide with long wavelength light. The dioxirane rearranged to phenyl trifluoroacetate when photolyzed at shorter wavelengths (Scheme 23) [52,53].

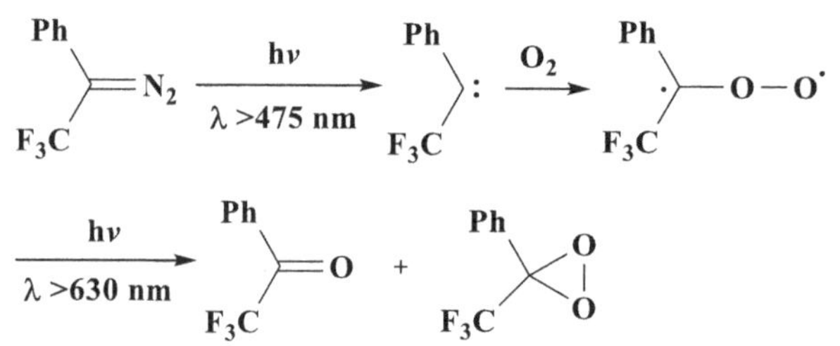

SCHEME 23

A study of synthetic approaches for dioxiranes is a little peculiar because the area is dominated by virtually a single procedure. There are alternative conditions that resulted in dioxirane system like the photolytic oxidation of diaryldiazoalkanes (Scheme 24) [53]; however, they have not been extensively adopted by synthetic laboratory. Actually, various procedures involving dioxiranes as reactive intermediates were designed to synthesize these species in situ, typically as a part of catalytic cycle and typically by oxidizing a ketone precursor. This has not only brought the advantage of requiring less than stoichiometric amounts of ketone substrates (a few of which were quite dear), but it also avoided the need to isolate, store, and frequently titrate mixtures of unstable dioxiranes. Nevertheless, sometimes it was appropriate or even required to work with isolated dioxirane reagents. The dilute solutions of dimethyldioxirane are accessible by reacting $(CH_3)_2CO$ with Oxone in an aqueous bicarbonate buffer at low temperature, after which the dioxirane was distilled at reduced pressure (Scheme 25). The obtained distillate possesses about a 0.1 M solution of dimethyldioxirane in $(CH_3)_2CO$ (toward which the dioxirane was stable [54]), and this was usually titrated before utilization [55]. The dilution factor could be problematic [56], although several practical modifications have been disclosed [57,58] including phase-transfer conditions, which could be employed even to large-scale preparations [59].

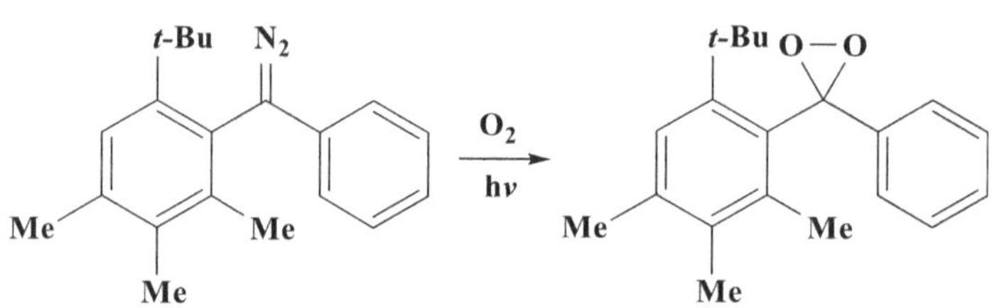

SCHEME 24

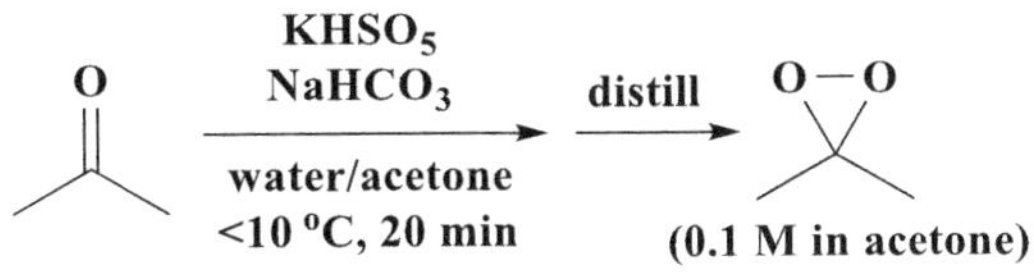

SCHEME 25

12.4 Photochemical synthesis of oxaziridines

Carmona et al. [60] reported that the irradiation products of α-styryl-*N*-phenylnitrone with UV light were azoxybenzene, cinnamaldehyde, and azobenzene. They proposed that the oxaziridine intermediate was unstable and these products were obtained by decomposition. A solution of α-styryl-*N*-phenylnitrone in deuterated DCM was photolyzed at a wavelength of 350 nm. The shifting of K-band and the higher E-max value for α-styryl-*N*-phenylnitrone could be because of extra conjugation present in the α-styryl-*N*-phenylnitrone. The α-styryl-*N*-phenylnitrone was irradiated for 30 min. During this time, the color of solution changed from bright yellow to only slightly yellow, and the ultraviolet spectrum showed the growth of 261 and 352 nm absorptions. Orange oil remained upon the removal of solvent. An analysis of nuclear magnetic resonance spectrum of this reaction product proved the presence of cinnamaldehyde as a major photolysis product along with unreacted nitrone and α-styryl-*N*-phenyloxaziridine (Scheme 26). In a consequent reaction, the α-styryl-*N*-phenylnitrone was irradiated for a shorter time (18 min) to provide the α-styryl-*N*-phenyloxaziridine as a major product as revealed by nuclear magnetic resonance study. Storage for 30 min at 36°C did not result in change in the nuclear magnetic resonance spectrum.

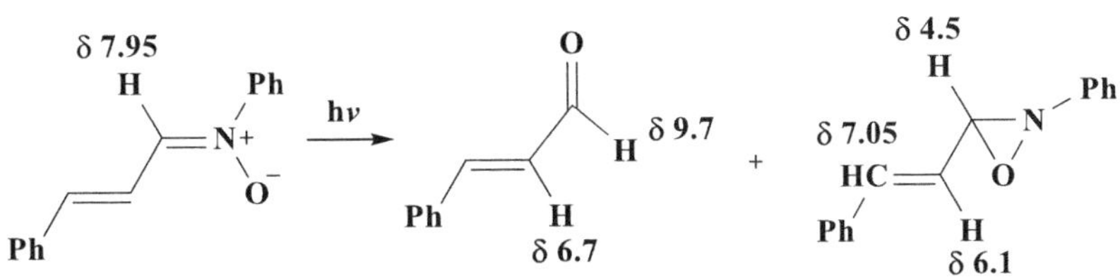

SCHEME 26

The α-styryl-*N*-*t*-butyloxaziridine was obtained by an irradiation of α-styryl-*N*-*t*-butyl nitrone for 15 min in deuterated CH_2Cl_2 solution. The structure of oxaziridine was verified by chemical and spectral data (Scheme 27) [61].

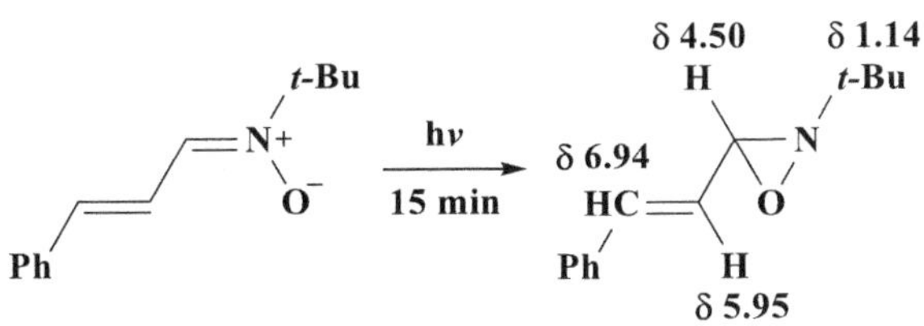

SCHEME 27

The stable α-styryl-*N*-benzyloxiziridine was synthesized by an irradiation of α-styryl-*N*-benzylnitrone for 15 min (Scheme 28). The nuclear magnetic resonance spectrum of this compound displayed an interesting doublet of doublets for diastereotopic *N*-benzyl protons. The nitrogen atom in the oxaziridine structure underwent an inversion only very slowly (or not at all), unlike the nitrogen in the starting nitrone in which the *N*-benzyl protons appeared as a singlet at δ 4.9 [61].

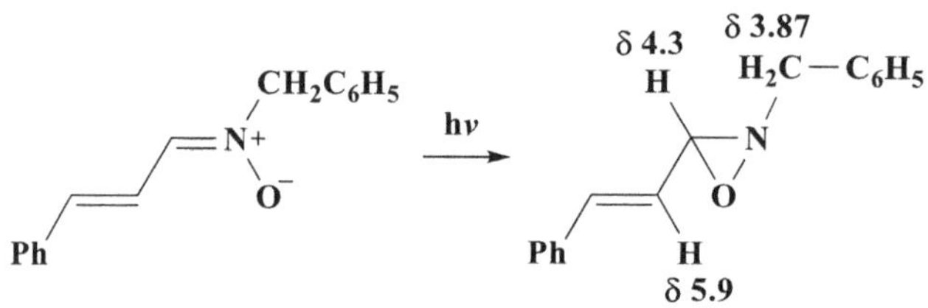

SCHEME 28

The α-phenyl-*N*-benzyloxiziridine was formed as a major product upon photochemical irradiation of α-phenyl-*N*-benzylnitrone for 15 min. The synthesis of oxaziridines was proved by the chemical analysis and spectral data (Scheme 29) [61].

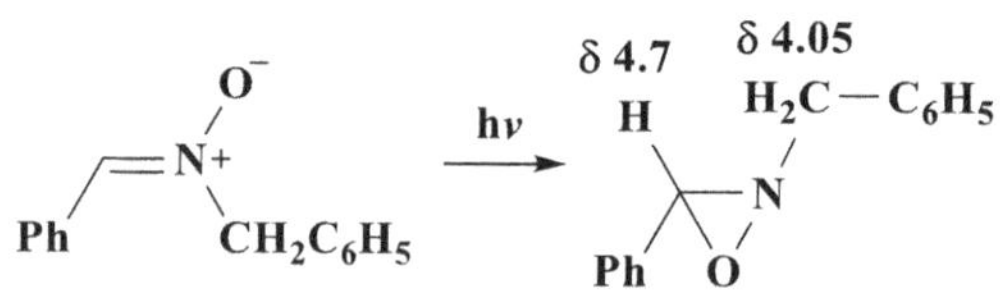

SCHEME 29

The nitrones were isomerized to oxaziridines photochemically. This reaction reversed thermally (Scheme 30) [62,63].

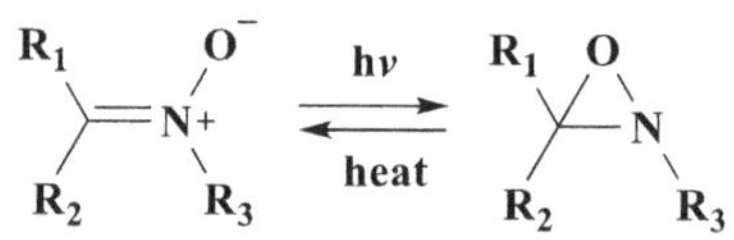

SCHEME 30

The trifluoromethylated hydroxyimines were formed diastereo- and chemoselectively when camphorquinone monoimines were reacted with (trifluoromethyl)trimethylsilane (Ruppert–Prakash reagent [64]). Unexpectedly, an oxidation of trifluoromethylated hydroxyimines with *m*-chloroperoxybenzoic acid provided nitrones instead of desired oxaziridines (Scheme 31) [65,66]. However, an oxaziridine derivative was synthesized by the photolysis of an isolated nitrone as a mixture of *exo*- and *endo*-isomers in one case.

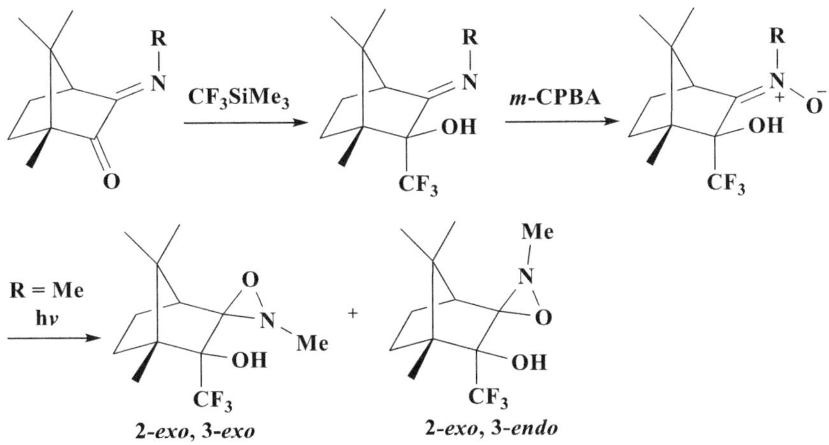

SCHEME 31

12.5 Photochemical synthesis of thiiranes

Green et al. [67] reported the first instance of allene episulfidation (Scheme 32). The ultraviolet photolysis of carbonyl sulfide using allene led to methylene thiirane as major product as verified by the mass spectrometry, nuclear magnetic resonance, and infrared spectroscopy data. Although the product persisted for many hours in the gas phase at rt, it polymerized during storage at temperatures as low as −196°C [68].

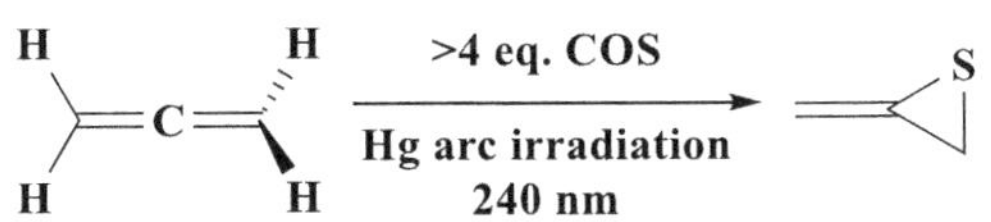

SCHEME 32

The episulfidation of methylallene afforded all three probable methylene thiiranes favoring the reaction of more substituted olefin (Scheme 33). The products also decomposed in less than a day at rt. A small degree of allenic carbon–hydrogen insertion at the CH_3 group was also reported. While these studies showed that allene episulfidation was possible, the results were not synthetically applicable [68,69].

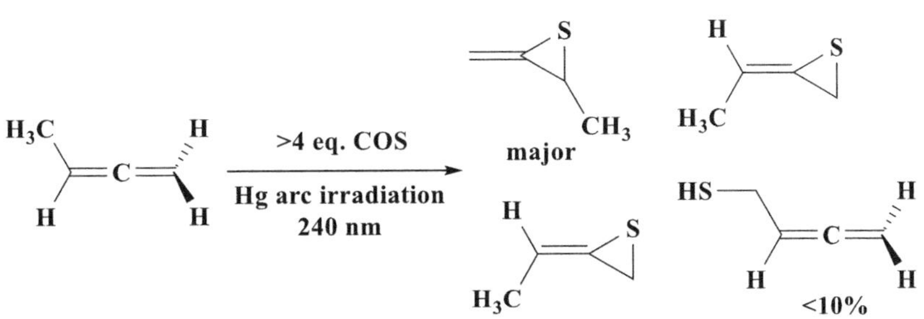

SCHEME 33

The reactions of triplet atomic sulfur were performed (Scheme 34). This radical-type atomic sulfur was added to (*Z*)-butene to synthesize the diradical, which on cyclization and intersystem crossing provided thiirane; an insertion into a carbon–hydrogen bond to synthesize the mercaptans was not noticed [70]. Because of the fast-intersystem crossing (heavy atom effect of sulfur) [71], the episulfidation with $S(^3P)$ took place under the retention of alkene configuration [72]. Since the trapping of $S(^3P)$ by a carbon–carbon double-bond substrate was very efficient, triplet atomic sulfur was also a good candidate for the stereoselective episulfidation of allenes (Scheme 35) [67]. For stoichiometric preparative episulfidation, the triplet atomic sulfur was utilized because of these reasons: (1) the photolysis of COS was conducted at around 254 nm wavelengths, and under these reaction conditions, the thiiranes did not persist sufficiently, since the small absorption coefficient of COS needed prolonged reaction times; and (2) the catenation tendency of triplet sulfur atoms synthesized S_8, which may only be suppressed at low COS pressure and becomes impractical at the preparative scale [73,74]. An approach to overcome these problems was to carry out the reaction in the liquid phase in solvents such as aromatic hydrocarbons [75], cyclohexene [76,77], alcohols, and CH_3CN [78]; nevertheless, the yields of episulfide were generally less than 20%. The utilization of isothiocyanates as direct substrate to triplet atomic sulfur [79–81] increased the yield of episulfide up to 45% at 76% conversion of the alkene as demonstrated for cyclohexene (Scheme 36) [82]. This was noteworthy as cyclohexene was quite unreactive toward direct episulfidation [50].

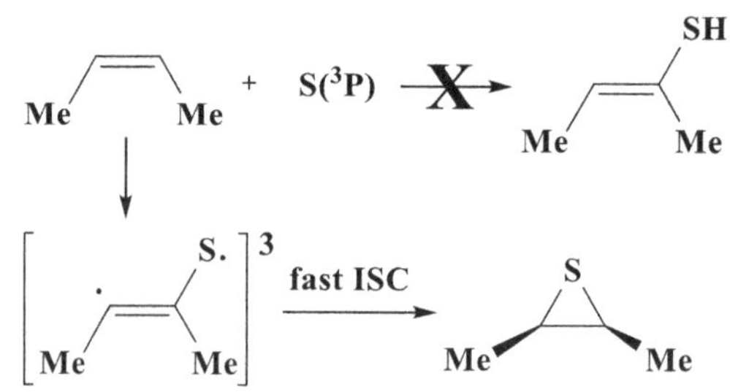

SCHEME 34

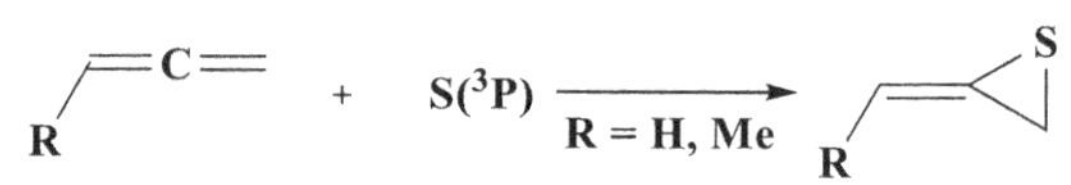

SCHEME 35

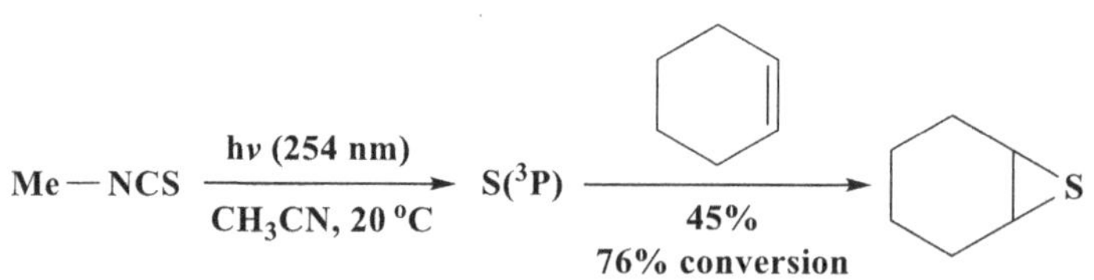

SCHEME 36

This method involved the in situ formation and used highly reactive thiocarbonyl-*S*-sulfides. Classically, the thiocarbonyl-*S*-sulfides were reacted as sulfur-rich 1,3-dipoles and trapped both thiocarbonyl precursors and electron-poor *C*,*C*-dipolarophiles providing five-membered cycloadducts. In some examples, the intermediate thiocarbonyl-*S*-sulfides and/or their cyclic isomers (i.e., dithiiranes) served as sulfur-transfer agents to afford the sulfur-rich heterocyclic compounds such as pentathiepanes and hexathiepanes (Scheme 37) [68].

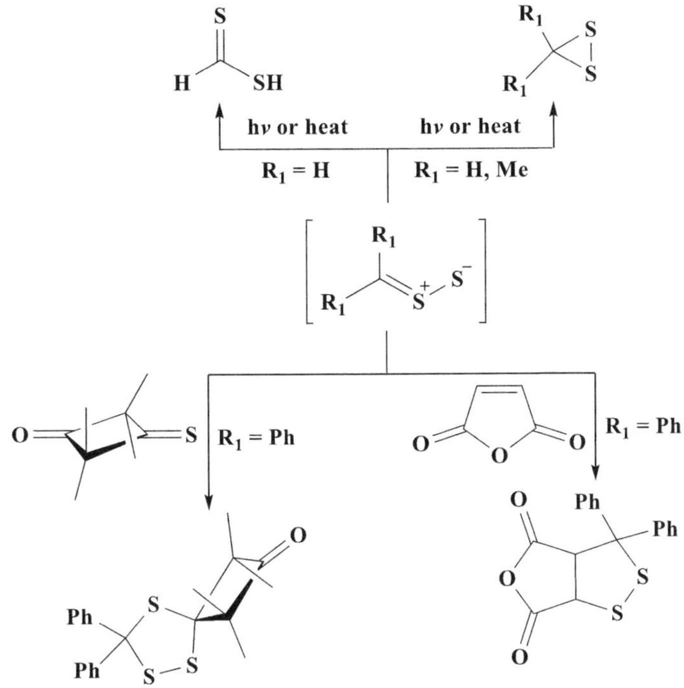

SCHEME 37

The chemistry of thiiranyl acetals has not been well studied. For the formation of thiiranyl acetals, only two approaches have been reported. Singh et al. [83,84] synthesized thiiranyl acetals by the photochemical addition of alcohols to thioketenes (Scheme 38), and Beiner et al. [85,86] synthesized 2-alkoxy thiiranes by reacting thioesters with diazo compounds (Scheme 39).

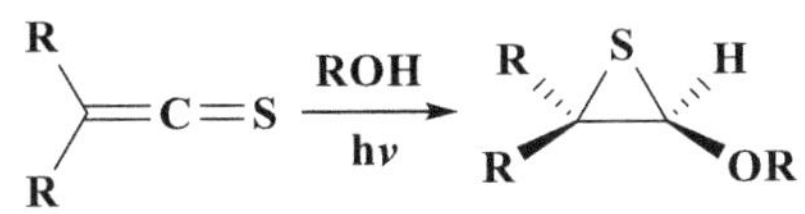

SCHEME 38

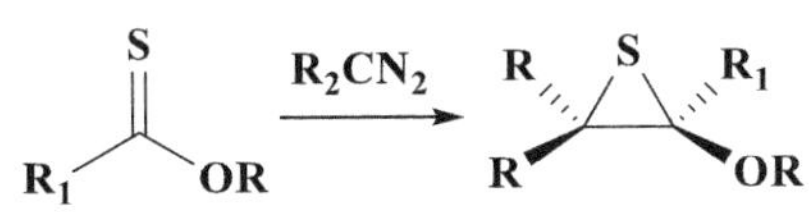

SCHEME 39

The photocycloelimination afforded a synthetic pathway for the formation of thiirane via photolysis of pyrazole (Scheme 40) [87]. The thiiranes were prepared from oxothiole ($R_1 = H$ or phenyl, $R_2 = H$ or methyl, $R_3 =$ various aryl or alkyl) by reacting with methyl iodide and nitro methane followed by ethanol–sodium ethoxide (Scheme 41) [88].

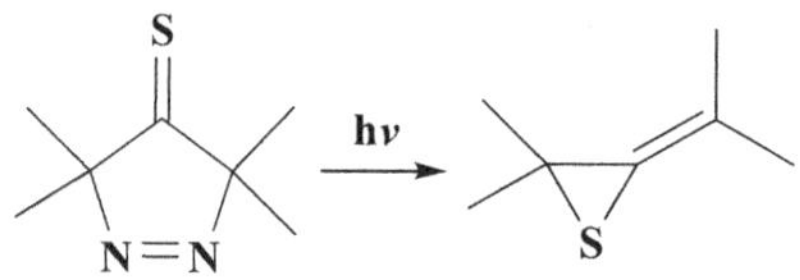

SCHEME 40

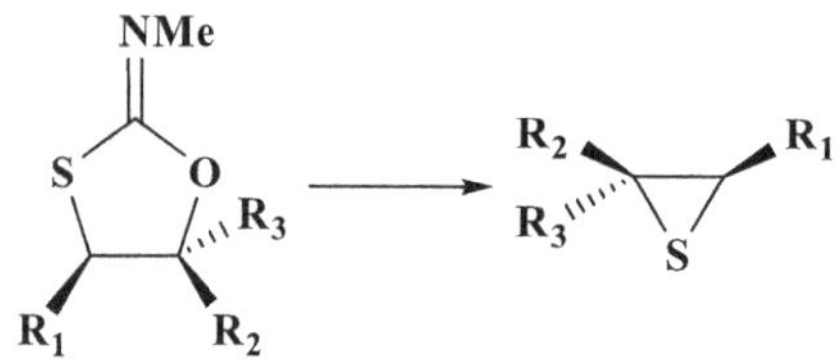

SCHEME 41

The valence isomerization provided fluorothiiranes. The perfluorotetramethylthiophene was isomerized to Dewar isomer in the vapor phase. The reaction of triplet sulfur atoms with fluoroalkynes and fluoroalkenes also gave thiiranes among other products (Scheme 42) [22,23].

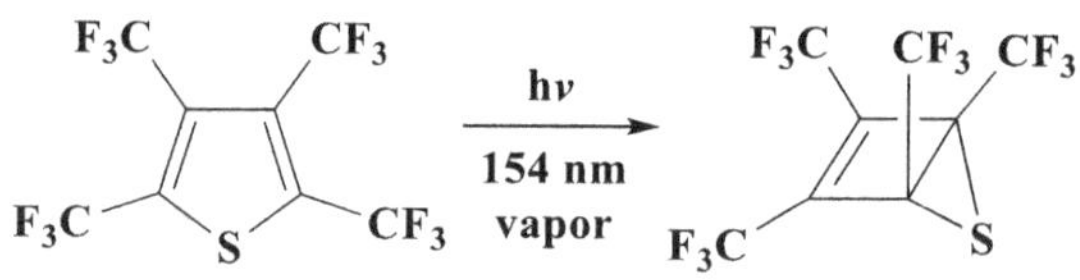

SCHEME 42

The Dewar isomers were formed in 8:1 ratio by an ultraviolet irradiation of 2,3-bis(trifluoromethyl) thiophene along with all four bis(trifluoromethyl)thiophenes. The reaction of mixture with 2,5-dimethylfuran has proved that Dewar isomers were thermally equilibrated with one another. Only one Dewar isomer underwent a Diels–Alder reaction with diene, providing adduct, but other Dewar isomer disappeared gradually with concomitant formation of more adduct (Scheme 43) [22,23].

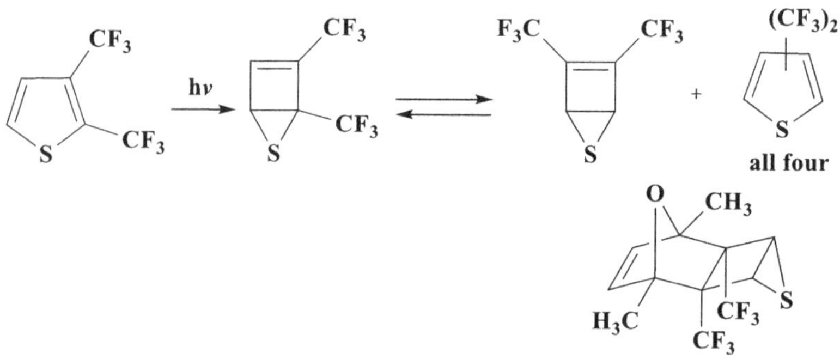

SCHEME 43

The formation and characterization of matrix-isolated thiirene (R = H) was reported [89]. For the existence of thiirene, three pieces of evidence were applied. First, the same labeled C_2H_2S species was synthesized by an irradiation of thiadiazoles (R = H) that carry a specific ^{13}C-label at either of the carbon atoms. Secondly, the photoisomerization of labeled C_2H_2S species formed $H_2C{=}C{=}S$ and $HC{\equiv}CSH$, each with randomized label, and thirdly, the infrared spectrum of thiirene was consistent with a cyclopropenol species (Scheme 44) [90].

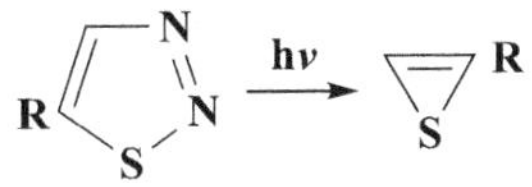

SCHEME 44

The photolysis and thermolysis of 1,2,3-benzothiazoles and 1,2,3-thiazoles resulted in the removal of nitrogen. Based on the substituents in the five- and four-positions, the fragments achieve stability to different degrees providing primarily thiirenes and/or thioketenes (Scheme 45) [91].

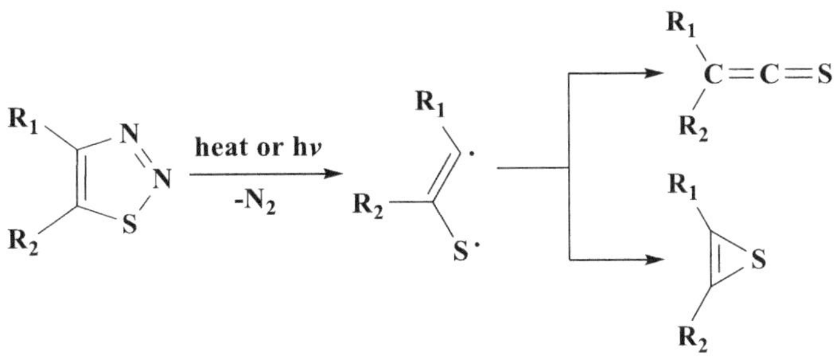

SCHEME 45

In medicinal chemistry and organic synthesis, the 1,2,3-thiadiazoles and their derivatives make an important class of molecules [92], and their rich photochemistry has been analyzed by the physical, chemical, and computational methods [93]. The light irradiation of 1,2,3-thiadiazoles resulted in the extrusion of N_2 and generation of intermediates such as thioketenes and thiirenes. Both species have been directly characterized in various studies in cryogenic matrices [89,94–97]. The theoretical intermediacy of thiocarbenes and 1,3-diradicals as substrates of thioketenes or thiirenes was not clear. The mechanistic details of photoreaction were studied by time-resolved spectroscopy for MCT (4-methyl-5-carboethoxy-1,2,3-thiadiazole) in solution phase (Scheme 46) [98]. The structure immediately collapsed to thiirene when N_2 extrusion occurred from 1,2,3-thiadiazole. The thioketene was synthesized in a hetero-Wolff rearrangement possibly in a concerted procedure without interference of thiocarbene [99].

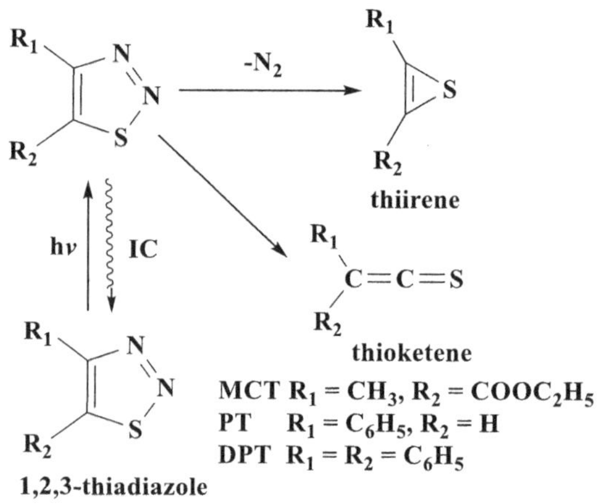

SCHEME 46

The *exo*-adducts were formed predominantly in most cases. The *endo*-derivatives also were prone to rearrange into *exo*-products on a silica gel column. The *exo*-adducts were usually photochemically inactive; therefore, focus was turned only on the *endo*-adducts. Surprisingly, the *exo*-3-benzoyl-2-thiabicyclo[2.2.1]hept-5-enes on an irradiation in Rayonet photoreactor with RPR-3500 lamps or UVLED-based illuminator with five 250 mW @ 365 nm Nichia chips were rearranged to provide the thiirane as a major product (Schemes 47 and 48) [100–102].

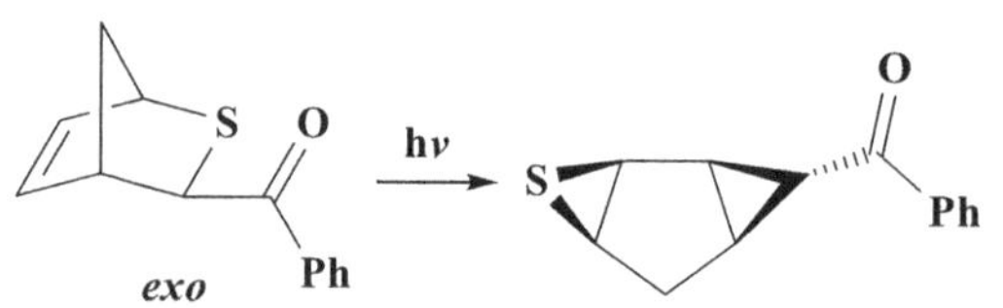

SCHEME 47

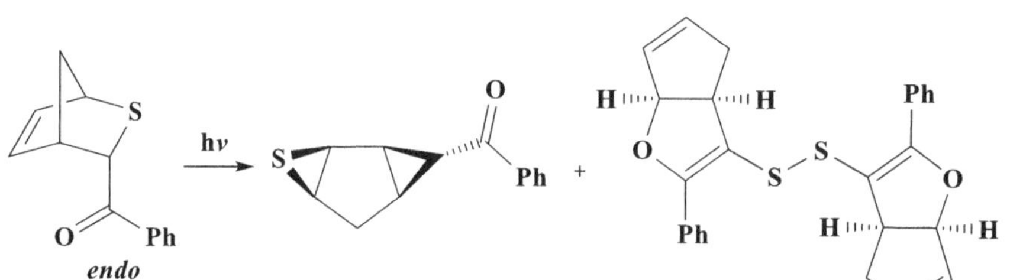

SCHEME 48

A plausible mechanism for the synthesis of thiiranes included a homolytic carbon–sulfur bond fragmentation providing two radicals, which were then reacted with double bond, forming the final product. The cyclopropyl and thiirane rings were in a *syn*-configuration with respect to each other (Scheme 49) [103].

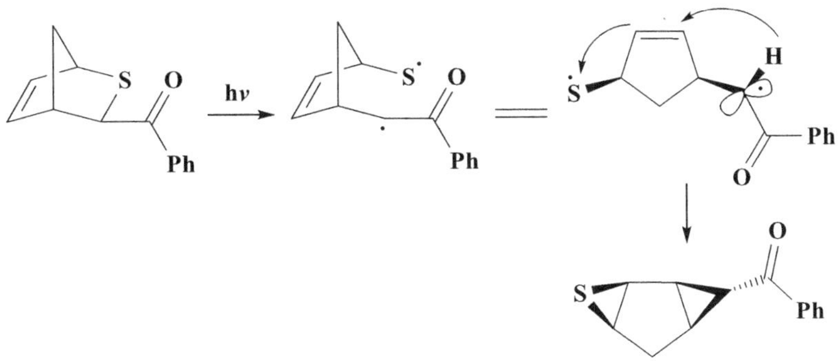

SCHEME 49

References

1. (a) N. Kaur, N. Ahlawat, Y. Verma, P. Grewal, P. Bhardwaj and N.K. Jangid. 2021. Crown ethers for the synthesis of heterocycles. Curr. Org. Chem. 25: 1270-1297. (b) N. Kaur. 2015. Polycyclic six-membered *N*-heterocycles: microwave-assisted synthesis. Synth. Commun. 45: 35-69. (c) N. Kaur. 2015. Application of microwave-assisted synthesis in the synthesis of fused six-membered heterocycles with *N*-heteroatom. Synth. Commun. 45: 173-201. (d) N. Kaur. 2015. Microwave-assisted synthesis of fused polycyclic six-membered *N*-heterocycles. Synth. Commun. 45: 273-299. (e) N. Kaur. 2015. Review on the synthesis of six-membered *N,N*-heterocycles by microwave irradiation. Synth. Commun. 45: 1145-1182. (f) N. Kaur. 2015. Review of microwave-assisted synthesis of benzo-fused six-membered *N,N*-heterocycles. Synth. Commun. 45: 300-330. (g) N. Kaur. 2014. Microwave-assisted synthesis of five-membered *O*-heterocycles. Synth. Commun. 44: 3483-3508. (h) N. Kaur. 2014. Microwave-assisted synthesis of five-membered *O,N*-heterocycles. Synth. Commun. 44: 3509-3537.
2. (a) N. Kaur, P. Bhardwaj and M. Gupta. 2021. Recent developments in the synthesis of five- and six-membered *N*-heterocycles from dicarbonyl compounds. Curr. Org. Chem. 25: 2765-2790. (b) N. Kaur. 2018. Recent developments in the synthesis of nitrogen-containing five-membered polyheterocycles using rhodium catalysts. Synth. Commun. 48: 2457-2474. (c) N. Kaur. 2017. Gold catalysts in the synthesis of five-membered *N*-heterocycles. Curr. Organocatal. 4: 122-154. (d) N. Kaur. 2019. Palladium acetate and phosphine assisted synthesis of five-membered *N*-heterocycles. Synth. Commun. 49: 483-514. (e) N. Kaur. 2019. Applications of palladium dibenzylideneacetone as catalyst in the synthesis of five-membered *N*-heterocycles. Synth. Commun. 49: 1205-1230. (f) N. Kaur, P. Grewal, P. Bhardwaj, M. Devi, N. Ahlawat and Y. Verma. 2020. Synthesis of five-membered *N*-heterocycles using silver metal. Synth. Commun. 49: 3058-3100. (g) N. Kaur. 2015. Synthesis of fused five-membered *N,N*-heterocycles using microwave irradiation. Synth. Commun. 45: 1379-1410. (h) N. Kaur. 2015. Role of microwaves in the synthesis of fused five-membered heterocycles with three *N*-heteroatoms. Synth. Commun. 45: 403-431.
3. (a) N. Kaur, P. Grewal and K. Poonia. 2021. Dicarbonyl compounds in *O*-heterocycle synthesis. Synth. Commun. 51: 2423-2444. (b) N. Kaur. 2019. Multiple nitrogen-containing heterocycles: metal and non-metal assisted synthesis. Synth. Commun. 49: 1633-1658. (c) N. Kaur. 2018. Synthesis of six- and seven-membered and larger heterocycles using Au and Ag catalysts. Inorg. Nano Met. Chem. 48: 541-568. (d) N. Kaur, Y. Verma, P. Grewal, P. Bhardwaj and M. Devi. 2019. Application of titanium catalysts for the syntheses of heterocycles. Synth. Commun. 49: 1847-1894. (e) N. Kaur. 2019. Cobalt-catalyzed C-N, C-O, C-S bond formation: synthesis of heterocycles. J. Iran. Chem. Soc. 16: 2525-2553. (f) N. Kaur. 2018.

Mercury-catalyzed synthesis of heterocycles. Synth. Commun. 48: 2715-2749. (g) N. Kaur. 2019. Ionic liquid assisted synthesis of six-membered oxygen heterocycles. SN Appl. Sci. 1: 1-20. (h) N. Kaur. 2019. Ionic liquid: an efficient and recyclable medium for the synthesis of fused six-membered oxygen heterocycles. Synth. Commun. 49: 1679-1707.

4. (a) A.L. Margolin. 1993. Enzymes in the synthesis of chiral drugs. Enzym. Microb. Technol. 15: 266-280. (b) D.E. Uehling, K.H. Donaldson, D.N. Deaton, C.E. Hyman, E.E. Sugg, D.G. Barrett, R.G. Hughes, B. Reitter, K.K. Adkison, M.E. Lancaster, F. Lee, R. Hart, M.A. Paulik, B.W. Sherman, T. True and C. Cowan. 2002. Synthesis and evaluation of potent and selective β3 adrenergic receptor agonists containing acylsulfonamide, sulfonylsulfonamide, and sulfonylurea carboxylic acid isosteres. J. Med. Chem. 45: 567-583. (c) R.N. Bream, S.V. Ley and P.A. Procopiou. 2002. Synthesis of the β2 agonist (*R*)-salmeterol using a sequence of supported reagents and scavenging agents. Org. Lett. 4: 3793-3796. (d) D.J. Buchanan, D.J. Dixon and B.E. Lookerb. 2005. A short stereoselective synthesis of (*R*)-salmeterol. Synlett 12: 1948-1950. (e) T. Izumi, K. Satou and K. Ono. 1996. Baker's yeast reduction of α-(alkoxycarbonylamino)acetophenones and lipase-catalysed resolution of 2-(alkoxycarbonylamino)-1-arylethanols. J. Chem. Technol. Biotechnol. 66: 233-242.
5. B.E. Rossiter, T. Katsuki and K.B. Sharpless. 1981. Asymmetric epoxidation provides shortest routes to four chiral epoxy alcohols which are key intermediates in syntheses of methymycin, erythromycin, leukotriene C-1, and disparlure. J. Am. Chem. Soc. 103: 464-465.
6. R.D. Fabio, C. Pietra, R.J. Thomas and L. Ziviani. 1995. The asymmetric synthesis of both enantiomers of eliprodil. Bioorg. Med. Chem. Lett. 5: 551-554.
7. A. Kamal, G.B.R. Khanna and R. Ramu. 2002. Chemoenzymatic synthesis of both enantiomers of fluoxetine, tomoxetine and nisoxetine: lipase-catalyzed resolution of 3-aryl-3-hydroxypropanenitriles. Tetrahedron: Asymmetry 13: 2039-2051.
8. (a) N. Kaur. 2019. Synthesis of three-membered and four-membered heterocycles with the assistance of photochemical reactions. J. Heterocycl. Chem. 56: 1141-1167. (b) N. Kaur. 2018. Photochemical-mediated reactions in five-membered *O*-heterocycles synthesis. Synth. Commun. 48: 2119-2149.
9. (a) N. Kaur. 2018. Photochemical reactions as key steps in five-membered *N*-heterocycles synthesis. Synth. Commun. 48: 1259-1284. (b) N. Kaur. 2017. Photochemical reactions: synthesis of six-membered *N*-heterocycles. Curr. Org. Synth. 14: 972-998.
10. (a) N. Kaur, Y. Verma, P. Grewal, N. Ahlawat, P. Bhardwaj and N.K. Jangid. 2020. Photochemical C-N bond forming reactions for the synthesis of five-membered fused *N*-heterocycles. Synth. Commun. 50: 1286-1334. (b) N. Kaur. 2018. Photochemical reactions for the synthesis of six-membered *O*-heterocycles. Curr. Org. Synth. 15: 298-320. (c) N. Kaur. 2018. Photochemical irradiation: seven- and higher-membered *O*-heterocycles. Synth. Commun. 48: 2935-2964.
11. (a) N. Kaur, P. Bhardwaj, M. Devi, Y. Verma and P. Grewal. 2019. Photochemical reactions in five and six-membered polyheterocycles synthesis. Synth. Commun. 49: 2281-2318. (b) N. Kaur. 2018. Synthesis of seven- and higher-membered nitrogen-containing heterocycles using photochemical irradiation. Synth. Commun. 48: 2815-2849.
12. G.K.S. Prakash, R.W. Ellis, J.D. Felberg and G.A. Olah. 1986. Onium ylide chemistry. 5. Formaldehyde *O*-methylide, [$CH_2{=}O^+{-}CH_2$]: the parent carbonyl ylide. J. Am. Chem. Soc. 108: 1341-1342.
13. P.C. Wong, D. Griller and J.C. Scaiano. 1982. A kinetic study of the reactions of carbonyl ylides formed by the addition of fluorenylidene to ketones. J. Am. Chem. Soc. 104: 6631-6635.
14. T.T. Tidwell. 2006. Ketenes II. John Wiley and Sons Inc., ISBN-13: 978-0-471-69282-9, ISBN-10: 0-471-69282-4.
15. G.W. Griffin, R.L. Smith and A. Manmade. 1976. New precursors for arylcarbenes. Photocycloelimination reactions of cyclic carbonates. J. Org. Chem. 41: 338-345.
16. R.C. White and T. Rix. 1987. Photolysis of *meso*- and D,L-hydrobenzoin carbonate. J. Org. Chem. 52: 2309-2311.

17. R.S. Givens, R.J. Olsen and P.L. Wylie. 1979. Mechanistic studies in photochemistry. 21. Photoextrusion of sulfur dioxide: general route to [2.2]cyclophanes. J. Org. Chem. 44: 1608-1613.
18. G.W. Griffin and A. Manmade. 1972. New precursors for arylcarbenes. Photocycloelimination reactions of cyclic sulfites. J. Org. Chem. 37: 2589-2600.
19. N.J. Turro, C.H. Tung, I.R. Gould, G.W. Griffin, R.L. Smith and A. Manmade. 1984. Cycloelimination reactions of cyclic aryl pinacol sulfites and carbonates via photosensitized reversible electron transfer. J. Photochem. 24: 265-270.
20. R.J. Olsen and F.M. Butler. 1982. Singlet-triplet reactivity in the photo extrusion reactions of cyclic sulphites. J. Chem. Res. Synop. 7: 175-175.
21. R.C. White, B.E. Arney and H. Ihmels. 2012. Photoreactions of cyclic sulfite esters: evidence for diradical intermediates. Beilstein J. Org. Chem. 8: 1208-1212.
22. T. Eicher and S. Hauptmann. 2003. The chemistry of heterocycles: structures, reactions, and applications. John Wiley and Sons: Chichester, UK, 5-16.
23. R.E. Banks and J.C. Tatlow. 1994. Organofluorine chemistry: nomenclature and history. R.E. Banks, B.E. Smart and J.C. Tatlow (Eds.). Organofluorine chemistry: principles and commercial applications. Plenum Press: New York, 2-4.
24. R. Okazaki, K. Tamura, Y. Hirabayashi and N. Inamoto. 1976. Photoreactions of diphosphines with benzophenone. J. Chem. Soc. Perkin Trans. 1 18: 1924-1926.
25. H. Tanaka, H. Nishikawa, T. Uchida and T. Katsuki. 2010. Photopromoted Ru-catalyzed asymmetric aerobic sulfide oxidation and epoxidation using water as a proton transfer mediator. J. Am. Chem. Soc. 132: 12034-12041.
26. F. Teply. 2011. Photoredox catalysis by $[Ru(bpy)_3]^{2+}$ to trigger transformations of organic molecules. Organic synthesis using visible-light photocatalysis and its 20th century roots. Collect. Czech. Chem. Commun. 76: 859-917.
27. E. Weitz and A. Scheffer. 1921. Über die einwirkung von alkalischem wasserstoffsuperoxyd auf ungesättigte verbindungen. Chem. Ber. 54: 2327-2344.
28. O.L. Chapman and T.C. Hess. 1979. Cyclopentadienone epoxide: synthesis and photochemistry. J. Org. Chem. 44: 962-964.
29. P. Galatsis and S.D. Millan. 1991. Use of iodobenzene diacetate for the synthesis of α-iodoepoxides. Tetrahedron Lett. 32: 7493-7496.
30. V.H. Rawal and S. Iwasa. 1992. Radical-induced epoxide fragmentation chemistry. Atom transfer cyclization of iodoepoxides to cyclopentanols. Tetrahedron Lett. 33: 4687-4690.
31. B. Heller, B. Sundermann, H. Buschmann, H.J. Drexler, J. You, U. Holzgrade, E. Heller and G. Oehme. 2002. Photocatalyzed [2+2+2]-cycloaddition of nitriles with acetylene: an effective method for the synthesis of 2-pyridines under mild conditions. J. Org. Chem. 67: 4414-4422.
32. B. Heller, B. Sundermann, C. Fischer, J. You, W. Chen, H.J. Drexler, P. Knochel, W. Bonrath and A. Gutnov. 2003. Facile and racemization-free conversion of chiral nitriles into pyridine derivatives. J. Org. Chem. 68: 9221-9225.
33. B.A. Aleiwi. 2010. Transition metal-mediated higher-order cycloaddition reactions: application to the total synthesis of complex natural products. PhD Thesis, Wayne State University.
34. O. Ando, H. Satake, K. Itoi, A. Sat, M. Nakajima, S. Takahashi, H. Haruyama, Y. Ohkuma, T. Kinoshita and R. Enokita. 1991. Trehazolin, a new trehalase inhibitor. J. Antibiot. 44: 1165-1168.
35. T. Damiano, D. Morton and A. Nelson. 2007. Photochemical transformations of pyridinium salts: mechanistic studies and applications in synthesis. Org. Biomol. Chem. 5: 2735-2735.
36. D.M. Jerina, D.R. Boyd and J.W. Daly. 1970. Photolysis of pyridine-*N*-oxide: an oxygen atom transfer model for enzymatic oxygenation, arene oxide formation, and the NIH shift. Tetrahedron Lett. 6: 457-460.

37. T.D. Avery, P.I. Macreadie, B.W. Greatrex, T.V. Robinson, D.K. Taylor and I.G. Macreadie. 2007. Design of endoperoxides with anti-candida activity. Bioorg. Med. Chem. 15: 36-42.
38. V.M. Dembitsky and M. Srebnik. 2002. Natural halogenated fatty acids: their analogues and derivatives. Prog. Lipid Res. 41: 315-367.
39. E. Fattorusso, S. Parapini, C. Campagnuolo, N. Basilico, O. Taglialatela-Scafati and D. Taramelli. 2002. Activity against *Plasmodium falciparum* of cycloperoxide compounds obtained from the sponge *Plakortis simplex*. J. Antimicrob. Chemother. 50: 883-888.
40. I.G. Macreadie, T.D. Avery, T.V. Robinson, P. Macreadie, M. Barraclough, D.K. Taylor and E.R.T. Tiekink. 2008. Design of 1,2-dioxines with anti-candida activity: aromatic substituted 1,2-dioxines. Tetrahedron 64: 1225-1232.
41. S.P. Gunasekera, M. Gunasekera, G.P. Gunawardana, P. McCarthy and N. Burres. 1990. Two new bioactive cyclic peroxides from the marine sponge *Plakortis angulospiculatus*. J. Nat. Prod. 53: 669-674.
42. D.W. Phillipson and K.L. Rinehart. 1983. Antifungal peroxide-containing acids from two *Caribbean* sponges. J. Am. Chem. Soc. 105: 7735-7736.
43. J.K. Rugutt and K.J. Rugutt. 2002. Relationships between molecular properties and antimycobacterial activities of steroids. Nat. Prod. Lett. 16: 107-113.
44. L.-W. Chen, M.-J. Cheng, C.-F. Peng and I.-S. Chen. 2010. Secondary metabolites and antimycobacterial activities from the roots of *Ficus nervosa*. Chem. Biodivers. 7: 1814-1821.
45. D.K. Taylor, T.D. Avery, B.W. Greatrex, E.R.T. Tiekink, I.G. Macreadie, P.I. Macreadie, A.D. Humphries, M. Kalkanidis, E.N. Fox, N. Klonis and L. Tilley. 2004. Novel endoperoxide antimalarials: synthesis, heme binding, and antimalarial activity. J. Med. Chem. 47: 1833-1839.
46. M. del Pilar Crespo, T.D. Avery, E. Hanssen, E. Fox, T.V. Robinson, P. Valente, D.K. Taylor and L. Tilley. 2008. Artemisinin and a series of novel endoperoxide antimalarials exert early effects on digestive vacuole morphology. Antimicrob. Agents Chemother. 52: 98-109.
47. A.O. Terentev, D.A. Borisov, V.A. Vil and V.M. Dembitsky. 2014. Synthesis of five- and six-membered cyclic organic peroxides: key transformations into peroxide ring-retaining products. Beilstein J. Org. Chem. 10: 34-114.
48. M.G. Matturro, R.P. Reynolds, R.V. Kastrup and C.F. Pictroski. 1986. Thioozonide decomposition. Sulfur and oxygen atom transfer. Evidence for the formation of a carbonyl *O*-sulfide intermediate. J. Am. Chem. Soc. 108: 2775-2776.
49. W. Adam and S. Weinkotz. 1996. Synthesis of thiiranes through direct sulfur transfer to cycloalkenes in the thermolysis of a thiophene endoperoxide. J. Chem. Soc. Chem. Commun. 2: 177-178.
50. W. Adam and R.M. Bargon. 2004. Synthesis of thiiranes by direct sulfur transfer: the challenge of developing effective sulfur donors and metal catalysts. Chem. Rev. 104: 251-262.
51. L.A. Paquette, F. Bellarny, M.C. Boehm and R. Gleiter. 1980. Electronic control of stereoselectivity. 6. Directionality of singlet oxygen addition to 1,4-dimethoxynaphthalenes laterally fused to bridged bicyclic systems. J. Org. Chem. 45: 4913-4921.
52. R.I. Fryer and A. Walser. 1991. Bicyclic diazepines: diazepines with an additional ring. R.I. Fryer (Ed.). The chemistry of heterocyclic compounds. John Wiley & Sons, ISBN 0471-52148-5, 5: 1-1019.
53. W.S.K. Schroeder. 2005. Sterically hindered carbonyl *O*-oxides and dioxiranes-(6-*tert*-butyl-2,3,4-trimethylphenyl)phenylcarbonyl *O*-oxide and (6-*tert*-butyl-2,3,4-trimethylphenyl)phenyldioxirane. Eur. J. Org. Chem. 3: 496-504.
54. K. Zeller, M. Kowallik and P. Schuler. 2005. Dimethyldioxirane and acetone are not in equilibrium. Eur. J. Org. Chem. 24: 5151-5153.
55. R.W. Murray and M. Singh. 1997. Synthesis of epoxides using dimethyldioxirane: *trans*-stilbene oxide. Org. Synth. 74: 91-91.

56. R.J. Duffy, K.A. Morris and D. Romo. 2005. Synthesis, structure, and reactivity of unexpectedly stable spiroepoxy-β-lactones obtained by epoxidation of 4-alkylidene-2-oxetanones. J. Am. Chem. Soc. 127: 16754-16755.
57. W. Adam, J. Bialas and L. Hadjiarapoglou. 1991. Kurzmitteilung/short communication a convenient preparation of acetone solutions of dimethyldioxirane. Chem. Ber. 124: 2377-2377.
58. W.C. Broshears, J.J. Esteb, J. Richter and A.M. Wilson. 2004. Simple epoxide formation for the organic laboratory using Oxone. J. Chem. Edu. 81: 1018-1019.
59. P. Kachasakul, S. Assabumrungrat, P. Praserthdam and U. Pancharoen. 2003. Extractive reaction for epoxidation of cyclohexene to cyclohexene oxide using dioxirane in ketone/Oxone® system. Chem. Eng. J. 92: 131-139.
60. I. Carmona, R.S. Schechter, W.H. Wade, U. Weerasooriya and V. Weerasooriya. 1983. Synthesis and performance of linear monoisomeric ethylene oxide sulfonate surfactants. J. Dispers. Sci. Technol. 4: 361-370.
61. O. Katkova. 2005. Photochemical isomerization and stereoselective thermal reactions of conjugated nitrones. Master Thesis, Graduate College of Bowling Green State University.
62. M.T.H. Liu. 1982. The thermolysis and photolysis of diazirines. Chem. Soc. Rev. 11: 127-140.
63. I.R. Likhotvorik, E.L. Tae, C. Ventre and M.S. Platz. 2000. A facile one-pot conversion of non-enolizable aldehydes to diazirines. Tetrahedron Lett. 41: 795-796.
64. G.K.S. Prakash and A.K. Yudin. 1997. Perfluoroalkylation with organosilicon reagents. Chem. Rev. 97: 757-786.
65. G. Mlostod, E. Obijalska, A. Linden and H. Heimgartner. 2008. Trifluoromethylation of camphorquinone and its monoimine derivatives. Tetrahedron: Asymmetry 19: 1676-1683.
66. G. Mlostod, E. Obijalska, A. Linden and H. Heimgartner. 2010. Synthesis and structure of nitrones derived from 2-trifluoromethyl bornane 3-imines. J. Fluorine Chem. 131: 578-583.
67. M. Green, B. Verkoczy, E.M. Lown and O.P. Strausz. 1985. The reactions of sulfur atoms with propadiene and 1,2-butadiene. Can. J. Chem. 63: 667-675.
68. C.S. Adams, C.D. Weatherly, E.G. Burke and J.M. Schomaker. 2014. The conversion of allenes to strained three-membered heterocycles. Chem. Soc. Rev. 43: 3136-3163.
69. W. Ando, N. Tokitoh and Y. Kabe. 1991. Synthesis and reaction of novel sulfur containing heterocyclic systems. Phosphorus, Sulfur, Silicon Relat. Elem. 58: 179-205.
70. K.S. Sidhu, E.M. Lown, O.P. Strausz and H.E. Gunning. 1966. The reactions of sulfur atoms. VI. The addition to C4 olefins. A stereospecific triplet-state reaction. J. Am. Chem. Soc. 88: 254-263.
71. E. Leppin and K. Gollnick. 1969. On the mechanism of stereospecific S (^3P) and O (^3P) additions to olefins. Tetrahedron Lett. 43: 3819-3824.
72. J. Joseph, R.K. Gosavi, A. Otter, G. Kotovych, E.M. Lown and O.P. Strausz. 1990. Stereospecificity of the sulfur (^3P_J) + butene-2 reaction and the NMR spectra of the 1,2-dimethylthiiranes: an experimental and theoretical study. J. Am. Chem. Soc. 112: 8670-8678.
73. O.P. Strausz. 1972. The addition of sulfur atoms to olefins. Adv. Chem. Ser. 110: 137-149.
74. H.E. Gunning and O.P. Strausz. 1966. The reactions of sulfur atoms. Adv. Photochem. 4: 143-194.
75. E. Leppin and K. Gollnick. 1971. Sensitized photolysis of carbonyl sulfide in solution. Photosensitization by aromatic hydrocarbons. Products and general phenomena. J. Am. Chem. Soc. 93: 2848-2853.
76. R. Okazaki, K. Okawa, S. Weijiki and N. Inamoto. 1971. Photolytic and thermal decompositions of 5-substituted 1,2,3,4-thiatriazoles in the presence of olefins. Reactions of sulfur atoms. Bull. Chem. Soc. Jpn. 44: 3167-3170.
77. C. Bertaına, R. Fellous, F. Lemaire and R. Stringat. 1985. Laser induced synthesis: selective formation of cyclohexene episulphide from molecular sulphur S_8. Tetrahedron Lett. 26: 5521-5522.

78. E. Leppin and K. Gollnick. 1970. Direkte photolyse von kohlenoxidsulfid in alkoholen und acetonitril in gegenwart von olefinen. Reaktionen von triplett-schwefelatomen mit norbornen in äthanol und 1.4-dioxane. Chem. Ber. 103: 2894-2901.
79. T.L. Peppard and F.R. Sharpe. 1980. Chemistry of hop constituents. Part 42. The formation and characterisation of *Sesquiterpene episulphides* in the essential oil of hops. J. Chem. Soc. Perkin Trans. 1 0: 311-313.
80. R. Jahn and U. Schmidt. 1978. Photochemische abspaltung von atomarem schwefel aus isothiocyanaten. Monatsh. Chem. 109: 161-164.
81. U. Schmidt, K. Kabitze, I. Boie and C. Osterroht. 1965. Lichtreaktionen mit carbonsäurederivaten, V. Eliminierung doppelt gebundenen schwefels durch bestrahlung: endioläther aus thionestern und isonitrile aus senfölen. Erzeugung atomaren schwefels in lösung. Chem. Ber. 98: 3819-3826.
82. W. Nau, G. Bucher and J.C. Scaiano. 1997. Absolute rate constants for the reactions of sulfur (^3P_J) atoms in solution. J. Am. Chem. Soc. 119: 1961-1970.
83. S. Singh and V. Ramamurthy. 1984. Upper excited state reactions of thioketenes: di-*tert*-butylthioketene. J. Org. Chem. 49: 393-395.
84. S. Singh, H. Nimmesgem, E. Schaumann and V. Ramamurthy. 1985. Photochemical behavior of thioketenes in solution: reaction from S_2. J. Org. Chem. 50: 4799-4805.
85. J.M. Beiner, D. Lecadet, D. Paquer, A. Thuillier and J. Vialle. 1973. Réaction du diazométhane avec composés thiocarbonylés; préparation de thiiranes. Bull. Chim. Soc. Fr. 6: 1979-1983.
86. J.M. Beiner, D. Lecadet, D. Paquer and A. Thuillier. 1973. Composes organiques sulfures. Reaction du diazoethane et du diazo-2-propane sur des composes thiocarbonyles; preparation de thiirannes (2 partie). Bull. Chim. Soc. Fr. 6: 1983-1987.
87. H. Quast and A. Fuss. 1981. Tetramethylcyclopropanon und isopropyliden(dimethyl)thiiran durch photolyse reaktionsträger 3,3,5,5-tetramethyl-1-pyrazoline; solvens- und temperatureinfluß auf die konkurrenz von [3+2]- und [4+1]-photocycloeliminierung. Angew. Chem. 93: 293-294.
88. D. Hoppe and R. Follmann. 1980. Metallierte stickstoffderivate der kohlensäure in der organischen synthese, XVII. Die umwandlung von 2-methylimino-1,3-oxathiolanen in thiirane. Liebigs Ann. Chem. 11: 1779-1785.
89. A. Krantz and J. Laureni. 1981. Characterization of matrix-isolated antiaromatic three-membered heterocycles. Preparation of the elusive thiirene molecule. J. Am. Chem. Soc. 103: 486-496.
90. B.A. Hess, L.J. Schaad and C.S. Ewig. 1980. An ab initio SCF study of the structure and vibrational spectrum of thiirene. J. Am. Chem. Soc. 102: 2507-2508.
91. B.D. Larsen, H. Eggert, N. Harrit and A. Holm. 1992. Photolysis of 1,2,3-thiadiazole. Formation of thiirene by secondary photolysis of thioketene. Acta Chem. Scand. 46: 482-486.
92. Y.Y. Morzherin, T.V. Glukhareva and V.A. Bakulev. 2003. Rearrangements and transformations of 1,2,3-thiadiazoles in organic synthesis (review). Chem. Heterocycl. Compd. 39: 679-706.
93. W. Kirmse. 2002. 100 Years of the Wolff rearrangement. Eur. J. Org. Chem. 14: 2193-2256.
94. E. Schaumann. 1988. The chemistry of thioketens. Tetrahedron 44: 1827-1871.
95. A. Krantz and J. Laureni. 1977. A methodology for the preparation and characterization of three-membered, potentially antiaromatic molecules. Preparation of matrix-isolated thiirene and selenirene. J. Am. Chem. Soc. 99: 4842-4844.
96. A. Krantz and J. Laureni. 1974. Matrix photolysis of 1,2,3-thiadiazole. Possible involvement of thiirene. J. Am. Chem. Soc. 96: 6768-6770.
97. M. Torres, A. Clement, J.E. Bertie, H.E. Gunning and O.P. Strausz. 1978. Low-temperature matrix isolation of thiirenes. J. Org. Chem. 43: 2490-2493.

98. G. Burdzinski, M. Sliwa, Y. Zhang and S. Delbaere. 2011. Early events in the photochemistry of 1,2,3-thiadiazole studied by ultrafast time-resolved UV-Vis and IR spectroscopies. J. Phys. Chem. A 115: 14300-14305.
99. G. Burdzinski, M. Sliwa, Y. Zhang, S. Delbaere, T. Pedzinski and J. Réhault. 2013. Photochemical formation of thiirene and thioketene in 1,2,3-thiadiazoles with phenyl substituents studied by time-resolved spectroscopy. Photochem. Photobiol. Sci. 12: 895-901.
100. D. Margetic, R.N. Warrener, G. Sun and D.N. Butler. 2007. The preparation of 7-substituted norbornadiene-2,3-dicarboxylic anhydrides and an experimental and theoretical study of their reactivity. Tetrahedron 63: 4338-4346.
101. R.N. Warrener, D. Margetic, G. Sun and R.A. Russell. 2003. Position-addressable nano-scaffolds. I. The preparation of *N,O-*, *N,C-* and *N,N*-bridged sesquinorbornadiene succinimides as compact, highly functionalized addressable building blocks. Aust. J. Chem. 56: 263-267.
102. K. Paulvannan. 2004. An atom-economical approach to conformationally constrained tricyclic nitrogen heterocycles via sequential and tandem Ugi/intramolecular Diels-Alder reaction of pyrrole. J. Org. Chem. 69: 1207-1214.
103. M. Bianchi, A. Butti, Y. Christidis, J. Perronnet, F. Barzaghi, R. Cesana and A. Nencioni. 1988. Gastric antisecretory, anti-ulcer and cytoprotective properties of substituted (*E*)-4-phenyl- and heteroaryl-4-oxo-2-butenoic acids. Eur. J. Med. Chem. 23: 45-52.

Index

'*Note:* Page numbers followed by "f" indicate figures and "t" indicate tables.'

C

D

E

F

G

H

U

V

9780443135835